CONTINUOUS ADVANCES IN QCD 2006

Proceedings of the Conference on

CONTINUOUS ADVANCES IN QCD 2006

William I. Fine Theoretical Physics Institute
Minneapolis, USA 11 – 14 May 2006

Editors

M. Peloso & M. Shifman
University of Minnesota, USA

World Scientific
NEW JERSEY • LONDON • SINGAPORE • BEIJING • SHANGHAI • HONG KONG • TAIPEI • CHENNAI

Published by

World Scientific Publishing Co. Pte. Ltd.

5 Toh Tuck Link, Singapore 596224

USA office: 27 Warren Street, Suite 401-402, Hackensack, NJ 07601

UK office: 57 Shelton Street, Covent Garden, London WC2H 9HE

British Library Cataloguing-in-Publication Data
A catalogue record for this book is available from the British Library.

CONTINUOUS ADVANCES IN QCD 2006

ISBN-13 978-981-270-552-5
ISBN-10 981-270-552-X

Printed in Singapore by World Scientific Printers (S) Pte Ltd

FROM THE EDITORS

The Seventh Workshop on Continuous Advances in QCD, "Continuous Advances in QCD 2006," was held at the William I. Fine Theoretical Physics Institute on May 11-14, 2006. It is the latest in a series which started in 1994 and eventually grew into a major international event in this area of physics, which attracts the most active researchers working at the cutting edge of QCD.

This year was special for two reasons. First, following the emerging trend in string theory to "return to its roots," we made a special effort to invite leading string theorists interested in applications to gauge theories and hadronic physics. This was a complete success. Sections on B-theory, AdS/CFT, and solitons, branes and strings in Yang–Mills theories (conveners: Z. Bern, A. Armoni, D. Tong) proved to be unprecedented in the breadth of coverage of their corresponding topics. They set the standard for future meetings of this kind. They attracted a record number of participants. Never before had we needed to run three parallel sessions to accommodate all talks selected by the Organizing Committee.

Traditional QCD was also well represented. A healthy mix of new ideas and developments was presented in sections on large-N methods, QCD at high temperatures/densities, interplay of weak and strong interactions, light-cone methods, and ideas on confinement (conveners: T. Cohen, E. Shuryak, A. Khodjamirian, J. Hiller, M. Shifman). Some topics, such as strongly coupled quark-gluon plasma or chiral symmetry restoration in high excitations, turned out to be especially intriguing and sparked heated debates which, remarkably, engaged not only QCD practitioners but string theorists as well.

The second new feature of the workshop was inclusion in the program of invited review talks which summarized major trends in the field and set guidelines for future developments. We had 10 plenary talks delivered by the leading world experts. Suffice it to mention Klebanov's talk on AdS/CFT, Strassler's talk on the Pomeron and gauge/string duality and Shuryak's talk on strongly coupled quark-gluon plasma.

By and large, it is fair to say that our attempt to unite QCD and string theorists was successful. The workshop stimulated "interdisciplinary" discussions in the most vibrant areas of high-energy theory and made a significant contribution towards cross-fertilization of ideas of string/brane theory on the one hand and gauge field theory on the other.

This volume presents write-ups of 65 talks given by the Workshop participants. Some of the speakers decided to limit their contributions to pdf files, which can be found at

http://www.ftpi.umn.edu/qcd_06/qcd06_program.html

In conjunction with the *Continuous Advances in QCD 2006* meeting, a related workshop, "Light-Cone QCD and Non-Perturbative Hadronic Physics," was held immediately after *Continuous Advances* (May 14–19). It was devoted to applications of light-cone quantization to the understanding of hadron structure and strong-interaction phenomena at a fundamental level. This arrangement allowed many participants of *Continuous Advances* to smoothly pass to the discussion of the light-cone methods.

Included in this volume is a special chapter entitled *Pages of the Past.* Here we present a memoir of Sasha Migdal entitled *Paradise Lost*, a note on early history of strong interactions written by Stephen Gasiorowicz, and Gasiorowicz's "Brief history of FTPI."

Finally, on behalf of the organizers, we would like to take this opportunity to express our gratitude and appreciation to Roxanne Keen and Anne Barthel for their dedicated efforts to the smooth running of the workshop. They were the driving force behind practically seamless organization which was noted by all participants.

M. Peloso
M. Shifman
September 28, 2006

CONTENTS

SECTION 1

PLENARY TALKS

STRONGLY COUPLED QUARK-GLUON PLASMA: THE STATUS REPORT

E.V. SHURYAK

Department of Physics and Astronomy, University at Stony Brook,
Stony Brook NY 11794 USA
E-mail: shuryak@tonic.physics.sunysb.edu

RHIC data have shown robust collective flows, strong jet and charm quenching, and charm flow. Recently "conical flows" from damped jets were seen. Non-Abelian classical strongly coupled plasmas were introduced and studied via molecular dynamics, with first results for its transport (diffusion and viscosity) reported. Quantum-mechanical studies reveal the survival for $T > T_c$ of the lowest binary states, including colored ones, and also of some manybody ones such as baryons. "Polymeric chains" $\bar{q}.g.g..q$ are also bound in some range of T, perhaps the progonitors of the QCD strings. AdS/CFT applications advanced to a completely new level of detail: they now include studies of thermal heavy quark motion, jet quenching and even conical flow. Confinement is however still beyond simply strong coupling: its de-facto inclusion the so called AdS/QCD approach is so far added as a model, while true understanding may probably only come from further insights into the monopole dynamics.

Keywords: Quark-gluon plasma, strong coupling, AdS/CFT correspondence, heavy ion collisions

1. Why strongly coupled?

A realization [1–3] that QGP at RHIC is not a weakly coupled gas but rather a strongly coupled liquid has lead to a paradigm shift in the field. It was extensively debated at the "discovery" BNL workshop in 2004 [4] (at which the abbreviation sQGP was established) and multiple other meetings since.

In the intervening three years we had to learn a lot, some new some from other branches of physics which happened to have some experience with strongly coupled systems. Those range from quantum gases to classical plasmas to string theory. In short, there seem to be not one but actually two difficult issues we are facing. One is to understand why QGP at $T \sim 2T_c$ is strongly coupled, and what exactly it means. The second large problem is to understand what happens at the deconfinement, at $|T - T_c| \ll T_c$, which may be a key to the famous confinement problem.

As usual, progress proceeds from catching/formulating the main concepts and qualitative pictures, to mastering technical tools, to final quantitative predictions: and now we are somewhere in the middle of this process. The work is going on at many fronts. At classical level, first studies of the transport properties of strongly coupled non-Abelian plasmas have been made. Quantum-mechanical studies of the bound states above T_c have revealed a lot of unusual states, including "polymeric chains". At the quantum field/string theory front, a surprisingly detailed uses of AdS/CFT correspondence has been made. And yet, to be honest, deep understanding is still missing: e.g. we don't know what the CFT plasma is made of.

The list of arguments explaining why we think QGP is strongly coupled at T above T_c is long and constantly growing. Let me start with its short version, as I see them today.

1.Collective phenomena observed at RHIC lead hydro practitioners to a conclusion that QGP as a "near perfect liquid", with unusually small *viscosity-to-entropy ratio* $\eta/s = .1 - .2 << 1$ [5] in striking contrast to pQCD predictions. Not only light jets, but also charmed ones are strongly quenched. Charm diffusion constant D_c deduced from its flow is an order of magnitude lower than pQCD estimates [6].

2. Combining lattice data on quasiparticle masses and interparticle potentials, one finds a lot of quasiparticle bound states [7]. The same approach explains why η_c, J/ψ remains bound till near $3T_c$, as was directly observed on the lattice [8] and perhaps experimentally at RHIC. The resulting resonances enhance transport cross sections [2,9] and may lead to a liquid-like behavior. Similar thing does happen for ultracold trapped atoms, due to Feshbach-type resonances at which the scattering length $a \to \infty$.

3.The interaction parameter $\Gamma \sim < potential\, energy > / < kinetic\, energy >$ in sQGP is obviously not small. Classical e/m plasmas at the comparable coupling $\Gamma \sim 1 - 10$ are good liquids too.

4. Exact correspondence between a conformal (CFT) $\mathcal{N}$=4 supersymmetric Yang-Mills theory at strong coupling and string theory in Anti-de-Sitter space (AdS) in classical SUGRA regime was conjectured by Maldacena [10]. The results obtained this way on the $g^2 N_c \to \infty$ regime of the CFT plasma are all close to what we know about sQGP. Indeed, it has a very similar thermodynamics and is a good liquid with record low viscosity as well. Recent works (see below) added to the list parametrically large jet quenching and small diffusion constant for heavy quarks.

5.[a] The N=2 SUSY YM ("Seiberg-Witten" theory) is a working example of confinement due to condensed monopoles[11]. If it is also true for QCD, at $T \to T_c$ magnetic monopoles must become light and weakly interacting at large distances due to U(1) beta function. Then the Dirac condition forces electric coupling g be large (in IR).

2. Collective flows in heavy ion collisions

This meeting is mostly theoretical in nature, and thus I would not go into details of heavy ion phenomenology. Collective flows, related with explosive behavior of hot matter, were observed at SPS and RHIC and are quite accurately reproduced by the ideal hydrodynamics. The flow affect different secondaries differently, yet their spectra are in quantitative agreement with the data for all of them, from π to Ω^-. At non-zero impact parameter the original excited system is deformed in the transverse plane, creating the so called elliptic flow. It is described by the parameter $v_2(s, pt, M_i, y, b, A)$ $=< cos(2\phi) >$, where ϕ is the azimuthal angle and the others stand for the collision energy, transverse momentum, particle mass, rapidity, centrality and system size. Hydrodynamics explains well all of those dependences, for about 99% of the particles[b].

New hydrodynamical phenomenon suggested recently [12], is the so called *conical* flow which is induced by jets quenched in sQGP. Although the QCD Lagrangian tells us that charges are coupled to gluons and thus it is gluons which are to be radiated, at strong coupling those are rapidly quenched. Effectively the jet energy is dumped into the medium and then it transformed into coherent radiation of sound waves, which unlike gluons are much less absorbed and can survive till freezout to be detected. As shown in Fig.1, this seem to be what indeed is observed.

Antinori and myself [13] suggested to test it further by b-quark jets, which can be tagged experimentally even if not ultrarelativistic: the Mach cone should then shrink, till it goes to zero at the critical velocity $v = c_s = 1/\sqrt{3}$. Gluon radiation behaves oppositely, expanding with *decreasing* v, and never shrinks to zero. Casalderrey and myself[14] have shown, using conservation of adiabatic invariants, that fireball expansion should greatly enhance the sonic boom[c].

---

[a]This part is presented at this conference for the first time.
[b]At large $p_t > 2GeV$ a different regime starts, related with jets.
[c]The reason is similar to enhancement of a sea wave such tsunami as it goes onshore.

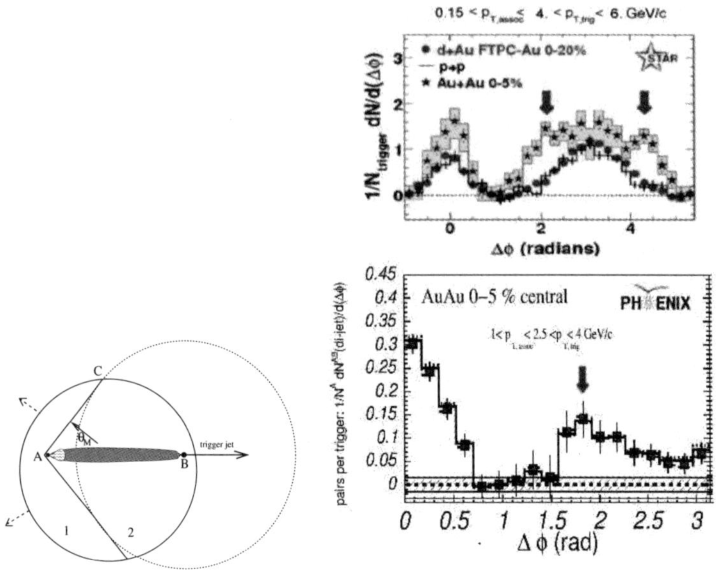

Fig. 1. (a) A schematic picture of flow created by a jet going through the fireball. The trigger jet is going to the right from the origination point B. The companion quenched jet is moving to the left, heating the matter (in shadowed area) and producing a shock cone with a flow normal to it, at the Mach angle $cos\theta_M = v/c_s$, where v, c_s are jet and sound velocities. (b)The background subtracted correlation functions from STAR and PHENIX experiments, a distribution in azimuthal angle $\Delta\phi$ between the trigger jet and associated particle. Unlike in pp and dAu collisions where the decay of the companion jet create a peak at $\Delta\phi = \pi$ (STAR plot), central $AuAu$ collisions show a minimum at that angle and a maximum corresponding to the Mach angle (downward arrows).

3. Classical strongly coupled non-Abelian plasmas

In the electromagnetic plasmas the term "strongly coupled" is expressed via dimensionless parameter $\Gamma = (Ze)^2/(a_{WS}T)$ characterizing the strength of the interparticle interaction. Ze, a_{WS}, T are respectively the ion charge, the Wigner-Seitz radius $a_{WT} = (3/4\pi n)^{1/3}$ and the temperature. Γ is convenient to use because it only involves the *input* parameters, such as the temperature and density. Extensive studies using both MD and analytical methods, have revealed the following regimes: **i.** a gas regime for $\Gamma < 1$; **ii.** a liquid regime for $\Gamma \approx 10$; **iii.** a glass regime for $\Gamma \approx 100$; **iv.** a solid regime for $\Gamma > 300$.

Gelman, Zahed and myself [15] proposed a model for the description of strongly interacting quarks and gluon quasiparticles as a classical and non-relativistic Non-Abelian Coulomb gas. The sign and strength of the interparticle interactions are fixed by the scalar product of their classical *color vectors* subject to Wong's equations. The EoM for the phase space coordinates follow from the usual Poisson brackets:

$$\{x^m_{\alpha i}, p^n_{\beta j}\} = \delta^{mn}\delta_{\alpha\beta}\delta_{ij} \quad \{Q^a_{\alpha i}, Q^b_{\beta j}\} = f^{abc}\, Q^c_{\alpha i} \tag{1}$$

For the color coordinates they are classical analogue of the SU(N_c) color commutators, with f^{abc} the structure constants of the color group. The classical color vectors are all adjoint vectors with $a = 1...(N_c^2 - 1)$. For the non-Abelian group SU(2) those are 3d vectors on a unit sphere, for SU(3) there are 8 dimensions minus 2 Casimirs=6 d.o.f.[d].

The model was studied using Molecular Dynamics (MD), which means solving numerically EoM for $n \sim 100$ particles. It also displays a number of phases as the Coulomb coupling is increased ranging from a gas, to a liquid, to a crystal with anti-ferromagnetic-like color ordering. There is no place for details here: in Fig.2 one can see the result for diffusion and viscosity vs coupling: note how different and nontrivial they are. When extrapolated to the sQGP suggest that the phase is liquid-like, with a diffusion constant $D \approx 0.1/T$ and a bulk viscosity to entropy density ratio $\eta/s \approx 1/3$. The second paper of the same group[15] discussed the energy and the screening at $\Gamma > 1$, finding large deviations from the Debye theory.

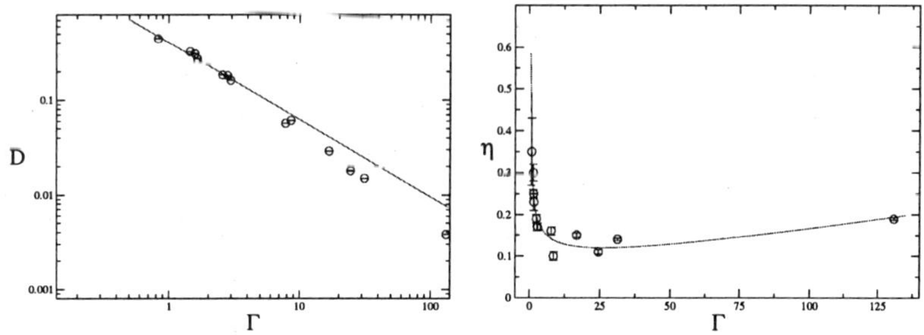

Fig. 2. The diffusion constant (a) and shear viscosity (b) of a one species cQGP as a function of the dimensionless coupling Γ. Blue points are the MD simulations; the red curve is the fit.

[d]Although color EoM do not look like the usual canonical relations between coordinates and momenta, they actually are pairs of conjugated variables, as can be shown via the so called Darboux parameterization.

4. Quantum mechanics of the quasiparticles

In the deconfined phase, at $T > T_c$, the basic objects are dressed quarks and gluons. Even perturbatively they get masses $M_{eff} \sim gT$ [16] and dispersion curve close to that of massive particle. However it follows from lattice measurements [17] (admittedly, with still poor accuracy) that (i) masses are very large, about $M_{eff} \approx (3-4)T$; and (ii) quarks and gluons have very close masses, in contrast to pQCD prediction. Thus, in first approximation, quasiparticles are rather non-relativistic.

As emphasized by Zahed and myself [7], a gas of such heavy quasiparticles cannot generate the pressure observed in other lattice works. The resolution of a puzzle may be found if there are multiple bound states of the quasiparticles, which also contribute to pressure.

The existence of bound states also follows from the interaction deduced for static quarks. For marginal states, with near-zero binding, those should be applicable. The most obvious state to think of is charmonium, which had a long story of a debate whether it will survive in QGP or not. The answer depends on which is the effective potential: in contrast to many earlier works we pointed out [2] that one has to remove the entropy term and use the "energy" potentials $V(T,r) = F - TdF/dT$, not the free energy ones $F(T,r)$ which are directly measured from Wilson/Polyakov lines. This leads effectively to deeper potentials and better binding, so J/ψ survives till $T = (2-3)T_c$. This was confirmed by direct calculation of spectral densities by maximal entropy method [8]. It also nicely correlates well with surprisingly small J/ψ suppression observed at RHIC, where $T < 2T_c$.

It was then pointed out in [7] that also multiple binary *colored* bound states should exist in the same T domain. Since QGP is a deconfined phase, there is nothing wrong with that, and the forces between say singlet $\bar{q}q$ and octet qg quasiparticle pairs are about the same. Liao and myself [18] have also found survival of the s-wave baryons $(N, \Delta ...)$ at $T < 1.6T_c$.

A particularly interesting objects are *multibody* [18] bound states, such as "polymer chains". Those can be "open strings" $\bar{q}gg..gq$ or closed chains of gluons (e.g. very robust ggg state we studied). Their binding *per bond* was proven to be the same as for light-quark mesons, and both are bound till about $1.5T_c$ or so. They have interesting AdS/CFT analogs (see below) and they also can be viewed as precursor to the formation of the QCD strings from the deconfined phases.

A curve of marginal stability (CMS) is not a thermodynamic singularity but it often indicates a change of physics. Zahed and myself [2] argued that resonances can strongly enhance transport cross section near multiple

CMS's and thus explain small viscosity. Rapp and van Hees [9] studied $\bar{q}c$ resonances, and found enhancement of charm stopping.

Similar phenomenon does happen for ultracold trapped atoms, which are extremely dilute but due to Feshbach-type resonances at which the scattering length $a \to \infty$ they behave like very good liquids with small viscosity, see[19].

One notable colored bound state is a diquark qq, the main player in color superconductivity at high density and low T. Diquarks are weaker bound than mesons[e] and are expected to melts right above the deconfinement transition. In my recent paper[20] I argued that may bring color superconductivity into strongly coupled regime as well. The usual BCS theory of superconductors are then inapplicable: it is also weak coupling theory.

Fortunately superfluidity of ultracold fermionic strongly coupled atoms have been studied recently experimentally. The system is known to enter the universal strongly coupled regime as their scattering length a gets large, and therefore it is possible to use some universal properties to get such properties as the slope of the critical line of color superconductivity, $dT_c/d\mu$. I also deduced limitations on the critical temperature of color superconductivity itself and conclude that it is limited by $T_{CS} < 70\, MeV$.

5. AdS/CFT correspondence at finite T

Thermodynamics of the CFT plasma was studied started from the early work[21], its result is that the free energy (pressure) of a plasma is

$$F(g, N_c, T)/F(g=0, N_c, T) = [(3/4) + O((g^2 N_c)^{-3/2})] \tag{2}$$

which compares well with the lattice value[f] of about 0.8.

Heavy-quark potentials in vacuum and then at finite T [22] were calculated by calculating the configuration of the static string, deformed by gravity into the 5-th dimension. Let me write the result schematically as

$$V(T, r, g) \sim \frac{\sqrt{g^2 N_c}}{r} exp(-\pi T r) \tag{3}$$

The Debye radius at strong coupling is unusual: unlike in pQCD it has no coupling constant. Although potential depends on distance r still as in the Coulomb law, $1/r$ (at $T = 0$ it is due to conformity), it is has a notorious square root of the coupling. Semenoff and Zarembo [23] noticed that summing

[e] Due to extra 1/2 in color Casimirs.

[f] Not too close to T_c, of course, but in the "conformal domain" of $T = few\, T_c$, in which p/T^4 and ϵ/T^4 are constant.

ladder diagrams one can explain $\sqrt{g^2 N_c}$, although not a numerical constant. Zahed and myself [3] pointed out that both static charges are color correlated during a parametrically small time $\delta t \sim r/(g^2 N_c)^{1/4}$: this explains [24] why a field of the dipole is $1/r^7$ at large distance[25], not $1/r^6$. Debye screening range can also be explained by resummation of thermal polarizations [3].

Zahed and myself [26] had also discussed the velocity-dependent forces , as well as spin-spin and spin-orbit ones, at strong coupling. Using ladder resummation for non-parallel Wilson lines with spin they concluded that all of them join into one common square root

$$V(T,r,g) \sim \sqrt{(g^2 N_c)[1 - \vec{v}_1 * \vec{v}_2 + (spin - spin) + (spin - orbit)]}/r \quad (4)$$

Here $\vec{v}_1, \vec{v}_2$ are velocities of the quarks: and the corresponding term is a strong coupling version of Ampere's interaction between two currents[g]. No results on that are known from a gravity side, to my knowledge.

Bound states Zahed and myself [3] looked for heavy quarks bound states, using a Coulombic potential with Maldacena's $\sqrt{g^2 N_c}$ and Klein-Gordon/Dirac eqns. There is no problem with states at large orbital momentum $J >> \sqrt{g^2 N_c}$, otherwise one has the famous "falling on a center" solutions[h]: we argued that a significant density of bound states develops, at all energies, from zero to $2M_{HQ}$.

And yet, a study of the gravity side [27] found that there is no falling. In more detail, the Coulombic states at large J are supplemented by two more families: Regge ones with the mass $\sim M_{HQ}/(g^2 N_c)^{1/4}$ and the lowest s-wave states (one may call $\eta_c, J/\psi$) with even smaller masses $\sim M_{HQ}/\sqrt{g^2 N_c}$. The issue of "falling" was further discussed by Klebanov, Maldacena and Thorn [24] for a pair of static quarks: they calculated the spectral density of states via a semiclassical quantization of string vibrations. They argued that their corresponding density of states should appear at exactly the same critical coupling as the famous "falling" in the Klein-Gordon eqn..

AdS/CFT also has multi-body states similar to "polymeric chains" $\bar{q}.g.g...q$ discussed above. For the endpoints being static quarks and the intermediate gluons conveniently replaced by adjoint scalars, Hong, Yoon and Strassler [28] have studied such states and even their formfactors.

Transport properties of the CFT plasma was a subject of recent breakthroughs [i]. The (already famous) work by Polykastro, Son and

---

[g] Note that in a quarkonium their scalar product is negative, increasing attraction.

[h] Note that all relativistic corrections mentioned above cannot prevent it from happening.

[i] The works which appeared between the conference and the time when this summary is written are included.

Starinets[29] have calculated viscosity (at infinite coupling) $\eta/s => 1/4\pi$ which is in the ballpark of the empirical RHIC value. It taught us that gravitons in the bulk at large distances are dual to phonons on the brane. Dual to a viscous sound absorption is thus interception of gravitons by the black hole.

Heavy quark diffusion constant has been calculated by Casalderrey-Solana and Teaney [30]: their result is

$$D_{HQ} = \frac{2}{\pi T \sqrt{g^2 N_c}} \tag{5}$$

which is parametrically smaller than an expression for the momentum diffusion $D_p = \eta/(\epsilon + p) \sim 1/4\pi T$. This work is methodically quite different from others in that Kruskal coordinates are used, which allows to consider the inside of the black hole and *two* Universes (with opposite time directions) simultaneously, see Fig.3a. This is indeed necessary [j] in any problems when a *probability* is evaluated, because that contains both an amplitude and a conjugated amplitude at the same time.

Jet quenching studies[32–36] have been reported by L.Yaffe, see his talk for details. The result for the drag force is

$$\frac{dP}{dt} = -\frac{\pi T^2 \sqrt{g^2 N_c v}}{2\sqrt{1 - v^2}} \tag{6}$$

Quite remarkably, the Einstein relation which relates the heavy quark diffusion constant (given above) to the drag force is actually fulfilled, in spite of quite different gravity settings shown in Fig.3, a and b.

This result is valid only for quarks heavy enough $M > M_{eff} \sim \sqrt{g^2 N_c} T$ and is obtained in a stationary setting, in which a quark is dragged with constant by "an invisible hand" via some rope through QGP, resulting in constant production of a string length per time, see Fig.3b . I have borrowed it from the paper by Friess et al [37], who have made the next (and technically much more difficult) step, namely solving the Einstein equation with this falling string as a source and found corrections to the metric $h_{\mu\nu}$ and thus the matter stress tensor on the brane. Quite remarkably, when they analyzed harmonics of this stress at small momenta they have seen the "conical flow"! And, as one can see from plots for "subsonic" $v < 1/\sqrt{3}$, the Mach cone disappear in this case, as argued in [13].

---

[j] One such problem is evaluation of the so called $\hat{q}$ parameter: two lines of the loop should also belong to *two* different Universes, not one as assumed in [31]. It remains unknown whether similar calculation in Kruskal geometry would produe the same result or not.

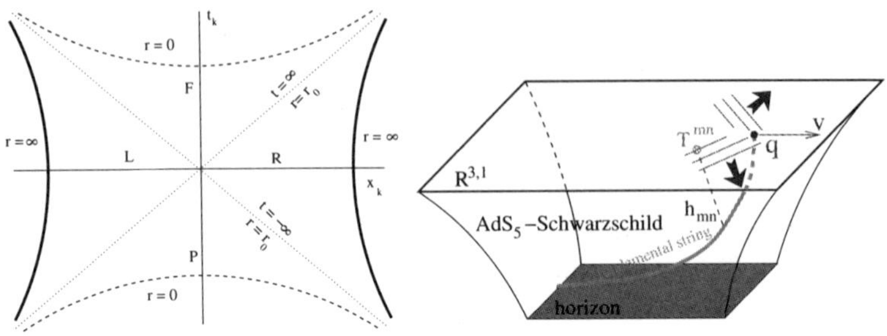

Fig. 3. (a) (from [30]): In Kruskal coordinates one can study two Universes at the same time, shown right and left, and the evaluated Wilson line contains static quarks on their boundaries. (b) (from [37]) The dragged quark trails a string into the five-dimensional AdS bulk, representing color fields sourced by the quark's fundamental charge and interacting with the thermal medium. The back gravity reaction describes how matter flows on the brane.

The ultimate goal would be a complete "gravity dual" to the whole RHIC collision process, in which thermalization and subsequent hydro explosion will be described via dynamical production of a black hole, as emphasized by Nastase [38]. Sin, Zahed and myself [39] further argued that exploding and cooling fireball on the brane is dual to departing black hole, formed by collision debris falling into the AdS center. Janik and Peschanski [40], found that the Bjorken expansion can be mapped into a metric with a departing *horizon*[k].

6. AdS/QCD

"Holographic" ideas have been also used for theories more resembling QCD. There are famous papers on it; my favorite is Sakai-Sugimoto model[49] with light quarks, deconfinement and chiral symmetry restoration. The closest to RHIC physics is nice paper by Peeters et al[50] which document how light mesons (e.g. ρ) get T-dependent masses after they survive deconfinement.

Quite intriguing is also a "bottom-up" approach, in which the violations of comformity of AdS/CFT is introduced explicitly. Confinement forbids QCD phenomena from exploring large distances. In the now popular "holographic" language, it means that all object we study are somehow prevented from going too far into the 5-th coordinate z into the IR. First attempts to model confinement (used mostly for QCD spectroscopy) used just such a cut-off, at some $z = z_0$.

[k]However they solve vacuum Einstein eqns without any matter: their departing horison is due to acausal time dependence of the central black hole (the stack of branes).

Recently three groups suggested different arguments that confinement induces a *quadratic* potential in z. Karch et al [41] have argued that this is needed to get correct dependence of the Regge trajectories on particle spin S and principal quantum number n. In my paper [42] the probe for confinement are instantons, and it was first argued that their size should be identified to the 5-th coordinate z. The proposed potential consists of two parts, related to asymptotic freedom[l] and confinement

$$V_{eff}(z) = V_{AF} + V_{conf} = -\beta_0 log(z) + 2\pi\sigma z^2 \qquad (7)$$

where $\beta_0 = (11/3)N - (2/3)N_f$ and the coefficient of the quadratic term was proposed previously in [43], it contains the string tension σ. The reason (and coefficient[m]) for quadratic behavior is not ad hoc, but because a it is related with a VEV of a *dual superconductor*; it is in excellent agreement with lattice datat on instanton size distribution.

Andreev and Zakharov [44] put the quadratic potential into metric, and calculated a number of string-based potentials and spatial Wilson lines. However if it is in metric, wrong Regge trajectories follows[41] and also one cannot have non-universal coupling just mentioned: so I think it should be put into some extra potential instead.

7. (Post)Confinement and monopoles

Here comes an old question: is there any progress in understanding confinement, as well as the deconfinement transition region, at $T \approx T_c$?

Recent lattice data have revealed a puzzling behavior of static $\bar{Q}Q$ potentials, which I call "postconfinement". At $T = 0$ we all know that a potential between heavy quarks is a sum of the Coulomb and a confining $\sigma(T = 0)r$ potential. At deconfinement $T = T_c$ the Wilson or Polyakov lines with a static quark pair has vanishing string tension; but this is the free energy $exp(-F(T,r)) =< W >$. Quite shockingly, if one calculates the *energy* or *entropy* separately (by $F = E - TS$, $S = -\partial F/\partial T$) one finds [45] a force between $\bar{Q}Q$ to be more than twice $\sigma(T = 0)$ till rather large distances. The total energy added to a pair is surprisingly large reaches

[l] I found it remarcable that both other groups igore asymptotic freedom and deviations from conformity in UV. The corresponding coupling constants for any operator is related to its perturbative anomalous dimension.

[m] It is similar but not the same as the one proposed in [41], which should not surprise us, as the constant depends on how strongly the object studied is coupled to confinement-related condensates. For example, for glueball Regge trajectories (including the Pomeron and the 2++ glueball) the Regge slope is already significantly different.

about $E(T = T_c, r \to \infty) = 3 - 4\,GeV$, and the entropy as large as $S(T = T_c, r \to \infty) \sim 10$. Since this energy of "associated matter" is about 20 times larger than T, any separation of two static quarks must be extremely suppressed by the Boltzmann factor exp(-E/T). (As T grows, this phenomenon disappears, and thus it is obviously related to the phase transition itself.)

Where all this energy and entropy may come from in the deconfined phase? It can only be long and complicated QCD string connecting two static quarks. We already mentioned that such strings can be explained by a "polymerization" of gluonic quasiparticles in sQGP.

Let us now add a twist to this story related with magnetic excitations, the monopoles[n]. According to t'Hooft-Mandelstamm scenario, confinement is supposed to be due to monopole condensation. Seiberg-Witten solution for the N=2 SYM is an example of how it is all supposed to work: it has taught us that as one approaches the deconfinement transition the electrically charged particles – quarks and gluons – are getting heavier while monopoles gets lighter and more numerous. Although I cannot go into details here, we do have hints from lattice studies of monopoles and related observables that this is happening in QCD as well.

Let us now think what all of it means for the sQGP close to T_c. Even at classical level, it means that one has a plasma with both type of charges – *electric and magnetic* – at the same time, with the former dominant at large T and the latter dominant close to T_c.

A binary dyon-dyon systems have been studied before , but not many-body ones. The first numerical studies of such systems (by molecular dynamics) are now performed by (Stony Brook student) Liao and myself [46]. We found that a monopole can be trapped by an electric static dipole, both classically and quantum mechanically. We also found that classical gas of monopoles leads to electric flux tubes[o] because monopoles scatter from the electric flux tube back into plasma, compressing it. Whether monopoles are condensed or not is not crucial.

Are ther bound states of electric and magnetic quasiparticles? Yes, there are a lot of them. A surprize is that even finite-T instantons can be viewed

---

[n]Recall that they appear naturally if there is an explicit Higgs VEV breaking of the color group. We cannot discuss in detail a QCD setting: the reader may simply imagine a generic finite-T configuration with a nonzero mean $< A_0 >$, an adjoint Higgsing leaving $N_c - 1$ U(1) massless gauge fields. These U(1)'s corresponds to magnetic charges of the monopoles. In AdS/CFT language one may simply considered N_c branes to be placed not at exactly the same point in the orthogonal space.

[o]Those are dual to magnetic flux tubes in solar classical plasmas.

as being made of N_c selfdual dyons [47], attracted to each other pairvise, electrically and magnetically. Not only such baryons-made-of-dyons have the same moduli space as instantons, the solutions can be obtained vis very interesting AdS/CFT brane construction [48]. Many more exotic bound states of those are surely waiting to be discoverd.

References

1. E.V.Shuryak, Prog. Part. Nucl. Phys. **53**, 273 (2004) [hep-ph/0312227].
2. E.V.Shuryak and I. Zahed, hep-ph/0307267, Phys. Rev. C **70**, 021901 (2004)
3. E.V.Shuryak and I. Zahed, Phys. Rev. **D69** (2004) 014011. [hep-th/0308073].
4. M. Gyulassy and L. McLerran, Nucl. Phys. A **750**, 30 (2005) [nucl-th/0405013]. E. V. Shuryak,Prog.Part.Nucl.Phys.53:273-303,2004, hep-ph/0312227 Nucl. Phys. A **750**, 64 (2005).
5. D. Teaney, Phys. Rev. C **68**, 034913 (2003) [arXiv:nucl-th/0301099].
6. G. D. Moore and D. Teaney, hep-ph/0412346.
7. E. V. Shuryak and I. Zahed, Phys. Rev. D **70**, 054507 (2004), hep-ph/0403127.
8. S. Datta, F. Karsch, P. Petreczky and I. Wetzorke, hep-lat/0208012. M. Asakawa and T. Hatsuda, Nucl. Phys. **A715** (2003) 863c; hep-lat/0308034;
9. H. van Hees, V. Greco and R. Rapp, arXiv:hep-ph/0601166.
10. J. M. Maldacena, Adv. Theor. Math. Phys. **2**, 231 (1998) [Int. J. Theor. Phys. **38**, 1113 (1999)] [arXiv:hep-th/9711200].
11. N. Seiberg and E. Witten, Nucl. Phys. B **426**, 19 (1994) [Erratum-ibid. B **430**, 485 (1994)] [arXiv:hep-th/9407087].
12. J. Casalderrey-Solana, E. V. Shuryak and D. Teaney, hep-ph/0411315. arXiv:hep-ph/0602183.
13. F. Antinori and E. V. Shuryak, nucl-th/0507046.
14. J. Casalderrey-Solana and E. V. Shuryak, arXiv:hep-ph/0511263.
15. B. A. Gelman, E. V. Shuryak and I. Zahed, Phys.Rev.C, in press, arXiv:nucl-th/0601029, nucl-th/0605046.
16. E. V. Shuryak, Zh. Eksp. Teor. Fiz. **74**, 408 (1978); Sov. Phys. JETP **47**, 212 (1978),Phys. Rept. **61**, 71 (1980).
17. P. Petreczky, F. Karsch, E. Laermann, S. Stickan, I. Wetzorke, Nucl. Phys. Proc. Suppl. **106** (2002) 513.
18. J. Liao and E. V. Shuryak, Nucl.Phys.A, in press, arXiv:hep-ph/0508035. Phys. Rev. D **73**, 014509 (2006) [arXiv:hep-ph/0510110].
19. B. A. Gelman, E. V. Shuryak and I. Zahed, Phys.Rev.A, in press, arXiv:nucl-th/0410067.
20. E. V. Shuryak, arXiv:nucl-th/0606046.
21. S.S.Gubser, I.R.Klebanov and A.A. Tseytlin, Nucl. Phys. **B534** (1998) 202
22. J. M. Maldacena, Phys. Rev. Lett. **80**, 4859 (1998) [arXiv:hep-th/9803002]. S. J. Rey and J. T. Yee, Eur. Phys. J. C **22**, 379 (2001) [arXiv:hep-th/9803001].
23. G. W. Semenoff and K. Zarembo, Nucl. Phys. Proc. Suppl. **108**, 106 (2002)

[arXiv:hep-th/0202156].
24. I. R. Klebanov, J. M. Maldacena and C. B. Thorn, JHEP **0604**, 024 (2006) [arXiv:hep-th/0602255].
25. C. G. . Callan and A. Guijosa, Nucl. Phys. B **565**, 157 (2000) [arXiv:hep-th/9906153].
26. E. V. Shuryak and I. Zahed, Phys. Lett. B **608**, 258 (2005) [arXiv:hep-th/0310031].
27. M. Kruczenski, D. Mateos, R. C. Myers and D. J. Winters, JHEP **0307**, 049 (2003) [arXiv:hep-th/0304032].
28. S. Hong, S. Yoon and M. J. Strassler, JHEP **0603**, 012 (2006) [arXiv:hep-th/0410080].
29. G. Policastro, D. T. Son and A. O. Starinets, Phys. Rev. Lett. **87** (2001) 081601.
30. J. Casalderrey-Solana and D. Teaney, hep-ph/0605199.
31. H. Liu, K. Rajagopal, and U. A. Wiedemann, hep-ph/0605178.
32. S.-J. Sin and I. Zahed, *Phys. Lett.* **B608** (2005) 265–273, hep-th/0407215.
33. C. P. Herzog, A. Karch, P. Kovtun, C. Kozcaz, and L. G. Yaffe, hep-th/0605158.
34. S. S. Gubser, hep-th/0605182.
35. A. Buchel, hep-th/0605178.
36. S.-J. Sin and I. Zahed, hep-ph/0606049.
37. J. J. Friess, S. S. Gubser, and G. Michalogiorgakis, hep-th/0605292.
38. H. Nastase, hep-th/0501068.
39. E. Shuryak, S. J. Sin and I. Zahed, arXiv:hep-th/0511199.
40. R. A. Janik and R. Peschanski, Phys. Rev. D **73**, 045013 (2006) arXiv:hep-th/0512162, arXiv:hep-th/0606149.
41. A. Karch, E. Katz, D. T. Son and M. A. Stephanov, arXiv:hep-ph/0602229.
42. E. Shuryak, arXiv:hep-th/0605219.
43. E. V. Shuryak, hep-ph/9909458 (unpublished).
44. O. Andreev and V. I. Zakharov, "Heavy-quark potentials and AdS/QCD," arXiv:hep-ph/0604204. "The spatial string tension, thermal phase transition, and AdS/QCD," arXiv:hep-ph/0607026.
45. O. Kaczmarek, S. Ejiri, F. Karsch, E. Laermann and F. Zantow, hep-lat/0312015.
46. J. Liao and E. V. Shuryak, in progress
47. T. C. Kraan and P. van Baal, Phys. Lett. B **428**, 268 (1998) [arXiv:hep-th/9802049].
48. K. M. Lee and P. Yi, Phys. Rev. D **56**, 3711 (1997) [arXiv:hep-th/9702107].
49. T. Sakai and S. Sugimoto, Prog. Theor. Phys. **114**, 1083 (2006) [arXiv:hep-th/0507073].
50. K. Peeters, J. Sonnenschein and M. Zamaklar, arXiv:hep-th/0606195.

NEW OPEN AND HIDDEN CHARM SPECTROSCOPY

P. COLANGELO*, F. DE FAZIO, R. FERRANDES, S. NICOTRI

Istituto Nazionale di Fisica Nucleare, Sezione di Bari, Bari, Italy

**E-mail: pietro.colangelo@ba.infn.it*

Many new results on open and hidden charm spectroscopy have been obtained recently. We present a short review of the experimental findings in the meson sector, of the theoretical interpretations and of the open problems, with a discussion on the possibility that some mesons are not quark-antiquark states.

Keywords: charmed mesons, quarkonium, nonstandard quark/gluon states

1. Introduction

Observation of a long list of new hadrons has been recently reported by experiments at e^+e^- and $p\bar{p}$ colliders, by fixed target experiments and by reanalyses of old data. We can use Leporello's words in Mozart's Don Giovanni: *Madamina, il catalogo é questo:* [a]

$D^*_{sJ}(2317)$, $D_{sJ}(2460)$, $D_{sJ}(2632)$, $D_{sJ}(2860)$, $D^*_0(2308)$, $D'_1(2440)$, h_c, η'_c, $X(3872)$, $X(3940)$, $Y(3940)$, $Z(3930)$, $Y(4260)$, $\Upsilon(1D)$, B_1, B_2, B_{s2}, $\Theta(1540)^+$, $\Theta_c(3099)$, $\Xi_{cc}(3518)$, ...

Not all the states in the list have been confirmed ($D_{sJ}(2632)$, $\Theta(1540)^+$, $\Theta_c(3099)$) and therefore we can ignore them. Other states ($\Xi_{cc}(3518)$) are baryons, deserving a dedicated analysis, and mesons with open (B_1, B_2, B_{s2}) or hidden beauty ($\Upsilon(1D)$), that we do not discuss here.

We only consider mesons with open and hidden charm. The wealth of information collected in recent years is impressive: not only the number of known states has nearly doubled, but a few experimental observations seem to challenge the current picture of mesons as simple quark-antiquark configurations. Therefore, it is important to search the signatures allowing us to assign a given state to a particular multiplet, so that the hints of exotic structures can be clearly interpreted. The next Sections are devoted to such

---

[a] My lady, this is the list:

a discussion, considering separately the case of open charm mesons, which at present can be classified according to known rules, and that of hidden charm states where a couple of mesons seem to escape simple classification schemes. [b]

2. Mesons with open charm

In QCD, for hadrons containing a single heavy quark Q and in the limit $m_Q \to \infty$, there is a spin-flavour symmetry due to the decoupling of the heavy quark from the dynamics of the light degrees of freedom (light quarks and gluons). Therefore, it is possible to classify states containing the heavy quark Q according to the total angular momentum s_ℓ of the light degrees of freedom. For mesons, states belonging to doublets with the same $s_\ell = s_{\bar{q}} + \ell$, with $s_{\bar{q}}$ the spin of the light antiquark and ℓ the orbital angular momentum relative to the heavy quark, are degenerate in mass in the large m_Q limit. In case of charm, $D^{0,+}$ and $D^{*0,+}$, D_s and D_s^* are the states in the $s_\ell^P = \frac{1}{2}^-$ $c\bar{u}(\bar{d}, \bar{s})$ doublet, corresponding to $\ell = 0$. The mass difference between the members of the doublet is $O(\frac{1}{m_c})$, and vanishes when $m_c \to \infty$.

For $\ell = 1$ there are two doublets with $s_\ell^P = \frac{1}{2}^+$ and $s_\ell^P = \frac{3}{2}^+$, for $\ell = 2$ two other doublets with $s_\ell^P = \frac{3}{2}^-$ and $s_\ell^P = \frac{5}{2}^-$, and so on. The spin-flavour symmetry is important not only for spectroscopy, but also for the classification of strong decay modes and for evaluating the rates, since decays involving heavy mesons belonging to the same doublets are related. For example, the decays of mesons belonging to the $s_\ell^P = \frac{3}{2}^+$ doublet in one light pseudoscalar and one heavy $s_\ell^P = \frac{1}{2}^-$ meson occur in $d-$wave, so that these states are expected, *ceteris paribus*, to be narrower than the states belonging to the doublet $s_\ell^P = \frac{1}{2}^+$, which decay to the same final states by $s-$wave transitions. These observations are at the basis of the analyses of the new mesons observed in $c\bar{q}$ and in $c\bar{s}$ systems. They must be used together with the consideration that $\frac{1}{m_Q}$ effects can be important in case of charm: for example, the two 1^+ states belonging to $s_\ell^P = \frac{1}{2}^+$ and $\frac{3}{2}^+$ doublets, due to the finite charm quark mass, could mix with a mixing angle θ_c to provide the physical axial vector mesons. Such effects must be investigated on the basis of the experimental observation.

[b]For other recent reviews on this subject see Refs. 1, 2.

2.1. $c\bar{q}$ mesons: $D_0^*(2308)$ and $D_1'(2440)$.

Information about broad $c\bar{q}$ mesons, one scalar and one axial vector charmed meson that can be interpreted as the states belonging to the $s_\ell^P = \frac{1}{2}^+$ $c\bar{u}$, $c\bar{d}$ doublets, comes from Cleo[3], Belle[4] and Focus[5] Collaborations. The resonance parameters are reported in Table 1; they are obtained observing that the $D\pi$ and $D^*\pi$ mass distributions, produced for example in $B \to D^{**}\pi$ with a D^{**} a generic $\ell = 1$ meson, require contributions with scalar or axial vector quantum numbers. Improved determinations of mass and width of the two other positive parity charmed states D_1 ($J^P = 1^+$) and D_2 ($J^P = 2^+$) have been obtained, together with a measurement of the mixing angle between the two 1^+ states. It is small: $\theta_c = -0.10 \pm 0.03 \pm 0.02 \pm 0.02$ rad ($\simeq -6^0$) [4].

Table 1. Mass and width of broad resonances observed in Dπ and D$^*\pi$.

		Belle[4]	Focus[5]
D_0^{*0}	M (MeV)	$2308 \pm 17 \pm 15 \pm 28$	$2407 \pm 21 \pm 35$
	Γ (MeV)	$276 \pm 21 \pm 18 \pm 60$	$240 \pm 55 \pm 59$
D_0^{*+}	M (MeV)		$2403 \pm 14 \pm 35$
	Γ (MeV)		$283 \pm 24 \pm 34$
		Belle[4]	Cleo[3]
$D_1'^0$	M (MeV)	$2427 \pm 26 \pm 20 \pm 15$	$2461^{+41}_{-34} \pm 10 \pm 32$
	Γ (MeV)	$384^{+107}_{-75} \pm 24 \pm 70$	$290^{+101}_{-79} \pm 26 \pm 36$

2.2. $c\bar{s}$ mesons: $D_{sJ}^*(2317)$, $D_{sJ}(2460)$ and $D_{sJ}(2860)$.

$D_{sJ}^*(2317)$ and $D_{sJ}(2460)$ were found at the B factories in $D_s\pi^0$ and $D_s^*\pi^0$, $D_s^*\gamma$ distributions, respectively, in e^+e^- continuum and in B decays[6]. Their widths are unresolved, and this has arisen doubts about their identification as the scalar and axial vector $s_\ell^P = \frac{1}{2}^+$ $c\bar{s}$ mesons (D_{s0} and D'_{s1}), forming, together with $D_{s1}(2536)$ and $D_{s2}(2573)$, the set of four low-lying $\ell = 1$ states. However, the masses of $D_{sJ}^*(2317)$ and $D_{sJ}(2460)$ are below their respective thresholds for strong decays, DK and D^*K, therefore the small width is natural. Moreover, analyses of radiative transitions, that probe the structure of hadrons, support the $c\bar{s}$ interpretation of the two states[7,8]. For example, by Light-Cone QCD sum rules one can compute the hadronic parameters d, g_1, g_2 and g_3 governing the $D_{sJ}^*(2317) \to D_s^*\gamma$ and

$D_{sJ}(2460) \to D_s^{(*)}\gamma$, $D_{sJ}^*(2317)\gamma$ decay amplitudes[9]:

$$\begin{aligned}
\langle\gamma(q,\lambda)D_s^*(p,\lambda')|D_{s0}(p+q)\rangle &= ed\,[(\varepsilon^*\cdot\tilde{\eta}^*)(p\cdot q)-(\varepsilon^*\cdot p)(\tilde{\eta}^*\cdot q)]\\
\langle\gamma(q,\lambda)D_s(p)|D'_{s1}(p+q,\lambda'')\rangle &= eg_1\,[(\varepsilon^*\cdot\eta)(p\cdot q)-(\varepsilon^*\cdot p)(\eta\cdot q)]\\
\langle\gamma(q,\lambda)D_s^*(p,\lambda')|D'_{s1}(p+q,\lambda'')\rangle &= i\,e\,g_2\,\varepsilon_{\alpha\beta\sigma\tau}\eta^\alpha\tilde{\eta}^{*\beta}\varepsilon^{*\sigma}q^\tau\\
\langle\gamma(q,\lambda)D_{s0}(p)|D'_{s1}(p+q,\lambda'')\rangle &= i\,e\,g_3\,\varepsilon_{\alpha\beta\sigma\tau}\varepsilon^{*\alpha}\eta^\beta p^\sigma q^\tau
\end{aligned} \tag{1}$$

($\varepsilon(\lambda)$ is the photon polarization vector and $\tilde{\eta}(\lambda')$, $\eta(\lambda'')$ the D_s^* and D'_{s1} polarization vectors). Considering the correlation functions[10,11]

$$F(p,q) = i\int d^4x\, e^{ip\cdot x}\langle\gamma(q,\lambda)|T[J_A^\dagger(x)J_B(0)]|0\rangle \tag{2}$$

of quark-antiquark currents $J_{A,B}$ having the same quantum number of the decaying and of the produced charmed mesons, and an external photon state of momentum q and helicity λ, and expanding on the light-cone, it

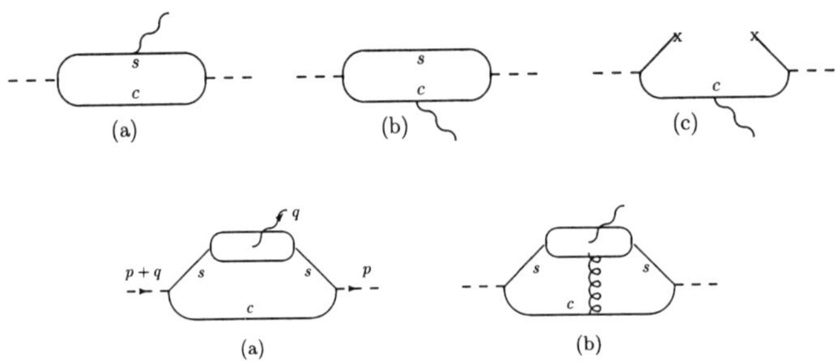

Fig. 1. Leading contributions to the correlation functions eq.(2) expanded on the light-cone: perturbative photon emission by the strange and charm quark ((a,b) in the first line) and two- and three-particle photon distribution amplitudes (second line); (c) corresponds to the strange quark condensate contribution.

is possible to express F in terms of the perturbative photon coupling to the strange and charm quarks, together with the contributions of the photon emission from the soft s quark, expressed as photon matrix elements of increasing twist[12], see fig.1 . The hadronic representation of the correlation function involves the contribution of the lowest-lying resonances, the current-vacuum matrix elements of which are computed by the same method[13], and a continuum of states treated invoking global quark-hadron duality. A Borel transformation introduces an external parameter M^2, the hadronic quantities being independent of it (fig. 2).

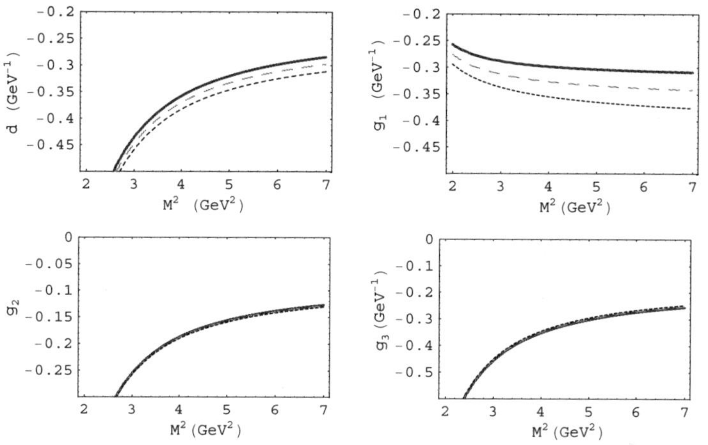

Fig. 2. Results for the hadronic parameters in eq.(1); M^2 is the Borel parameter.

Looking at the results, collected in Table 2, one sees that the rate of $D_{sJ}(2460) \to D_s\gamma$ is the largest one among the radiative $D_{sJ}(2460)$ rates, and this is confirmed by experiment, as reported in Table 3. Quantitative understanding of the ratios in Table 3 requires a precise knowledge of the widths of the isospin violating transitions $D_{s0} \to D_s\pi^0$ and $D'_{s1} \to D_s^*\pi^0$. In the description of these transitions based on the mechanism of $\eta - \pi^0$ mixing [7,8] the accurate determination of the strong $D_{s0}D_s\eta$ and $D'_{s0}D_s^*\eta$ couplings for finite heavy quark mass and including $SU(3)$ corrections is required.

If there are no reasons to consider $D_{sJ}^*(2317)$ and $D_{sJ}(2460)$ as exotic mesons, the same conclusion seems mandatory for $D_{sJ}(2860)$, a state recently observed by BaBar[15] in the DK system inclusively produced in $e^+e^- \to DKX$. The parameters of the resonance are: $M(D_{sJ}(2860)) = 2856.6 \pm 1.5 \pm 5.0$ MeV and $\Gamma(D_{sJ}(2860) \to DK) = 48 \pm 7 \pm 10$ MeV, (where $DK = D^0K^+$ and D^+K_S). In the same set of data and range of

Table 2. Radiative decay widths (in keV) of $D_{sJ}^*(2317)$ and $D_{sJ}(2460)$ obtained by Light-Cone sum rules (LCSR), Vector Meson Dominance (VMD) and constituent quark model (QM).

Initial state	Final state	LCSR [9]	VMD [8]	QM [7]	QM [14]
$D_{sJ}^*(2317)$	$D_s^*\gamma$	4-6	0.85	1.9	1.74
$D_{sJ}(2460)$	$D_s\gamma$	19-29	3.3	6.2	5.08
	$D_s^*\gamma$	0.6-1.1	1.5	5.5	4.66
	$D_{sJ}^*(2317)\gamma$	0.5-0.8	—	0.012	2.74

Table 3. Measurements and 90% CL limits of ratios of $D^*_{sJ}(2317)$ and $D_{sJ}(2460)$ decay widths.

	Belle	BaBar	CLEO
$\Gamma\left(D^*_{sJ}(2317)\to D^*_s\gamma\right)/\Gamma\left(D^*_{sJ}(2317)\to D_s\pi^0\right)$	< 0.18	—	< 0.059
$\Gamma\left(D_{sJ}(2460)\to D_s\gamma\right)/\Gamma\left(D_{sJ}(2460)\to D^*_s\pi^0\right)$	0.45 ± 0.09	0.30 ± 0.04	< 0.49
$\Gamma\left(D_{sJ}(2460)\to D^*_s\gamma\right)/\Gamma\left(D_{sJ}(2460)\to D^*_s\pi^0\right)$	< 0.31	—	< 0.16
$\Gamma\left(D_{sJ}(2460)\to D^*_{sJ}(2317)\gamma\right)/\Gamma\left(D_{sJ}(2460)\to D^*_s\pi^0\right)$	—	< 0.23	< 0.58

mass no structures seem to appear in the D^*K distribution, while a broad contribution seems to be present in the DK distribution at smaller mass.

It is interesting to discuss this new meson in some detail[16]. A possible quantum number assignment for a $c\bar{s}$ meson decaying to DK is either $s_\ell^P = \frac{3}{2}^-$ $J^P = 1^-$, or $s_\ell^P = \frac{5}{2}^-$ $J^P = 3^-$, in both cases corresponding to $\ell = 2$ and lowest radial quantum number ($n = 0$). Another possibility is that $D_{sJ}(2860)$ is a radial excitation ($n = 1$) of already observed $c\bar{s}$ mesons: the $J^P = 1^-$ $s_\ell^P = \frac{1}{2}^-$ state (the first radial excitation of D^*_s), the $J^P = 0^+$ $s_\ell^P = \frac{1}{2}^+$ state (radial excitation of $D^*_{sJ}(2317)$) or the $J^P = 2^+$ $s_\ell^P = \frac{3}{2}^+$ state (radial excitation of $D_{s2}(2573)$). In the absence of the helicity distribution of the final state, arguments can be provided to support a particular assignment of J^P considering the observed mass, the decay modes and width.

A piece of information comes from the DK width. Using an effective QCD Lagrangian incorporating spin-flavour heavy quark symmetry and light quark chiral symmetry, an estimate is possible of the ratios $\frac{\Gamma(D_{sJ}(2860)\to D^*K)}{\Gamma(D_{sJ}(2860)\to DK)}$ and $\frac{\Gamma(D_{sJ}(2860)\to D_s\eta}{\Gamma(D_{sJ}(2860)\to DK)}$ for various quantum number assignments to $D_{sJ}(2860)$ (Table 4). Non observation (at present) of a D^*K signal implies that the production of D^*K is not favoured, and therefore the assignments $s_\ell^P = \frac{1}{2}^-$, $J^P = 1^-$, $n = 1$, and $s_\ell^P = \frac{3}{2}^+$,

Table 4. Predicted $\frac{\Gamma(D_{sJ}\to D^*K)}{\Gamma(D_{sJ}\to DK)}$ and $\frac{\Gamma(D_{sJ}\to D_s\eta)}{\Gamma(D_{sJ}\to DK)}$ for various assignment of quantum numbers to $D_{sJ}(2860)$. The sum $DK = D^0K^+ + D^+K_S$ is understood.

s_ℓ^p, J^P, n	$D_{sJ}(2860)\to DK$	$\frac{\Gamma(D_{sJ}\to D^*K)}{\Gamma(D_{sJ}\to DK)}$	$\frac{\Gamma(D_{sJ}\to D_s\eta)}{\Gamma(D_{sJ}\to DK)}$
$\frac{1}{2}^-$, 1^-, 1	p-wave	1.23	0.27
$\frac{1}{2}^+$, 0^+, 1	s-wave	0	0.34
$\frac{3}{2}^+$, 2^+, 1	d-wave	0.63	0.19
$\frac{3}{2}^-$, 1^-, 0	p-wave	0.06	0.23
$\frac{5}{2}^-$, 3^-, 0	f-wave	0.39	0.13

$J^P = 2^+$, $n = 1$ can be excluded. The assignment $s_\ell^P = \frac{3}{2}^-$, $J^P = 1^-$, $n = 0$ can also be excluded, since the width $\Gamma(D_{sJ} \to DK)$ would be naturally large for a $p-$wave $D_{sJ} \to DK$ transition. [c]

In the case of the assignment $s_\ell^P = \frac{1}{2}^+$, $J^P = 0^+$, $n = 1$ the decay $D_{sJ} \to D^*K$ is forbidden and $D_{sJ} \to DK$ occurs in $s-$wave: this is the assignment proposed in Ref. 18. However, for the state with the lowest radial quantum number $n = 0$ the computed coupling costant $g_{D_{sJ}DK}$ is in agreement with observation[13,1]; using it one would obtain $\Gamma(D_{sJ} \to DK) \simeq$ 1.4 GeV. Although it is reasonable to suppose that the coupling of radial excitation is smaller, a large signal would be expected in any case in the $D_s\eta$ channel. Moreover, the spin partner with $J^P = 1^+$ ($s_\ell^p = \frac{1}{2}^+$, $n = 1$) would decay to D^*K with a small width, $\simeq 40$ MeV, a rather easy signal to observe. Therefore, to explain the absence of the D^*K signal one must invoke a mechanism favouring the production of the 0^+ state and inhibiting that of 1^+ state in $e^+e^- \to DKX$, a mechanism discriminating the first radial excitation from the case $n = 0$.

For $s_\ell^P = \frac{5}{2}^-$, $J^P = 3^-$, $n = 0$ the small DK width is mainly due to kinematics ($\Gamma \propto q_K^7$). A smaller but non negligible signal in the D^*K mode is predicted, and a small signal in the $D_s\eta$ mode is also expected. The coupling constant is similar to the couplings of the other doublets to light pseudoscalars. If $D_{sJ}(2860)$ has $J^P = 3^-$, it is not expected to be produced in non leptonic B decays such as $\bar{B}^0 \to D_{sJ}(2860)^- D^+$, so that the quantum number assignment can be confirmed by studies of D_{sJ} production in B transitions. $D_{sJ}(2860)$ can be one of the predicted high mass, high spin and relatively narrow $c\bar{s}$ states [19,16]; its non-strange partner D_3 is also expected to be narrow: $\Gamma(D_3^+ \to D^0\pi^+) \simeq 37$ MeV, and can be produced in semileptonic and non leptonic B decays, such as $\bar{B}^0 \to D_3^+ \ell^- \bar{\nu}_\ell$ and $\bar{B}^0 \to D_3^+ \pi^-$.

We conclude this Section showing in fig. 3 a tentative classification of the known $c\bar{s}$ mesons. Confirmation of this classification and the search for the missing states is a task for current and future investigations.

3. Hidden charm mesons

While the results in the open charm sector can be organized in a well-established scheme, the situation in the hidden charm sector is more com-

---

[c]A natural candidate for this assignment is the resonance $D_{sJ}(2715)$ observed very recently by Belle[17] in B decays with $M = 2715 \pm 11^{+11}_{-14}$ MeV, $\Gamma = 115 \pm 20^{+36}_{-32}$ MeV and $J^P = 1^-$.

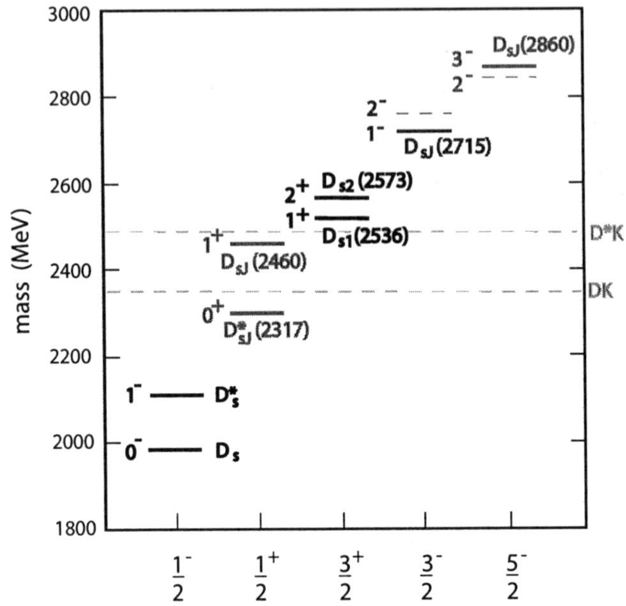

Fig. 3. Possible classification of the known $c\bar{s}$ states according to s_ℓ^P.

plex. A few new results, in particular those concerning η_c' and h_c, essentially agree with the expectations, although some particular aspects deserve investigations. Others, namely those concerning $X(3940)$, $Y(3940)$ and $Z(3930)$, could be organized according to generally accepted schemes with some caveat. The observations concerning $Y(4260)$ and $X(3872)$ have puzzling aspects: in particular, these states present features that could be expected for non standard quark-antiquark mesons, as we briefly discuss below.

3.1. h_c *and* η_c'

The observation of h_c ($J^{PC} = 1^{+-}$) by Cleo[20] in $\psi' \to \pi^0 h_c$, with h_c decaying to $\eta_c\gamma$, completes the set of four low-lying charmonium states with $\ell = 1$. The mass: $M(h_c) = 3524.4 \pm 0.6 \pm 0.4$ MeV deviates by less than 1 MeV from the center of gravity of the χ_{cJ} states. The strategy of searching h_c in B decays[21] has not been successful, yet, since the branching fraction of $B \to K h_c$ is smaller than estimated by the methods that reproduce the measured $B(B \to K\chi_{c0})$ [22].

Also the observation of η_c', made by Belle, Cleo and BaBar in B decays:

$B \to K\eta_c'$, in $e^+e^- \to J/\psi\eta_c'$ and in $\gamma\gamma \to \eta_c' \to K_S K^\pm \pi^\mp$ [23] completes the doublet of the first radial excitations of $(\eta_c, J/\psi)$. The parameters of the resonance are: $M(\eta_c') = 3638 \pm 4$ MeV (thus the hyperfine splitting is 48 MeV) and $\Gamma(\eta_c') = 14 \pm 7$ MeV. The observations are in agreement with the expectations, with some difficulty with the $\gamma\gamma$ rate of η_c' which is smaller than estimated[24]: the $c\bar{c}$ spectrum below the open charm threshold can be reproduced by a one-gluon-exchange short-distance potential, a scalar linearly confining potential and spin-spin and spin-orbit interactions[25]. However, when the energy increases, the theoretical determination of the meson properties, in particular of the spectrum, cannot ignore the open charm thresholds, starting from $D^0\bar{D}^0$, an old problem for which there is no model-independent solution, yet. Mass shifts of $20 - 40$ MeV have been estimated for states close to the thresholds [26]. This type of effects must be considered in the discussion of $X(3940)$, $Y(3940)$ and $Z(3930)$.

3.2. *X*(3940), *Y*(3940) *and* *Z*(3930).

For $X(3940)$, found by Belle[27] in the hadronic system recoiling against J/ψ in e^+e^- annihilation, with $M = 3943 \pm 6 \pm 6$ MeV, $\Gamma < 52$ MeV and decays into $D^*\bar{D}$, two interpretations are possible: i) the 3^1S_0 partner of 3^3S_1 ($\psi(4040)$), an assignment that could be confirmed by observation of the state in $\gamma\gamma$; ii) the first radial excitation of χ_{c1}, with the difficulty that χ_{c1} has not been found in the same set of data; moreover, another candidate, $Y(3940)$, is available for the same assignment.

Indeed, $Y(3940)$ was also found by Belle[28] in the $J/\psi\omega$ system produced in $B \to KJ/\psi\omega$. Its parameters are: $M = 3943 \pm 11 \pm 13$ MeV and $\Gamma = 85 \pm 22 \pm 26$ MeV; decays to open charm mesons have not been found, so far. The possible assignment as 2^3P_1 (χ_{c1}') implies that it should be observed in DD^*, even though the phase space for such a mode is small.

$Z(3930)$ is the last state in this region of mass found by Belle[29] in $\gamma\gamma \to D\bar{D}$, with $M = 3941 \pm 4 \pm 2$ MeV and $\Gamma = 20 \pm 8 \pm 3$ MeV. The helicity distribution in the final state is consistent with a $J = 2$ state, therefore it can be identified as the 2^3P_2 (χ_{c2}') meson.

In spite of the uncertainties in the quantum number assignment, the three states can be arranged in the $c\bar{c}$ spectrum, as shown in fig. 4. The case of $Y(4260)$ and $X(3872)$ is more difficult.

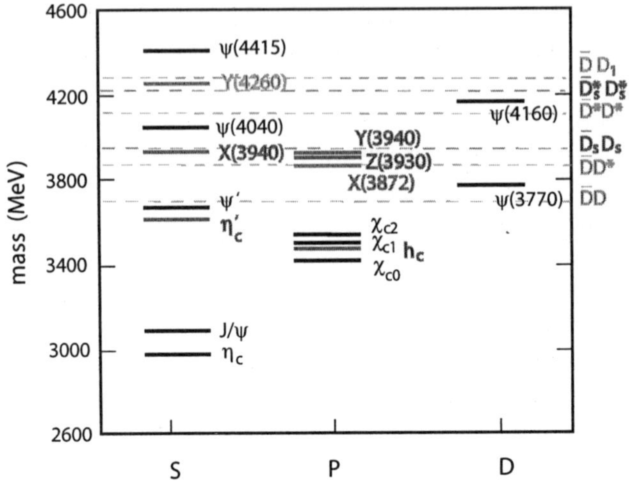

Fig. 4. Spectrum of $c\bar{c}$ states together with the thresholds for decays to open charm mesons. Possible positions of $X(3872)$ and $Y(4260)$ are shown.

3.3. $Y(4260)$

$Y(4260)$ is the first meson in the list of states seeming to escape ordinary classifications. It was found by BaBar[30] in $B^- \to K^-\pi\pi J/\psi$ and in radiative return analyses $e^+e^- \to \gamma_{ISR}\pi\pi J/\psi$, and confirmed by Cleo[31] in $e^+e^- \to \gamma_{ISR}Y$, with Y observed in $\pi^+\pi^- J/\psi$, $\pi^0\pi^0 J/\psi$ and $K^+K^- J/\psi$. The properties of the resonance are: $M = 4259 \pm 8 \pm 4$ MeV, $\Gamma = 88 \pm 23 \pm 5$ MeV and $J^{PC} = 1^{--}$. Moreover, the dipion mass distribution is consistent with a s-wave structure, so that a decay through $f_0(980)$ can be supposed. The problem with a $c\bar{c}$ interpretation is that a 1^{--} meson can be either a $\ell = 0$ state, a radial excitation between $\psi(4040)$ ($\psi(3S)$) and $\psi(4415)$ (at present interpreted as $\psi(4S)$), or a $\ell = 2$ state above $\psi(4159)$ (interpreted as $\psi(2D)$), with mass not predicted by any theoretical determination. Therefore, the meson looks as an extra state with respect to the 1^{--} levels, a state with a large coupling to $\pi\pi J/\psi$ and without (so far) observed decays in open charm mesons. Its mass is just above the $\bar{D}_s^* D_s^*$ threshold and below the $\bar{D}D_1$ threshold, $D_1(2420)$ being the narrow $c\bar{q}$ axial vector state. Among various interpretations[32], the one suggesting that $Y(4260)$ is a $\bar{c}Gc$ 1^{--} hybrid[33] emphasizes the agreement of the observation with some expectations. Indeed, charmed hybrids in this range of mass are conjectured, namely on the basis of lattice QCD simulations, with large couplings to J/ψ and light (η, η') mesons and with decays in open charm mesons with different orbital angular momentum (a decay in $DD_1'(2440)$ is possible, due to the broad width of D_1'). Noticeably, other hybrids with different quantum numbers are extected in the same range of mass; their observation would open a new chapter of the hadron spectroscopy.

3.4. $X(3872)$

We have left $X(3872)$ as the last meson to discuss, since it presents the most puzzling aspects. The observations can be summarized as follows:

(i) the X resonance has been found in $J/\psi\pi^+\pi^-$ distribution by four experiments, both in B decays ($B^{-(0)} \to K^{-(0)}X$), both in $p\bar{p}$ annihilation[34]. The mass is $M = 3871.9 \pm 0.6$ MeV while the width remains unresolved: $\Gamma < 2.3$ MeV (90 % CL);

(ii) there is no evidence of resonances in the charged mode $J/\psi\pi^{\pm}\pi^0$ or in $J/\psi\eta$ [35];

(iii) the state is not observed in e^+e^- annihilation;

(iv) for X produced in B decays the ratio $\dfrac{B(B^0 \to K^0X)}{B(B^+ \to K^+X)} = 0.61 \pm 0.36 \pm 0.06$ is obtained[35];

(v) the dipion spectrum in $J/\psi\pi^+\pi^-$ is peaked at large mass;

(vi) the decay in $J/\psi\pi^+\pi^-\pi^0$ is observed[36] with $\dfrac{B(X \to J/\psi\pi^+\pi^-\pi^0)}{B(X \to J/\psi\pi^+\pi^-)} = 1.0 \pm 0.4 \pm 0.3$: this implies G-parity violation;

(vii) the radiative mode $X \to J/\psi\gamma$ is found[36,37] with $\dfrac{B(X \to J/\psi\gamma)}{B(X \to J/\psi\pi^+\pi^-)} = 0.19 \pm 0.07$, therefore charge conjugation of the state is C=+1;

(viii) the angular distribution of the final state is compatible with the spin-parity assignment $J^P = 1^+$ [38];

(ix) there is a signal in $D^0\bar{D}^0\pi^0$ with $\frac{B(X \to D^0\bar{D}^0\pi^0)}{B(X \to J/\psi\pi^+\pi^-)} = 9 \pm 4$ [39] .

All the measurements are thus compatible with the assignment $J^{PC} = 1^{++}$. If the 2^3P_2 is identified with $Y(3940)$, there is overpopulation of 1^{++} $c\bar{c}$ mesons.

Noticeably, the mass of the resonance coincides with that of the $D^{*0}\bar{D}^0$ pair; this suggests that the state could be a realization of the molecular quarkonium[40], a bound state of two mesons, D^{*0} and $\bar{D}^0$, with small binding energy[41]. The absence of a $D^{*+}D^-$ molecule can be interpreted in this scheme observing that, being heavier by 7 MeV, such a state can rapidly decay in $D^{*0}\bar{D}^0$. In this description, the wave function of X(3872) has various components[42]:

$$|X(3872)>= a\,|D^{*0}\bar{D}^0 + \bar{D}^{*0}D^0> +b\,|D^{*+}D^- + D^{*-}D^+> + \dots \quad (3)$$

allowing to explain a few observations and to make predictions:

(i) the state has no definite isospin;
(ii) the decay $X \to J/\psi\pi^0\pi^0$ is forbidden;
(iii) since the decays of the resonance are mainly due to the decays of its components, the radiative transition in neutral mesons $X \to D^0\bar{D}^0\gamma$ should be dominant with respect to $X \to D^+D^-\gamma$;
(iv) a resonance $X_b(10604)$ is expected as a bound state of $\bar{B}^0$ and B^{*0};
(v) if the molecular binding mechanism is provided by a single pion exchange, this model explains the absence of $D\bar{D}$ molecular states.

The description of $X(3872)$ in the simple charmonium scheme, leaving unsolved the issue of the overpopulation of 1^{++} states, presents alternative arguments to the molecular description[43]. First, the molecular binding mechanism cannot be a single π^0 exchange, since this produces an attractive potential which is a delta function in space:

$$V(r) = -\frac{1}{3}\,g^2_{D^*D\pi}\,\epsilon'\cdot\epsilon\,\delta(r) + \dots \quad (4)$$

($g_{D^*D\pi}$ is the coupling constant of the $D^*D\pi$ vertex, ϵ and ϵ' the D^* polarization vectors) and therefore it does not give rise to a bound state in three spatial dimensions. Concerning the isospin (G-parity) violation, to correctly interpret the large value of the ratio $\dfrac{B(X \to J/\psi\pi^+\pi^-\pi^0)}{B(X \to J/\psi\pi^+\pi^-)}$ one has to consider that the phase space effects in two and three pion modes are very different. The amplitude ratio is rather small: $\dfrac{A(X \to J/\psi\rho^0)}{A(X \to J/\psi\omega)} \simeq 0.2$, so that the isospin violating amplitude is 20% of the isospin conserving one, an effect that could be related to another isospin violating effect, the mass difference between neutral and charged D mesons, considering the contribution of DD^* intermediate states to X decays. Finally, also the eventual dominance of $X \to D^0\bar{D}^0\gamma$ with respect to $X \to D^+D^-\gamma$ could be interpreted invoking standard mechanisms. Notice that a prediction of the charmonium description is that the rates of $B^0 \to XK^0$ and $B^- \to XK^-$ are nearly equal; the measurements are not conclusive on this point.

Our conclusion is that, at present, there are no compelling arguments allowing to exclude an interpretation in favour of others. Further analyses are requested to solve the issue of $X(3872)$.

4. Conclusions

In this short review of the new charm meson spectroscopy we have attempted to schematically describe the experimental observations, various interpretations and the main open problems. We do not want to emphasize how interesting the present situation is, and how much work is needed, both on the experimental, both on the theory side, to elaborate the information collected so far. We prefer to borrow the conclusion from another review on charm, written about 30 years ago: "It is easy to see the time when the charmed particles will be studied in detail... so that we look for new enjoyment and surprises." [44]

Acknowledgments

We are grateful to M.A.Shifman for invitation to the QCD Workshop. We thank A. Palano and T.N. Pham for discussions, and we acknowledge partial support from the EC Contract No. HPRN-CT-2002-00311 (EURIDICE).

References

1. P. Colangelo, F. De Fazio, R. Ferrandes, Mod. Phys. Lett. A **19** (2004) 2083.
2. E. S. Swanson, Phys. Rept. **429** (2006) 243.
3. S. Anderson *et al.* [CLEO Collaboration], Nucl. Phys. A**663** (2000) 647.
4. K. Abe *et al.* [Belle Collaboration], Phys. Rev. D **69** (2004) 112002.
5. J. M. Link *et al.* [FOCUS Collaboration], Phys. Lett. B **586** (2004) 11.
6. B. Aubert *et al.* [BABAR Collaboration], Phys. Rev. Lett. **90** (2003) 242001; D. Besson *et al.* [CLEO], Phys. Rev. D**68** (2003) 032002; Y. Mikami *et al.* [Belle Collaboration], Phys. Rev. Lett. **92** (2004) 012002; P. Krokovny *et al.* [Belle Collaboration], Phys. Rev. Lett. **91** (2003) 262002; A. Drutskoy *et al.* [Belle Collaboration], Phys. Rev. Lett. **96** (2005) 061802; E. W. Vaandering [FOCUS Collaboration], hep-ex/0406044; B. Aubert *et al.* [BABAR Collaboration], Phys. Rev. Lett. **93** (2004) 181801; B. Aubert *et al.* [BABAR Collaboration], Phys. Rev. D **69** (2004) 031101; B. Aubert *et al.* [BABAR Collaboration], hep-ex/0408067.
7. S. Godfrey, Phys. Lett. B **568** (2003) 254.
8. P. Colangelo and F. De Fazio, Phys. Lett. B **570** (2003) 180.
9. P. Colangelo, F. De Fazio and A. Ozpineci, Phys. Rev. D **72** (2005) 074004.
10. V. M. Belyaev *et al.*, Phys. Rev. D **51** (1995) 6177, and references therein.
11. For a review see: P. Colangelo and A. Khodjamirian, in "QCD sum rules: A modern perspective," 'At the Frontier of Particle Physics/Handbook of QCD', ed. by M. Shifman (World Scientific, Singapore, 2001), page 1495, arXiv:hep-ph/0010175.
12. P. Ball, V. M. Braun and N. Kivel, Nucl. Phys. B **649** (2003) 263.
13. P. Colangelo *et al.*, Phys. Rev. D **52** (1995) 6422; P. Colangelo and F. De

Fazio, Eur. Phys. J. C **4** (1998) 503;
14. W. A. Bardeen, E. J. Eichten and C. T. Hill, Phys. Rev. D **68** (2003) 054024.
15. B. Aubert [BABAR Collaboration], arXiv:hep-ex/0607082.
16. P. Colangelo, F. De Fazio and S. Nicotri, arXiv:hep-ph/0607245.
17. K. Abe *et al.*, [Belle Collaboration], arXiv:hep-ex/0608031.
18. E. van Beveren and G. Rupp, arXiv:hep-ph/0606110.
19. P. Colangelo, F. De Fazio and G. Nardulli, Phys. Lett. B **478** (2000) 408.
20. J. L. Rosner *et al.* [CLEO Collaboration], Phys. Rev. Lett. **95** (2005) 102003.
21. P. Colangelo, F. De Fazio and T. N. Pham, Phys. Rev. D **69** (2004) 054023.
22. F. Fang *et al.* [Belle Collaboration], arXiv:hep-ex/0605007.
23. S. K. Choi *et al.* [Belle Collaboration], Phys. Rev. Lett. **89** (2002) 102001 [Erratum-ibid. **89** (2002) 129901]; D. M. Asner *et al.* [CLEO Collaboration], Phys. Rev. Lett. **92** (2004) 142001; B. Aubert *et al.* [BABAR Collaboration], Phys. Rev. Lett. **92** (2004) 142002.
24. J. P. Lansberg and T. N. Pham, arXiv:hep-ph/0603113.
25. T. Barnes, S. Godfrey and E. S. Swanson, Phys. Rev. D **72** (2005) 054026.
26. E. J. Eichten, K. Lane and C. Quigg, Phys. Rev. D **73** (2006) 014014 [Erratum-ibid. D **73** (2006) 079903].
27. K. Abe *et al.* [Belle Collaboration], arXiv:hep-ex/0507019.
28. K. Abe *et al.* [Belle Collaboration], Phys. Rev. Lett. **94** (2005) 182002.
29. S. Uehara *et al.* [Belle Collaboration], Phys. Rev. Lett. **96** (2006) 082003.
30. B. Aubert *et al.* [BABAR Collaboration], Phys. Rev. Lett. **95** (2005) 142001.
31. T. E. Coan *et al.* [CLEO Collaboration], Phys. Rev. Lett. **96** (2006) 162003.
32. L. Maiani, V. Riquer, F. Piccinini and A. D. Polosa, Phys. Rev. D **72** (2005) 031502; S. L. Zhu, Phys. Lett. B **625** (2005) 212; E. van Beveren and G. Rupp, arXiv:hep-ph/0605317.
33. F. E. Close and P. R. Page, Phys. Lett. B **628** (2005) 215; E. Kou and O. Pene, Phys. Lett. B **631** (2005) 164.
34. S. K. Choi *et al.* [Belle Collaboration], Phys. Rev. Lett. **91** (2003) 262001; D. Acosta *et al.* [CDF II Collaboration], Phys. Rev. Lett. **93** (2004) 072001; V. M. Abazov *et al.* [D0 Collaboration], Phys. Rev. Lett. **93** (2004) 162002; B. Aubert *et al.* [BABAR Collaboration], Phys. Rev. D **71** (2005) 071103.
35. D. Bernard, talk at the International Conference on QCD and Hadronic Physics, June 16-20 2005, Beijing, China.
36. K. Abe *et al.* [Belle Collaboration] arXiv:hep-ex/0505037.
37. B. Aubert *et al.* [BABAR Collaboration], arXiv:hep-ex/0607050.
38. K. Abe *et al.* [Belle Collaboration] arXiv:hep-ex/0505038.
39. G. Gokhroo et al., [Belle Collaboration], BELLE-CONF-0568 (2005); P. Pakhlov, talk at ICHEP 2006, 26/7-2/8 2006, Moskow, Russia.
40. M. B. Voloshin and L. B. Okun, JETP Lett. **23** (1976) 333.
41. Other proposals are discussed in Ref. 2.
42. M. B. Voloshin, Phys. Lett. B **579** (2004) 316; Int. J. Mod. Phys. A **21** (2006) 1239.
43. M. Suzuki, Phys. Rev. D **72** (2005) 114013.
44. V. A. Novikov, L. B. Okun, M. A. Shifman, A. I. Vainshtein, M. B. Voloshin and V. I. Zakharov, Phys. Rept. **41** (1978) 1.

PLANAR EQUIVALENCE — AN UPDATE

ADI ARMONI

Department of Physics, University of Wales Swansea,
Singleton Park, Swansea, SA2 8PP, UK

We review the recent developments in the area of planar equivalence for orientifold field theories.

1. Introduction and background

The most interesting regime of QCD is the strong coupling (non-perturbative) regime. In particular, chiral symmetry breaking and confinement are non-perturbative phenomena. Despite of the importance of the problem, no analytical methods of calculations at strong coupling exist.

In supersymmetric theories the situation is much better: certain quantities (F-terms) can be calculated *exactly*. In particular, it is possible to calculate the exact value of the gluino condensate in pure $\mathcal{N} = 1$ Super Yang-Mills.

The basic idea behind planar equivalence is to approximate QCD by a supersymmetric theory !

The history of planar equivalence is as follows: in 1998 soon after the seminal AdS/CFT paper of Maldacena [1], Kachru and Silverstein [2] suggested a class of non-supersymmetric large-N conformal gauge theories. The candidate theories were the duals of $AdS_5 \times S^5/\Gamma$ and therefore named 'orbifold field theories'. Although it turns out that these theories are in fact not conformal (even not perturbatively, see [3,4]) their conjecture led to a more subtle conjecture by Strassler [5]. A refined version of Strassler's conjecture is 'planar equivalence for orientifold field theories'. In contrast to various conjectures the latter can be actually proved [6,7]. This is the subject of this talk.[a]

The statement of planar equivalence for (the minimal) orientifold field

[a]For a more detailed review on the subject, see Ref. [8].

theory is as follows: **at large-N, in a certain well defined bosonic sector, $SU(N)$ $\mathcal{N}=1$ SYM is equivalent to an $SU(N)$ gauge theory with Dirac fermions in the two-index antisymmetric representation.** The same statement holds if the fermions are in the symmetric representation.

Although planar equivalence is an extremely interesting statement by itself, since it relates a supersymmetric gauge theory with a non-supersymmetric gauge theory, it is a very useful tool for QCD. Let us make a simple observation: for $SU(3)$ a Dirac fermion in the antisymmetric representation is equivalent to a Dirac fermion in the fundamental representation. Therefore the $SU(3)$ version of 'orientifold field theory' is one-flavor QCD ! Thus, we can approximate one-flavor QCD by supersymmetric Yang-Mills and evaluate non-perturbative quantities in QCD. In particular planar equivalence will enable us to calculate the quark condensate in one-flavor QCD by using the value of the gluino-condensate in super Yang-Mills.

2. Planar equivalence - a proof

Let us present the main ingredients of the proof. Namely that at large-N the supersymmetric and the non-supersymmetric coincide. A detailed rigorous proof is presented in [7].

Let us start with a perturbative proof. We wish to show that the planar graphs of the two theories coincide. It is useful to use 't Hooft's notation. In this notation the adjoint representation is denoted by two parallel lines with arrows pointing in opposite directions, whereas the antisymmetric(symmetric) representation is denoted by two parallel lines with arrows pointing in the same direction. The Feynman rules of the two theories are depicted in figure (1).

We argue that the direction of the arrows does not affect the value of planar graphs. To see that imagine that we color every pair of fermionic lines by blue and red. Accordingly gluonic lines will be either both red or both blue. A typical planar graphs will be divided to blue regions and red regions — separated by fermionic loops. An example is given in figure (2). Now imagine that we reverse the arrows on the red lines. In this way we map a planar graph of one theory into a planar graph of the other theory. But this action does not change the value of the graph. This argument completes the perturbative proof.

The complete non-perturbative proof is much more involved [7]. The main ingredients are as follows: define, for a generic Dirac fermion in the repre-

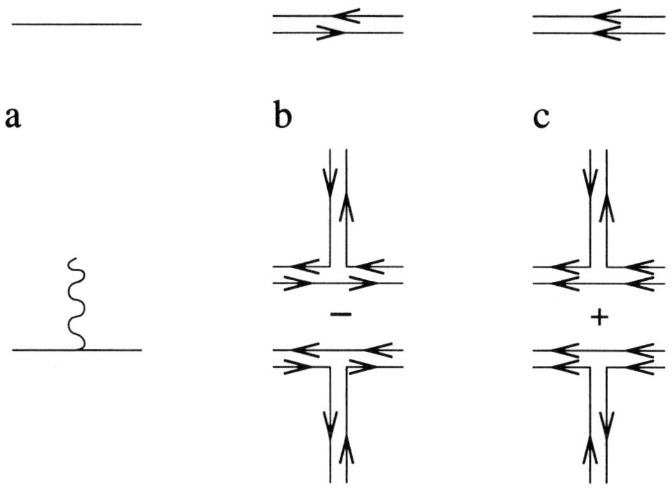

Fig. 1. a. The Quark-Gluon vertex. b. In $\mathcal{N}=1$ SYM. c. In 'orientifold field theory'

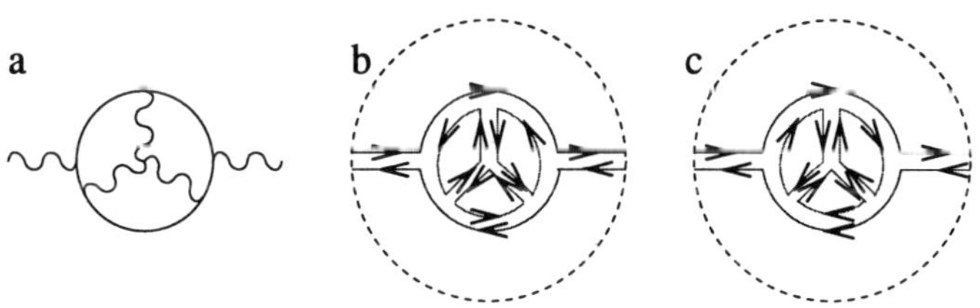

Fig. 2. A typical planar graph in SYM and the 'orientifold field theory'.

sentation r, the generating functional,

$$e^{-\mathcal{W}_r(J_{\rm YM},\,J_\Psi)} = \int DA_\mu\, D\Psi\, D\bar\Psi\, e^{-S_{\rm YM}[A,J_{\rm YM}]}\, \exp\left\{\bar\Psi\left(i\not\partial + \not A^a\, T^a_r + J_\Psi\right)\Psi\right\}\,, \tag{1}$$

Integrate out the fermions to arrive at

$$e^{-\mathcal{W}_r(J_{\rm YM},\,J_\Psi)} = \int DA_\mu\, e^{-S_{\rm YM}[A,J_{\rm YM}]+\Gamma_r[A,J_\Psi]}\,, \tag{2}$$

where

$$\Gamma_r[A,J_\Psi] = \log\,\det\left(i\not\partial + \not A^a\, T^a_r + J_\Psi\right)\,. \tag{3}$$

For what follows it is convenient to write the effective action $\Gamma_r[A, J_\Psi]$ in the world-line formalism as an integral over (super-)Wilson loops

$$\Gamma_r[A, J_\Psi] = -\frac{1}{2}\int_0^\infty \frac{dT}{T}$$

$$\times \int \mathcal{D}x\mathcal{D}\psi \,\exp\left\{-\int_\epsilon^T d\tau \left(\frac{1}{2}\dot{x}^\mu\dot{x}^\mu + \frac{1}{2}\psi^\mu\dot{\psi}^\mu - \frac{1}{2}J_\Psi^2\right)\right\}$$

$$\times \operatorname{Tr}\mathcal{P}\exp\left\{i\int_0^T d\tau \left(A_\mu^a\dot{x}^\mu - \frac{1}{2}\psi^\mu F_{\mu\nu}^a\psi^\nu\right)T_r^a\right\}, \quad (4)$$

Thus, the generating functional of a theory with matter in the antisymmetric/adjoint is very similar. The dependence on the representation enters through the Wilson loops. The latter can be written as follows

$$W_{\rm AS} = \frac{1}{2}\left((\operatorname{Tr}U)^2 - \operatorname{Tr}U^2\right) + (U \to U^\dagger), \quad (5)$$

$$W_{\rm adjoint} = \operatorname{Tr}U\operatorname{Tr}U^\dagger - 1 + (U \to U^\dagger) = 2\left(\operatorname{Tr}U\operatorname{Tr}U^\dagger - 1\right), \quad (6)$$

where U (resp. $U^\dagger$) represents the same group element in the *fundamental* (resp. *antifundamental*) representation of $SU(N)$.

To complete the proof [7], one has to show that at large-N we can use

$$W_{\rm AS} \sim \frac{1}{2}(\operatorname{Tr}U)^2 + \frac{1}{2}(\operatorname{Tr}U^\dagger)^2, \quad (7)$$

$$W_{\rm adjoint} \sim 2\operatorname{Tr}U\operatorname{Tr}U^\dagger, \quad (8)$$

and that U can be replaced by $U^\dagger$ everywhere. The factor 2 in (8) is cancelled by a factor $\frac{1}{2}$, since the adjoint representation is realized by Majorana fermions.

The implications of non-perturbative planar equivalence are that the non-supersymmetric orientifold field theory exhibits many supersymmetric properties. Although the planar theory consists of bosons only (it is impossible to form fermionic color-singlets), they will exhibit an even/odd parity degeneracy — as in the supersymmetric theory. In addition, a quark condensate $\langle\bar{\Psi}\Psi\rangle$ will form and its value will be identical to the value of the gluino condensate in $\mathcal{N}=1$ SYM. Other important properties are the NSVZ beta-function, the domain walls spectrum and gluonic Green-functions [6,8].

3. The orientifold large-N expansion

There are various options of generalizing QCD to $SU(N)$ gauge theory. Ultimately we are interested in a large-N approximation. A generalization makes sense only if the large-N theory is controlled by a planar theory. It is even tempting to say that a large-N theory makes sense only if it is dual to a string theory.

In the original 't Hooft large-N expansion both g^2N and the number of flavors N_f is kept fixed (this is realized in the modern gauge/string duality by keeping the number of flavor branes fixed). In the Veneziano large-N expansion (the topological expansion) the ratio N_f/N is kept fixed together with g^2N (it can be achieved by placing branes on orbifold singularities, at a certain region of the moduli space). The advantage of the latter expansion is that quark loops are not suppressed at large N and hence flavor physics is captured by the approximation. The η' mass, for example, is non-vanishing even when $N \to \infty$, namely it is part of the planar theory.

While both expansions are interesting and useful, the planar approximation is still hard and difficult to use for predictions in QCD.

Let us present a new large-N expansion [9]. It will lead to *quantitative* predictions for QCD. We start with an $SU(3)$ theory with N_f flavors in the fundamental representation ('multi-flavor QCD'). Since for $SU(3)$ a Dirac fermion in the fundamental representation in equivalent to a Dirac fermion in the antisymmetric representation, we have the option of generalizing the theory to an $SU(N)$ gauge theory with N_f antisymmetric Dirac fermions (see figure (3)).The next step is to consider the large-N limit of the theory while keeping N_f fixed. This large-N approximation is somewhat similar to the topological expansion since quark loops are not suppressed with respect to glue loops. In fact, as we shall see, for $N_f = 1$ they are equally important.

Fig. 3. Antisymmetric/fundamental representation in $SU(3)$

4. Applications for one-flavor QCD

As we explained in the previous section, we can approximate one-flavor QCD by a planar theory with an antisymmetric fermion. This theory is planar equivalent to $\mathcal{N} = 1$ SYM. We can therefore make several *quantitative* predictions about the non-perturbative regime of the one-flavor QCD !

The first prediction is about the spectrum of the theory. As we discussed at the end of section (2), the color-singlet spectrum of the 'orientifold field theory' will exhibit an odd/even parity degeneracy. Thus, we expect a similar degeneracy in the spectrum of one-flavor QCD, within a $1/N$ error,

$$\frac{M_-^S}{M_+^S} = 1 + \mathcal{O}(1/N)\,, \tag{9}$$

where M_-^S is a color-singlet bosonic degree of freedom with spin S and *odd* parity and M_+^S is a color-singlet bosonic degree of freedom with spin S and *even* parity. Clearly it will be simple to check this prediction in a lattice simulation.

Another prediction is the value of the quark condensate in one-flavor QCD. The analysis was carried out in [10] and checked recently in a lattice simulation by DeGrand et.al.[11] .

The idea is to start with a RGI definition of the gluino condensate and the quark condensate.

$$\langle \bar{\Psi}\Psi \rangle_{\mathrm{RGI}} \equiv (g^2)^{\gamma/\beta} \langle \bar{\Psi}\Psi \rangle \tag{10}$$

The RGI value of the gluino condensate is

$$\langle \lambda\lambda \rangle = -\frac{N^2}{2\pi^2}\Lambda^3\,. \tag{11}$$

Non-perturbative planar equivalence implies the equality of the orientifold quark condensate and the gluino condensate at infinite N. Moreover, since we know that for $N = 2$ the antisymmetric representation is equivalent to the singlet we can make the following 'educated guess' for the value of the quark condensate at any N

$$\langle \bar{\Psi}\Psi \rangle = -\left(1 - \frac{2}{N}\right) \frac{N^2}{2\pi^2}\Lambda^3\,. \tag{12}$$

The evaluation of the quark condensate for $N = 3$, namely for one-flavor QCD, at 2 GeV (assuming the 't Hooft coupling is 0.115) yields

$$\langle \bar{q}q \rangle_{2\mathrm{GeV}}^{\mathrm{orientifold}} = -(262 \text{ Mev})^3 \ \pm 30\% \tag{13}$$

this value can be compared with a recent lattice evaluation by DeGrand et.al. [11]

$$\langle \bar{q}q \rangle_{2\mathrm{GeV}}^{\mathrm{lattice}} = -(269(9) \text{ Mev})^3 \tag{14}$$

The agreement is more than satisfactory.

5. Applications for three-flavor QCD

It is possible to use planar equivalence to calculate non-perturbative quantities in real QCD, namely in three-flavor QCD. Consider an $SU(N)$ gauge theory with one Dirac fermion in the antisymmetric Ψ and two extra Dirac fermions in the fundamental representation χ^i. For $SU(3)$ this model reduces to three flavor QCD. When $N \to \infty$ the fundamental flavors can be neglected and our model is planar equivalent to $\mathcal{N}=1$ SYM. Thus the model interpolates between QCD for $SU(3)$ and SYM at large-N.

Several subtleties arise in the consideration of this model. Note that because of chiral symmetry breaking there will be Goldstone bosons in the model, at any finite N. Therefore, in the attempt to match quantities of this theory and $\mathcal{N}=1$ SYM, one has to choose sources which do not couple to the Goldstones.

A detailed analysis of the model [12] leads to

$$\langle\bar{\Psi}\Psi\rangle_{\rm RGI}/\Lambda^3 = -\left(1-\frac{2}{N}\right)\frac{N^2}{2\pi^2}\,, \qquad (15)$$

as in the previous case. Note however, that in this model $\beta_0 = 3N$, and hence the running coupling is different than in one-flavor QCD. Therefore we find

$$\langle\bar{\Psi}\Psi\rangle^{\rm orientifold}_{2\,{\rm GeV}} \quad = \quad -\,(317\,\pm\,30\,\pm\,36\ {\rm MeV})^3\,, \qquad (16)$$

(the errors are due to the 30% uncertainty and the experimental uncertainty of the 't Hooft coupling at 2 Gev.)

Our prediction should be compared with a recent lattice analysis by McNeile [13]

$$\langle\bar{\Psi}\Psi\rangle^{\rm lattice}_{2\,{\rm GeV}} \quad = \quad -\,(259\,\pm\,27\ {\rm MeV})^3\,. \qquad (17)$$

The orientifold prediction and the lattice simulation result are presented in figure (4).

6. Sagnotti's model and the gauge/string correspondence

Orientifold field theories originate in string theory. The starting point is 10d type 0B string theory. By adding the orientifold $\Omega' \equiv \Omega(-)^{f_R}$ and 32 D9 branes we end up with a non-supersymmetric non-tachyonic string theory [14,15]. The low energy spectrum of the closed string modes consists of the dilaton, the graviton and a set of RR field. There are no fermions (NSR sector). The open string sector consists of a ten dimensional $U(32)$ gauge theory with an antisymmetric fermion. The model is free of RR tadpoles.

In order to obtain a realization of the 4d orientifold field theory one can use a Hanany-Witten brane configuration in type 0A, namely a set of N D4 branes and O'4 plane suspended between rotated NS5 branes [6]. An alternative realization [16] is via fractional D3 branes placed on a $C^3/Z_2 \times Z_2$ orbifold singularity in type 0'B. The latter description is useful for the gauge/gravity correspondence [17]. Since at $g_{st} = 0$ the bosonic gravity modes of type 0'B and their interactions are identical to those of type IIB, the gauge/gravity correspondence (provided that it holds) provides an additional evidence in favor of planar equivalence: if the bosonic sector of two gauge theories are described by the same bosonic sector of two string theories at $g_{st} = 0$ then the two gauge theories must be equivalent at infinite N.

The gauge/gravity correspondence for 'orientifold field theories' was used recently [17] to make predictions about the theories at finite-N. In contrast to the supersymmetric type IIB background that contains N units of RR flux, the type 0B background contains $N-2$ units of RR flux, due to the presence of the $O'5$ plane that shifts the flux by -2. Certain quantities are sensitive to this shift. For example, it was found that the ratio between the lowest pseudo-scalar and scalar color-singlets is

$$M_-/M_+ \sim (N-2)/N \,. \tag{18}$$

This ratio is in agreement with an earlier prediction based on an effective action approach [18].

7. Recent developments and outlook

Planar equivalence was used in both formal works and in phenomenology. Papers on the subject appeared on all theoretical high-energy archives: hep-th, hep-ph and hep-lat. We wish to mention a few.

The lattice works concern the verification of planar equivalence. A formal strong-coupling and large mass proof was provided by Patella [19]. The paper by DeGrand et.al. [11] confirms the prediction of the quark condensate in one flavor QCD.

The phenomenological papers, mainly by Sannino and collaborators [20,21] were devoted to the constructions of technicolor models based on orientifold field theories with symmetric matter. In another recent work predictions about one-flavor QCD were used for 'beyond the standard model phenomenology' [22].

Among the more formal aspects, we wish to mention the work of Di Vecchia et.al. [16] who studied realizations of 'orientifold field theories' in type 0' string theory and tree level string amplitudes in those models.

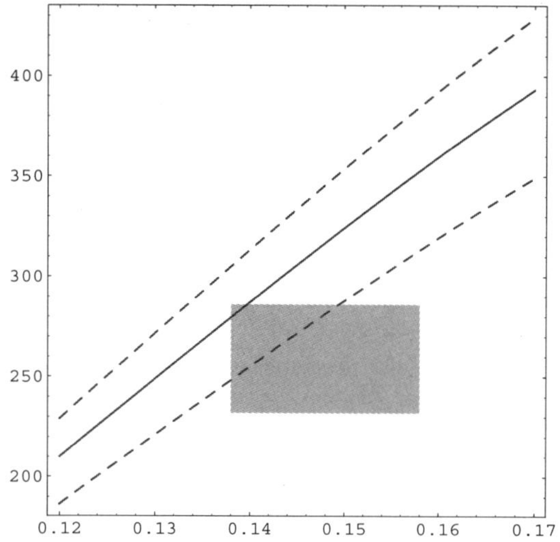

Fig. 4. The quark condensate expressed as $-(y \text{ MeV})^3$ as a function of the 't Hooft coupling λ. The solid line represent he prediction of planar equivalence. The two dashed lines represent the $\pm 30\%$ error. The $\pm 1\sigma$ range of the coupling, $0.138 < \lambda < 0.158$ and the lattice estimate $-(259 \pm 27 \text{ MeV})^3$ define the shaded region.

A partial list of other related works is given in refs.[23–26].

In conclusion planar equivalence is a new useful tool to calculate non-perturbative quantities. It already led to a couple of applications, both in QCD, string theory, AdS/CFT, lattice gauge theory and beyond the standard model phenomenology. We believe that further study is needed in order to exploit the potential of this new method. In particular to find new planar-equivalent pairs and to learn about one of them from the other.

ACKNOWLEDGMENTS

I wish to thank my collaborators Dr. Emiliano Imeroni, Prof. Mikhail Shifman, Prof. Graham Shore and Prof. Gabriele Veneziano. I am supported by a PPARC advanced fellowship research award.

References

1. J. M. Maldacena, "The large N limit of superconformal field theories and supergravity," Adv. Theor. Math. Phys. **2**, 231 (1998) [Int. J. Theor. Phys. **38**, 1113 (1999)] [hep-th/9711200].

2. S. Kachru and E. Silverstein, "4d conformal theories and strings on orbifolds," Phys. Rev. Lett. **80**, 4855 (1998) [hep-th/9802183].
3. A. Armoni, E. Lopez and A. M. Uranga, "Closed strings tachyons and non-commutative instabilities," JHEP **0302**, 020 (2003) [arXiv:hep-th/0301099].
4. A. Dymarsky, I. R. Klebanov and R. Roiban, "Perturbative gauge theory and closed string tachyons," JHEP **0511**, 038 (2005) [arXiv:hep-th/0509132].
5. M. J. Strassler, "On methods for extracting exact non-perturbative results in non-supersymmetric gauge theories," hep-th/0104032.
6. A. Armoni, M. Shifman and G. Veneziano, "Exact results in non-supersymmetric large N orientifold field theories," Nucl. Phys. B **667**, 170 (2003) [hep-th/0302163].
7. A. Armoni, M. Shifman and G. Veneziano, "Refining the proof of planar equivalence," Phys. Rev. D **71**, 045015 (2005) [hep-th/0412203].
8. A. Armoni, M. Shifman and G. Veneziano, "From super-Yang-Mills theory to QCD: Planar equivalence and its implications," hep-th/0403071.
9. A. Armoni, M. Shifman and G. Veneziano, "SUSY relics in one-flavour QCD from a new 1/N expansion," Phys. Rev. Lett. **91**, 191601 (2003) [hep-th/0307097].
10. A. Armoni, M. Shifman and G. Veneziano, "QCD quark condensate from SUSY and the orientifold large-N expansion," Phys. Lett. B **579**, 384 (2004) [hep-th/0309013].
11. T. DeGrand, R. Hoffmann, S. Schaefer and Z. Liu, "Quark condensate in one-flavor QCD," hep-th/0605147.
12. A. Armoni, G. Shore and G. Veneziano, "Quark condensate in massless QCD from planar equivalence," Nucl. Phys. B **740**, 23 (2006) [hep-ph/0511143].
13. C. McNeile, "An estimate of the chiral condensate from unquenched lattice QCD," Phys. Lett. B **619**, 124 (2005) [hep-lat/0504006].
14. A. Sagnotti, "Some properties of open string theories," hep-th/9509080.
15. A. Sagnotti, "Surprises in open-string perturbation theory," Nucl. Phys. Proc. Suppl. **56B**, 332 (1997) [hep-th/9702093].
16. P. Di Vecchia, A. Liccardo, R. Marotta and F. Pezzella, "Brane-inspired orientifold field theories," JHEP **0409**, 050 (2004) [hep-th/0407038].
17. A. Armoni and E. Imeroni, "Predictions for orientifold field theories from type 0' string theory," Phys. Lett. B **631**, 192 (2005) [hep-th/0508107].
18. F. Sannino and M. Shifman, "Effective Lagrangians for orientifold theories," Phys. Rev. D **69**, 125004 (2004) [hep-th/0309252].
19. A. Patella, "A proof of orientifold planar equivalence on the lattice," hep-lat/0511037.
20. F. Sannino and K. Tuominen, "Techniorientifold," Phys. Rev. D **71**, 051901 (2005) [hep-ph/0405209].
21. D. K. Hong, S. D. H. Hsu and F. Sannino, "Composite Higgs from higher representations," Phys. Lett. B **597**, 89 (2004) [hep-ph/0406200].
22. M. J. Strassler and K. M. Zurek, "Echoes of a hidden valley at hadron colliders," hep-ph/0604261.
23. A. Feo, P. Merlatti and F. Sannino, "Information on the super Yang-Mills spectrum," Phys. Rev. D **70**, 096004 (2004) [hep-th/0408214].

24. J. L. F. Barbon and C. Hoyos, "Small volume expansion of almost supersymmetric large N theories," JHEP **0601**, 114 (2006) [hep-th/0507267].
25. G. Veneziano and J. Wosiek, "Planar quantum mechanics: An intriguing supersymmetric example," JHEP **0601**, 156 (2006) [hep-th/0512301].
26. S. Bolognesi, "Baryons and Skyrmions in QCD with quarks in higher representations," hep-th/0605065.

NUCLEONS ON THE LIGHT CONE: THEORY AND PHENOMENOLOGY OF BARYON DISTRIBUTION AMPLITUDES

V. M. BRAUN

Institut für Theoretische Physik,
Universität Regensburg,
D-93040 Regensburg, Germany
E-mail: vladimir.braun@physik.uni-regensburg.de

This is a short review of the theory and phenomenology of baryon distribution amplitudes, including recent applications to the studies of nucleon form factors at intermediate momentum transfers using the light-cone sum rule approach.

Keywords: hard exclusive processes, distribution amplitudes, form factors

1. Introduction

In the next generation of experiments in hadron physics there is a tendency to go for more and more exclusive channels. One main reason for this is that one has understood only in recent years how much one can learn from reactions like deeply virtual Compton scattering (DVCS) about the internal hadron structure and especially the spin structure. All future plans also call for very high luminosity and would therefore be perfectly suited for the investigation of exclusive and semi-exclusive reactions with and without polarization.

Main question which has to be addressed at this stage is whether studies of hard exclusive processes can be made fully quantitative. The classical theoretical framework for the calculation of hard exclusive processes in QCD is based on QCD factorization [1–3]. This approach introduces a concept of hadron distribution amplitudes (DAs) as fundamental nonperturbative functions describing the hadron structure in rare parton configurations with a fixed number of Fock constituents at small transverse separation. DAs are ordered by increasing twist. For example, the leading-twist-2 meson DA $\phi_{2;P}$ describes the momentum distribution of the valence quarks in the meson P and is related to the meson's Bethe–

Salpeter wave function $\phi_{P,BS}$ by an integral over transverse momenta: $\phi_{2;P}(x,\mu) = Z_2(\mu)\int^{|k_\perp|<\mu} d^2k_\perp\, \phi_{P,BS}(x,k_\perp)$. Here x is the quark momentum fraction, Z_2 is the renormalization factor (in the light-cone gauge) for the quark-field operators in the wave function, and μ denotes the renormalization scale. Higher-twist DAs are much more numerous and describe either contributions of "bad" components in the wave function, or contributions of transverse motion of quarks (antiquarks) in the leading-twist components, or contributions of higher Fock states with additional gluons and/or quark–antiquark pairs. Within the hard-rescattering picture, the corresponding contributions to the hard exclusive reactions are suppressed by a power (or powers) of the large momentum Q and usually have received less attention.

The distribution amplitudes are equally important and to a large extent complementary to conventional parton distributions which correspond to one-particle probability distributions for the parton momentum fraction in an *average* configuration. They are, however, much less studied: The direct experimental information is only available for the pion distribution amplitude and comes from the CLEO measurement [4] of the $\gamma^*\gamma\pi$ transition form factor. In this talk I present a short review of the theoretical status of baryon DAs, mainly those of the nucleon. I describe the basic theoretical framework and discuss how the nucleon DAs can be related to the experimental measurements of form factors at accessible momentum transfers within the light-cone sum rule (LCSR) framework.

2. General framework

2.1. *Definitions*

Nucleon DAs are most conveniently defined as nucleon-to-vacuum transition matrix elements of nonlocal light-ray three-quark operators. In order to facilitate the power counting it is usually convenient to choose a light-like vector z_μ orthogonal to the large momentum q_μ involved in the problem: $q\cdot z = 0\,, \qquad z^2 = 0\,.$ The nucleon momentum $P_\mu, P^2 = m_N^2$ can be used to introduce the second light-like vector $p_\mu = P_\mu - \frac{1}{2}\, z_\mu \frac{m_N^2}{P\cdot z}$, $p^2 = 0$, so that $P \to p$ if the nucleon mass can be neglected, $m_N \to 0$.

The nucleon leading-twist-three DA $\Phi_N(x_i)$ can be defined as [5–8]

$$\langle 0|\varepsilon^{ijk}\left(u_i^\uparrow(a_1 z) C \not{z} u_j^\downarrow(a_2 z)\right)\not{z} d_k^\uparrow(a_3 z)|N(P)\rangle = \\ = -\frac{1}{2} pz \not{z} N^\uparrow(P)\int \mathcal{D}x e^{-ipz\sum x_i a_i}\Phi_N(x_i) \qquad (1)$$

where $\int \mathcal{D}x = \int_0^1 dx_1\, dx_2\, dx_3\, \delta\,(1 - x_1 - x_2 - x_3)$, x_i correspond to quark momentum fractions, C is the charge-conjugation matrix, $N(P)$ is the Dirac spinor and the arrows correspond to the helicity projections $q^{\uparrow(\downarrow)} = (1 \pm \gamma_5)/2\, q$.

The definition in (1) is equivalent to the following representation for the three-quark component of the proton wave function [3]

$$|p\uparrow\rangle = \int \frac{\mathcal{D}x\, \Phi_N(x_i)}{2\sqrt{24 x_1 x_2 x_3}} \left\{ |u^\uparrow(x_1) u^\downarrow(x_2) d^\uparrow(x_3)\rangle - |u^\uparrow(x_1) d^\downarrow(x_2) u^\uparrow(x_3)\rangle \right\}, \tag{2}$$

where the standard relativistic normalization of spinors is implied. One often writes $\Phi_N(1,2,3) = V_1(1,2,3) - A_1(1,2,3)$, where $V_1(x_1,x_2,x_3) = V_1(x_2,x_1,x_3,)$ and $A_1(x_1,x_2,x_3) = -A_1(x_2,x_1,x_3,)$ correspond to the symmetric and the antisymmetric part of $\Phi_N(x_i)$ w.r.t. the interchange of the u-quark momenta, respectively. The definitions of the leading-twist DAs of other baryons of the octet can be found in Ref. 9.

For higher twists, there exists an important conceptual difference between mesons and baryons. For mesons, all effects of the transverse motion of quarks in the valence quark-antiquark state can be rewritten in terms of higher Fock state contributions by using QCD equations of motion (EOM). Since quark-antiquark-gluon admixture in meson wave functions turns out to be numerically small, the transverse momentum contributions are small as well, and the higher-twist contributions to hard exclusive reactions involving mesons are dominated in most cases by meson mass corrections, see e.g. Ref. 10. For baryons, EOM are not sufficient to eliminate higher-twist three-quark DAs in favor of the components with extra gluons, so that the former present genuine new degrees of freedom. A systematic classification of such contributions is carried out in Ref. 8. One finds that to the twist-four accuracy there exist three independent DAs:

$$\begin{aligned}
&\langle 0|\varepsilon^{ijk}\left(u_i^\uparrow(a_1 z) C \not{z} u_j^\downarrow(a_2 z)\right) \not{p} d_k^\uparrow(a_3 z)|N(P)\rangle = \\
&\qquad = -\frac{1}{2} pz\, \not{p} N^\uparrow(P) \int \mathcal{D}x\, e^{-ipz\sum x_i a_i} \Phi_4(x_i)\,, \\
&\langle 0|\varepsilon^{ijk}\left(u_i^\uparrow(a_1 z) C \not{z}\gamma_\perp \not{p} u_j^\downarrow(a_2 z)\right) \gamma_\perp \not{z} d_k^\downarrow(a_3 z)|N(P)\rangle = \\
&\qquad = -m_N\, pz\, \not{z} N^\uparrow(P) \int \mathcal{D}x\, e^{-ipz\sum x_i a_i} \Psi_4(x_i)\,, \\
&\langle 0|\,\varepsilon^{ijk}\left(u_i^\uparrow(a_1 z) C \not{p} \not{z} u_j^\uparrow(a_2 z)\right) \not{z} d_k^\uparrow(a_3 z)\,|N(P)\rangle = \\
&\qquad = \frac{1}{2} m_N pz\, \not{z} N^\uparrow(P) \int \mathcal{D}x\, e^{-ipz\sum x_i a_i} \Xi_4(x_i)\,.
\end{aligned} \tag{3}$$

In addition, there exist three twist-5 and one twist-6 three-quark DA, which do not involve new parameters to this accuracy, and can be expressed in terms of twist-3,4 DAs.

Note that in the approach of Ref. 8 the higher-twist DAs are introduced as matrix elements of light-ray operators involving "minus" components of the quark field operators. All transverse degrees of freedom are eliminated. There exists an alternative approach [11] in which only "plus" components are involved, but the transverse momentum dependence is retained. Both techniques are probably equivalent but the precise connection has not been worked out yet.

2.2. *Conformal expansion*

A convenient tool to study DAs is provided by conformal expansion [12–18]. The underlying idea is similar to partial-wave decomposition in quantum mechanics and allows one to separate transverse and longitudinal variables in the Bethe–Salpeter wave function. The dependence on transverse coordinates is traded for the scale dependence of the relevant operators and is governed by renormalization-group equations, the dependence on the longitudinal momentum fractions is described in terms of irreducible representations of the corresponding symmetry group, the collinear conformal group SL(2,$\mathbb{R}$). The conformal partial-wave expansion is explicitly consistent with the equations of motion since the latter are not renormalized. It thus makes maximum use of the symmetry of the theory to simplify the dynamics.

To construct the conformal expansion for an arbitrary multiparticle distribution, one first has to decompose each constituent field into components with fixed Lorentz-spin projection onto the light-cone. Each such component has conformal spin $j = \frac{1}{2}(l+s)$, where l is the canonical dimension and s the (Lorentz-) spin projection. In particular, $l = 3/2$ for quarks and $l = 2$ for gluons. The quark field is decomposed as $\psi_+ \equiv \Lambda_+\psi$ and $\psi_- = \Lambda_-\psi$ with spin projection operators $\Lambda_+ = \frac{\not p \not z}{2pz}$ and $\Lambda_- = \frac{\not z \not p}{2pz}$, corresponding to $s = +1/2$ and $s = -1/2$, respectively. Note that the "minus" components of quark fields that contribute to higher-twist DAs correspond to the negative spin projection and thus lower conformal spin. The three-particle states built of quark with definite Lorentz-spin projection can be expanded in irreducible representations of SL(2,$\mathbb{R}$) with increasing conformal spin. The explicit expression for the DA with the lowest possible conformal spin $j = j_1 + j_2 + j_3$, the so-called asymptotic DA, is

$$\phi_{as}(x_1, x_2, x_3) = \frac{\Gamma(2j_1 + 2j_2 + 2j_3)}{\Gamma(2j_1)\Gamma(2j_2)\Gamma(2j_3)}\, x_1^{2j_1-1} x_2^{2j_2-1} x_3^{2j_3-1}. \qquad (4)$$

For the leading twist DAs $j_1 = j_2 = j_3 = 1$ reproducing the familiar result

$$\Phi_N^{\rm as}(x_1, x_2, x_3) = 120 x_1 x_2 x_3 \,. \tag{5}$$

The nucleon DA can be expanded in the sum over irreducible representations with higher spin $N+3$. For example for leading twist

$$\Phi_N(x_i) = \Phi_N^{\rm as}(x_i) \sum_{N=0}^{\infty} \sum_{n=0}^{N} \varphi_{N,n}(\mu) \Psi_{N,n}^{(12)3}(x_i) \tag{6}$$

where [7]

$$\Psi_{N,n}^{(12)3}(x_i) = (x_1 + x_2)^n P_{N-n}^{(2n+3,1)}(x_3 - x_1 - x_2) C_n^{3/2}\left(\frac{x_1 - x_2}{x_1 + x_2}\right) \tag{7}$$

where the constraint $\sum_{k=1}^{3} x_k = 1$ is implied and $C_n^{3/2}(x)$ and $P_n^{\alpha,\beta}(x)$ are Gegenbauer and Jacoby polynomials, respectively. The superscript (12)3 stands for the order in which the conformal spins of the three quarks are summed to form the total spin $N+3$: First the u-quark spins are summed to the total spin $n+2$, and then the d-quark spin is added. This order is of course arbitrary; the functions $\Psi_{N,n}^{(12)3}$ and e.g. $\Psi_{N,n}^{1(23)}$ are related with each other through the Racah $6j$ symbols of the $SL(2)$ group, see Ref. 7 and Appendix B in Ref. 19 for details. The basis functions in (7) are mutually orthogonal w.r.t. the conformal scalar product (4) and are more convenient than Appell polynomials used in earlier studies [20,12,14,21,15,22,23].

The explicit expression for the conformal expansion of the leading-twist proton DA (1) to the next-to-leading conformal spin accuracy ($N = 0, 1$) reads

$$\Phi_N(x_i, \mu) = 120 x_1 x_2 x_3 \left[\phi_3^0 + (x_1 - x_2)\phi_3^- + \phi_3^+(1 - 3x_3)\right] , \tag{8}$$

and the twist-4 DAs (3) to the same accuracy are given by

$$\begin{aligned} \Phi_4(x_i) &= 24 x_1 x_2 \left[\phi_4^0 + \phi_4^-(x_1 - x_2) + \phi_4^+(1 - 5x_3)\right] , \\ \Psi_4(x_i) &= 24 x_1 x_3 \left[\psi_4^0 + \psi_4^-(x_1 - x_3) + \psi_4^+(1 - 5x_2)\right] , \\ \Xi_4(x_i) &= 24 x_2 x_3 \left[\xi_4^0 + \xi_4^-(x_2 - x_3) + \xi_4^+(1 - 5x_1)\right] , \end{aligned} \tag{9}$$

The twelve coefficients $\phi_3^0 \ldots \xi_4^+$ can be expressed in terms of eight independent non-perturbative parameters $f_N, \lambda_1, \lambda_2, f_1^u, f_1^d, f_2^d, A_1^u, V_1^d$ corresponding to matrix elements of local operators. One obtains [8]

$$\phi_3^0 = f_N \,, \quad \phi_3^- = \frac{21}{2} f_N A_1^u \,, \quad \phi_3^+ = \frac{7}{2}(1 - 3V_1^d) \tag{10}$$

for the leading twist, and

$$\begin{aligned}
\phi_4^0 &= \frac{1}{2}\left(\lambda_1 + f_N\right), \quad \xi_4^0 = \frac{1}{6}\lambda_2, \quad \psi_4^0 = \frac{1}{2}\left(f_N - \lambda_1\right) \\
\phi_4^- &= \frac{5}{4}\left(\lambda_1(1 - 2f_1^d - 4f_1^u) + f_N(2A_1^u - 1)\right), \\
\phi_4^+ &= \frac{1}{4}\left(\lambda_1(3 - 10f_1^d) - f_N(10V_1^d - 3)\right), \\
\psi_4^- &= -\frac{5}{4}\left(\lambda_1(2 - 7f_1^d + f_1^u) + f_N(A_1^u + 3V_1^d - 2)\right), \\
\psi_4^+ &= -\frac{1}{4}\left(\lambda_1(-2 + 5f_1^d + 5f_1^u) + f_N(2 + 5A_1^u - 5V_1^d)\right), \\
\xi_4^- &= \frac{5}{16}\lambda_2(4 - 15f_2^d), \qquad \xi_4^+ = \frac{1}{16}\lambda_2(4 - 15f_2^d). \qquad (11)
\end{aligned}$$

for the twist-four DAs, respectively. Note that the truncation of the conformal expansion at first order tacitly implies an assumption that this expansion is well convergent at least as a distribution in mathematical sense: after convolution with a smooth test function.

2.3. *Scale dependence and Complete Integrability*

The scale dependence of the nonperturbative coefficients $\varphi_{N,n}(\mu)$ in (6) is obtained by the diagonalization of the mixing matrix for the three-quark operators

$$B_{k_1,k_2,k_3} = (D_+^{k_1}q)(D_+^{k_2}q)(D_+^{k_3}q); \quad k_1 + k_2 + k_3 = N \qquad (12)$$

As well known, conformal symmetry allows one to resolve the mixing with operators containing total derivatives [12–18]. In particular, the coefficients $\varphi_{N,n}(\mu)$ with different values of N (related to the total conformal spin $J = N+3$) do not mix with each other by the one-loop evolution. The conformal symmetry is not sufficient, however, to solve the evolution equations: the coefficients $\varphi_{N,n}(\mu)$ with the same N but different n do mix, producing a nontrivial spectrum of anomalous dimensions, see Fig. 1. The corresponding multiplicatively renormalizeable contributions to the DA are given by linear combinations of the conformal polynomials

$$P_{N,q}(x_i) = \sum_{n=0}^{N} c_{N,n}^{(q)} \Psi_{N,n}^{(12)3}(x_i) \qquad (13)$$

with the coefficients $c_{N,n}^{(q)}$ and anomalous dimensions $\gamma_{N,q}$ that have to be found by the diagonalization of the mixing matrix.

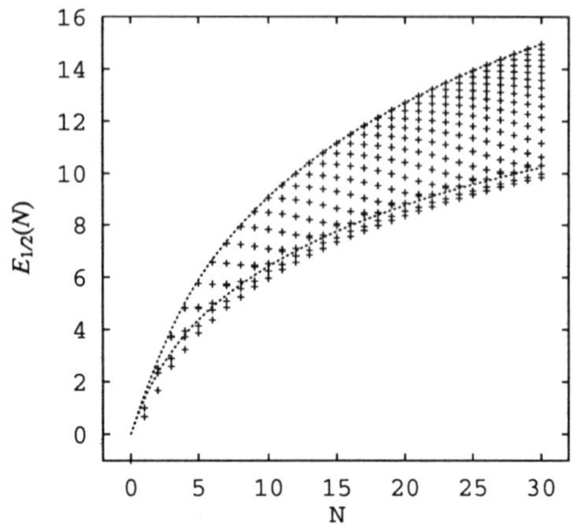

Fig. 1. The spectrum of anomalous dimensions $\gamma_N \equiv (1+1/N_c)E_N + 3/2C_F$ for the baryon distribution amplitudes with helicity $\lambda = 1/2$. The lines of the largest and the smallest eigenvalues for $\lambda = 3/2$ are indicated by dots for comparison.

It turns out [24] that the the index q that enumerates the solutions can be identified with an eigenvalue of a certain conserved charge. The physical interpretation is that one is able to find a new 'hidden' quantum number that distinguishes between partonic components in the proton with different scale dependence.

To explain this result, we have to introduce the so-called Hamiltonian approach [25], in which the evolution kernels are rewritten in terms of the $SL(2)$ generators. It is instructive to consider two cases separately, corresponding to helicity $\lambda = 3/2$ and $\lambda = 1/2$ operators related to the evolution of the Δ-isobar DA and the nucleon, respectively. The corresponding evolution kernels can be written in the following compact form [24,7]:

$$H_{3/2} = 2\left(1+\frac{1}{N_c}\right)\sum_{i<k}\Big[\psi(J_{ik}) - \psi(2)\Big] + \frac{3}{2}C_F\,, \tag{14}$$

$$H_{1/2} = H_{3/2} - 2\left(1+\frac{1}{N_c}\right)\left[\frac{1}{J_{12}(J_{12}-1)} + \frac{1}{J_{23}(J_{23}-1)}\right]. \tag{15}$$

Here $\psi(x)$ is the logarithmic derivative of the Γ-function and $J_{ik}, i,k = 1,2,3$ are defined in terms of the two-particle Casimir operators of the $SL(2,R)$ group

$$J_{ik}(J_{ik}-1) = L_{ik}^2 \equiv (\vec{L}_i + \vec{L}_k)^2\,, \tag{16}$$

with $\vec{L}_i$ being the group generators acting on the i-th quark, which have to be taken in the adjoint representation [7]:

$$
\begin{aligned}
L_{k,0}P(x_i) &= (x_k\partial_k + 1)P(x_i)\,,\\
L_{k,+}P(x_i) &= -x_k P(x_i)\,,\\
L_{k,-}P(x_i) &= (x_k\partial_k^2 + 2\partial_k)P(x_i)\,.
\end{aligned} \tag{17}
$$

Solution of the evolution equations corresponds in this language to solution of the Schrödinger equation

$$
HP_{N,q}(x_i) = \gamma_{N,q}P_{N,q}(X_i) \tag{18}
$$

with $\gamma_{N,q}$ being the anomalous dimensions. The $SL(2,R)$ invariance of the evolution equations implies that the generators of conformal transformations commute with the 'Hamiltonians'

$$
[H, L^2] = [H, L_\alpha] = 0\,, \tag{19}
$$

where $L^2 = (\vec{L}_1+\vec{L}_2+\vec{L}_3)^2$ and $L_\alpha = L_{1,\alpha}+L_{3,\alpha}+L_{3,\alpha}$, so that the polynomials $P_{N,q}(x_i)$ corresponding to multiplicatively renormalizable operators can be chosen simultaneously to be eigenfunctions of L^2 and L_0:

$$
L^2P_{N,q} = (N+3)(N+2)P_{N,q}\,,\quad L_0P_{N,q} = (N+3)P_{N,q}\,,\quad L_-P_{N,q} = 0\,. \tag{20}
$$

The third condition in (20) ensures that the operators do not contain overall total derivatives.

Main finding of Ref. 24 is that the Hamiltonian $H_{3/2}$ possesses an additional integral of motion (conserved charge):

$$
Q = \frac{i}{2}[L_{12}^2, L_{23}^2] = i(\partial_1-\partial_2)(\partial_2-\partial_3)(\partial_3-\partial_1)x_1x_2x_3\,,\quad [H_{3/2}, Q] = 0 \tag{21}
$$

The evolution equation for baryon distribution functions with maximum helicity is, therefore, completely integrable. The premium is that instead of solving a Schrödinger equation with a complicated nonlocal Hamiltonian, it is sufficient to solve a much simpler equation

$$
QP_{N,q}(x_i) = qP_{N,q}(x_i)\,. \tag{22}
$$

Once the eigenfunctions are found, the eigenvalues of the Hamiltonian (anomalous dimensions) are obtained as algebraic functions of N, q.

The Hamiltonian in (14) is known as the Hamiltonian of the so-called $XXX_{s=-1}$ Heisenberg spin magnet. The same Hamiltonian was encountered before in the interactions between reggeized gluons in QCD [26,27].

The equation in (22) cannot be solved exactly, but a wealth of analytic results can be obtained by means of the $1/N$ expansion [28]. One general

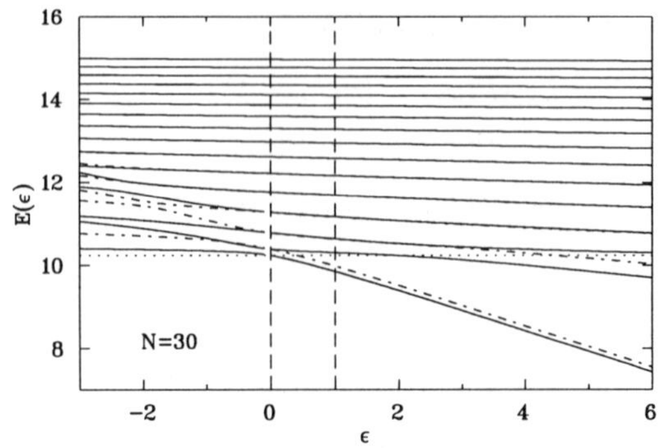

Fig. 2. The flow of energy eigenvalues for the Hamiltonian $H(\epsilon)$ for $N = 30$. The solid and the dash-dotted curves show the parity-even and parity-odd levels, respectively. The two vertical dashed lines indicate $H_{3/2} \equiv H(\epsilon = 0)$ and $H_{1/2} \equiv H(\epsilon = 1)$, respectively (up to the color factors). The horizontal dotted line shows position of the unperturbed 'ground state' given by Eq. (24).

consequence of complete integrability is that all anomalous dimensions are double degenerate except for the lowest ones for each *even* N, corresponding to the solution with $q = 0$. The corresponding eigenfunctions have a very simple form [7]

$$x_1x_2x_3P^{\lambda=3/2}_{N,q=0}(x_i) = x_1(1-x_1)C^{3/2}_{N+1}(1-2x_1) + x_2(1-x_2)C^{3/2}_{N+1}(1-2x_2) + x_3(1-x_3)C^{3/2}_{N+1}(1-2x_3) \tag{23}$$

and the anomalous dimension is equal to

$$\gamma_{N,q=0} = (1+1/N_c)\Big[4\psi(N+3)+4\gamma_E-6\Big]+3/2C_F\,. \tag{24}$$

The asymptotic expansions for the charge q and the anomalous dimensions at large N are available to the order $1/N^8$ [28,7] and give very accurate results.

The additional term in $H_{1/2}$ (the nucleon) spoils integrability but can be considered as a (calculable) small correction for all of the spectrum except for two lowest levels [7]. To illustrate this, consider the flow of energy levels for the Hamiltonian $H(\epsilon) = \sum_{i<k}\Big[\psi(J_{ik})-\psi(2)\Big]-\epsilon\Big[1/L^2_{12}+1/L^2_{23}\Big]$ (cf. (14), (15)) as a function of an auxiliary parameter ϵ, see Fig. 2. It is seen that the two lowest levels decouple from the rest of the spectrum and are separated from it by a finite mass gap. As shown in Ref. 7, this phenomenon can be interpreted as binding of the two quarks with opposite helicity and forming a scalar "diquark". The effective Hamiltonian for the low-lying levels can

be constructed and turns out to be a generalization of the famous Kroning-Penney problem for a particle in a δ-function type periodic potential. The value of the mass gap between the lowest and the next-to-lowest anomalous dimensions at $N \to \infty$ can be calculated combining the small-ϵ and the large-ϵ expansions and is equal to $\Delta\gamma = 0.32 \cdot (1 + 1/N_c)$ in agreement with the direct numerical calculations. The corresponding contributions to the nucleon DA are of the form [7]

$$P^{\lambda=1/2}_{N,q=0}(x_i) \stackrel{\ln N \to \infty}{=} P^{(1,3)}_N(1-2x_3) \pm P^{(1,3)}_N(1-2x_1)\,, \qquad (25)$$

where $P^{(1,3)}_N(x)$ are Jacobi polynomials.

The approach based on complete integrability can be used to obtain parts of the two-loop evolution kernels for baryon operators beyond the leading order [29], but a complete calculation to the two-loop accuracy is so far absent.

3. Nonperturbative parameters

To the leading-order accuracy in the conformal spin expansion, the leading-twist-3 DA involves one, f_N, and the twist-4 DAs two, λ_1 and λ_2, nonperturbative parameters, To the next-to-leading accuracy in the conformal spin there are two additional parameters for twist-3, A_1^u and V_1^d, and three parameters for twist-4, f_1^u, f_1^d and f_2^d, cf. Eqs. (10),(11). The number of parameters proliferates rapidly if higher spins are included, and their estimates become increasingly complicated and unreliable. Hence I stop at the first nontrivial order and summarize the existing estimates in Table 1 and Table 2 for the leading and the higher twist DAs, respectively.

Most of the estimates are obtained using QCD sum rules. The quoted numbers correspond to the sum rules to the leading order accuracy in the QCD coupling. The NLO radiative corrections are known for λ_1 [30,31] but not for other cases, to my knowledge. The effect of such corrections can be substantial, see e.g. Ref. 32.

The calculations presented in Refs. 6, 9, 33, 34 make use of the same sum rule and are, therefore, not entirely independent. The errors are difficult to quantify, but are probably of the size of the spread in the quoted values. The result for f_N is expected to be rather reliable although the quoted error might well be underestimated. The parameter λ_1 is also well known to QCD sum rule practitioners and corresponds to the nucleon coupling to the so-called Ioffe current [37]. Note that although an overall sign in the couplings $f_N, \lambda_{1,2}$ is arbitrary and can be readjusted by the phase factor in

Table 1. Parameters of the leading-twist nucleon distribution amplitude (8), (10) at the scale 1 GeV. The constant f_N is given in units of 10^{-3} GeV^2

Method	f_N	V_1^d	A_1^u	Ref.
asymptotic	–	1/3	0	
QCDSR	5.3 ± 0.5	0.220	0.480	[6]
QCDSR	5.0 ± 0.3	0.229	0.387	[9]
QCDSR	5.1 ± 0.3	0.240	0.340	[33]
QCDSR	–	0.236	0.490	[34]
Model	–	0.310	0.071	[35]
LCSR	–	0.300	0.130	[36]

Table 2. Parameters of the twist-four nucleon distribution amplitudes (9), (11) at the scale 1 GeV. The constants λ_1 and λ_2 are given in units of 10^{-3} GeV^2

Method	λ_1	λ_2	f_1^d	f_2^d	f_1^u	Ref.
asymptotic	–	–	3/10	4/15	1/10	
QCDSR	-27 ± 5	54 ± 19	0.40 ± 0.05	0.22 ± 0.05	0.07 ± 0.05	[36]
LCSR	–	–	0.33	0.25	0.09	[36]

the nucleon wave function, the relative sign is physical and important for the applications.

Alternatively, there exists a phenomenological model for the leading-twist DA [35] which was obtained by modelling the soft contribution to electromagnetic form factors by a convolution of light-cone wave functions. Estimates of the higher-twist DAs in the same technique are not available. Finally, I quote the parameters obtained in Ref. 36 from the fit of the light-cone sum rules to the experimental data on the nucleon form factors. This approach will be explained below. In future, one should expect that a few lowest order parameters in the conformal expansion of baryon DAs can be calculated on the lattice, cf. Ref. 38. Main technical problem on this way seems to be the necessity to use nonperturbative renormalization of three-quark operators.

4. Light-Cone Sum Rules

Main problem that does not allow to extract the information on baryon DAs from experiment is that nature does not provide us with point-like three-quark currents. Baryon number conservation implies that physical processes always involve baryons in pairs. Hence one has to deal with the convolution of two baryon DAs, and also the so-called "soft" contributions to the form factors which cannot expressed in terms of DAs prove to be numerically significant at present energies. A (partial) remedy is suggested by

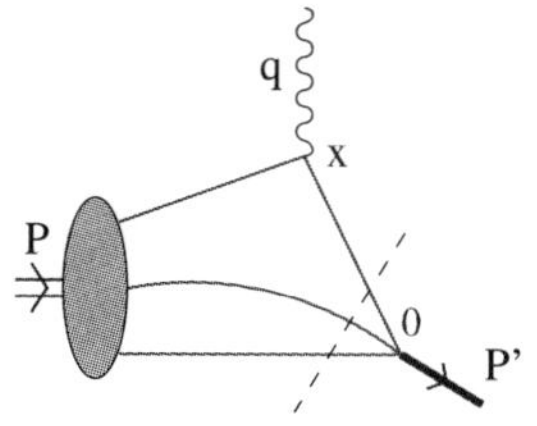

Fig. 3. Schematic structure of the light-cone sum rule for baryon form factors.

the approach known as light-cone sum rules (LCSRs) [39–41]. This technique is attractive because in LCSRs "soft" contributions to the form factors are calculated in terms of the same DAs that enter the pQCD calculation and there is no double counting. Thus, the LCSRs provide one with the most direct relation of the hadron form factors and distribution amplitudes that is available at present, with no other nonperturbative parameters.

The basic object of the LCSR approach is the correlation function

$$\int dx\, e^{-iqx} \langle 0|T\{\eta(0)j(x)\}|N(P)\rangle$$

in which j represents the electromagnetic (or weak) probe and η is a suitable operator with nucleon quantum numbers. The other (in this example, initial state) nucleon is explicitly represented by its state vector $|N(P)\rangle$, see a schematic representation in Fig. 3. When both the momentum transfer Q^2 and the momentum $(P')^2 = (P-q)^2$ flowing in the η vertex are large and negative, the asymptotics of the correlation function is governed by the light-cone kinematics $x^2 \to 0$ and can be studied using the operator product expansion (OPE) $T\{\eta(0)j(x)\} \sim \sum C_i(x)\mathcal{O}_i(0)$ on the light-cone $x^2 = 0$. The x^2-singularity of a particular perturbatively calculable short-distance factor $C_i(x)$ is determined by the twist of the relevant composite operator $\mathcal{O}_i$, whose matrix element $\langle 0|\mathcal{O}_i(0)|N(P)\rangle$ is given by an appropriate moment of the nucleon DA. Next, one can represent the answer in form of the dispersion integral in $(P')^2$ and define the nucleon contribution by the cut-off in the quark-antiquark invariant mass, the so-called interval of duality s_0 (or continuum threshold). The main role of the interval of duality is that it does not allow large momenta $|k^2| > s_0$ to flow through the η-vertex; to the lowest order $O(\alpha_s^0)$ one obtains a purely soft contribution to the form factor as a sum of terms ordered by twist of the relevant operators and hence including both the leading- and the higher-twist nucleon DAs. Note that, in difference to the hard mechanism, the contribution of higher-twist DAs is only suppressed by powers of $|(P')^2| \sim 1-2$ GeV^2 (which is trans-

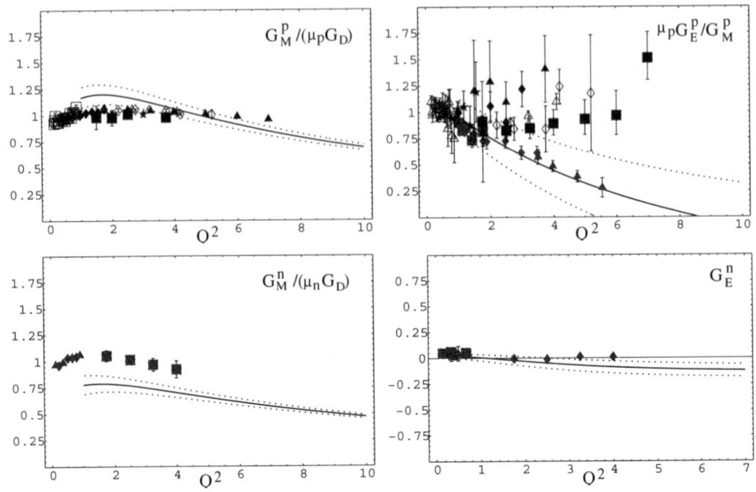

Fig. 4. LCSR results (solid curves) for the electromagnetic form factors of the nucleon, obtained using the model of the nucleon DAs with parameters from Tables 1,2. The dotted curves show the effect of the variation of the ratio f_N/λ_1 by 30%. For the identification of the data points and details of the calculation see Ref. 36.

lated to the suppression by powers of the Borel parameter after applying the usual QCD sum rule machinery), but not by powers of Q^2. This feature is in agreement with the common wisdom that soft contributions are not constrained to small transverse separations.

The LCSR expansion also contains terms generating the asymptotic pQCD contributions. They appear at proper order in α_s, i.e., in the $O(\alpha_s)$ term for the pion form factor, at the α_s^2 order for the nucleon form factors, etc. In the pion case, it was explicitly demonstrated [42,43] that the contribution of hard rescattering is correctly reproduced in the LCSR approach as a part of the $O(\alpha_s)$ correction. It should be noted that the diagrams of LCSR that contain the "hard" pQCD contributions also possess "soft" parts, i.e., one should perform a separation of "hard" and "soft" terms inside each diagram. As a result, the distinction between "hard" and "soft" contributions appears to be scale- and scheme-dependent [42]. During the last years there have been numerous applications of LCSRs to mesons, see Refs. 44, 45 for a review. Following the work Ref. 46 nucleon form factors were further considered in this framework in Refs. 47, 48, 49, 36 and the weak decays $\Lambda_b \to p\ell\nu_\ell$, $\Lambda_c \to \Lambda\ell\nu_\ell$ in Refs. 50, 51. The generalization to the $N\gamma\Delta$ transition form factor was worked out in Ref. 52.

The net outcome of these studies is that all nucleon form factors (with an exception of the magnetic $N\Delta\gamma$ transition) can be reproduced to roughly 20% accuracy by using the parameters of the proton DA summarized in Tables 1,2 above, which are roughly in the middle of the range between asymptotic DAs and the QCD sum rule predictions, see Fig. 4. This conclusion is preliminary, however. More studies are needed and in particular radiative corrections to the sum rules have to be calculated.

5. Conclusions

Baryon distribution amplitudes are fundamental nonperturbative functions describing the hadron structure in configurations with a fixed number of Fock constituents at small transverse separation. They are equally important and to a large extent complementary to conventional parton distributions which correspond to one-particle probability distributions for the parton momentum fraction in an average configuration. The theory of baryon DAs has reached a certain degree of maturity. In particular, their scale dependence is well understood and reveals a beautiful hidden symmetry of QCD which is not seen at the level of the QCD Lagrangian. The basic tool to describe DAs is provided by the conformal expansion combined with EOM (for higher twists) that allows one to obtain parameterizations with the minimum number of nonperturbative parameters. There are indications that the conformal expansion is converging sufficiently rapidly so that only a few terms are needed for most of the practical purposes. A qualitative picture inspired by the QCD sum rule calculations [6] seems to be that the valence quark with the spin parallel to that of the proton carries most of its momentum. It is timely to make this picture quantitative; combination of LCSRs and lattice calculations should allow one to determine momentum fractions carried by the three valence quarks with 5-7% precision within a few years. Further progress will depend decisively on whether studies of hard exclusive processes can be made fully quantitative. High quality data are needed in the $Q^2 \sim 10$ GeV^2 range, and one has to develop a consistent theoretical framework for the treatment of end-point contributions.

Acknowledgements

The author is grateful to I. Balitsky, P. Ball, S. Derkachov, R. Fries, G. Korchemsky, A. Lenz, A. Manashov, N. Mahnke, D. Mueller, A. Peters, G. Peters, A. Radyushkin, E. Stein and M. Wittmann for the collaboration on the subject of this review. Special thanks are to the organizers of CAQCD06 for the invitation and hospitality.

References

1. V. L. Chernyak and A. R. Zhitnitsky, JETP Lett. **25**, 510 (1977); Sov. J. Nucl. Phys. **31**, 544 (1980); V. L. Chernyak, A. R. Zhitnitsky and V. G. Serbo, JETP Lett. **26**, 594 (1977); Sov. J. Nucl. Phys. **31**, 552 (1980).
2. A. V. Radyushkin, JINR report R2-10717 (1977), [arXiv:hep-ph/0410276]; A. V. Efremov and A. V. Radyushkin, Theor. Math. Phys. **42**, 97 (1980) Phys. Lett. B **94**, 245 (1980).
3. G. P. Lepage and S. J. Brodsky, Phys. Lett. B **87**, 359 (1979); Phys. Rev. Lett. **43**, 545 (1979) [Erratum-ibid. **43**, 1625 (1979)]; Phys. Rev. D **22**, 2157 (1980); S. J. Brodsky, G. P. Lepage and S. A. A. Zaidi, Phys. Rev. D **23**, 1152 (1981).
4. J. Gronberg *et al.* [CLEO Collaboration], Phys. Rev. D **57**, 33 (1998).
5. V. L. Chernyak and I. R. Zhitnitsky, Nucl. Phys. B **246**, 52 (1984).
6. V. L. Chernyak and A. R. Zhitnitsky, Phys. Rept. **112**, 173 (1984).
7. V. M. Braun, S. E. Derkachov, G. P. Korchemsky and A. N. Manashov, Nucl. Phys. B **553**, 355 (1999).
8. V. Braun, R. J. Fries, N. Mahnke and E. Stein, Nucl. Phys. B **589**, 381 (2000) [Erratum-ibid. B **607**, 433 (2001)].
9. V. L. Chernyak, A. A. Ogloblin and I. R. Zhitnitsky, Z. Phys. C **42**, 583 (1989); Z. Phys. C **42**, 569 (1989).
10. P. Ball and V. M. Braun, Nucl. Phys. B **543**, 201 (1999).
11. X. d. Ji, J. P. Ma and F. Yuan, Nucl. Phys. B **652**, 383 (2003); Eur. Phys. J. C **33**, 75 (2004).
12. S. J. Brodsky, Y. Frishman, G. P. Lepage and C. T. Sachrajda, Phys. Lett. B **91**, 239 (1980); Phys. Rev. D **33**, 1881 (1986); S. J. Brodsky, Y. Frishman and G. P. Lepage, Phys. Lett. B **167**, 347 (1986).
13. Y. M. Makeenko, Sov. J. Nucl. Phys. **33**, 440 (1981).
14. T. Ohrndorf, Nucl. Phys. B **198**, 26 (1982).
15. V. M. Braun and I. E. Filyanov, Z. Phys. C **48**, 239 (1990).
16. D. Mueller, Phys. Rev. D **49**, 2525 (1994).
17. A. V. Belitsky and D. Mueller, Nucl. Phys. B **537**, 397 (1999).
18. V. M. Braun, G. P. Korchemsky and D. Mueller, Prog. Part. Nucl. Phys. **51**, 311 (2003).
19. V. M. Braun, G. P. Korchemsky and A. N. Manashov, Nucl. Phys. B **603**, 69 (2001).
20. M. E. Peskin, Phys. Lett. B **88**, 128 (1979).
21. K. Tesima, Nucl. Phys. B **202**, 523 (1982).
22. S. L. Nyeo, Z. Phys. C **54**, 615 (1992).
23. N.G. Stefanis, Acta Phys. Polon. **B**25 (1994) 1777; Dissertation, JINR, 1997 (unpublished); M. Bergmann, W. Schroers and N. G. Stefanis, Phys. Lett. B **458**, 109 (1999).
24. V. M. Braun, S. E. Derkachov and A. N. Manashov, Phys. Rev. Lett. **81**, 2020 (1998).
25. A. P. Bukhvostov, G. V. Frolov, L. N. Lipatov and E. A. Kuraev, Nucl. Phys. B **258**, 601 (1985).
26. L. N. Lipatov, JETP Lett. **59**, 596 (1994).

27. L. D. Faddeev and G. P. Korchemsky, Phys. Lett. B **342**, 311 (1995).
28. G.P. Korchemsky, Nucl. Phys. B **443**, 255 (1995); *ibid.* B **462**, 333 (1996); Preprint LPTHE-Orsay-97-62 [hep-ph/9801377].
29. A. V. Belitsky, G. P. Korchemsky and D. Mueller, Phys. Rev. Lett. **94**, 151603 (2005); Nucl. Phys. B **735**, 17 (2006).
30. M. Jamin, Z. Phys. C **37**, 635 (1988).
31. A. A. Ovchinnikov, A. A. Pivovarov and L. R. Surguladze, Sov. J. Nucl. Phys. **48**, 358 (1988); Int. J. Mod. Phys. A **6**, 2025 (1991).
32. V. A. Sadovnikova, E. G. Drukarev and M. G. Ryskin, Phys. Rev. D **72**, 114015 (2005).
33. I. D. King and C. T. Sachrajda, Nucl. Phys. B **279**, 785 (1987).
34. M. Gari and N. G. Stefanis, Phys. Rev. D **35**, 1074 (1987).
35. J. Bolz and P. Kroll, Z. Phys. A **356**, 327 (1996).
36. V. M. Braun, A. Lenz and M. Wittmann, Phys. Rev. D **73**, 094019 (2006).
37. B. L. Ioffe, Nucl. Phys. B **188**, 317 (1981) [Erratum-ibid. B **191**, 591 (1981)].
38. G. Martinelli and C. T. Sachrajda, Phys. Lett. B **217**, 319 (1989).
39. I. I. Balitsky, V. M. Braun and A. V. Kolesnichenko, Nucl. Phys. B **312**, 509 (1989).
40. V. M. Braun and I. E. Filyanov, Z. Phys. C **44**, 157 (1989).
41. V. L. Chernyak and I. R. Zhitnitsky, Nucl. Phys. B **345**, 137 (1990).
42. V. M. Braun, A. Khodjamirian and M. Maul, Phys. Rev. D **61**, 073004 (2000).
43. J. Bijnens and A. Khodjamirian, Eur. Phys. J. C **26**, 67 (2002).
44. V. M. Braun, arXiv:hep-ph/9801222.
45. P. Colangelo and A. Khodjamirian, arXiv:hep-ph/0010175.
46. V. M. Braun, A. Lenz, N. Mahnke and E. Stein, Phys. Rev. D **65**, 074011 (2002).
47. Z. G. Wang, S. L. Wan and W. M. Yang, Phys. Rev. D **73**, 094011 (2006).
48. Z. G. Wang, S. L. Wan and W. M. Yang, Eur. Phys. J. C **47**, 375 (2006).
49. A. Lenz, M. Wittmann and E. Stein, Phys. Lett. B **581**, 199 (2004).
50. M. Q. Huang and D. W. Wang, Phys. Rev. D **69**, 094003 (2004).
51. M. Q. Huang and D. W. Wang, arXiv:hep-ph/0608170.
52. V. M. Braun, A. Lenz, G. Peters and A.V. Radyushkin, Phys. Rev. D **73**, 034020 (2006).

SOLITONS IN SUPERSYMMETRIC GAUGE THEORIES: MODULI MATRIX APPROACH

MINORU ETO[†], YOUICHI ISOZUMI, MUNETO NITTA[††], KEISUKE OHASHI and NORISUKE SAKAI[*]

Department of Physics, Tokyo Institute of Technology
[†]*Inst. of Physics, University of Tokyo*
[††]*Department of Physics, Hiyoshi, Keio University*

We review our recent works on solitons in $U(N_C)$ gauge theories with $N_F(\geq N_C)$ Higgs fields in the fundamental representation, which possess eight supercharges. The moduli matrix is proposed as a crucial tool to exhaust all BPS solutions, and to characterize all possible moduli parameters. Since vacua are in the Higgs phase, we find domain walls (kinks) and vortices as the only elementary solitons. Stable monopoles and instantons can exist as composite solitons with vortices attached. Webs of walls are also found as another composite soliton. The moduli space of all these elementary as well as composite solitons are found in terms of the moduli matrix. The total moduli space of walls is given by the complex Grassmann manifold $SU(N_F)/[SU(N_C) \times SU(N_F - N_C) \times U(1)]$ and is decomposed into various topological sectors corresponding to boundary conditions specified by particular vacua. We found charges characterizing composite solitons contribute negatively (either positively or negatively) in Abelian (non-Abelian) gauge theories. Effective Lagrangians are constructed on walls and vortices in a compact form. The power of the moduli matrix is illustrated by an interaction rule of monopoles, vortices, and walls, which is difficult to obtain in other methods. More thorough description of the moduli matrix approach can be found in our review article[1] (hep-th/0602170).

Keywords: Soliton; Higgs phase; Supersymmetry; Moduli.

1. Discrete Vacua in Higgs Phase

Solitons have been playing a central role in understanding nonperturbative effects. The solitons are classified by their codimensions. Kinks (domain walls), vortices, monopoles and instantons are well-known typical solitons with codimensions one, two, three and four, respectively. They carry topological charges classified by certain homotopy groups according to their

[*]Speaker at the conference.

codimensions. Moreover, they are also important to construct models of the brane-world, where our four-dimensional world is realized on a topological defect in higher-dimensional spacetime[2–4]. These topological defects are preferably solitons as a solution of field equations.

When energy of solitons saturates a bound from below, which is called the Bogomol'nyi bound, they are the most stable among all possible configurations with the same boundary condition, and automatically satisfy field equations. They are called Bogomol'nyi-Prasad-Sommerfield (BPS) solitons[5]. If a part of supersymmetry (SUSY) is preserved in supersymmetric theories, the field configuration becomes a BPS state[6]. The representation theory of SUSY shows that they are non-perturbatively stable. With this fact non-perturbative effects have been established in SUSY gauge theories and string theory[7]. In supersymmetric theories, BPS solitons often have parameters, which are called moduli. When they are promoted to fields on the world volume of solitons, they become massless fields of the low-energy effective theory.

We are primarily interested in $U(N_{\rm C})$ gauge theory with $N_{\rm F}$ flavors in the fundamental representation, which can be made SUSY theories with eight supercharges.

Bosonic components of a vector multiplet are a gauge field $W_M, M = 0, 1, \cdots, d-1$ and a real adjoint scalar field $\Sigma_p, p = d, \cdots, 5$ in the adjoint representation. Matter fields are represented by hypermultiplets containing two $N_{\rm C} \times N_{\rm F}$ matrices of complex Higgs (scalar) fields H^1, H^2 as bosonic components. The theory contains a common gauge coupling g for $SU(N_{\rm C})$ and $U(1)$ and the Fayct-Iliopoulos parameter[8] c. Since one of the hypermultiplet scalar $H^2 = 0$ in all of our BPS solutions if $c > 0$, we ignore it: $H^1 \equiv H$, $H^2 = 0$. Our (bosonic part of) the Lagrangian is given by

$$\mathcal{L} = \mathcal{L}_{\rm kin} - V, \tag{1}$$

$$\mathcal{L}_{\rm kin} = {\rm Tr}\left(-\frac{1}{2g^2}F_{\mu\nu}F^{\mu\nu} + \frac{1}{g^2}\mathcal{D}_\mu\Sigma_p\mathcal{D}^\mu\Sigma_p + \mathcal{D}^\mu H\left(\mathcal{D}_\mu H\right)^\dagger\right), \tag{2}$$

where the covariant derivatives and field strengths are defined as $\mathcal{D}_\mu\Sigma_p = \partial_\mu\Sigma_p + i[W_\mu, \Sigma_p]$, $\mathcal{D}_\mu H = (\partial_\mu + iW_\mu)H$, $F_{\mu\nu} = -i[\mathcal{D}_\mu, \mathcal{D}_\nu]$. Our convention for the metric is $\eta_{\mu\nu} = {\rm diag}(+, -, \cdots, -)$. The scalar potential V is given in terms of diagonal mass matrices M_p and a real parameter c as

$$V = {\rm Tr}\Big[\frac{g^2}{4}\left(c\mathbf{1}_{N_{\rm C}} - HH^\dagger\right)^2 + (\Sigma_p H - HM_p)(\Sigma_p H - HM_p)^\dagger\Big]. \tag{3}$$

To obtain domain walls, we need real mass parameters $M = {\rm diag}(m_1, m_2, \cdots, m_{N_{\rm F}})$. Therefore we consider $d = 5$ with a single adjoint scalar Σ. For simplicity, we choose fully non-degenerate mass: $m_A > m_{A+1}$. Then the flavor symmetry is broken to $U(1)_{\rm F}^{N_{\rm F}-1}$. Let us note that a common mass can be absorbed into the adjoint scalar Σ. Because of non-degenerate masses, we obtain discrete supersymmetric vacua, labeled by $N_{\rm C}$ flavors $\langle A_1 A_2 \cdots A_{N_{\rm C}} \rangle$, which are called color-flavor locking vacua:

$$H^{rA} = \sqrt{c}\,\delta^{A_r}{}_A, \qquad \Sigma = \mathrm{diag}(m_{A_1}, \cdots, m_{A_{N_{\rm C}}}). \tag{4}$$

The number of these vacua increases exponentially as the number of colors $N_{\rm C}$ and flavors $N_{\rm F}$ increases:

$$\frac{N_{\rm F}!}{(N_{\rm F}-N_{\rm C})!N_{\rm C}!} \sim e^{N_{\rm F}\log(x^{-x}(1-x)^{-(1-x)})}, \quad x \equiv N_{\rm C}/N_{\rm F}. \tag{5}$$

Since the Higgs H charged under $U(N_{\rm C})$ gauge group have nonvanishing values, the vacua are in the Higgs phase. In the Higgs phase, only walls and vortices are elementary solitons, whereas the instantons, monopoles, and (wall-)junctions appear as composite solitons.

2. 1/2 BPS Walls

To obtain domain wall solutions we assume that all fields depend on one spatial coordinate, say $y \equiv x^4$ with $3+1$ dimensional Poincaré invariance. The 1/2 **BPS equations for walls** are obtained by requiring the following direction ε^i of SUSY to be preserved[9]: $\gamma^4\varepsilon^i = -i(\sigma^3)^i{}_j\varepsilon^j$,

$$\mathcal{D}_y H = -\Sigma H + HM, \qquad \mathcal{D}_y\Sigma = \frac{g^2}{2}\left(c\mathbf{1}_{N_{\rm C}} - HH^\dagger\right). \tag{6}$$

The topological sector of (multi-)BPS wall configurations is labeled by vacua $\langle A_1 A_2 \cdots A_{N_{\rm C}} \rangle$ at $y=\infty$ and $\langle B_1 B_2 \cdots B_{N_{\rm C}} \rangle$ at $y=-\infty$ as shown in Fig. 1.

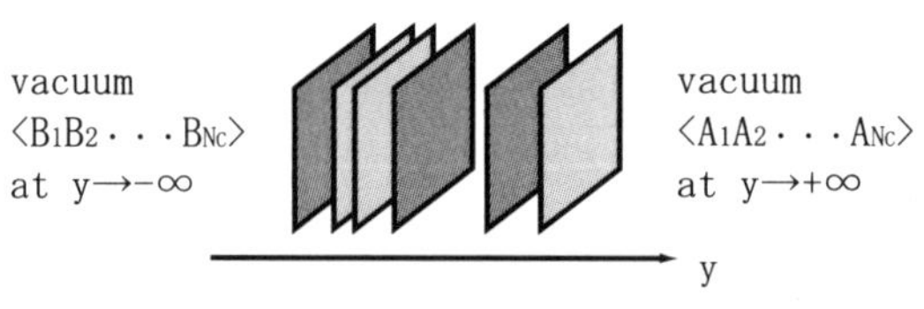

Fig. 1. Multi-wall connecting vacua $\langle A_1 A_2 \cdots A_{N_{\rm C}} \rangle$ and $\langle B_1 B_2 \cdots B_{N_{\rm C}} \rangle$.

The BPS equations for hypermultiplet (the first of Eq.(6)) can be solved[9] by defining an element $S(y)$ of a complexified gauge group $GL(N_{\rm C}, \mathbf{C})$ as

$$\Sigma + iW_y \equiv S^{-1}(y)\partial_y S(y), \qquad H(y) = S^{-1}(y)H_0 e^{My}. \tag{7}$$

We call the constant $N_C \times N_F$ matrix H_0 " **moduli matrix**".

With the above solution, we can rewrite the vector multiplet BPS equation into the following equation in terms of the gauge invariant quantity $\Omega \equiv SS^\dagger$

$$\partial_y\left(\Omega^{-1}\partial_y\Omega\right) = g^2 c\left(\mathbf{1}_{N_C} - \Omega^{-1}\Omega_0\right), \qquad \Omega_0 \equiv c^{-1}H_0 e^{2My}H_0^\dagger. \quad (8)$$

We call this equation "**master equation**". The index theorem[10] shows that the number of moduli parameters contained in the moduli matrix H_0 is just enough, implying that the solution of the master equation exists and is unique for a given Ω_0. The existence and uniqueness have been proved rigorously for the case of $U(1)$ gauge theory[11].

Since the solution $S(y)$ of Eq.(7) has N_C^2 integration constants, two sets (S, H_0) and (S', H_0') give the same $H = S^{-1}H_0 e^{My}$, if they are related by the following global $GL(N_C, \mathbf{C})$ transformation V (called the V-transformation),

$$S \to S' = VS, \quad H_0 \to H_0' = VH_0, \qquad V \in GL(N_C, \mathbf{C}). \quad (9)$$

Therefore the genuine moduli parameters of domain walls are given by the equivalence class defined by the V-transformation. We thus find that the **total moduli space** for (multi-)wall solutions is the complex **Grassmann manifold**[9]:

$$\begin{aligned}\mathcal{M}_{N_F,N_C} &= \{H_0 | H_0 \sim VH_0, V \in GL(N_C, \mathbf{C})\} \equiv G_{N_F,N_C} \\ &\simeq \frac{SU(N_F)}{SU(N_C)\times SU(N_F - N_C)\times U(1)}.\end{aligned} \quad (10)$$

This is a compact (closed) set of complex dimension $N_C\tilde{N}_C = N_C(N_F - N_C)$. We did not put any boundary conditions at $y \to \pm\infty$ to get the moduli space (10). Therefore it contains configurations with all possible boundary conditions, and can be decomposed into the sum of topological sectors

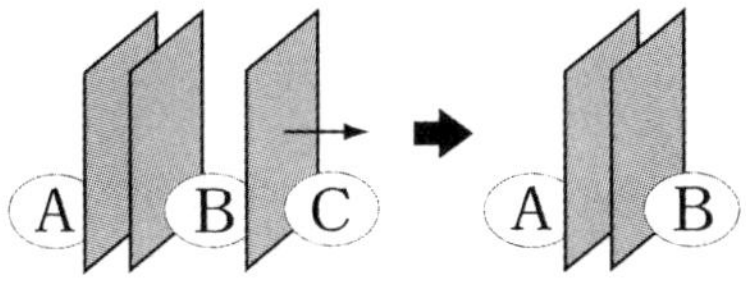

Fig. 2. A three wall solution connecting vacuum A to C through B (left). By letting the right-most wall to infinity, we obtain a two wall solution connecting vacuum A to B.

$$\mathcal{M}_{\rm wall}^{\rm total} = \sum_{\rm BPS} \mathcal{M}^{\langle A_1,\cdots,A_{N_C}\rangle \leftarrow \langle B_1,\cdots,B_{N_C}\rangle}. \quad (11)$$

As shown in Fig. 2, by sending one of the wall to infinity, we obtain one less walls. Namely the boundaries of a topological sector consists of topological sectors with one less wall. It is interesting to observe that this natural compactification of the moduli space of walls leads to the compact total moduli space (10) for the 1/2 BPS solutions, if we add vacua as points to compactify the manifold.

Components of the moduli matrix H_0 represent weights of the vacua. For

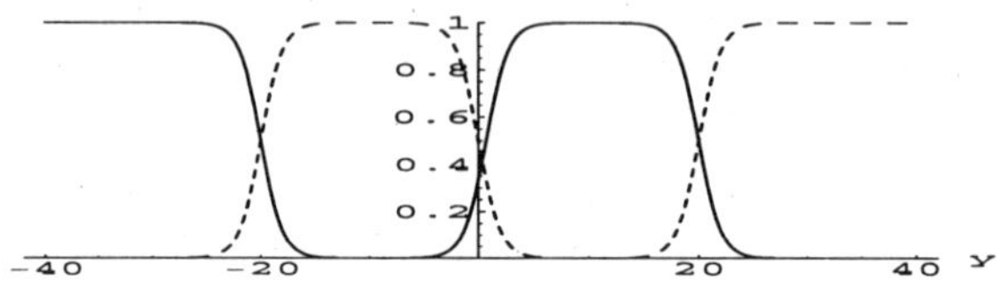

Fig. 3. Rapid change of hypermultiplets indicates the positions of walls.

instance, the moduli matrix for the $U(1)$ gauge theory can be parametrized by $H_0 = (e^{r_1}, e^{r_2}, \cdots, e^{r_{N_F}})$. Then the hypermultiplets are given by

$$H = S^{-1} H_0 e^{My} = S^{-1} (e^{r_1 + m_1 y}, \cdots, e^{r_{N_F} + m_{N_F} y}) \tag{12}$$

We see that wall separating i- and $i+1$-th vacua is located where the magnitudes of the i- and $i+1$-th components become equal as illustrated in Fig. 3. The wall position y is

$$\mathrm{Re} r_i + m_i y \sim \mathrm{Re} r_{i+1} + m_{i+1} y \rightarrow y = -\mathrm{Re}(r_i - r_{i+1})/(m_i - m_{i+1}). \tag{13}$$

The imaginary part $\mathrm{Im}(r_i - r_{i+1})$ gives the relative phase of the two adjacent vacua. We see that there are $N_F - 1$ walls maximally. Similarly, the number of walls in non-Abelian $U(N_C)$ gauge theory is given by $N_C(N_F - N_C)$, and each wall carries two moduli, position and relative phase of adjacent vacua.

The low-energy effective Lagrangian on domain walls is given by promoting the moduli parameters in the moduli matrix H_0 to fields on the world volume of the soliton and by assuming the weak dependence on the world volume coordinates. We assume the slow-movement of moduli fields compared to the two typical mass scales $g\sqrt{c}$ and Δm of the wall of hypermultiplets

$$\lambda \ll \min(\Delta m, g\sqrt{c}). \tag{14}$$

We find it extremely useful to use the superfield formalism maintaining the preserved four SUSY manifest. The **effective Lagrangian** for the 1/2 BPS domain walls is given in terms of the solution $\Omega_{\rm sol}(y,\phi,\phi^*)$ of the master equation

$$\mathcal{L} = -T_{\rm w} + \int d^4\theta K(\phi,\phi^*) + \text{higher derivatives}, \tag{15}$$

where $T_{\rm w}$ is the tension of the domain wall, and K is the **Kähler potential** of moduli fields ϕ,ϕ^*, given by[12]

$$K(\phi,\phi^*) = \int dy \left[c \log \det\Omega + c{\rm Tr}\left(\Omega_0\Omega^{-1}\right) + \frac{1}{2g^2}{\rm Tr}\left(\Omega^{-1}\partial_y\Omega\right)^2\right]\Big|_{\Omega=\Omega_{\rm sol}} \tag{16}$$

One of the merit of our superfield formulation is that Kähler potential can be obtained directly without going through Kähler metric and integrating it. We find that the Kähler potential serves as the action for Ω to obtain the master equation (8).

Let us make some comments. If we take the strong gauge coupling limit $g^2c/(\Delta m)^2 \gg 1$, the model becomes a nonlinear sigma model[13] and the master equation (8) can be solved algebraically

$$\Omega = \Omega_0 \equiv c^{-1}H_0 e^{2My}H_0^\dagger. \tag{17}$$

The domain wall configuration in our system can be realized as a bound state of kinky Dp-brane and D$(p+4)$-branes in the type II string theories[14]. By doing so ample dynamics of walls have been uncovered. We have found that the moduli space of domain walls is generally the Lagrangian submanifold of the vacuum manifold of corresponding massless model[15].

3. 1/2 BPS Vortices

Vortices can exist in $5+1$ dimensions or lower. In particular they carry non-Abelian orientational moduli in massless theory[16,17] as instantons. For simplicity let us consider the case of $N_{\rm F} = N_{\rm C} = N$. Taking the Lagrangian (2) in $5+1$ dimensions, and requiring the half of SUSY to be preserved, we obtain the 1/2 BPS equations for vortices as

$$0 = \mathcal{D}_1 H + i\mathcal{D}_2 H, \quad 0 = F_{12} + \frac{g^2}{2}(c\mathbf{1}_N - HH^\dagger). \tag{18}$$

Hypermultiplet BPS equation can be easily solved in terms of a complexified gauge transformation $S(z,\bar{z}) \in GL(N_{\rm C},\mathbf{C})$ and the holomorphic moduli matrix $H_0(z)$[18,19]

$$H = S^{-1}H_0(z), \quad W_1 + iW_2 = -i2S^{-1}\bar{\partial}_z S, \quad z \equiv x^1 + ix^2. \tag{19}$$

The vector multiplet BPS equation can be transformed to the following master equation

$$\partial_z(\Omega^{-1}\bar{\partial}_z\Omega) = \frac{g^2}{4}(c\mathbf{1}_N - \Omega^{-1}H_0H_0^\dagger). \tag{20}$$

The solutions of the BPS equation saturate the BPS bound for the energy density for **vorticity** $k \in \mathbf{Z}_{\geq 0}$

$$T \equiv -c\int d^2x\ \mathrm{Tr}F_{12} = 2\pi ck = -i\frac{c}{2}\oint dz\ \partial\log(\det H_0) + \text{c.c.}, \tag{21}$$

with the boundary condition $\det(H_0) \sim z^k$ at $z \to \infty$. The moduli matrices $H_0(z)$ related by the V-transformation give identical physical fields : $H_0 \to VH_0$, $S \to VS$, $V = V(z) \in GL(N, \mathbf{C}), \det V = \text{const.} \neq 0$. Therefore, the moduli space for vortices is found as

$$\mathcal{M}_{k,N} = \frac{\{H_0(z)|H_0(z) \in M_N, \deg\det(H_0(z)) = k\}}{\{V(z)|V(z) \in M_N, \det V(z) = \text{const.} \neq 0\}}. \tag{22}$$

The generic points of moduli space has $\dim(\mathcal{M}_{N,k}) = 2kN$ and can be represented by

$$H_0 = \begin{pmatrix} \mathbf{1}_{N-1} & -\vec{R}(z) \\ 0 & P(z) \end{pmatrix}, \quad P(z) = \prod_{i=1}^{k}(z - z_i). \tag{23}$$

Moduli space of a single vortex $k = 1$ is given[16,17] by $\mathcal{M}_{N,k=1} \simeq \mathbf{C}\times\mathbf{C}P^{N-1}$ and is represented by the moduli matrix (23) with $\vec{R}^T = (b_1, \cdots, b_{N-1})$ and $P(z) = z - z_0$. Moduli space of k separated vortices is given by a symmetric product $\left(\mathbf{C} \times \mathbf{C}P^{N-1}\right)^k / \mathfrak{S}_k$. The orbifold singularities of this are appropriately resolved in the full moduli space[19]. We also find that the Kähler quotient construction[16] can also be obtained from our moduli matrix by a change of basis. Superfield formulation with the slow-movement expansion readily yields the effective Lagrangian on the world volume of vortices[12]. The duality between vortices and walls has been discussed[20].

4. 1/4 BPS Webs of Domain Walls

The direction of BPS walls are related to the phase of the hypermultiplet masses. If we have **complex masses** $\mu_A = m_A + in_A$, we can obtain two or more non-parallel walls, which can lead to **wall junctions**[21]. We consider the Lagrangian (2) in 3+1 dimensions, since complex masses can be realized in $3+1$ dimensions or lower

$$M_1 = \mathrm{diag}\left(m_1, m_2, \cdots, m_{N_\mathrm{F}}\right), \quad M_2 = \mathrm{diag}\left(n_1, n_2, \cdots, n_{N_\mathrm{F}}\right). \tag{24}$$

The wall junctions can be realized as a solution of the following 1/4 BPS equations[22]

$$F_{12} = i\left[\Sigma_1, \Sigma_2\right], \quad \mathcal{D}_1\Sigma_2 = \mathcal{D}_2\Sigma_1, \quad \mathcal{D}_\alpha H = HM_\alpha - \Sigma_\alpha H, \tag{25}$$

$$\mathcal{D}_1\Sigma_1 + \mathcal{D}_2\Sigma_2 = \frac{g^2}{2}(c\mathbf{1}_{N_C} - HH^\dagger). \tag{26}$$

The solutions saturate the Bogomol'nyi bound for the energy density

$$\mathcal{E} \geq \mathcal{Y} + \mathcal{Z}_1 + \mathcal{Z}_2 + \sum_{\alpha=1,2} \partial_\alpha J_\alpha, \quad J_\alpha \equiv \mathrm{Tr}\Big[H(M_\alpha H^\dagger - H^\dagger \Sigma_\alpha)\Big], \tag{27}$$

$$\mathcal{Y} \equiv \frac{2}{g^2}\partial_\alpha \mathrm{Tr}\left(\epsilon^{\alpha\beta}\Sigma_2\mathcal{D}_\beta\Sigma_1\right), \quad \mathcal{Z}_1 \equiv c\partial_1\mathrm{Tr}\Sigma_1, \quad \mathcal{Z}_2 \equiv c\partial_2\mathrm{Tr}\Sigma_2. \tag{28}$$

The first two equations in Eq.(25) assures the integrability of the last one in Eq.(25), which is solved by[22]

$$H = S^{-1}H_0 e^{M_1x^1 + M_2x^2}, \quad W_\alpha - i\Sigma_\alpha = -iS^{-1}\partial_\alpha S, \quad \alpha = 1, 2. \tag{29}$$

The remaining BPS equation (26) can be rewritten in terms of the gauge invariant quantity $\Omega \equiv SS^\dagger$ as the master equation

$$\sum_{\alpha=1,2} \partial_\alpha\left(\partial_\alpha\Omega\Omega^{-1}\right) = cg^2\left(\mathbf{1}_{N_C} - c^{-1}H_0 e^{2(M_1x^1 + M_2x^2)}H_0^\dagger\Omega^{-1}\right), \tag{30}$$

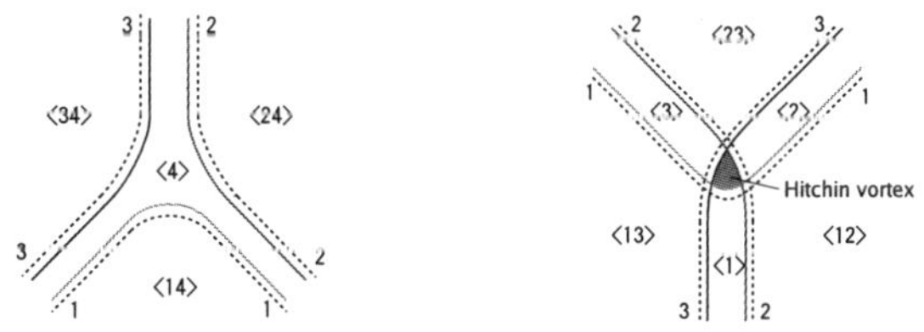

(a) Abelian junction (b) non-Abelian junction

Fig. 4. Internal structures of the junctions with $g\sqrt{c} \ll |\Delta m + i\Delta n|$

The total moduli space of 1/4 BPS equations (25), (26) can be decomposed into 1/4, 1/2, and 1/1 BPS sectors

$$\begin{aligned} \mathcal{M}^{\rm webs}_{\rm tot} &\simeq G_{N_F,N_C} = \{H_0 \mid H_0 \sim VH_0,\ V \in GL(N_C, \mathbf{C})\} \\ &= \mathcal{M}^{\rm webs}_{1/4} \bigcup \mathcal{M}^{\rm walls}_{1/2} \bigcup \mathcal{M}^{\rm vacua}_{1/1}. \end{aligned} \tag{31}$$

We find that Abelian gauge theory gives only Abelian junctions with the negative junction charge $\mathcal{Y} < 0$, whereas non-Abelian gauge theory gives non-Abelian junction with positive junction charge $\mathcal{Y} > 0$ in addition to

Abelian junction. Physical interpretation of the positive junction charge is the presence of the Hitchin vortex residing at the junction as illustrated in Fig. 4. We also find that there are cases with vanishing junction charge $\mathcal{Y} = 0$ corresponding to the intersections of penetrable walls.

We find that the **normalizable moduli** of web of walls are given by **loops** in web as shown in Fig. 5. The moduli matrix of $N_{\rm F} = 4, N_{\rm C} = 1$

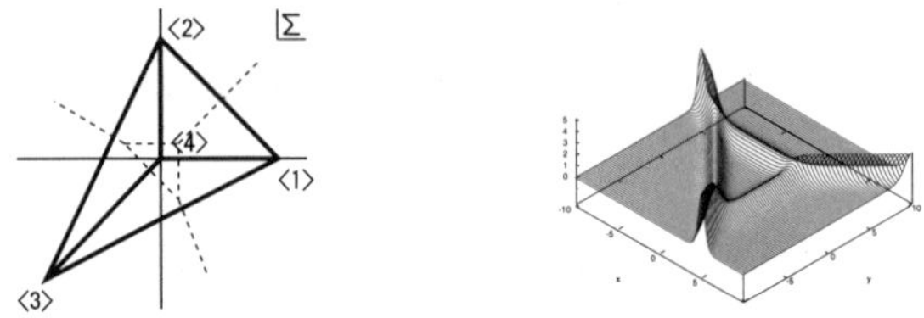

(a) grid diagram (b) energy density ($g^2 \to \infty$)

Fig. 5. Abelian junction with 1 loop and 3 external walls in $N_{\rm C} = 1, N_{\rm F} = 4$ model.

case with $M = \mathrm{diag}(1, i, -1-i, 0)$ can be parametrized by

$$H_0 = \sqrt{c}\left(e^{a_1+ib_1}, e^{a_2+ib_2}, e^{a_3+ib_3}, e^{a_4+ib_4}\right), \tag{32}$$

with $a_j + ib_j, j = 1, 2, 3$ as external wall moduli, and $a_4 + ib_4$ as the loop moduli, corresponding to the normalizable mode. Grid diagrams are found to be useful to specify the moduli of the web of walls as illustrated in Fig. 5. A brane configuration is proposed in [23].

5. 1/4 BPS Monopoles (Instantons) inside a vortex

Vacua outside of monopoles are in the Coulomb phase with unbroken $U(1)$ gauge group. In our $U(N_{\rm C})$ gauge theory with $N_{\rm F}(\geq N_{\rm C})$ flavors of hypermultiplets in the fundamental representation, vacua are in the Higgs phase. If we place a monopole in the Higgs phase, magnetic flux emanating from the monopole is squeezed into vortices, as illustrated in Fig. 6. Therefore

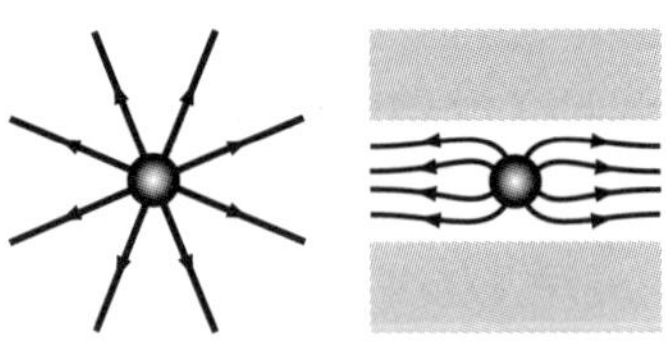

Fig. 6. Monopole in Coulomb phase (left) and in Higgs phase (right).

monopoles in the Higgs phase become composite of monopoles and vortices[24]. Monopoles depend on x^1, x^2, x^3 coordinates and preserve 1/2 SUSY defined by $\gamma^{123}\varepsilon^i = \varepsilon^i$. Vortices along x^3-axis preserve another 1/2 SUSY defined by $\gamma^{12}(i\sigma_3)^i{}_j\varepsilon^j = \varepsilon^i$. If they coexist as a monopole in the Higgs phase, 1/4 of SUSY is preserved. $\gamma^3(i\sigma_3)^i{}_j\varepsilon^j = \varepsilon^i$. We found that this 1/4 SUSY precisely allows domain walls perpendicular to the vortices[18].

Let us consider $U(N_C)$ gauge theory in 4 + 1 dimensions and assume field configurations of monopole-vortex-wall composite to depend on $x^m \equiv (x^1, x^2, x^3)$ and Poincaré invariance in x^0, x^4 space. With the above preserved 1/4 SUSY, we obtain the 1/4 BPS equations :

$$\mathcal{D}_3\Sigma = \frac{g^2}{2}\left(c\mathbf{1}_{N_C} - HH^\dagger\right) + F_{12}, \quad \mathcal{D}_3 H = -\Sigma H + HM, \tag{33}$$

which amounts to the contribution of vortex magnetic field F_{12} added to the wall BPS equation. These are supplemented by the BPS equations for vortices

$$0 = \mathcal{D}_1 H + i\mathcal{D}_2 H, \quad 0 = F_{23} - \mathcal{D}_1\Sigma, \quad 0 = F_{31} - \mathcal{D}_2\Sigma. \tag{34}$$

The solutions of these BPS equations saturate the **BPS bound** of the energy density

$$\mathcal{E} \geq t_{\rm w} + t_{\rm v} + t_{\rm m} + \partial_m J_m, \tag{35}$$

where J_m is the current that does not contribute to the topological charge, and $t_{\rm w}$, $t_{\rm v}$ and $t_{\rm m}$ are energy densities for **walls, vortices** and **monopoles**

$$t_{\rm w} = c\partial_3 {\rm Tr}(\Sigma), \quad t_{\rm v} = -c{\rm Tr}(F_{12}), \quad t_{\rm m} = \frac{2}{g^2}\partial_m {\rm Tr}(\frac{1}{2}\epsilon^{mnl}F_{nl}\Sigma). \tag{36}$$

Integrability condition $[\mathcal{D}_1 + i\mathcal{D}_2, \mathcal{D}_3 + \Sigma] = 0$ coming from the second and third equations in (34) assures the existence of an invertible complex matrix function $S(x^m) \in GL(N_C, \mathbf{C})$ defined by[18]

$$(\mathcal{D}_3 + \Sigma)S^{-1} = 0 \rightarrow \Sigma + iW_3 \equiv S^{-1}\partial_3 S, \tag{37}$$

$$(\mathcal{D}_1 + i\mathcal{D}_2)S^{-1} = 0 \rightarrow W_1 + iW_2 \equiv -2iS^{-1}\bar{\partial}S, \tag{38}$$

where $z \equiv x^1 + ix^2$, and $\bar{\partial} \equiv \partial/\partial z^*$. With this matrix function, the BPS Eq. for hypermultiplet is solved by

$$H = S^{-1}(z, z^*, x^3)H_0(z)e^{Mx^3}, \tag{39}$$

where the **moduli matrix** $H_0(z)$: $N_C \times N_F$ matrix is a holomorphic function of z. The remaining BPS equation is rewritten into a master equation for $\Omega \equiv SS^\dagger$ with $\Omega_0 \equiv H_0\, e^{2My} H_0{}^\dagger$ as input data[18]

$$4\partial(\Omega^{-1}\bar{\partial}\Omega) + \partial_3(\Omega^{-1}\partial_3\Omega) = g^2\left(c - \Omega^{-1}\Omega_0\right). \tag{40}$$

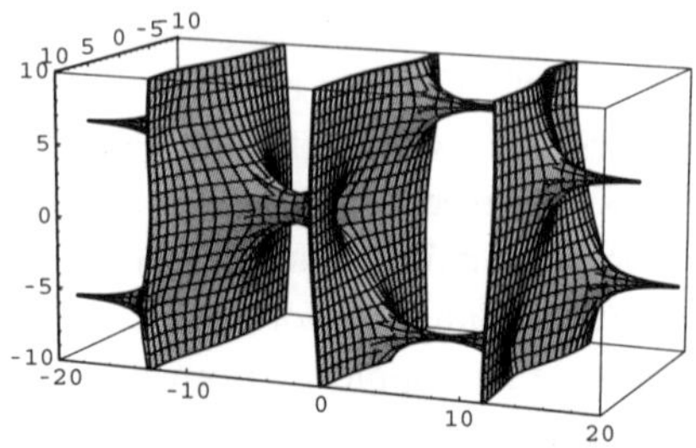

Fig. 7. Surfaces defined by the same energy density with $t_{\rm w} + t_{\rm v} = 0.5c$.

We can obtain exact solutions at strong coupling limit: $g^2 \to \infty$, since the master equation reduces to an algebraic equation

$$\Omega = \Omega_0 \equiv c^{-1} H_0 e^{2My} H_0^\dagger. \tag{41}$$

Our construction produces rich contents, even for $U(1)$ gauge theory whose moduli matrix is given by

$$H_0(z) = \sqrt{c}\left(f^1(z), \ldots, f^{N_{\rm F}}(z)\right), \quad \Omega = \sum_{A=1}^{N_{\rm F}} |f^A(z)|^2 e^{2m_A x^3}. \tag{42}$$

Nonconstant $f^A(z)$ can be interpreted as wall positions to depend on z. In particular, walls are bent to form **vortices**, if $f^A(z)$ has **zeroes**. If $f^A(z) \propto (z - z_\alpha^A)^{k_\alpha^A}$, we obtain vorticity k_α^A at $z = z_\alpha^A$ on the A-th wall. An illustrative configuration of monopoles-vortices-walls is given in Fig. 7. A **monopole in Higgs phase** is realized as a **kink on a vortex**, whereas **instantons inside a vortex** is realized as a **vortex on a vortex**[25].

The moduli matrix approach is powerful enough to establish interaction rules of monopoles, vortices, and walls. In $U(2)$ gauge theory with $N_{\rm F} = 3$ flavors, we can list up all possible moduli matrices for a single vortex in both sides of a wall ($M = {\rm diag}(m_1, m_2, m_3)$ ordered as $m_1 > m_2 > m_3$)

$$\begin{pmatrix} z - z_2 & a_2(z - z_3) & 0 \\ 0 & 0 & 1 \end{pmatrix} e^{Mx^3}, \quad \begin{pmatrix} 1 & a_3 & 0 \\ 0 & 0 & z - z_4 \end{pmatrix} e^{Mx^3}, \tag{43}$$

$$\begin{pmatrix} 1 & a_1 & b \\ 0 & 0 & z - z_1 \end{pmatrix} e^{Mx^3} \sim \begin{pmatrix} \frac{1}{a_1}(z - z_1) & z - z_1 & 0 \\ \frac{1}{b} & \frac{a_1}{b} & 1 \end{pmatrix} e^{Mx^3}. \tag{44}$$

Moduli matrices in (43) are essentially those in $U(1)$ gauge theory. The first one has two vortices at $z = z_2, z_3$ stretching to opposite directions as in the left of Fig. 8. The second one represents a single vortex penetrating the wall as in the right of Fig. 8. We see that vortices can end on a wall in different

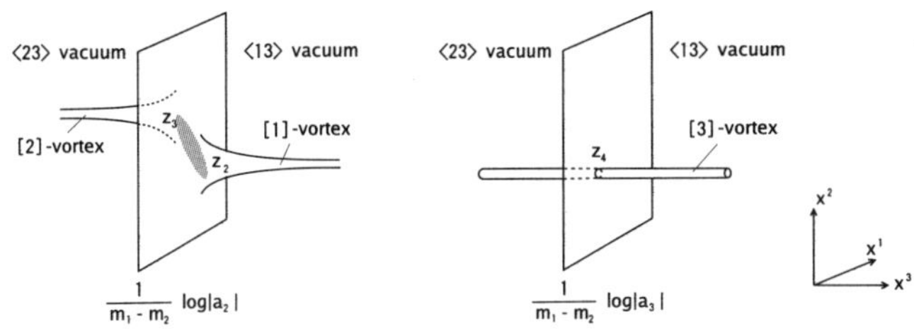

Fig. 8. The vortex positions can separate on domain wall

positions as long as monopole does not sit on any of the vortices. The moduli matrix in Eq.(44) is intrinsically non-Abelian and gives the configuration depicted in Fig. 9. We find that monopole can penetrate through a wall as long as the position of vortices on both sides of the wall coincide. Vortices on a wall can separate only when the monopole is removed to infinity.

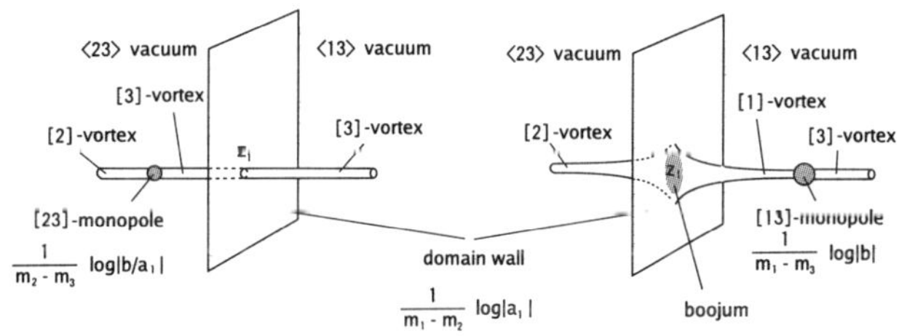

Fig. 9. Monopole can go through wall only if vortex positions on the wall coincide.

We find that these 1/4 BPS composite solitons are related by the Scherk-Schwarz dimensional reduction from $5+1$ dimensions to $4+1$ or $3+1$ dimensions as the table below. Dyonic extension of these solitons are also discussed [26].

dim \ charge	positive	negative
$d = 5, 6$ instanton	Instanton inside vortex	Intersecton[25]
$d = 4, 5$ monopole	Monopole attached by vortices	Boojum[18,10]
$d = 3, 4$ Hitchin	Non-Abelian wall junction	Abelian w. j. [22]

6. Conclusion

(1) The BPS solitons are constructed in SUSY $U(N_C)$ gauge theories with N_F hypermultiplets in the fundamental representation.

(2) Total moduli space of the non-Abelian walls is given by a compact

complex Grassmann manifold described by the moduli matrix H_0.

(3) A general formula for the effective Lagrangian is obtained.

(4) Webs of domain walls are obtained. There are Abelian and non-Abelian junctions of walls in non-Abelian gauge theory. Normalizable moduli of the web of walls are associated with loops of walls.

(5) Composite 1/4 BPS solitons in Higgs phase are systematically obtained by Scherk-Schwarz dimensional reduction: instanton-vortex-vortex, wall-vortex-monopole, webs of walls.

(6) All possible 1/4 BPS solutions are obtained exactly and explicitly in the strong gauge coupling limit.

Acknowledgements

We would like to thank Toshiaki Fujimori, Kazutoshi Ohta, Yuji Tachikawa, David Tong, and Yisong Yang for collaborations in various stages. This work is supported in part by Grant-in-Aid for Scientific Research from the Ministry of Education, Culture, Sports, Science and Technology, Japan No.17540237 (N. S.). The work of K. O. (M. E. and Y. I.) is supported by Japan Society for the Promotion of Science under the Post-doctoral (Pre-doctoral) Research Program.

References

1. M. Eto, Y. Isozumi, M. Nitta, K. Ohashi and N. Sakai, J. Phys. A **39** (2006) R315 [arXiv:hep-th/0602170].
2. P. Horava and E. Witten, Nucl. Phys. B **475** (1996) 94 [arXiv:hep-th/9603142].
3. N. Arkani-Hamed, S. Dimopoulos and G. R. Dvali, Phys. Lett. B **429** (1998) 263 [arXiv:hep-ph/9803315]; I. Antoniadis, N. Arkani-Hamed, S. Dimopoulos and G. R. Dvali, Phys. Lett. B **436** (1998) 257 [arXiv:hep-ph/9804398].
4. L. Randall and R. Sundrum, Phys. Rev. Lett. **83** (1999) 3370 [arXiv:hep-ph/9905221]; Phys. Rev. Lett. **83** (1999) 4690 [arXiv:hep-th/9906064].
5. E. B. Bogomolny, Sov. J. Nucl. Phys. **24** (1976) 449 [Yad. Fiz. **24** (1976) 861]; M. K. Prasad and C. M. Sommerfield, Phys. Rev. Lett. **35** (1975) 760.
6. E. Witten and D. I. Olive, Phys. Lett. B **78** (1978) 97.
7. N. Seiberg and E. Witten, Nucl. Phys. B **426** (1994) 19 [Erratum-ibid. B **430** (1994) 485] [arXiv:hep-th/9407087]; Nucl. Phys. B **431** (1994) 484 [arXiv:hep-th/9408099].
8. P. Fayet and J. Iliopoulos, Phys. Lett. B **51** (1974) 461.
9. Y. Isozumi, M. Nitta, K. Ohashi and N. Sakai, Phys. Rev. Lett. **93** (2004) 161601 [arXiv:hep-th/0404198]; Phys. Rev. D **70** (2004) 125014 [arXiv:hep-th/0405194].
10. N. Sakai and D. Tong, JHEP **0503** (2005) 019 [arXiv:hep-th/0501207].

11. N. Sakai and Y. Yang, Comm. Math. Phys. (in press) [arXiv:hep-th/0505136].
12. M. Eto, Y. Isozumi, M. Nitta, K. Ohashi and N. Sakai, Phys. Rev. D **73** (2006) 125008 [arXiv:hep-th/0602289].
13. M. Arai, M. Naganuma, M. Nitta and N. Sakai, Nucl. Phys. B **652** (2003) 35 [arXiv:hep-th/0211103].
14. M. Eto, Y. Isozumi, M. Nitta, K. Ohashi, K. Ohta and N. Sakai, Phys. Rev. D **71** (2005) 125006 [arXiv:hep-th/0412024].
15. M. Eto, Y. Isozumi, M. Nitta, K. Ohashi, K. Ohta, N. Sakai and Y. Tachikawa, Phys. Rev. D **71** (2005) 105009 [arXiv:hep-th/0503033].
16. A. Hanany and D. Tong, JHEP **0307** (2003) 037 [arXiv:hep-th/0306150].
17. R. Auzzi, S. Bolognesi, J. Evslin, K. Konishi and A. Yung, Nucl. Phys. B **673** (2003) 187 [arXiv:hep-th/0307287].
18. Y. Isozumi, M. Nitta, K. Ohashi and N. Sakai, Phys. Rev. D **71** (2005) 065018 [arXiv:hep-th/0405129].
19. M. Eto, Y. Isozumi, M. Nitta, K. Ohashi and N. Sakai, Phys. Rev. Lett. **96** (2006) 161601 [arXiv:hep-th/0511088].
20. M. Eto, T. Fujimori, Y. Isozumi, M. Nitta, K. Ohashi, K. Ohta and N. Sakai, Phys. Rev. D **73** (2006) 085008 [arXiv:hep-th/0601181].
21. G. W. Gibbons and P. K. Townsend, Phys. Rev. Lett. **83** (1999) 1727 [arXiv:hep-th/9905196]; S. M. Carroll, S. Hellerman and M. Trodden, Phys. Rev. D **61** (2000) 065001 [arXiv:hep-th/9905217]; H. Oda, K. Ito, M. Naganuma and N. Sakai, Phys. Lett. B **471** (1999) 140 [arXiv:hep-th/9910095]; K. Ito, M. Naganuma, H. Oda and N. Sakai, Nucl. Phys. B **586** (2000) 231 [arXiv:hep-th/0004188]; Nucl. Phys. Proc. Suppl. **101** (2001) 304 [arXiv:hep-th/0012182]; A. Gorsky and M. A. Shifman, Phys. Rev. D **61** (2000) 085001 [arXiv:hep-th/9909015]; M. Naganuma, M. Nitta and N. Sakai, Phys. Rev. D **65** (2002) 045016 [arXiv:hep-th/0108179]; K. Kakimoto and N. Sakai, Phys. Rev. D **68** (2003) 065005 [arXiv:hep-th/0306077].
22. M. Eto, Y. Isozumi, M. Nitta, K. Ohashi and N. Sakai, Phys. Rev. D **72** (2005) 085004 [arXiv:hep-th/0506135]; Phys. Lett. B **632** (2006) 384 [arXiv:hep-th/0508241].
23. M. Eto, Y. Isozumi, M. Nitta, K. Ohashi, K. Ohta and N. Sakai, AIP Conf. Proc. **805** (2006) 354 [arXiv:hep-th/0509127].
24. D. Tong, Phys. Rev. D **69** (2004) 065003 [arXiv:hep-th/0307302]; R. Auzzi, S. Bolognesi, J. Evslin and K. Konishi, Nucl. Phys. B **686** (2004) 119 [arXiv:hep-th/0312233]; M. Shifman and A. Yung, Phys. Rev. D **70** (2004) 045004 [arXiv:hep-th/0403149]; A. Hanany and D. Tong, JHEP **0404** (2004) 066 [arXiv:hep-th/0403158].
25. M. Eto, Y. Isozumi, M. Nitta, K. Ohashi and N. Sakai, Phys. Rev. D **72** (2005) 025011 [arXiv:hep-th/0412048].
26. K. Lee and H. U. Yee, Phys. Rev. D **72** (2005) 065023 [arXiv:hep-th/0506256]; M. Eto, Y. Isozumi, M. Nitta and K. Ohashi, Nucl. Phys. B (in press) [arXiv:hep-th/0506257].

SECTION 2
AdS/QCD

Convener A. Armoni

EMERGING HOLOGRAPHY: ADS FROM QCD*

JOSHUA ERLICH

Department of Physics, College of William and Mary, Williamsburg, VA 23187

We systematically derive an AdS/CFT correspondence for the vector current in QCD, beginning with a Padé approximation to the vector current two-point function. 5D Anti-de Sitter space emerges naturally in this approach.

1. Introduction

Recent phenomenological applications of the AdS/CFT correspondence[2–4] include the proposals of a simple holographic dual to low energy QCD[5–8]. These models have allowed for the calculation of properties of mesons and baryons, given relatively few input parameters, with results that so far are remarkably consistent with experimental data. Chiral symmetry breaking and confinement are input for the model, following the AdS/CFT dictionary. The discrete spectrum of Kaluza-Klein excitations of bulk fields become the composites of QCD. These models are referred to as AdS/QCD models; several extensions and simplifications of the AdS/QCD models have been proposed[9–13]. For some of the earlier attempts to apply AdS/CFT to QCD, see [14–17].

Shifman[18] and Voloshin[19] recently pointed out that the results of a 1970's calculation of vector meson masses in large N_c QCD by way of a Padé approximation to QCD current-current correlators[20,21] is reminiscent of predictions via the AdS/CFT correspondence. We will make explicit the relation between the Padé approximation and the AdS/CFT correspondence, and show that both approaches give identical leading order predictions of meson masses and decay constants[1]. We will also show how the radial direction of 5D Anti-de Sitter space emerges systematically, by analogy with deconstructed extra dimensions[22,23].

---

*This talk was based on work done with G. Kribs and I. Low[1].

2. The AdS/CFT correspondence

In the supergravity limit of the AdS/CFT correspondence, the basic dictionary relating correlation functions in the CFT to supergravity on an AdS background was described in [3,4]. Each operator in the CFT corresponds to a field in Anti-de Sitter space of one dimension higher. If we consider the AdS metric in the form,

$$ds^2 = \frac{1}{z^2}\left(\eta_{\mu\nu}dx^\mu dx^\nu - dz^2\right), \tag{1}$$

where $\eta_{\mu\nu} = \text{Diag}(1,-1,-1,-1)$, the AdS/CFT correspondence is the statement that,

$$\langle e^{i\int d^4x\, \phi_0(x)\mathcal{O}(x)}\rangle_{\text{CFT}} = e^{iS_{\text{AdS}}[\phi^{\text{cl}}]}\Big|_{\phi^{\text{cl}}(x,z=0)=\phi_0}, \tag{2}$$

where $S_{\text{AdS}}[\phi^{\text{cl}}]$ is the classical action in the AdS space and ϕ^{cl} is a solution to the equation of motion whose boundary value is fixed to be the source $\phi_0(x)$ (up to a conformal rescaling). For a vector current such as $J^\mu = \bar{q}\gamma^\mu q$ in the CFT, there is a corresponding bulk gauge field $A_M(x,z)$ whose boundary value is the source for J^μ. The 5D action in AdS space is

$$S_{\text{AdS}} = -\int d^4x dz \sqrt{-g}\,\frac{1}{4g_5^2}F_{MN}F^{MN}, \tag{3}$$

where the capital roman letters $M,N = 0,1,2,3,z$. According to the AdS/CFT correspondence to calculate the current-current correlator, we calculate the 5D action on a solution to the equations of motion for the corresponding gauge field such that the 5D gauge field at the UV boundary has the profile of the 4D source of the current. We will consider a finite AdS space with an infrared boundary at $z = z_0$. The IR boundary corresponds to confinement in the now broken CFT. The profile of the 5D gauge field satisfying those boundary conditions is the bulk-to-boundary propagator, which we call $V(q,z)$. Varying the action (twice) with respect to the boundary source gives the current-current correlator. We impose the $A_z = 0$ gauge and Fourier-transform the gauge field in four dimensions:

$$A_\mu(q,z) = \frac{1}{V(q,\epsilon)}\widetilde{A}_\mu(q)V(q,z), \tag{4}$$

where $\widetilde{A}_\mu(q)$ is the Fourier-transformed current source. The boundary condition $A_\mu(q,\epsilon) = \widetilde{A}_\mu(q)$ is built into (4). The boundary condition at $z = z_0$ is not completely predetermined, but for definiteness we assume Neumann boundary conditions there, $\partial_z V(q,z_0) = 0$, corresponding to the gauge invariant condition $F_{\mu z}(x,z_0) = 0$. The equations of motion for the transverse part of the gauge field are,

$$z\,\partial_z\left(\frac{1}{z}\partial_z V(q,z)\right)+q^2\,V(q,z)=0, \tag{5}$$

which, given the boundary conditions, leads to

$$V(q,z)=qz\left(Y_0(qz_0)J_1(qz)-J_0(qz_0)Y_1(qz)\right). \tag{6}$$

Evaluating the action on the solution leaves only the boundary term at the UV

$$S_{\mathrm{AdS}}=-\frac{1}{2g_5^2}\int d^4q\,\widetilde{A}_\mu(q)\widetilde{A}^\mu(-q)\left(\frac{1}{z}\frac{\partial_z V(q,z)}{V(q,z)}\right)_{z=\epsilon}. \tag{7}$$

If we write the Fourier-transformed vector current two-point function as,

$$\int d^4x\,e^{iq\cdot x}\,\langle J_\mu(x)J_\nu(0)\rangle=\left(g_{\mu\nu}-\frac{q_\mu q_\nu}{q^2}\right)\Sigma(q^2), \tag{8}$$

then functionally differentiating the action (7) with respect to the source $\widetilde{A}_\mu(q)$ yields the vector current-current correlator determined by AdS/CFT:

$$\begin{aligned}\Sigma(q^2) &= -\frac{1}{g_5^2}\frac{1}{z}\frac{\partial_z V(q,z)}{V(q,z)}\bigg|_{z=\epsilon\to 0}\\ &= -\frac{1}{g_5^2}\frac{1}{z}\,q\,\frac{Y_0(qz_0)J_0(qz)-J_0(qz_0)Y_0(qz)}{Y_0(qz_0)J_1(qz)-J_0(qz_0)Y_1(qz)}\bigg|_{z=\epsilon}\\ &\to \frac{q^2}{g_5^2}\frac{\log(q\epsilon)J_0(qz_0)-(\pi/2)Y_0(qz_0)}{J_0(qz_0)},\end{aligned} \tag{9}$$

where we have retained the leading non-vanishing contribution from each of the Bessel functions in the limit $q\epsilon\to 0$. If we compare this to a one-loop QCD calculation at large Euclidean momentum, we can identify $g_5^2=\frac{12\pi^2}{N_c}$ in units of the AdS curvature[7,8].

3. The Padé approximation and vector mesons

Two-point correlators of operators with arbitrary conformal dimension in the UV generally involve expressions of the form,

$$\Sigma_\nu(q^2)=(q^2)^\nu\log\frac{q^2}{\mu^2}. \tag{10}$$

The scale μ is a renormalization scale and $\nu=\Delta-2$, where Δ is the conformal dimension of the operator involved. For a conserved vector current, $\Delta=3$ and $\nu=1$, which is consistent with (9). We will consider the dimensionless quantity $f_0(t)=\log t$, so that $f_0(q^2/\mu^2)=(q^2)^{-\nu}\Sigma_\nu(q^2)$. Migdal proposed a systematic procedure for calculating masses of mesons in large N_c QCD[20,21], which approximates the two-point function in the deep Euclidean region by a rational function and then analytically continues into the infrared domain. Migdal proposed that one find a ratio of two polynomials of degree N respectively, which reproduces the first $2N+1$ terms in the Taylor expansion of $f_0(t)$ with respect to an arbitrary subtraction point, which we choose to be $t=-1$, *i.e.* $q^2=-\mu^2$.

Our goal is then to determine two polynomials $R_N(t)$ and $Q_N(t)$ of degree N that satisfy,

$$f_0(t) - \frac{R_N(t)}{Q_N(t)} = \mathcal{O}\left((t+1)^{2N+1}\right). \tag{11}$$

Since $R_N(t)$ is of degree N, it follows that $Q_N(t)$ satisfies,

$$\left.\frac{d^m}{dt^m} Q_N f_0(t)\right|_{t=-1} = 0, \quad m = N+1, \cdots, 2N. \tag{12}$$

To solve for Q_N we use the Cauchy integral formula along the path in Fig 1. The integral over the large and small circles vanish (for the orders

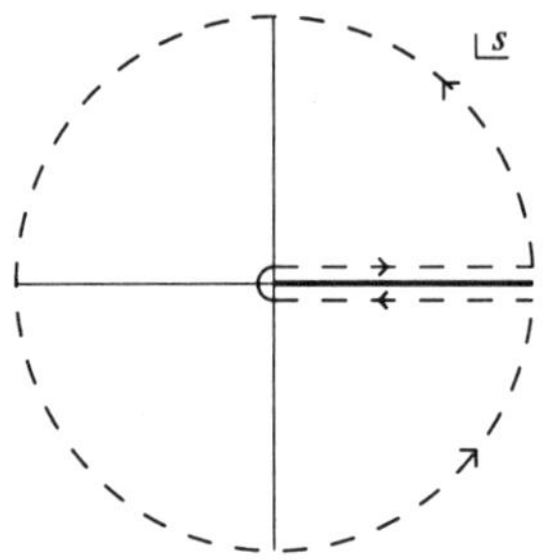

Fig. 1. Contour used to evaluate derivatives of $Q_N f_0(t)$ in (13).

of derivative we are considering), and the integrals above and below the branch cut cancel except for the change in the imaginary part of f_0. What remains is given by,

$$\left.\frac{d^m}{dt^m} Q_N f_0(t)\right|_{t=-1} = \frac{m!}{2\pi}\int_0^\infty ds\, \frac{Q_N\, \Delta \mathrm{Im} f_0(s)}{(s+1)^{m+1}} = 0, \quad m = N+1, \cdots, 2N, \tag{13}$$

where $\Delta \mathrm{Im} f_0(t) = \mathrm{Im} f_0(t+i\epsilon) - \mathrm{Im} f_0(t-i\epsilon)$ is the change in $f_0(t)$ across the branch cut.

In our case, $\Delta \mathrm{Im} f_0 = -2\pi$, and from (12) and (13) we obtain the following equations for Q_N:

$$\int_0^\infty ds \frac{Q_N}{(s+1)^{m+1}} = 0, \qquad m = N+1, \cdots, 2N\,. \tag{14}$$

Using the orthogonality of Legendre polynomials, we can identify,

$$Q_N(s) = (s+1)^N P_N\left(\frac{1-s}{1+s}\right), \tag{15}$$

up to an overall normalization. The normalization does not affect the computation of the two-point function since it is canceled between the numerator and the denominator in (11).

Next, we take $N \to \infty$ and also the low energy limit $q^2 \ll \mu^2$. Using the relation for Jacobi polynomials[26],

$$\lim_{N\to\infty} N^{-a} P_N^{(a,b)}\left(\cos\frac{z}{N}\right) = \lim_{N\to\infty} N^{-a} P_N^{(a,b)}\left(1 - \frac{z^2}{2N^2}\right) = \left(\frac{z}{2}\right)^{-a} J_a(z), \tag{16}$$

and $P_N(x) = P_N^{(0,0)}(x)$ we obtain,

$$Q_N(t) \to J_0(2N\sqrt{t}) \equiv Q_\infty(t). \tag{17}$$

The scale μ/N is not determined by the Padé approximation, just as the location of the IR wall in AdS was not predetermined. We take the limit, $\mu \to \infty$, $N \to \infty$, $\mu_{ir} \equiv \mu/N =$ fixed. In Sec. 4 we will give these scales a concrete physical interpretation in the deconstructed picture.

The polynomial $R_N(t)$ is determined by consistency with (11):

$$R_N = Q_N f_0(t) - \int_0^\infty ds\, \frac{1}{2\pi} \frac{Q_N\, \Delta\mathrm{Im} f_0(s)}{s-t} \left(\frac{t+1}{s+1}\right)^{N+1}. \tag{18}$$

The second term on the right hand side cancels the higher order terms in the Taylor expansion of $Q_N f_0(t)$ so that the result is a polynomial of degree N. Taking the limit $N \to \infty$ we obtain,

$$\begin{aligned} R_\infty(t) &\equiv Q_\infty(t) f_0(t) - \frac{1}{2\pi}\int_0^\infty ds\, \frac{Q_\infty(s)\,\Delta\mathrm{Im} f_0(s)}{s-t} \\ &= Q_\infty(t) f_0(t) - \frac{1}{2\pi}\int_0^\infty ds\, \frac{(-2\pi) I_0(2N\sqrt{-s})}{s-t} \\ &= J_0(2N\sqrt{t})(\log t) - \pi\, Y_0(2N\sqrt{t}). \end{aligned} \tag{19}$$

Up to an overall constant which has been factored out of (10), the Padé approximation for the polarization $\Sigma(q^2)$ becomes,

$$\Sigma(q^2) \propto q^2 \frac{J_0(2Nq/\mu)\log(q^2/\mu^2) - \pi\, Y_0(2Nq/\mu)}{J_0(2Nq/\mu)}. \tag{20}$$

Finally, identifying $\epsilon = 1/\mu$, $z_0 = 2/\mu_{ir}$, we recover (9) as promised. We have obtained the same results as AdS/CFT for the two-point function, without reference to the Anti-de Sitter space or holography.

4. The emergence of AdS

At finite order, the Padé approximation gives rise to a finite set of massive modes that approximate the KK spectrum of a gauge field in a slice of AdS_5. We propose that finite N in the Padé approximation is equivalent to N hidden local symmetries[27,28], which becomes equivalent to an N-site deconstruction [22,23,29–31] of the AdS space at large N. In the context of QCD, the idea of considering a large number of hidden local symmetries was discussed in [14] and found to qualitatively and in some cases quantitatively agree with low energy data. Here our observation is that the poles in the Padé approximation can be interpreted as vector resonances. Using hidden local symmetry, each massive vector is reinterpreted as a massive gauge boson of a broken gauge group. Hence, N massive vector resonances can be represented as a product gauge theory with N broken gauge groups. Furthermore, the residues at the poles are positive[20], and are related to the decay constants of the resonances.

In order to connect the spectrum of modes as determined by the Padé approximation to Anti-de Sitter space, we can construct a moose model such that the spectrum of gauge bosons is precisely that of the Padé approximation. As the number of gauge groups increases the theory approaches that of a deconstructed warped extra dimension. By restricting the number of gauge fields, the Padé approximation in this way provides a gauge invariant regularization of the 5D theory. Anti-de Sitter space, latticized in one dimension, thereby appears naturally.

Acknowledgments

This work was inspired by comments of M. Shifman and M. Voloshin. We benefitted from conversations with Johannes Hirn, Juan Maldacena, Veronica Sanz, Martin Schmaltz, and Lior Silberman. We are also grateful to Chris Carone for collaboration at early stages of this work. This work was supported in part by the National Science Foundation under grant PHY-0504442 and the Jeffress Memorial Trust under grant J-768 (JE), and by the Department of Energy under grants DE-FG02-96ER40969 (GDK) and DE-FG02-90ER40542 (IL). Part of this work was performed at the Aspen Center for Physics.

References

1. J. Erlich, G. D. Kribs and I. Low, Phys. Rev. D **73**, 096001 (2006) [arXiv:hep-th/0602110].
2. J. M. Maldacena, Adv. Theor. Math. Phys. **2**, 231 (1998) [arXiv:hep-th/9711200].
3. S. S. Gubser, I. R. Klebanov and A. M. Polyakov, Phys. Lett. B **428**, 105 (1998) [arXiv:hep-th/9802109].
4. E. Witten, Adv. Theor. Math. Phys. **2**, 253 (1998) [arXiv:hep-th/9802150].
5. T. Sakai and S. Sugimoto, Prog. Theor. Phys. **113**, 843 (2005) [arXiv:hep-th/0412141].
6. G. F. de Teramond and S. J. Brodsky, Phys. Rev. Lett. **94**, 201601 (2005) [arXiv:hep-th/0501022].
7. J. Erlich, E. Katz, D. T. Son and M. A. Stephanov, arXiv:hep-ph/0501128.
8. L. Da Rold and A. Pomarol, Nucl. Phys. B **721**, 79 (2005) [arXiv:hep-ph/0501218].
9. J. Hirn and V. Sanz, JHEP **0512**, 030 (2005) [arXiv:hep-ph/0507049].
10. T. Sakai and S. Sugimoto, Prog. Theor. Phys. **114**, 1083 (2006) [arXiv:hep-th/0507073].
11. L. Da Rold and A. Pomarol, arXiv:hep-ph/0510268.
12. K. Ghoroku, N. Maru, M. Tachibana and M. Yahiro, Phys. Lett. B **633**, 602 (2006) [arXiv:hep-ph/0510334].
13. J. Hirn, N. Rius and V. Sanz, arXiv:hep-ph/0512240.
14. D. T. Son and M. A. Stephanov, Phys. Rev. D **69**, 065020 (2004) [arXiv:hep-ph/0304182].
15. S. J. Brodsky and G. F. de Teramond, Phys. Lett. B **582**, 211 (2004) [arXiv:hep-th/0310227].
16. N. J. Evans and J. P. Shock, Phys. Rev. D **70**, 046002 (2004) [arXiv:hep-th/0403279].
17. S. Hong, S. Yoon and M. J. Strassler, arXiv:hep-th/0409118.
18. M. Shifman, arXiv:hep-ph/0507246.
19. M. B. Voloshin, private communication.
20. A. A. Migdal, Annals Phys. **109**, 365 (1977).
21. A. A. Migdal, Annals Phys. **110**, 46 (1978).
22. N. Arkani-Hamed, A. G. Cohen and H. Georgi, Phys. Rev. Lett. **86**, 4757 (2001) [arXiv:hep-th/0104005].
23. C. T. Hill, S. Pokorski and J. Wang, Phys. Rev. D **64**, 105005 (2001) [arXiv:hep-th/0104035].
24. M. A. Shifman, A. I. Vainshtein and V. I. Zakharov, Nucl. Phys. B **147**, 385 (1979).
25. M. A. Shifman, A. I. Vainshtein and V. I. Zakharov, Nucl. Phys. B **147**, 448 (1979).
26. See, for example, G. Szego, "Orthogonal Polynomials," American Mathematical Society Colloquium Publications, Volume XXIII, 1975.
27. M. Bando, T. Kugo, S. Uehara, K. Yamawaki and T. Yanagida, Phys. Rev. Lett. **54**, 1215 (1985).
28. M. Bando, T. Kugo and K. Yamawaki, Phys. Rept. **164**, 217 (1988).
29. K. Sfetsos, Nucl. Phys. B **612**, 191 (2001) [arXiv:hep-th/0106126].
30. A. Falkowski and H. D. Kim, JHEP **0208**, 052 (2002) [arXiv:hep-ph/0208058].
31. L. Randall, Y. Shadmi and N. Weiner, JHEP **0301**, 055 (2003) [arXiv:hep-th/0208120].

A HOLOGRAPHIC MODEL OF HADRONS

M.A. STEPHANOV

Physics Department, University of Illinois, Chicago, IL 60607-7059
E-mail: misha@uic.edu

This short talk is based on the work with J. Erlich, E. Katz and D. Son, hep-ph/0501128. Inspired by ideas of gauge/string duality, we propose a five-dimensional framework for modeling low energy properties of QCD. The model naturally incorporates properties of QCD dictated by chiral symmetry, which we demonstrate by deriving the Gell-Mann-Oakes-Renner relationship for the pion mass. The couplings and masses of the infinite towers of vector and axial vector mesons described by the model automatically obey QCD sum rules. The phenomenon of vector-meson dominance is a straightforward consequence of the model.

1. Introduction

The discovery of AdS/CFT correspondence [1] has revived the hope that QCD can be reformulated as a solvable string theory. So far, theories which can be solved using AdS/CFT techniques differ substantially from QCD, most notably by the strong coupling in the ultraviolet (UV) regime and the lack of asymptotic freedom. Nevertheless, certain important properties of QCD such as confinement and chiral symmetry breaking are present in many of these theories, and the gravity/gauge duality provides a new approach to studying the resulting dynamics. An important development in the prototypical example of $\mathcal{N} = 4$ super Yang-Mills (SYM) theory has been the introduction of fundamental quarks using probe D7 branes [2]. The mesons that appear in these theories behave in many ways similarly to the mesons in QCD [3,4].

Inspired by the gravity/gauge duality we propose the following complementary approach. Rather than deform the SYM theory to obtain QCD [5], we start from QCD and attempt to construct its five-dimensional (5D) holographic dual. Such an approach is similar in spirit to the construction of the QCD moose theory in Ref. 6, where the holographic description arises in the continuum limit of infinitely many hidden local symmetries. As in

Ref. 6, vector meson dominance and QCD sum rules are natural consequences of our model. Hence, the success of the model is not coincidental, but a result of linking several proven approaches through the AdS/CFT correspondence. We expect the success of our model to diminish above roughly the scale given by the mass of the lightest isospin-carrying spin-2 resonance, namely the a_2 (1318 MeV [7]). In particular, we are completely neglecting stringy physics which becomes important at higher energies, and we have not included in our description any modes with spin larger than one (see, however, Ref. 8). At this stage, we also neglect running of the QCD coupling, which is likely a poor approximation for a larger range of energies. While our model is too simple to provide a complete dual description of QCD, its success seems to suggest that there is a quantitatively useful reformulation of QCD as a string theory in a higher-dimensional curved space.

2. Field content

Table 1 illustrates the field content of our model. The choice of the 5D fields is dictated by a principle of the AdS/CFT correspondence: each operator $\mathcal{O}(x)$ in the 4D field theory corresponds to a field $\phi(x,z)$ in the 5D bulk theory. The 5D theory dual to QCD should, therefore, contain an infinite number of fields corresponding to the infinite number of operators in QCD. There is, however, a small number of operators that are important in the chiral dynamics: the left- and right-handed currents corresponding to the $\mathrm{SU}(N_f)_L \times \mathrm{SU}(N_f)_R$ chiral flavor symmetry, and the chiral order parameter (see Table 1). We shall include in our model only the 5D fields which correspond to these operators and neglect all other fields.

Table 1. The fields of the model

4D: $\mathcal{O}(x)$	5D: $\phi(x,z)$	p	Δ	$(m_5)^2$
$\bar{q}_L\gamma^\mu t^a q_L$	$A^a_{L\mu}$	1	3	0
$\bar{q}_R\gamma^\mu t^a q_R$	$A^a_{R\mu}$	1	3	0
$\bar{q}^\alpha_R q^\beta_L$	$(2/z)X^{\alpha\beta}$	0	3	-3

The 5D masses m_5 of the fields $A^a_{L\mu}$, $A^a_{R\mu}$, and X are determined via the relation [9,10] $(\Delta - p)(\Delta + p - 4) = m_5^2$, where Δ is the dimension of the corresponding p-form operator – see Table 1. We assumed here that these operators keep their canonical dimensions, which is true only for the conserved currents. However, for the field X we could easily incorporate corrections to its classical dimension. The factor $1/z$ in Table 1 is dictated by the dimension of the operator $\bar{q}q$, while the factor of 2 is of no physical significance and is chosen for later convenience.

We shall choose the simplest possible metric for our model, namely a slice of the Anti-de Sitter (AdS) metric,

$$ds^2 = \frac{1}{z^2}(-dz^2 + dx^\mu dx_\mu), \quad 0 < z \le z_m. \tag{1}$$

The fifth coordinate z corresponds to the energy scale, as higher energy (or momentum transfer Q^2) QCD physics is reflected by the behavior of the fields closer to the AdS boundary $z = 0$: $Q \sim 1/z$. By virtue of the conformal isometry of the AdS space, in such a model the running of the QCD gauge coupling is neglected in a window of scales until an infrared (IR) scale $Q_m \sim 1/z_m$. To make the theory confining, one introduces an IR cutoff in the metric at $z = z_m$ where spacetime ends. We shall call $z = z_m$ the "infrared brane" and impose certain boundary conditions on the fields at $z = z_m$. Certainly, this is only a crude model of confinement. Indeed, our model requires two dimensionful parameters related to chiral symmetry breaking, whereas in QCD there is only one. In addition, an UV cutoff can be provided by setting the boundary to $z = \epsilon$ instead of $z = 0$.

3. 5D action and chiral symmetry breaking

The action of the theory in the bulk is:

$$S = \int d^5x \sqrt{g}\, \mathrm{Tr}\left\{|DX|^2 + 3|X|^2 - \frac{1}{4g_5^2}(F_L^2 + F_R^2)\right\} \tag{2}$$

where $D_\mu X = \partial_\mu X - iA_{L\mu}X + iXA_{R\mu}$, $A_{L,R} = A^a_{L,R}t^a$, and $F_{\mu\nu} = \partial_\mu A_\nu - \partial_\nu A_\mu - i[A_\mu, A_\nu]$. As usual, the gauge invariance in the 5D theory corresponds to the conservation of the global symmetry current in the 4D theory.

At the IR brane we must impose some gauge invariant boundary conditions, and we make the simplest choice: $(F_L)_{z\mu} = (F_R)_{z\mu} = 0$. In the following we shall be using the gauge $A_z = 0$. In this case our boundary conditions are simply Neumann.

The expectation value of the field X is determined by the classical solution satisfying the UV boundary condition $(2/\epsilon)X(\epsilon) = M$ for quark mass matrix M:

$$X_0(z) = \frac{1}{2}Mz + \frac{1}{2}\Sigma z^3, \tag{3}$$

The matrix Σ is determined by the IR boundary condition on X. Instead of specifying this condition we shall choose Σ as an input parameter of the model. The meaning of Σ in QCD can be found by calculating the variation of the vacuum energy w.r.t. M [11]: $\Sigma^{\alpha\beta} = \langle \bar{q}^{\alpha} q^{\beta} \rangle$. We shall assume, as usual, $\Sigma = \sigma \mathbf{1}$ and take $M = m_q \mathbf{1}$.

At this stage the model has four free parameters: m_q, σ, z_m and g_5. The gauge coupling g_5 will be fixed by the QCD operator product expansion (OPE) for the product of currents, leaving three adjustable parameters.

We will focus on the $N_f = 2$ lightest flavors and neglect effects of $O(m_q^2)$. Therefore, in Table 1, $\alpha, \beta = 1, 2$; $a, b = 1, 2, 3$ and $t^a = \sigma^a/2$, where σ^a are the Pauli matrices.

4. Matching the 5D gauge coupling

We will use the holographic duality to relate the 5D coupling g_5 in (2) to the number of colors N_c in QCD. The precise sense of the holographic correspondence is the equivalence between the generating functional of the connected correlators in the 4D theory $W_{4D}[\phi_0(x)]$ and the effective action of the 5D theory $S_{5D,\text{eff}}[\phi(x,z)]$, with UV boundary values of the 5D bulk fields set to the value of the sources in 4D theory:

$$W_{4D}[\phi_0(x)] = S_{5D,\text{eff}}[\phi(x,z)] \quad \text{at} \quad \phi(x,\epsilon) = \phi_0(x). \tag{4}$$

QCD Green's functions can therefore be obtained by differentiating the 5D effective action with respect to the sources. In the case that stringy effects can be neglected, $S_{5D,\text{eff}}$ is simply given by Eq. (2). The action is evaluated on solutions to the 5D equations of motion subject to the condition that the value of each bulk field at the boundary $z = \epsilon \to 0$ be given by the source ϕ_0 of the corresponding 4D operator $\mathcal{O}$ (see Table 1).

We may now fix the 5D gauge coupling by comparing the result for the vector current two-point function obtained from the above prescription with that of QCD. Introducing the vector field as $V = (A_L + A_R)/2$, one finds, in the $V_z(x,z) = 0$ gauge, the equation of motion for the transverse part of the gauge field:

$$\left(\partial_z \left(\frac{1}{z} \partial_z V_\mu^a(q,z) \right) + \frac{q^2}{z} V_\mu^a(q,z) \right)_\perp = 0. \tag{5}$$

Here $V_\mu^a(q,z)$ is the 4D Fourier transform of $V_\mu^a(x,z)$. If $V_0^{\mu a}(q)$ is the Fourier transform of the source of the vector current $J_\mu^a = \bar{q}\gamma_\mu t^a q$ at the boundary then letting $V^\mu(q,z) = V(q,z) V_0^\mu(q)$, we require that $V(q,\epsilon) = 1$. Differentiating twice with respect to the source V_0, we arrive at the vector current two-point function ($Q^2 \equiv -q^2$),

$$\int_x e^{iqx}\langle J^a_\mu(x)J^b_\nu(0)\rangle = \delta^{ab}(q_\mu q_\nu - q^2 g_{\mu\nu})\Pi_V(Q^2), \tag{6a}$$

$$\Pi_V(-q^2) = -\frac{1}{g_5^2 Q^2}\frac{\partial_z V(q,z)}{z}\bigg|_{z=\epsilon}, \tag{6b}$$

For large Euclidean Q^2 we only need to know $V(q,z)$ near the boundary, $V(Q,z) = 1 + \frac{Q^2 z^2}{4}\ln(Q^2 z^2) + \cdots$ which up to contact terms gives

$$\Pi_V(Q^2) = -\frac{1}{2g_5^2}\ln Q^2. \tag{7}$$

On the other hand, we can compute Π_V from QCD by evaluating Feynman diagrams [12]. The diagram dominating at large Q is the quark bubble,

$$\Pi_V(Q^2) = -\frac{N_c}{24\pi^2}\ln Q^2. \tag{8}$$

This leads to the identification

$$g_5^2 = \frac{12\pi^2}{N_c}. \tag{9}$$

5. Hadrons

A ρ-meson mode, $\psi_\rho(z)$, is a solution to Eq. (5) for an arbitrary component of V_μ with $q^2 = m_\rho^2$, subject to $\psi_\rho(\epsilon) = 0$, $\partial_z\psi_\rho(z_m) = 0$ and normalized as $\int(dz/z)\,\psi_\rho(z)^2 = 1$. The function $V(q,z)$ can be expanded in the basis of $\psi_\rho(z)$: $V(q,z) = \sum_\rho a_\rho(q)\psi_\rho(z)$. The coefficients $a_\rho(q)$ can be found using orthogonality of ψ_ρ and the fact the $V(q,z)$ is a solution of (5). Substituting the expansion into (6b) we find:

$$\Pi_V(-q^2) = -\frac{1}{g_5^2}\sum_\rho \frac{[\psi_\rho{}'(\epsilon)/\epsilon]^2}{(q^2 - m_\rho^2)m_\rho^2}. \tag{10}$$

This allows us to extract the weak decay constants F_ρ:

$$F_\rho^2 = \frac{1}{g_5^2}[\psi_\rho{}'(\epsilon)/\epsilon]^2 = \frac{1}{g_5^2}[\psi_\rho{}''(0)]^2, \tag{11}$$

where F_ρ is defined by $\langle 0|J^a_\mu|\rho^b\rangle = F_\rho\delta^{ab}\varepsilon_\mu$ for a ρ meson with polarization ε_μ. Eqs. (7), (10) are the holographic version of the QCD sum rules.

In the axial sector (a_1 and π mesons), the action to quadratic order is

$$S = \int d^5x\left[-\frac{1}{4g_5^2 z}F_A^a F_A^a + \frac{v(z)^2}{2z^3}(\partial\pi^a - A^a)^2\right], \tag{12}$$

where we have defined $v(z) = m_q z + \sigma z^3$, $A = (A_L - A_R)/2$, and $X = X_0 \exp(i2\pi^a t^a)$. In the $A_z = 0$ gauge, the resulting equations of motion for the 4D-transverse components of A_μ ($A_\mu = A_{\mu\perp} + \partial_\mu \varphi$) are:

$$\left(\partial_z \left(\frac{1}{z}\partial_z A_\mu^a\right) + \frac{q^2}{z} A_\mu^a - \frac{g_5^2 v^2}{z^3} A_\mu^a\right)_\perp = 0; \tag{13}$$

The a_1, being a spin-1 particle, is the solution to Eq.(13) with $\psi_{a_1}(0) = \partial_z \psi_{a_1}(z_m) = 0$. The a_1 decay constant, F_{a_1}, is given by an expression similar to Eq.(11), but with ρ replaced by a_1.

Our theory has all the consequences of chiral symmetry built in. Using the holographic recipe [cf. Eq. (6)],

$$f_\pi^2 = -\frac{1}{g_5^2} \left.\frac{\partial_z A(0,z)}{z}\right|_{z=\epsilon}, \tag{14}$$

where $A(0,z)$ is the solution to Eq.(13) with $q^2 = 0$, satisfying $A'(0, z_m) = 0$, $A(0,\epsilon) = 1$. One can also derive the Gell-Mann–Oakes–Renner (GOR) relation for the pion mass, applying the rules of holographic correspondence:

$$m_\pi^2 f_\pi^2 = (m_u + m_d)\langle \bar{q}q \rangle = 2 m_q \sigma. \tag{15}$$

6. Meson interactions and VMD

The meson interactions can be read from the non-bilinear terms in the 5D action. For example, we find that the π-ρ coupling is given by:

$$g_{\rho\pi\pi} = g_5 \int dz\, \psi_\rho(z) \left(\frac{\phi'(z)^2}{g_5^2\, z} + \frac{v(z)^2 (\pi - \phi)^2}{z^3}\right). \tag{16}$$

The normalization of π is fixed by the pion kinetic term: integrating the function in parentheses in Eq.(16) gives 1.

The well-known phenomenon of vector-meson dominance follows naturally in these types of models from the fact that 3-meson couplings are given by the overlap integral (16). The overlap of the "wave-function" $\psi_\rho(z)$ with the pion "wave-function" squared decreases rapidly with increasing ρ excitation number due to increasing oscillation frequency of ψ_ρ. As a result the coupling of excited ρ mesons to pions is small.

7. Comparison to experiment

The model described above in some ways mimics the approach of QCD sum rules, and we naturally expect our model to work about as well, at roughly the 20% level. As one can see from Table 2, the agreement is re-

Table 2. Model vs data comparison. Parameters marked with asterisks are fitted.

Observable	Measured (MeV)	Model (MeV)
m_π	139.6±0.0004 [7]	139.6*
m_ρ	775.8±0.5 [7]	775.8*
m_{a_1}	1230±40 [7]	1363
f_π	92.4±0.35 [7]	92.4*
$F_\rho^{1/2}$	345±8 [13]	329
$F_{a_1}^{1/2}$	433±13 [14,6]	486
$g_{\rho\pi\pi}$	6.03±0.07 [7]	4.48

markably good. The largest disagreement is in the value of $g_{\rho\pi\pi}$, the only coupling which involves trilinear terms in the action and therefore, expected to receive contribution from terms such as F^3, neglected in our exploratory study.

References

1. J. M. Maldacena, Adv. Theor. Math. Phys. **2**, 231 (1998).
2. A. Karch and E. Katz, J. High Energy Phys. 06 (2002) 043.
3. M. Kruczenski, D. Mateos, R. C. Myers, and D. J. Winters, J. High Energy Phys. 07 (2003) 049.
4. S. Hong, S. Yoon, and M. J. Strassler, J. High Energy Phys. 04 (2004) 046; hep-th/0409118; hep-th/0410080.
5. C. Csaki, H. Ooguri, Y. Oz, and J. Terning, J. High Energy Phys. 01 (1999) 017; L. Girardello, M. Petrini, M. Porrati, and A. Zaffaroni, Nucl. Phys. B **569**, 451 (2000); J. Babington, J. Erdmenger, N. Evans, Z. Guralnik, and I. Kirsch, Phys. Rev. D **69**, 066007 (2004); M. Kruczenski, D. Mateos, R. C. Myers, and D. J. Winters, J. High Energy Phys. 05 (2004) 041; H. Boschi-Filho and N. R. F. Braga, J. High Energy Phys. 05 (2003) 009; G. F. de Teramond and S. J. Brodsky, hep-th/0501022.
6. D. T. Son and M. A. Stephanov, Phys. Rev. D **69**, 065020 (2004).
7. S. Eidelman *et al.* [Particle Data Group Collaboration], Phys. Lett. B **592**, 1 (2004).
8. A. Karch, E. Katz, D. T. Son and M. A. Stephanov, Phys. Rev. D **74**, 015005 (2006) [arXiv:hep-ph/0602229].
9. S. S. Gubser, I. R. Klebanov, and A. M. Polyakov, Phys. Lett. B **428**, 105 (1998).
10. E. Witten, Adv. Theor. Math. Phys. **2**, 253 (1998).
11. I. R. Klebanov and E. Witten, Nucl. Phys. B **556**, 89 (1999).
12. M. A. Shifman, A. I. Vainshtein, and V. I. Zakharov, Nucl. Phys. B **147**, 385 (1979).
13. J. F. Donoghue, E. Golowich, and B. R. Holstein, *Dynamics of the Standard Model* (Cambridge University Press, Cambridge 1992)
14. N. Isgur, C. Morningstar, and C. Reader, Phys. Rev. D **39**, 1357 (1989).

GAUGE–STRING DUALITY, SPIN CHAINS AND 2-D EFFECTIVE ACTIONS

A.A. TSEYTLIN

Imperial College,
London, SW7 2AZ, U.K.
and
The Ohio State University,
Columbus, OH, USA

We review some recent advances in understanding the correspondence between anomalous dimensions of gauge-invariant operators of planar N=4 SYM theory and string energies in $AdS_5 \times S^5$.

Keywords: Super Yang-Mills, spin chains, anomalous dimensions, string theory

1. Introduction

AdS/CFT duality remains a focus of attention for already 8 years, but only during the last few years there appeared a hope of understanding it a quantitative level and eventually proving it. That is bound to have important implications for the attempts to find string theory description of strongly coupled gauge theories.

The canonical example of duality between $\mathcal{N} = 4$ SYM at $N = \infty$ and type IIB superstrings in $AdS_5 \times S^5$ has the following parameter: $\lambda = g^2_{_{YM}} N$ related to string tension $2\pi T = \frac{R^2}{\alpha'} = \sqrt{\lambda}$ while $g_s = \frac{\lambda}{4\pi N} \to 0$. The main statement is that string energies should be equal to dimensions of gauge-invariant operators $E(\sqrt{\lambda}, J, m, ...) = \Delta(\lambda, J, m, ...)$, where J are global charges of $SO(2,4) \times SO(6)$, i.e. spins S_1, S_2; J_1, J_2, J_3, while m stands for windings, folds, cusps, oscillation numbers, etc., of the strings. The gauge operators are built of SYM fields, symbolically, $\text{Tr}(\Phi_1^{J_1} \Phi_2^{J_2} \Phi_3^{J_3} D_+^{S_1} D_\perp^{S_2} ... F_{mn} ... \Psi ...)$. To solve SYM/string theory amounts, as a first step, to show that $E = \Delta$ for any λ (and any J,m). To solve non-trivial susy 4-d CFT or dual string in curved R-R background is a remarkable well-defined problem of mathematical physics. Its solution should help us to learn more about less supersymmetric theories, e.g., about the role of integrability and string picture in perturbative / large energy QCD.

The obvious complication is that the erturbative expansions are opposite: $\lambda \gg 1$ in perturbative string theory and $\lambda \ll 1$ in perturbative planar gauge theory. That suggests using "constructive" approach: use perturbative results on both sides and other properties (integrability, supersymmetry,...) as guides to guesses of exact answers (Bethe ansatz,...). This approach may have its limitations: guessing solution of string theory in terms of a Bethe ansatz (instead of finding, e.g., some free-field like representation for world-sheet theory) may not help in constructing string vertex operators and computing correlation functions.

There was a remarkable recent progress in uncovering the structure of string spectrum and comparing it to the structure of gauge-invariant operators. This was achieved by concentrating on "semiclassical" states with large quantum numbers ($J \gg 1$) dual to "long" gauge operators. It was found that in this limit E and Δ have the same dependence on J, m with coefficients being interpolating functions of λ. Also, a remarkable connection to spectrum of integrable spin chains was uncovered and important advances in understanding the structure of underlying Bethe ansatz were made.

One of the central questions is about the general properties of interpolating functions $f(\lambda)$. They are expected to be given by sign-alternating series with finite radius of convergence, to match the 't Hooft prediction about the properties of perturbation series in large N gauge theory.

2. Gauge theory spectrum

Below we shall concentrate on a particular sector of states of operators in $SU(2)$ sector corresponding to strings moving in S^3 of S^5. They are built out of SYM scalars $\Phi_1 = \phi_1 + i\phi_2$, $\Phi_2 = \phi_3 + i\phi_4$: $\mathrm{Tr}(\Phi^{J_1}\Phi^{J_2}) + ...$, $J = J_1 + J_2$. Their dimensions are eigenvalues of spin chain Hamiltonian related to the fact that the planar 1-loop dilatation operator of $\mathcal{N} = 4$ SYM. is Hamiltonian of ferromagnetic Heisenberg $\mathrm{XXX}_{1/2}$ spin chain (Minahan, Zarembo, 2002): $H_1 = \frac{\lambda}{(4\pi)^2}\sum_{l=1}^{J}(I - \vec{\sigma}_l \cdot \vec{\sigma}_{l+1})$ At higher orders (Beisert,Kristjansen,Staudacher, 2003; Beisert, 2004) one finds $H_2 = \frac{\lambda^2}{(4\pi)^4}\sum_{l=1}^{J}(-3+4\vec{\sigma}_l\cdot\vec{\sigma}_{l+1} - \vec{\sigma}_l\cdot\vec{\sigma}_{l+2})$, while H_3 contains $\vec{\sigma}_l\cdot\vec{\sigma}_{l+3}$ but also $(\vec{\sigma}_l \cdot \vec{\sigma}_{l+1})(\vec{\sigma}_{l+2} \cdot \vec{\sigma}_{l+3})$, etc. As a result, one finds a "long-range" ferromagnetic spin chain with "multi-spin" interactions. Interestingly, its H is an effective spin chain Hamiltonian related (at least to 3 loop order) to Hubbard model (Rej, Serban, Staudacher, 2005).

The spin chain appears to be integrable, and then one may hope to find the spectrum via the Bethe ansatz. The main expectation is that spectrum is to have qualitatively same structure at any λ, i.e. energies of states should change smoothly with λ (at least for large J_1, J_2). This is indeed supported by the remarkable correspondence between the semiclassical string ($\lambda \gg 1$) and weak-coupling gauge ($\lambda \ll 1$) results.

3. Structure of spectrum at large J and small λ

The qualitative structure of the spectrum appears to be the same at small λ (XXX chain or its BDS extension) and at large λ (semiclassical string). In 1-loop Heisenberg model $E = J + \lambda E_1 + O(\lambda^2)$ where E_1 has different scaling behavior at large J depending on the state in the spectrum:

$E_1 = 0$: ferromagnetic vacuum – BPS operator Tr Φ^J (dual to point-like string)

$E_1 \sim \frac{1}{J^2}$: magnons – BMN states $\mathrm{Tr}([\Phi_1....\Phi_1]\Phi_2[\Phi_1....\Phi_1]\Phi_2...)$ (dual to "short" fast strings with c.o.m. along S^5 geodesic $J_1 \gg J_2$)

$E_1 \sim \frac{1}{J}$: low-energy spin waves found in "thermodynamic" limit ($J_1 \sim J_2 \gg 1$); these are long "locally BPS" operators $\mathrm{Tr}([\Phi_1....\Phi_1][\Phi_2...\Phi_2][\Phi_1....\Phi_1][\Phi_2...\Phi_2]...)$ (dual to long fast strings, with nearly-null world surface, (Frolov, AT, 2002))

$E_1 \sim 1$: "intermediate" states or Bethe strings – bound states of finite number of magnons (dual to long rotating strings with $J_1 \gg J_2 \gg 1$; see Hofman, Maldacena, 2006; Dorey, 2006)

$E_1 \sim J$: anti-ferromagnetic state (dual to long slowly-rotating string $J_1 = J_2$, Roiban, Tirziu, AT, 2006)

Again, the gauge and string side correspond to different limits: (i) perturbative semiclassical string side: $\lambda \gg 1$, $\frac{J}{\sqrt{\lambda}}$=fixed; (ii) perturbative gauge side: $\lambda \ll 1$, $J \gg 1$. Still, in some cases few leading coefficients match exactly (e.g., for BMN and fast strings): this should be related to susy protection and peculiarities of large J limit. The general pattern is strong-weak coupling interpolation in λ.

4. Low-energy states: "fast" 2-spin strings

Perturbative string energies contain classical + quantum $\alpha' \sim \frac{1}{\sqrt{\lambda}}$ corrections. The limit one takes is first large λ or large J at fixed $\tilde{\lambda} \equiv \frac{\lambda}{J^2}$, then one may expand in $\tilde{\lambda}$. In perturbative SYM theory one first takes small λ, then expands in large J.

Remarkably, in these two different expansions one gets same structure and same coefficients in $E(\lambda, J)$ at first two leading orders (Frolov, AT,

2003; Beisert,Minahan,Staudacher,Zarembo, 2003). Starting with "3-loop" order one needs, however, an interpolating function of λ. The reason is that quantum string expansion near fast strings contains "non-analytic" terms with explicit factors of $\sqrt{\lambda}$ (Beisert, AT, 2005)

$$E = J\Big[1 + \tilde{\lambda}(a_0 + \frac{a_1}{J} + ...) + \tilde{\lambda}^2(b_0 + \frac{b_1}{J} + ...) + \ \tilde{\lambda}^3(f(\lambda) + ...) + ...\Big]$$

where $\tilde{\lambda} \equiv \frac{\lambda}{J^2}$ and the interpolating function is:

$$f(\lambda \gg 1) = c_0 + \frac{c_1}{\sqrt{\lambda}} + ... \ , \qquad f(\lambda \ll 1) = h_1 + h_2\lambda + ... \ , \qquad c_0 \neq h_1$$

The main issue is how to compute similar "interpolating functions"? They should follow from the underlying Bethe ansatz which in turn should be determined by the superstring world-sheet S-matrix.

5. All-order Bethe ansatz

Strong indications of integrability of both string and gauge theory suggest to expect a Bethe ansatz description of spectrum for any λ

$$e^{ip_k J} = \prod_{j\neq k}^{M} S(p_k, p_j; \lambda) \ , \qquad S(p_k, p_j; \lambda) = S_1(p_k, p_j; \lambda) \ e^{i\theta(p_k, p_j; \lambda)}$$

where $S_1 = \frac{u_k - u_j + i}{u_k - u_j - i}$. Here S is the scattering matrix of elementary excitations (magnons) with 1-d momenta p_j and rapidities u_j (Beisert, Dippel, Staudacher, 2004; Staudacher, 2005)

$$u_j = \tfrac{1}{2} \cot \tfrac{p_j}{2} \ \sqrt{1 + \tfrac{\lambda}{\pi^2} \sin^2 \tfrac{p_j}{2}}$$

This dispersion relation is justified on supersymmetry grounds (Beisert, 2005). One is supposed to find p_j for bound states with $\sum_{k=1}^{M} p_k = 2\pi m$ (cyclicity of the trace), and then energies of states are given by

$$E = \sum_{j=1}^{M} \Big(\sqrt{1 + \tfrac{\lambda}{\pi^2} \sin^2 \tfrac{p_j}{2}} - 1 \Big) \ .$$

The asymptotic (large J) gauge ansatz of BDS implies: $J \to \infty$, up to λ^J order: $S = S_1$, $\theta \to 0$. But to match string theory results we need non-trivial phase θ (Arutyunov, Frolov, Staudacher, 2004). This θ is common to all sectors, and its structure is fixed by symmetries (Beisert, 2005)

$$\theta(p', p; \lambda) = \sum_{r=2}^{\infty} \sum_{s=r+1}^{\infty} c_{rs}(\lambda) \big[q_s(p) q_r(p') - q_s(p') q_r(p)\big]$$

$$q_{r+1}(p) = \frac{2}{r} \sin \tfrac{rp}{2} \Big(\frac{\sqrt{1 + 4\bar\lambda \sin^2 \frac{p}{2}} - 1}{\bar\lambda \sin \frac{p}{2}} \Big)^r , \qquad \bar\lambda = \frac{\lambda}{(4\pi)^2}$$

$$c_{rs}(\lambda) = \bar\lambda^{\frac{r+s-1}{2}} \left[\delta_{r,s-1} + \frac{1}{\sqrt{\bar\lambda}} a_{rs} + \frac{1}{(\sqrt{\bar\lambda})^2} b_{rs} + ... \right]$$

Matching to classical string implies $(c_{rs})_{\lambda \to \infty} \to \lambda^{\frac{r+s-1}{2}} \delta_{r,s-1}$ (AFS). At small λ we should expect $c_{rs} \to 0$.

$AdS_5 \times S^5$ superstring action (Metsaev, AT, 98) leads to 1-loop corrections to spinning string energies (Frolov, AT, 2003) whose structure implies that $a_{rs} \neq 0$ (Beisert, AT, 2005). This in turn confirms the presence of interpolating functions in fast string energy starting with "3-loop" order. More generally, 1-loop string results translate into (Hernandez, Lopez, 2006)

$$a_{rs} = \frac{2}{\pi} (1 - (-1)^{r+s}) \frac{(r-1)(s-1)}{(r-1)^2 - (s-1)^2}$$

This result is consistent (Arutyunov, Frolov, 2006) with crossing-type condition (Janik, 2006) on the S-matrix.

An outstanding problem is how to compute the phase θ directly from string theory. The string sigma model in conformal gauge suggests to relate S to scattering matrix of integrable 2d field theory whose effective excitations correspond to spin chain magnons (Polchinski, Mann, 2005; Gromov, Kazakov, Sakai, Viera, 2006; Kloze, Zarembo, 2006). The relevant S-matrix should however correspond to scattering of magnons with "non-relativistic" dispersion relation. That raises the question of how to identify and compute same object on string side. The upshot is that this should be an effective S-matrix of "positive-energy" parts of BMN-type string modes (Roiban, Tirziu, AT, 2006). The key conceptual role here is played by non-relativistic "Landau-Lifshitz" type effective action.

6. Effective field theory approach

One may consider two microscopically consistent (UV finite) theories – the spin chain and the superstring – and compute the 2d effective actions that describe the lower part of the spectrum, i.e. slow modes at large J. This leads to non-relativistic "Landau-Lifshitz" (LL) 2d action (Kruczenski, 2003; Kruczenski, Ryzhov, AT, 2004). This works not only classically but also at the quantum level (Minahan, Tirziu, AT, 2005). One finds the $\lambda \gg 1$

to $\lambda \ll 1$ interpolation between "string" and "gauge" effective actions and the corresponding "spin chains".

The LL action is coherent-state action for low-energy excitations of spin chain (determined by the dilatation operator), and also the "fast-string" limit of the classical string action. It depends (in $SU(2)$ sector) on a unit 3-vector $\vec{n}$ which represents transverse position of a string in S^3 or spin coherent state $U^\dagger \vec{\sigma} U = \vec{n}$, $\vec{n}^2 = 1$. In continuum limit one finds classical Landau-Lifshitz equations of motion $\partial_t n_i = \frac{1}{2}\tilde{\lambda}\epsilon_{ijk} n_j \partial_\sigma^2 n_k$, $\tilde{\lambda} \equiv \frac{\lambda}{J^2}$, describing lower part of spectrum with energies $E \sim J\tilde{\lambda} = \frac{\lambda}{J}$. The LL model is an integrable system: it admits a Lax pair representation, finite gap solution, etc.

Beyond the leading order the LL-type effective actions from gauge-theory spin chain and string theory can be represented as $S = \int dt \int_0^J d\sigma\ L$, $\bar{\lambda} = \frac{\lambda}{(2\pi)^2}$, where

$$L = \vec{C}(n) \cdot \partial_t \vec{n} - \frac{1}{4}\vec{n}\left(\sqrt{1 - \bar{\lambda}\partial_\sigma^2} - 1\right)\vec{n} - \frac{3\bar{\lambda}^2}{128}(\partial_\sigma \vec{n})^4$$

$$- \frac{\bar{\lambda}^3}{64}\Big[-\frac{7}{4}(\partial_\sigma \vec{n})^2(\partial_\sigma^2 \vec{n})^2 + \mathrm{b}(\lambda)\,(\partial_\sigma \vec{n}\partial_\sigma^2 \vec{n})^2 + \mathrm{c}(\lambda)\,(\partial_\sigma \vec{n})^6\Big] + \ldots$$

The quadratic in $\vec{n}$ part is exact: it reproduces the BMN dispersion relation for small ("magnon") fluctuations near BPS vacuum $\vec{n} = (0,0,1)$. It describes half of usual massive modes, cf. $\partial_t^2 - \partial_\sigma^2 + m^2 \to (i\partial_t - \sqrt{m^2 - \partial_\sigma^2})(-i\partial_t - \sqrt{m^2 - \partial_\sigma^2})$. At orders $\bar{\lambda}$ and $\bar{\lambda}^2$ one finds agreement between the two effective actions, and that explains observed agreement of energies of states, integrable structures, etc. The "3-loop" coefficients in the string and the gauge theory expressions are interpolating functions:

$\lambda \gg 1: \quad \mathrm{b} = -\frac{25}{2} + O(\frac{1}{\sqrt{\lambda}})$, $\quad \mathrm{c} = \frac{13}{16} + O(\frac{1}{\sqrt{\lambda}})$,

$\lambda \ll 1: \quad \mathrm{b} = -\frac{23}{2} + O(\lambda)$, $\quad \mathrm{c} = \frac{12}{16} + O(\lambda)$

7. Field theory S-matrix for "magnons"

Staring with LL action one can compute the corresponding S-matrix (Kloze, Zarembo, 2006; Roiban, Tirziu, Tseytlin, 2006). Writing the LL action in terms of complex scalar ϕ: $n_s = 2\sqrt{1 - z^2}\, z_s$, $\quad \phi \equiv z_1 + iz_2$, one gets

$S = \int dt \int_0^J dx\ \{\phi^* \left[i\partial_t - (\sqrt{1 - \bar{\lambda}\partial_x^2} - 1)\right]\phi \ -\ V(\phi, \phi^*)\}$

where $V = V_4 + V_6 + \ldots$, $\quad V_{2n} \sim \sum_{k=1}^\infty \bar{\lambda}^k \partial_x^{2k}(\phi^*\phi)^n$ and

$V_4 = \frac{\bar{\lambda}}{4}(\phi^{*2}\phi'^2 + c.c) - \frac{\bar{\lambda}^2}{8}\left[\frac{1}{2}|\phi|^2(\phi''''\phi^* + c.c.) + 6|\phi''|^2|\phi|^2 - 3|\phi'|^4\right] + \ldots$

One can thenn establish matching of the effective LL model S-matrix with the BDS/AFS Bethe ansatz S-matrix in the limit $p \to 0$, $u(p) \to \frac{e(p)}{p} = \frac{1}{p}\sqrt{1 + \bar{\lambda}p^2}$

$$S_{\rm BDS} = \frac{u(p')-u(p)+i}{u(p')-u(p)-i} \quad \rightarrow \quad \tilde{S}_{\rm BDS} = \frac{1+\frac{ipp'}{pe(p')-p'e(p)}}{1-\frac{ipp'}{pe(p')-p'e(p)}}$$

At tree level: $(\tilde{S}_{\rm BDS})_{\rm tree} = \frac{2ipp'}{pe(p')-p'e(p)}$, $(\tilde{S}_{\rm AFS})_{\rm tree} = \frac{2iF(p,p')}{p\ e(p')-p'\ e(p)}$, where

$$F(p,p') = pp' + \tfrac{1}{2}[p\ e(p') - p'\ e(p)]\ \tilde{\theta}_{\rm AFS}$$
$$= \left(pp' - \bar{\lambda}^{-1}[e(p)-1][e(p')-1]\right)\left[1+\tfrac{1}{4}\left(\bar{\lambda}pp' - [e(p)-1][e(p')-1]\right)\right]$$

One can then extend this to full string theory, e.g., in bosonic S^3 sector. To compare spin chain phase shift (for magnons near ferromagnetic vacuum with "non-relativistic" first-order dispersion relation) one should re-organize the string-theory side S-matrix for BMN-type modes (which originally have relativistic dispersion relation) into the S-matrix for an effective field theory of the positive-energy modes. The classical string action on $R \times S^3$ is $ds^2 = -dt^2 + [d\alpha + C(n)]^2 + d\vec{n}d\vec{n}$ and in "uniform" gauge: $t = \tau,\ \ \tilde{\alpha} = \frac{J}{\sqrt{\lambda}}\sigma$, i.e. $p_\alpha = \frac{J}{\sqrt{\lambda}}$=const

$$L = C_t - \sqrt{[1-\tfrac{1}{4}(\partial_t\vec{n})^2][1+\tfrac{\bar{\lambda}}{4}(\partial_x\vec{n})^2] + \tfrac{\bar{\lambda}}{16}(\partial_t\vec{n}\cdot\partial_x\vec{n})^2}\ .$$

Expanding near $\vec{n} = (0,0,1)$ gives

$$L = i\phi^*\partial_t\phi - \tfrac{1}{2}\phi^*(\partial_t^2 - \bar{\lambda}\partial_x^2)\phi + \tfrac{1}{4}[\phi^{*2}(\dot{\phi}^2 - \bar{\lambda}\phi'^2) + c.c.]$$
$$+ \tfrac{1}{8}(\dot{\phi}^{*2} - \bar{\lambda}\phi'^{*2})(\dot{\phi}^2 - \bar{\lambda}\phi'^2) + O(\phi^6)$$

and solving for the negative-energy modes one gets indeed agreement (Roiban, Tirziu, AT) with the corresponding term in the AFS S-matrix, $(S^{\rm SU(2)}_{\rm string})_{\rm tree} = \frac{2iF(p,p')}{pe(p')-p'e(p)}$. The remaining problem is how to extend this string computation of S-matrix to the string quantum level.

Acknowledgements

I would like to thank J. Minahan, R. Roiban and A. Tirziu for collaboration on the work described here. This work was supported in part by the DOE grant DE-FG02-91ER40690, INTAS 03-51-6346 and the RS Wolfson award.

Most of the original papers mentioned in the text may be found in the lists of references in the papers cited below.

References

1. N. Beisert and A. A. Tseytlin, "On quantum corrections to spinning strings and Bethe equations," Phys. Lett. B **629**, 102 (2005) [hep-th/0509084].
2. J. A. Minahan, A. Tirziu and A. A. Tseytlin, "$1/J^2$ corrections to BMN energies from the quantum long range Landau-Lifshitz model," JHEP **0511**, 031 (2005) [hep-th/0510080].
3. R. Roiban, A. Tirziu and A. A. Tseytlin, "Asymptotic Bethe ansatz S-matrix and Landau-Lifshitz type effective 2-d actions," hep-th/0604199.

LINEAR CONFINEMENT AND ADS/QCD

A. KARCH

Department of Physics, University of Washington,
Seattle, Washington 98195, USA

E. KATZ

Department of Physics, Boston University,
Boston, Massachusetts 002215, USA

D. T. SON

Institute for Nuclear Theory, University of Washington,
Seattle, Washington 98195, USA
E-mail: son@phys.washington.edu

M. A. STEPHANOV

Department of Physics, University of Illinois,
Chicago, Illinois 60607, USA

In a theory with linear confinement, such as QCD, the masses squared $m^2_{n,S}$ of mesons with high spin S or high radial excitation number n are expected, from semiclassical arguments, to grow linearly with S and n. We show that this behavior can be reproduced within a putative 5-dimensional theory holographically dual to QCD (AdS/QCD). With the assumption that such a dual theory exists and describes highly excited mesons as well, we show that asymptotically linear m^2 spectrum translates into a strong constraint on the *infrared* behavior of that theory. In the simplest model which obeys such a constraint we find $m^2_{n,S} \sim (n + S)$.

Keywords: AdS/QCD

1. Introduction

Over the recent years gauge/gravity correspondence[1] has been be used to extract information about four-dimensional strongly coupled gauge theories by mapping them onto gravitational theories in five dimensions. The term AdS/QCD is often used to describe the efforts to apply a five-dimensional theory on an anti-de Sitter (AdS) gravity background to learn something about QCD. Although for QCD the exact form of the gravity dual is not yet known, there are two complementary approaches to the problem. One is to start from a string theory, choosing the background in such a way as to reproduce such essential ingredients of QCD as confinement or matter in the fundamental representation, and study the resulting QCD-like theories. Another, bottom-up, approach is to begin with QCD and attempt to determine or constrain the dual theory properties by matching them to known properties of QCD using gauge/gravity correspondence. From a practical point of view, one can model experimental data surprisingly well[2,3] by a local effective theory on a cutoff AdS space.

One criticism that has been brought against this program is that it so far appeared to be unable to describe correctly either (radially) excited rho mesons or higher spin mesons. The meson spectrum in AdS/QCD is determined by solving for the eigenmodes of a 5d gauge field living on the cutoff AdS. With the simplest cutoff — the hard IR wall — the spectrum of squared masses m_n^2 is similar to that of a Schrödinger equation for a particle in a box, i.e., for high excitation number, $n \gg 1$, m_n^2 grow as n^2. On the other hand, data shows growth consistent with $m_n^2 \sim n$. Furthermore, for large n semiclassical quantization of an oscillating flux tube gives $m_n^2 \sim n$. Such a behavior is also observed in the 1+1 dimensional 't Hooft model where linear confinement can be demonstrated analytically.

Here[4] we point out that the asymptotic behavior of the spectrum of highly excited mesons $m_n^2 \sim n^2$ is by no means an intrinsic property of AdS/QCD. We wish to emphasize that, contrary to thus far rather common assumption, the spectrum of the highly excited mesons is not determined by the *ultraviolet* behavior of the AdS/QCD (which is already constrained to be asymptotically AdS). Rather, it crucially depends on the details of the *infrared* region. We shall give an explicit example of the IR wall that gives the desired growth of the masses at large n and spin S,

$$m_n^2 \sim n + S. \tag{1}$$

At this point, we can give no explicit example of a background in an *ab initio* string theory that behaves in the way that we propose. At least, however, our construction can be thought of as a better phenomenological model of QCD than the hard-wall model.

2. Background geometry and overview

The gravitational backgrounds we are interested in can still be thought of as cutoff AdS spaces, but instead of the hard-wall IR cutoff we shall look at spacetimes that smoothly cap off. The only background fields we are considering are the dilaton Φ and the metric g_{MN}. The mesons are described by 5d fields propagating on this background with the action given by

$$I = \int d^5x \sqrt{g}\, e^{-\Phi}\, \mathcal{L}, \tag{2}$$

where $\mathcal{L}$ is the Lagrangian density and $g = |\det g_{MN}|$. We shall begin by considering a generic background parameterized by two functions $A(z)$ and $\Phi(z)$ such that:

$$g_{MN}\, dx^M dx^N = e^{2A(z)}(dz^2 + \eta_{\mu\nu} dx^\mu dx^\nu); \tag{3}$$

$$\Phi = \Phi(z); \tag{4}$$

where $\eta_{\mu\nu} = \text{diag}(-1, 1, 1, 1)$. We shall then determine the conditions that the background $A(z)$ and $\Phi(z)$ should obey to reproduce the Regge-like behavior of the mass spectrum 1. By considering the spectrum of radial ρ excitations only we conclude that the linear combination $\Phi - A$ must behave as z^2 at large z. In addition, conformal symmetry in the UV demands that $\Phi - A \sim \log z$ at small z. The simplest solution to both these constraints is $\Phi - A = z^2 + \log z$. It has the advantage that the spectrum of excited ρ masses can be determined exactly: $m_n^2 = 4(n+1)$.

In order to determine A and Φ functions separately we then consider higher spin mesons. We find that the behavior as in Eq. (1) requires that the metric function A does not have any contribution growing as z^2 at large z. In the simplest case obeying this constraint, $A = -\log z$, $\Phi = z^2$, the spectrum can be found exactly: $m_{n,S}^2 = 4(n+S)$.

3. Rho mesons

Following[2,3] let us consider the action

$$I = \int d^5x\, e^{-\Phi(z)} \sqrt{g} \left\{ -|DX|^2 + 3|X|^2 - \frac{1}{4g_5^2}(F_L^2 + F_R^2) \right\} \tag{5}$$

with $g_5^2 = 12\pi^2/N_c$. The boundary condition on the gauge fields A_L and A_R at $z = 0$ as required by the holographic correspondence, is given by the value of the sources of the currents J_L and J_R in 4d theory. The IR boundary condition now, in the case of the smooth wall extending to $z = \infty$, is simply that the action is finite. The ambiguity of the choice of the IR boundary condition in a theory with hard wall is not present in the theory with the smooth wall cutoff.

To determine the spectrum of the ρ mesons we need only the quadratic part of the action for the vector-like gauge field $V = A_L + A_R$. We use the gauge invariance of the action to go to the axial gauge $V_z = 0$ [2]. The equation for the 4d-transverse components V_μ^T ($\partial^\mu V_\mu^T = 0$) has normalizable solutions, v_n, only for discrete values of 4d momentum q^2 equal to m_n^2:

$$\partial_z \left(e^{-B}\partial_z v_n\right) + m_n^2 e^{-B} v_n = 0, \tag{6}$$

where $B = \Phi(z) - A(z)$. Via the substitution

$$v_n = e^{B/2}\psi_n \tag{7}$$

this equation can be brought into the form of a Schrödinger equation

$$-\psi_n'' + V(z)\psi_n = m_n^2 \psi_n, \tag{8}$$

$$V(z) = \frac{1}{4}(B')^2 - \frac{1}{2}B''. \tag{9}$$

In the particular case of $B = \Phi - A = z^2 + \log z$, we have $V = z^2 + 3/(4z^2)$ and the Schrödinger equation (8) is exactly solvable. More generally, for the quantum mechanical system

$$-\psi'' + \left[z^2 + \frac{m^2 - 1/4}{z^2}\right]\psi = E\psi \tag{10}$$

the eigenvalues are $(n = 0, 1, 2, \ldots)$

$$E = 4n + 2m + 2 \tag{11}$$

and the corresponding normalized eigenfunctions are

$$\psi_n(z) = e^{-\frac{z^2}{2}} z^{m+1/2} \sqrt{\frac{2n!}{(m+n)!}}\, L_n^m\left(z^2\right). \tag{12}$$

where L_n^m are associated Laguerre polynomials. The ρ meson mode equation (8) is of this form with $m = 1$. We can easily read off the squared masses of the ρs from this:

$$m_n^2 = 4(n+1). \tag{13}$$

The scale of the masses is fixed by the coefficient of the z^2 term in B, which violates explicitly the scale invariance. This coefficient should proportional to the QCD string tension σ. We measure masses in units in which this coefficient is equal to 1.

Undoing the change of variables (7), we get the original mode functions $v_n = e^{z^2/2}\sqrt{z}\,\psi_n$, hence

$$v_n(z) = z^2\sqrt{\frac{2n!}{(1+n)!}}\,L_n^1\left(z^2\right). \tag{14}$$

From the analytic form of the wavefunction we also can read off the corresponding decay constants [2]:

$$F_{\rho_n}^2 = \frac{1}{g_5^2}\left[v_n''(0)\right]^2 = \frac{8(n+1)}{g_5^2}, \tag{15}$$

whose large n behavior is also in accord with semiclassical QCD arguments[5]. It is interesting to note that, since $F_{\rho_n}^2/m_n^2 = 2/g_5^2$ is n-independent, our simplest choice of the background $B = z^2 + \log z$ reproduces the *ad hoc* resonance model of duality discussed in[6].

Note that in order to get the correct $m_n^2 \sim n$ behavior for large n it was crucial that the analog Schrödinger potential describes essentially a harmonic oscillator at large z. This is easy to see applying the WKB approximation for large n. The distance between successive levels m_n^2 of the Schrödinger equation (8) is given by the frequency of the classical oscillation in the potential V:

$$\frac{dm_n^2}{dn} = \pi\left[\int_{z_1}^{z_2}\frac{dz}{\sqrt{m_n^2 - V(z)}}\right]^{-1}, \tag{16}$$

where $z_{1,2}$ are the turning points. For large m_n, i.e., large z_2, and $z_1 \to 0$, the integral is dominated by large $z \sim z_2 \sim m_n$. This matches the expected growth of the size of the highly excited mesons in QCD – $L \sim m_n$.

By choosing a different function $V(z)$ (i.e., a different background $B(z)$) one can adjust the constant $O(1)$ term in m_n^2, but as long as $V(z) \sim z^2$ for large z, the spectrum will remain equidistant at large n.

Matching only the spectrum of the ρ mesons we are only able to constrain the linear combination $B = \Phi - A$ of the dilaton and the metrics background functions Φ and A. In the next section we shall see that all the z^2 asymptotics must all be in Φ and none in A.

4. Higher spin mesons

In order to create higher spin mesons we need to act with a higher spin current on the vacuum. Just like for the vector mesons, on the gravity side, we have to introduce a higher spin field whose normalizable modes determine meson masses and decay constants. In the free theory in the far

UV the corresponding current becomes a conserved twist 2 current, so the higher spin field has to become a massless higher spin field whose equations of motion are uniquely fixed by gauge and coordinate invariance.

Since we will only be concerned with the spectrum of the higher spin mesons, we shall not consider the full 5d action, but only its quadratic (free) part. It is known that simultaneously gauge and general-coordinate invariant action can be written for a higher spin field in a space with vanishing Weyl tensor. The background we consider (3) is conformally flat and obeys this condition.

The gauge field of spin S is represented by a tensor $\phi_{M_1 \dots M_S}$ of rank S totally symmetric over its indices. We require the action to be invariant with respect to the gauge transformation with gauge parameter $\xi_{M_2 \dots M_S}$ itself a symmetric rank $S-1$ tensor:

$$\delta \phi_{M_1 \dots M_S} = \nabla_{(M_1} \xi_{M_2 \dots M_S)}, \tag{17}$$

where ∇ is (general coordinate) covariant derivative and parentheses denote index symmetrization. We utilize the gauge invariance to go over to the axial gauge $\phi_{z \dots} = 0$. In this gauge, the part of the action involving the transverse and traceless part of the field ϕ ($\partial_\mu \phi^\mu{}_{\dots} = 0$ and $\phi^\nu{}_{\nu \dots} = 0$) decouples. As argued in[4], the quadratic action in this gauge must have the form

$$I = \frac{1}{2} \int d^5x \, e^{5A} \, e^{-\Phi} \left\{ e^{4(S-1)A} \, e^{-2A(1+S)} \partial_N \tilde{\phi}_{\mu_1 \dots \mu_S} \partial_N \tilde{\phi}_{\mu_1 \dots \mu_S} \right\}. \tag{18}$$

where $\tilde{\phi}$ is the rescaled field

$$\phi_{\dots} = e^{2(S-1)A} \, \tilde{\phi}_{\dots} \tag{19}$$

The equation for the modes $\tilde{\phi}_n$ of the transverse traceless field $\tilde{\phi}_{\dots}$ can be now easily derived from the action (18):

$$\partial_z \left(e^{(2S-1)A} e^{-\Phi} \partial_z \tilde{\phi}_n \right) + m_n^2 \, e^{(2S-1)A} e^{-\Phi} \, \tilde{\phi}_n = 0 \tag{20}$$

which has the form (6) with $B = \Phi - (2S-1)A$.

Converting to Schrödinger form using the procedure from Section 3 we see that the only way to have the slope dm_n^2/dn independent of S is to keep all z^2 asymptotics in Φ and none in A. For $A = -\log z$ and $\Phi = z^2$ the Schrödinger potential reads

$$V(z) = z^2 + 2(S-1) + \frac{S^2 - 1/4}{z^2}. \tag{21}$$

This has the same form as the potential in Eq. (10). The eigenvalues corresponding to the squared masses of the mesons now can easily be read off using (11)

$$m_{n,S}^2 = 4(n+S), \tag{22}$$

which generalizes our result (13) to higher S.

We can also read off the UV conformal dimension of the operators $\mathcal{O}^{\cdots}$ in QCD dual to the higher spin field $\phi_{\dots}$ from the behavior of the ϕ near the boundary. We find the dimension is $2+S$, i.e., the twist is indeed equal to 2.

5. Conclusions

In this paper we demonstrated that under the assumption of a local 5d bulk description of QCD there is a smoothing of the IR wall (asymptotically unique) that gives the right large n and large S behavior for highly radially or orbitally excited mesons characteristic of linear confinement $m_{n,S}^2 \sim (n+S)$. It is interesting to observe that in such a background the slopes of n and S trajectories automatically coincide:

$$\frac{dm_{n,S}^2}{dn} = \frac{dm_{n,S}^2}{dS}\,. \tag{23}$$

This matches the expectation from QCD if one considers highly excited mesons as semiclassically oscillating ($n \gg 1$) or rotating ($S \gg 1$) strings — the confining flux tubes connecting quark and antiquark. Indeed, the frequencies of the classical oscillatory and rotational motions of the relativistic Nambu-Goto open string at the same energy coincide.

It is not clear whether the background we considered can arise in any string-theory context. A conservative point of view on the 5d Lagrangian would be to consider it as a phenomenologically driven approach, along the line of Refs. [2,3], intelligently interpolating between the low-energy and high-energy limits of QCD. This approach, while being inspired by the AdS/CFT correspondence, may or may not have any direct relationship to the latter.

References

1. J. M. Maldacena, Adv. Theor. Math. Phys. **2**, 231 (1998).
2. J. Erlich, E. Katz, D. T. Son, and M. A. Stephanov, Phys. Rev. Lett **95**, 261602 (2005).
3. L. Da Rold and A. Pomarol, Nucl. Phys. B **721**, 79 (2005).
4. A. Karch, E. Katz, D. T. Son and M. A. Stephanov, Phys. Rev. D **74**, 015005 (2006).
5. M. Shifman, hep-ph/0507246.
6. M. Shifman, hep-ph/0009131.

INSTANTONS ON D7 BRANE PROBES AND ADS/CFT WITH FLAVOR

J. ERDMENGER

Max Planck-Institut für Physik (Werner Heisenberg-Institut),
Föhringer Ring 6, D-80805 München, Germany
E-mail: jke@mppmu.mpg.de

Recent work on adding flavor to the generalized AdS/CFT correspondence is reviewed. In particular, we consider instanton configurations on two coincident D7 brane probes. These are matched to the Higgs branch of the dual field theory. In $AdS_5 \times S^5$, the instanton generates a flow of the meson spectrum. For non-supersymmetric gravity backgrounds, the Higgs branch is lifted by a potential, which has non-trivial physical implications. In particular these configurations provide a gravity dual description of Bose-Einstein condensation and of a thermal phase transition.

Keywords: AdS/CFT correspondence, Flavour physics, Thermal field theory.

1. Introduction

D7 brane probes have proved a versatile tool for including quark fields into the AdS/CFT correspondence. Strings stretching between the D7s and the D3 branes of the original AdS/CFT construction provide $\mathcal{N} = 2$ fundamental hypermultiplets. The open string sector on the world-volume of a probe D7 brane is holographically dual to quark–anti-quark bilinears $\bar{\psi}\psi$ [1]. Moreover, the embedding of D7 brane probes into non-supersymmetric gravity backgrounds induces chiral symmetry breaking [2,3], with the symmetry breaking geometrically displayed by the D7 brane's bending which breaks an explicit symmetry of the space.

In scenarios involving two or more D7 probes, the Higgs branch spanned by squark vevs $\langle \bar{q}q \rangle$ can be identified with instanton configurations on the D7 world-volume [4,5]. These configurations are the standard four-dimensional instanton solutions living in the four directions of the D7 world-volume transverse to the D3 branes. The scalar Higgs vev in the field theory is identified with the instanton size on the supergravity side. In the case of a probe in AdS space, there is a moduli space for the magnitude of the instanton size or the scalar vev. In [5], the meson spectrum associated with a particular fluctuation about the instanton background is calculated. The spectrum exhibits a non-trivial spectral flow.

In less supersymmetric gravity backgrounds, the moduli space is expected to be lifted by a potential. This potential may have either a stable vacuum selecting a particular scalar vev, or a run-away behaviour. In [6], the Higgs branch of the $\mathcal{N} = 4$ gauge theory at finite temperature and density is analyzed. For the finite temperature case we find a stable minimum for the squark vev which undergoes a first order phase transition as a function of the temperature (or equivalently of the quark mass). On the other hand, in the presence of a chemical potential, the squark potential leads to an instability indicating Bose-Einstein condensation.

Moreover the potential obtained from evaluating the D7 probe action on a static instanton configuration may be used to obtain information about some aspects of the stability of brane embeddings into non-supersymmetric gravity backgrounds [7].

On a related issue, we have recently calculated the mass of mesons involving a light and a heavy quark from an effective field theory derived from strings stretched between two separated D7 brane probes [8]. From this model we are able to predict the B meson mass at the 20% level.

2. Higgs branch AdS/CFT dictionary

Consider a probe of two coincident D7 branes in $AdS_5 \times S^5$. This corresponds to two fundamental hypermultiplets in the dual $\mathcal{N} = 2$ gauge theory. The metric of $AdS_5 \times S^5$ is given by

$$ds^2 = H^{-1/2}(r)\eta_{\mu\nu}dx^\mu dx^\nu + H^{1/2}(r)(d\vec{y}^{\,2} + d\vec{z}^{\,2})\,, \tag{1}$$

$$H(r) = \frac{L^4}{r^4}, \quad r^2 = \vec{y}^{\,2} + \vec{z}^{\,2}, \quad L^4 = 4\pi g_s N_c(\alpha')^2, \quad \vec{y}^{\,2} = \sum_{m=4}^{7} y^m y^m,$$

$$C^{(4)}_{0123} = H^{-1}, \quad \vec{z}^{\,2} = (z^8)^2 + (z^9)^2, \quad e^\phi = e^{\phi_\infty} = g_s.$$

Two D7-branes are embedded into this geometry according to $z^8 = 0$, $z^9 = (2\pi\alpha')m$. This leads to the induced metric

$$\begin{aligned} ds^2_{\mathrm{D7}} &= H^{-1/2}(r)\eta_{\mu\nu}dx^\mu dx^\nu + H^{1/2}(r)d\vec{y}^{\,2}, \\ y^2 &\equiv y^m y^m\,, \quad r^2 = y^2 + (2\pi\alpha')^2 m^2\,. \end{aligned} \tag{2}$$

The parameter m corresponds to the mass of the fundamental hypermultiplets in the dual $\mathcal{N} = 2$ theory.

The effective action describing D7-branes in a curved background is

$$S = T_7 \int \sum_r C^{(r)} \wedge \operatorname{tr} e^{2\pi\alpha' F} - T_7 \int d^8\xi \sqrt{g} \, \frac{(2\pi\alpha')^2}{2} \operatorname{tr}\left(F_{\alpha\beta}F^{\alpha\beta}\right) + \cdots , \tag{3}$$

where we have not written terms involving fermions and scalars. This action is the sum of the Wess-Zumino and Yang-Mills terms, plus corrections at higher orders in α' indicated by $\cdots$ in (3). These corrections may be neglected at leading order in the AdS/CFT large 't Hooft coupling limit. Constraints on unknown higher order terms arising from the existence of instanton solutions, as well as from the exactly known metric on the Higgs branch, are discussed in [4]. - Since we need to consider at least two flavors (two D7's) in order to have a Higgs branch, the DBI action is non-Abelian.

At leading order in α', field strengths which are self dual with respect to the flat four-dimensional metric $ds^2 = \sum_{m=4}^{7} dy^m dy^m$ solve the equations of motion, due to a conspiracy between the Wess-Zumino and Yang-Mills term. Inserting the explicit AdS background values (1) for the metric and Ramond-Ramond four-form into the action for D7-branes embedded as given below (1), with non trivial field strengths only in the directions y^m, gives

$$\begin{aligned} S &= -\frac{T_7(2\pi\alpha')^2}{4} \int d^4x \, d^4y \, H(r)^{-1} \left(-\frac{1}{2}\epsilon_{mnrs}F_{mn}F_{rs} + F_{mn}F_{mn} \right) \\ &= -\frac{T_7(2\pi\alpha')^2}{2} \int d^4x \, d^4y \, H(r)^{-1} F_-^2 , \end{aligned} \tag{4}$$

where $F^-_{mn} = \frac{1}{2}(F_{mn} - \frac{1}{2}\epsilon_{mnrs}F_{rs})$. Field strengths $F^-_{mn} = 0$, which are self-dual with respect to the flat metric $dy^m dy^m$, manifestly solve the equations of motion. These solutions correspond to points on the Higgs branch of the dual $\mathcal{N} = 2$ theory. Strictly speaking, these are points on a mixed Coulomb-Higgs branch if $m \neq 0$ (see [5]). In order to neglect the back-reaction, we are considering a portion of the moduli space for which the instanton number k is fixed in the large N_c limit.

For $m = 0$, the AdS geometry (1) together with the embedding (2), is invariant under $SO(2,4) \times SU(2)_L \times SU(2)_R \times U(1)_R \times SU(2)_f$. The combination $SU(2)_L \times SU(2)_R$ acts as $SO(4)$ rotations of the coordinates y^m. The $SO(2,4)$ factor is the conformal symmetry of the dual gauge theory. The $SU(2)_L$ factor corresponds to a global symmetry of the dual gauge theory, while $SU(2)_R \times U(1)_R$ corresponds to the R symmetries. Finally $SU(2)_f$ is the gauge symmetry of the two coincident D7-branes which, at the AdS boundary, corresponds to the flavor symmetry of the dual gauge theory.

For $m \neq 0$, the symmetry is broken to $SO(1,3) \times SU(2)_L \times SU(2)_R \times SU(2)_f$. This is broken further if there is an instanton background on the D7-branes. We focus on that part of the Higgs branch, which is dual to a single instanton centered at the origin $y^m = 0$. The instanton, in "singular gauge," is given by

$$A_\mu = 0, \qquad A_m = \frac{2Q^2 \bar{\sigma}_{nm} y_n}{y^2(y^2+Q^2)}, \tag{5}$$

where Q is the instanton size and y_m denote the four coordinate directions parallel to the D7 branes but perpendicular to the D3 branes. Moreover $\bar{\sigma}_{mn} \equiv \frac{1}{4}(\bar{\sigma}_m \sigma_n - \bar{\sigma}_n \sigma_m)$, $\sigma_{mn} \equiv \frac{1}{4}(\sigma_m \bar{\sigma}_n - \sigma_n \bar{\sigma}_m)$, $\sigma_m \equiv (i\vec{\tau}, 1_{2\times 2})$, with $\vec{\tau}$ the three Pauli-matrices. We choose the singular gauge, as opposed to the regular gauge in which $A_n = 2\sigma_{mn} y^m/(y^2+Q^2)$, for improved asymptotic behaviour at large y. In the AdS setting, the Higgs branch should correspond to a normalizable deformation of the background at the origin of the moduli space. The singularity of (5) at $y^m = 0$ is not problematic for computations of physical (gauge invariant) quantities. The instanton (5) breaks the symmetries to $SO(1,3) \times SU(2)_L \times \text{diag}(SU(2)_R \times SU(2)_f)$ and corresponds to a point on the Higgs branch $q_{i\alpha} = v\,\varepsilon_{i\alpha}$, $v = \frac{Q}{2\pi\alpha'}$, where $q_{i\alpha}$ are scalar components of the fundamental hypermultiplets, labeled by a $SU(2)_f$ index $i = 1,2$, and a $SU(2)_R$ index $\alpha = 1,2$. All the broken symmetries are restored in the ultraviolet (large r), where the theory becomes conformal.

In [5] we have discussed fluctuations about the instanton background in view of calculating the associated meson masses. We found that in the limit of infinite instanton size, the levels of the meson spectrum are shifted by two with respect to the meson spectrum without instantons. We explain this shift, which corresponds to a spectral flow, by using a large gauge transformation applied to the instanton in the infinite size limit.

3. Higgs potential for non-supersymmetric backgrounds - thermal phase transition

The gravitational dual of $\mathcal{N} = 4$ gauge theory at large 't Hooft coupling and finite temperature is given by the AdS-Schwarzschild black-hole background [9]. The latter belongs to a general class of supergravity solutions which, in a choice of coordinates convenient for our purposes, have the form

$$ds^2 = f(r)(d\vec{x}^2 + g(r)d\tau^2) + h(r)(\sum_{m=4}^{7} dy^m dy^m + \sum_{i=8}^{9} dZ^i dZ^i), \tag{6}$$

$$e^{-\Phi} = \phi(r), \qquad r^2 = y^m y^m + Z^i Z^i,$$

$$F^{(5)} = 4R^4(V_{S^5} +^* V_{S^5}) = dC_{(4)}, \quad C_{(4)}|_{0123} = s(r)\, dx^0 \wedge dx^1 \wedge dx^2 \wedge dx^3 .$$

For the AdS-Schwarzschild solution, we have

$$f(r) = \frac{4r^4 + b^4}{4r^2R^2}, \quad g(r) = \left(\frac{4r^4 - b^4}{4r^4 + b^4}\right)^2, \qquad s(r) = \frac{r^4}{R^4}\left(1 + \frac{b^8}{16r^8}\right),$$
$$h(r) = \frac{R^2}{r^2}. \tag{7}$$

The coordinates $\vec{x}$ are the spatial coordinates of the dual gauge theory and τ is the Euclidean time direction, which is compactified on a circle of radius b^{-1}, corresponding to the inverse temperature. Note that the temperature $T \sim b$ only enters to the fourth power. The D7 embedding in this background is given by $Z^9 = 0$, $Z^8 = z(y)$.

The potential generated on the Higgs branch for this background was calculated in [6]. Specifically, the action is evaluated on the space of field strengths which are self-dual with respect to the induced metric in the directions transverse to $\tau, \vec{x}$;

$$V = \frac{T_7(2\pi\alpha')^2}{2}\left(\frac{1}{g_s}\int d^4y \operatorname{tr} C^{(4)} \wedge F \wedge F - \frac{1}{2}\int d^4y \sqrt{-\det G} \operatorname{tr} F^2\right), \tag{8}$$

where F_{mn} is self-dual with respect to the metric $ds_\perp^2 = h(r)\left((1 + z'(y)^2)dy^2 + y^2 d\Omega_3^2\right)$. This metric is conformally flat. With new coordinates $\tilde{y}(y)$ such that $ds^2 = \alpha(\tilde{y})(d\tilde{y}^2 + \tilde{y}^2 d\Omega_3^2)$, the instanton configurations (self-dual field strengths) take the usual form.

To compute $V(Q)$ in general requires knowledge of the embedding function $z(y)$, which has been computed by a numerical shooting technique in [2] and is displayed in Figure 1. Imposing boundary conditions for the large y behaviour, and requiring smooth behaviour in the interior, such that an RG flow interpretation is possible, leads to a dependence of the chiral quark condensate $\langle\bar{\psi}\psi\rangle$ on the quark mass m and on the temperature. Depending on the ratio m/b, there are two types of solutions, which differ by the topology of the D7-branes. At large r (or y) the geometry of the D7-branes is $AdS_5 \times S^3$ and the topology of the $r \to \infty$ boundary is $S^1 \times R^3 \times S^3$. For sufficiently large m/b, the S^3 component of the D7-geometry contracts to zero size at finite $r > b$. In this case the D7-brane "ends" before reaching the horizon at $r = b$. However, for sufficiently small m/b, the D7-brane ends at the horizon, at which point the thermal S^1 contracts to zero size. Both these types of solutions are plotted in figure 2. There is a first order phase transition at the critical value of $m/b \approx 0.92$ where the two types of solution meet [2,10]. The $\langle\bar{\psi}\psi\rangle$ condensate is non-zero on both sides of this transition, although there is a discontinuous jump in its value.

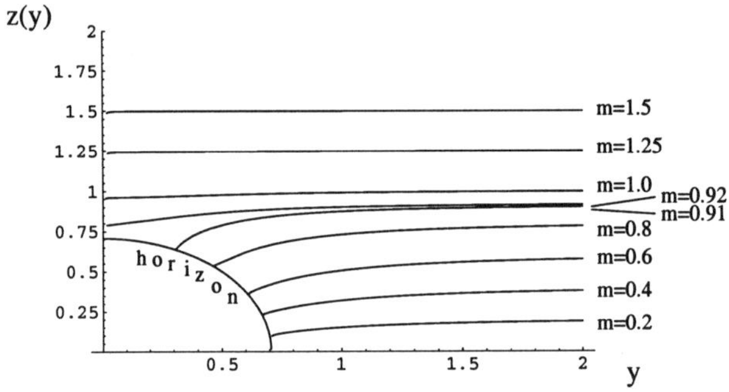

Figure 1. Brane embeddings in AdS-Schwarzschild for different values of the quark mass (with $b = 1$).

This same phase transition is also observed in the Higgs potential as shown in Figure 2. For $m = \infty$, the potential is flat as in the AdS case. For smaller and smaller values of m, a minimum forms at $Q = 0$, until at a critical value of m, the minimum of the potential moves to a finite value of Q.

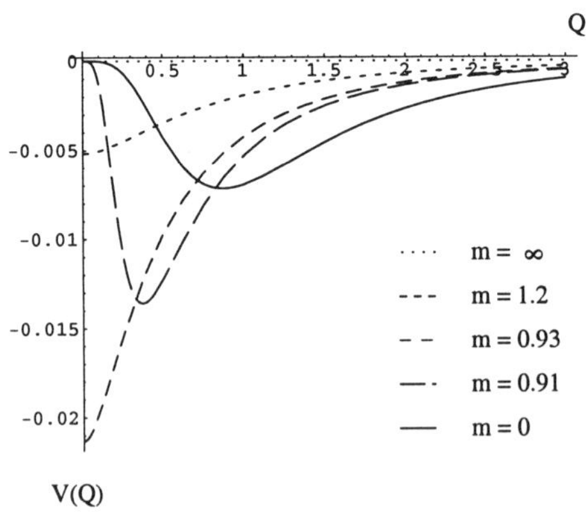

Figure 2. Potential $V(Q)$ as a function of the instanton size / Higgs VEV Q for various values of the quark mass m.

In Figure 3 the Higgs vev Q_0, for which the Higgs potential is minimised, is plotted versus the quark mass. This clearly displays the first order nature of the phase transition. Q_0 is a suitable order parameter for this transition.

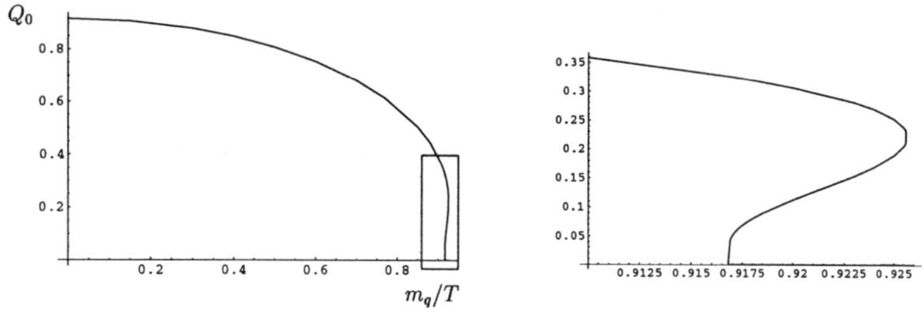

Figure 3. Position of the minimum of the potential Q_0 versus the bare quark mass m, zoom of the critical region.

Acknowledgements: I am grateful to my collaborators R. Apreda, N. Evans, Z. Guralnik, J. Große and I. Kirsch. The work of R. A. and of Z. G. has been supported by the Deutsche Forschungsgemeinschaft (DFG), grants ER301/1-4 and ER301/2-1.

References

1. A. Karch and E. Katz, JHEP **0206** (2002) 043 [arXiv:hep-th/0205236]
2. J. Babington, J. Erdmenger, N. J. Evans, Z. Guralnik and I. Kirsch, Phys. Rev. D **69** (2004) 066007 [arXiv:hep-th/0306018].
3. M. Kruczenski, D. Mateos, R. C. Myers and D. J. Winters, JHEP **0307** (2003) 049 [arXiv:hep-th/0304032]; M. Kruczenski, D. Mateos, R. C. Myers and D. J. Winters, JHEP **0405** (2004) 041 [arXiv:hep-th/0311270]; T. Sakai and S. Sugimoto, Prog. Theor. Phys. **113** (2005) 843 [arXiv:hep-th/0412141]; T. Sakai and J. Sonnenschein, JHEP **0309** (2003) 047 [arXiv:hep-th/0305049]; K. Ghoroku and M. Yahiro, Phys. Lett. B **604** (2004) 235 [arXiv:hep-th/0408040].
4. Z. Guralnik, S. Kovacs and B. Kulik, JHEP **0503**, 063 (2005) [arXiv:hep-th/0405127], Z. Guralnik, S. Kovacs and B. Kulik, Fortsch. Phys. **53**, 480 (2005) [arXiv:hep-th/0501154]; Z. Guralnik, [arXiv:hep-th/0412074].
5. J. Erdmenger, J. Große and Z. Guralnik, JHEP **0506**, 052 (2005) [arXiv:hep-th/0502224].
6. R. Apreda, J. Erdmenger, N. Evans and Z. Guralnik, Phys. Rev. D **71**, 126002 (2005) [arXiv:hep-th/0504151].
7. R. Apreda, J. Erdmenger and N. Evans, JHEP **0605** (2006) 011 [arXiv:hep-th/0509219].
8. J. Erdmenger, N. Evans and J. Große, Heavy-light mesons from the AdS/CFT correspondence, hep-th/0605241.
9. E. Witten, Adv. Theor. Math. Phys. **2** (1998) 505 [arXiv:hep-th/9803131].
10. I. Kirsch, Fortsch. Phys. **52** (2004) 727 [arXiv:hep-th/0406274].

MAPPING STRING STATES INTO PARTONS: FORM FACTORS AND THE HADRON SPECTRUM IN AdS/QCD

G. F. DE TÉRAMOND

Universidad de Costa Rica,
San José, Costa Rica
E-mail: gdt@asterix.crnet.cr

New developments in holographic QCD are described in this talk in the context of the correspondence between string states in AdS and light-front wavefunctions of hadronic states in physical space-time.

The AdS/CFT correspondence[1] gives unexpected connections between seemingly different theories which represent the same observables. On the bulk side it describes the propagation of weakly coupled strings, where physical quantities are computed using an effective gravity approximation. The duality provides a non-perturbative definition of quantum gravity in a $(d+1)$-dimensional AdS spacetime in terms of a d-dimensional conformally-invariant quantum field theory at the anti–de Sitter (AdS) boundary[2].

The AdS/CFT duality has the potential for understanding fundamental properties of quantum chromodynamics such as confinement and chiral symmetry breaking which are inherently non-perturbative. As shown by Polchinski and Strassler[3], the AdS/CFT duality, modified to incorporate a mass scale, provides a non-perturbative derivation of dimensional counting rules[4] for the leading power-law fall-off of hard scattering. The modified theory generates the hard behavior expected from QCD, instead of the soft behavior characteristic of strings.

In its original formulation[1], a correspondence was established between the supergravity approximation to Type IIB string theory on a curved background asymptotic to the product space of $AdS_5 \times S^5$ and the large N_C, $\mathcal{N} = 4$, super Yang-Mills (SYM) gauge theory in four dimensional space-time. The group of conformal transformations $SO(2,4)$ which acts at the AdS boundary, is isomorphic with the group of isometries of AdS space, and S^5 corresponds to the $SU(4) \sim SO(6)$ global symmetry which rotates the particles present in the SYM supermultiplet. The supergravity duality requires a large AdS radius R corresponding to a large value of the 't Hooft parameter $g_s N_C$, where $R = (4\pi g_s N_C)^{1/4} \alpha_s'^{1/2}$ and $\alpha_s'^{1/2}$ is the string scale. The classical approximation corresponds to the stiff limit where the string tension $T = R^2/2\pi\alpha' \to \infty$, effectively suppressing string fluctuations.

QCD is fundamentally different from SYM theories where all the matter fields transform in adjoint multiplets of $SU(N_C)$. QCD is also a confining theory in the infrared with a mass gap $\Lambda_{\rm QCD}$ and a well-defined spectrum of color singlet states. Its fundamental string dual is unknown. The duality should be extended to include different boundary conditions and non conformal quantum field theories. We may expect that a dual gravitational description would emerge in the strong coupling regime of QCD. Indeed, the string dual should remain well defined also in a highly curved space where the AdS radius become small compared to the string size[5].

In practice, we can deduce salient properties of the QCD dual theory by studying its general behavior, such as its ultraviolet limit at the conformal AdS boundary $z \to 0$, as well as the large-z infrared region, characteristic of strings dual to confining gauge theories. The fifth dimension in the anti-de Sitter metric corresponds to the scale transformations of the quantum field theory, thus incorporating the renormalization group flow of the boundary theory. This approach, which can be described as a bottom-up approach, has been successful in obtaining general properties of scattering processes of QCD bound states[3,6], the low-lying hadron spectra[7,8], hadron couplings and chiral symmetry breaking[8,9], quark potentials in confining backgrounds[10] and pomeron physics[11].

In contrast to the simple bottom-up approach described above, a top-bottom approach consists in studying the full supergravity equations to compute the glueball spectrum[12] or the introduction of additional higher dimensional branes to the $AdS_5 \times S^5$ background[13], as a prescription for the introduction of flavor with quarks in the fundamental representation and the computation of the meson spectrum.

It has been shown recently that the string amplitude $\Phi(z)$ describing hadronic modes in AdS_5 can be precisely mapped to the light-front wavefunctions $\psi_{n/h}$ of hadrons in physical space-time[14]. Indeed, there is an exact correspondence between the holographic variable z and an impact variable ζ which represents the measure of the transverse separation of the constituents within the hadrons. This remarkable holographic feature follows from the fact that current matrix element in AdS space can be mapped to the exact Drell-Yan-West formula at the asymptotic AdS boundary[14]. It was also found that effective Schrödinger equations describing hadronic bound states can be expressed as $3+1$ QCD light-front wave equations[14].

The boost invariant light-front wavefunctions (LFWFs) in the Fock expansion at fixed light-cone time $x^+ = x^0 + x^3$ of any hadronic system $\widetilde{\psi}_{n/h}(x_i, \mathbf{b}_{\perp i}, \lambda_i)$, encode all its bound-state quark and gluon properties and their behavior in high-momentum transfer reactions[15]. The light-cone momentum fractions $x_i = k_i^+/P^+$ and the impact position variables $\mathbf{b}_{\perp i}$ represent the relative coordinates of constituent i in Fock state n, and λ_i the helicity along the $\mathbf{z}$ axis.

In the case of a two-parton constituent bound state the correspondence between the string amplitude $\Phi(z)$ and the light-front wave function $\widetilde{\psi}(x, \mathbf{b})$ is expressed in closed form[14]

$$\left|\widetilde{\psi}(x,\zeta)\right|^2 = \frac{R^3}{2\pi}\, x(1-x)\, e^{3A(\zeta)}\, \frac{|\Phi(\zeta)|^2}{\zeta^4}, \tag{1}$$

where $\zeta^2 = x(1-x)\mathbf{b}_\perp^2$. The variable ζ, $0 \le \zeta \le \Lambda_{\rm QCD}^{-1}$, represents the invariant separation between point-like constituents, and it is also the holographic variable z in AdS; *i.e.*, we can identify $\zeta = z$. In the "hard wall" approximation[3] the nonconformal metric factor $e^{3A(z)}$ is a step function.

The short-distance behavior of a hadronic state is characterized by its twist (dimension minus spin) $\tau = \Delta - \sigma$, where σ is the sum over the constituent's spin $\sigma = \sum_{i=1}^n \sigma_i$. Twist is also equal to the number of partons $\tau = n$. Matching the boundary behavior of string modes $\phi(z)$ with the twist of the boundary interpolating operators we find, upon the substitution $\phi(z) = z^{-3/2}\Phi(z)$ in the wave equations in AdS space, an effective Schrödinger equation as a function of the weighted impact variable ζ

$$\left[-\frac{d^2}{d\zeta^2} + V(\zeta)\right]\phi(\zeta) = \mathcal{M}^2\phi(\zeta), \tag{2}$$

with the effective potential $V(\zeta) \to -(1-4L^2)/4\zeta^2$ in the conformal limit[14]. The solution to (2) is $\phi(z) = z^{-\frac{3}{2}}\Phi(z) = Cz^{\frac{1}{2}}J_L(z\mathcal{M})$. Its lowest stable state is determined by the Breitenlohner-Freedman bound[16]. Its eigenvalues are obtained from the boundary conditions at $\phi(z = 1/\Lambda_{\rm QCD}) = 0$, and are given in terms of the roots of the Bessel functions: $\mathcal{M}_{L,k} = \beta_{L,k}\Lambda_{\rm QCD}$. Normalized LFWFs $\widetilde{\psi}_{L,k}$ follow from (1)

$$\widetilde{\psi}_{L,k}(x,\zeta) = B_{L,k}\sqrt{x(1-x)}J_L\left(\zeta\beta_{L,k}\Lambda_{\rm QCD}\right)\theta\big(z \le \Lambda_{\rm QCD}^{-1}\big), \tag{3}$$

where $B_{L,k} = \pi^{-\frac{1}{2}}\Lambda_{\rm QCD}/J_{1+L}(\beta_{L,k})$. The spectrum of the light mesons is compared in Figure 1 with the data listed by the PDG[17].

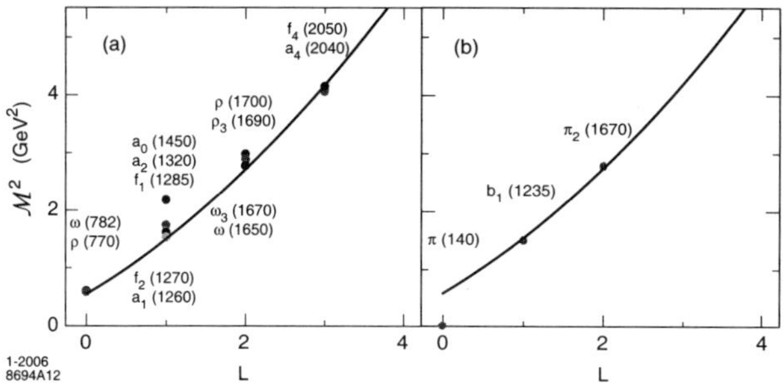

Fig. 1. Light meson orbital states for $\Lambda_{\rm QCD} = 0.32$ GeV: (a) vector mesons and (b) pseudoscalar mesons.

A different approach consists on matching AdS results following Migdal procedure for the regularization of UV conformal correlators, using Padé approximants to build the spectrum with poles of zeros of Bessel functions[18]. This has been discussed recently for two- and three-point functions[19].

Consider the twist-three, dimension $\frac{9}{2}+L$, baryon operators $\mathcal{O}_{(9/2)+L} = \psi D_{\{\ell_1} \ldots D_{\ell_q} \psi D_{\ell_{q+1}} \ldots D_{\ell_m\}} \psi$. Since we are taking a product of operators at the same point, we match the dependence of the corresponding AdS spin-$\frac{1}{2}$ or $\frac{3}{2}$ modes to the boundary operators at the ultraviolet $Q \to \infty$ or $z \to 0$ limit. A three-quark baryon is described by wave equation[20]

$$\left[z^2\, \partial_z^2 - 3z\, \partial_z + z^2\mathcal{M}^2 - \mathcal{L}_\pm^2 + 4\right] \psi_\pm(z) = 0 \tag{4}$$

with $\mathcal{L}_+ = L+1$, $\mathcal{L}_- = L+2$, and solution

$$\Psi(x,z) = Ce^{-iP\cdot x}\left[\psi(z)_+\; u_+(P) + \psi(z)_-\; u_-(P)\right], \tag{5}$$

with $\psi_+(z) = z^2 J_{1+L}(z\mathcal{M})$ and $\psi_-(z) = z^2 J_{2+L}(z\mathcal{M})$. The constant C in (5) is determined by the normalization $R^3 \int \frac{dz}{z^3}\frac{1}{2}\left[|\psi_+(z)|^2 + |\psi_-(z)|^2\right] = 1$ and is given by $C = \sqrt{2} R^{-\frac{3}{2}} \Lambda_{\rm QCD}/J_0(\beta_{1,1})$. The physical string solutions have plane waves and chiral spinors $u_\pm(P)$ along the Poincaré coordinates and hadronic invariant mass states $P_\mu P^\mu = \mathcal{M}^2$. Similar solutions follow from the Rarita-Schwinger AdS modes Ψ^μ in the $\Psi_z = 0$ gauge. In the large P^+ limit $\psi_\pm$ are the light-cone $\pm$ components along the $\mathbf{z}$ axis: $\psi_+ = \psi^\uparrow$, $\psi_- = \psi^\downarrow$. The four-dimensional spectrum follows from $\psi_\pm(z = 1/\Lambda_{\rm QCD}) = 0$: $\mathcal{M}^+_{\alpha,k} = \beta_{\alpha,k}\Lambda_{\rm QCD}$, $\mathcal{M}^-_{\alpha,k} = \beta_{\alpha+1,k}\Lambda_{\rm QCD}$, with a scale independent mass ratio[7]. Figure 2(a) shows the predicted orbital spectrum of the nucleon states and Fig. 2(b) the Δ orbital resonances. The data is from [17]. The internal parity of states is determined from the SU(6) spin-flavor symmetry.

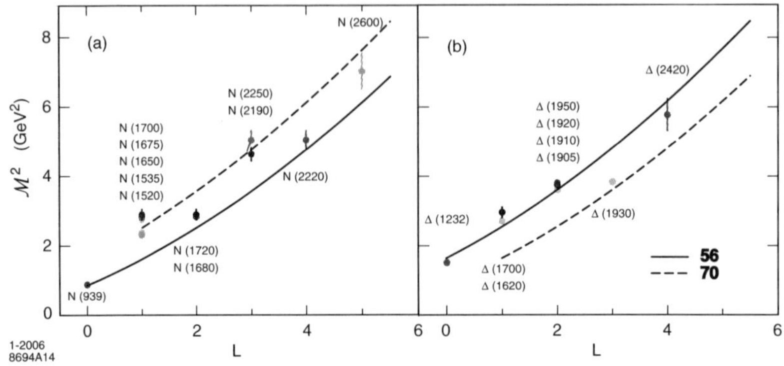

Fig. 2. Predictions for the light baryon orbital spectrum for $\Lambda_{QCD} = 0.25$ GeV. The **56** trajectory corresponds to L even $P = +$ states, and the **70** to L odd $P = -$ states. The only parameter is the value of $\Lambda_{\rm QCD}$ which is fixed by the proton mass.

The predictions for the lightest hadrons are improved relative to the results of [7] using the boundary conditions determined by twist instead of conformal dimensions. The model is remarkably successful in organizing the hadron spectrum, although it underestimates the spin-orbit splittings of the $L = 1$ states. A better understanding of the relation between chiral symmetry breaking and confinement is required to describe successfully the pion. This would probably need a description of quark spin-flip mechanisms at the wall.

We now consider the spin non-flip nucleon form factors in the hard wall model. The effective charges are determined from the spin-flavor structure of the theory. We choose the struck quark to have $s^z = +1/2$. The two AdS solutions ψ_+ and ψ_- correspond to nucleons with $J^z = +1/2$ and $-1/2$. For $SU(6)$ spin-flavor symmetry[20]

$$F_1^p(Q^2) = R^3 \int \frac{dz}{z^3} J(Q,z) \, |\psi_+(z)|^2, \tag{6}$$

$$F_1^n(Q^2) = -R^3 \, \frac{1}{3} \int \frac{dz}{z^3} J(Q,z) \left[|\psi_+(z)|^2 - |\psi_-(z)|^2 \right], \tag{7}$$

where $J(Q,z)$ is a solution to the AdS wave equation for the external electromagnetic current polarized along the Minkowski coordinates, $A_\mu = \epsilon_\mu e^{-iQ\cdot x} J(Q,z)$, $A_z = 0$, subject to the boundary conditions $J(Q = 0, z) = J(Q, z = 0) = 1$ and is given by $J(Q,z) = zQK_1(zQ)^3$. The conditions $F_1^p(0) = 1$ and $F_1^n(0) = 0$ follow from the identity $\int_0^1 x dx \left[J_\alpha^2(x\beta) - J_{\alpha+1}^2(x\beta)\right] = J_\alpha(\beta)J_{\alpha+1}(\beta)/\beta$. Figure 3 compares the predictions for the Dirac nucleon form factors with the experimental data[21].

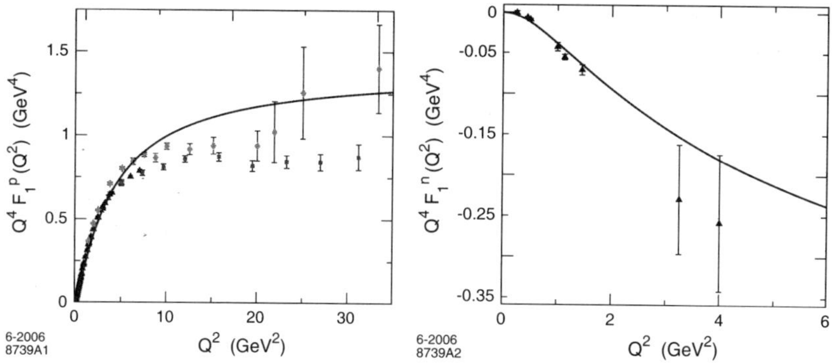

Fig. 3. Prediction for $Q^4 F_1^p(Q^2)$ and $Q^4 F_1^n(Q^2)$ in the valence approximation for $\Lambda_{QCD} = 0.21$ GeV. Analysis of the data is from Diehl. Data from Sill (solid boxes in red) and superimposed data from Kirk (solid diamonds in green).

We have shown how the string amplitude $\Phi(z)$ defined on the fifth dimension in AdS_5 space can be precisely mapped to the light-front wavefunctions of hadrons in physical spacetime[14]. This specific correspondence provides an exact holographic mapping in the conformal limit at all energy scales between string modes in AdS and boundary states with well-defined number of partons. Consequently, the AdS string mode $\Phi(z)$ can be regarded as the probability amplitude to find n-partons at transverse impact separation $\zeta = z$. Its eigenmodes determine the hadronic mass spectrum. Although major dynamical questions remain to be solved for extending the duality from large to small 't Hooft coupling, the string-parton correspondence described in [14] suggests that basic features of QCD can be understood in terms of a higher dimensional dual gravity theory which holographically encodes multi-parton boundary states into string modes and allows the computation of physical observables at strong coupling.

Acknowledgements

This work was done in collaboration with Stan Brodsky. We thank Misha Shifman for his invitation to CAQCD 2006 and Joe Polchinski for encouraging comments.

References

1. J. M. Maldacena, Adv. Theor. Math. Phys. **2**, 231 (1998).
2. S. S. Gubser, I. R. Klebanov and A. M. Polyakov, Phys. Lett. B **428**, 105 (1998); E. Witten, Adv. Theor. Math. Phys. **2**, 253 (1998).

3. J. Polchinski and M. J. Strassler, Phys. Rev. Lett. **88**, 031601 (2002); JHEP **0305**, 012 (2003).
4. S. J. Brodsky and G. R. Farrar, Phys. Rev. Lett. **31**, 1153 (1973); V. A. Matveev, R. M. Muradian and A. N. Tavkhelidze, Lett. Nuovo Cim. **7**, 719 (1973).
5. G. T. Horowitz and J. Polchinski, arXiv:gr-qc/0602037.
6. S. J. Brodsky and G. F. de Teramond, Phys. Lett. B **582**, 211 (2004).
7. G. F. de Teramond and S. J. Brodsky, Phys. Rev. Lett. **94**, 201601 (2005).
8. J. Erlich, E. Katz, D. T. Son and M. A. Stephanov, Phys. Rev. Lett. **95**, 261602 (2005); E. Katz, A. Lewandowski and M. D. Schwartz, arXiv:hep-ph/0510388; A. Karch, E. Katz, D. T. Son and M. A. Stephanov, arXiv:hep-ph/0602229; J. P. Shock and F. Wu, arXiv:hep-ph/0603142; N. Evans and T. Waterson, arXiv:hep-ph/0603249.
9. S. Hong, S. Yoon and M. J. Strassler, arXiv:hep-ph/0501197; L. Da Rold and A. Pomarol, Nucl. Phys. B **721**, 79 (2005); JHEP **0601**, 157 (2006); J. Hirn, N. Rius and V. Sanz, Phys. Rev. D **73**, 085005 (2006); J. Hirn and V. Sanz, JHEP **0512**, 030 (2005); K. Ghoroku, N. Maru, M. Tachibana and M. Yahiro, Phys. Lett. B **633**, 602 (2006).
10. H. Boschi-Filho, N. R. F. Braga and C. N. Ferreira, Phys. Rev. D **73**, 106006 (2006); O. Andreev and V. I. Zakharov, arXiv:hep-ph/0604204.
11. H. Boschi-Filho, N. R. F. Braga and H. L. Carrion, Phys. Rev. D **73**, 047901 (2006); R. C. Brower, J. Polchinski, M. J. Strassler and C. I. Tan, arXiv:hep-th/0603115.
12. E. Witten, Adv. Theor. Math. Phys. **2**, 505 (1998); C. Csaki, H. Ooguri, Y. Oz and J. Terning, JHEP **9901**, 017 (1999); R. C. Brower, S. D. Mathur and C. I. Tan, Nucl. Phys. B **587**, 249 (2000); E. Caceres and C. Nunez, JHEP **0509**, 027 (2005).
13. A. Karch and E. Katz, JHEP **0206**, 043 (2002); M. Kruczenski, D. Mateos, R. C. Myers and D. J. Winters, JHEP **0307**, 049 (2003); T. Sakai and J. Sonnenschein, JHEP **0309**, 047 (2003); J. Babington, J. Erdmenger, N. J. Evans, Z. Guralnik and I. Kirsch, Fortsch. Phys. **52**, 578 (2004); M. Kruczenski, L. A. P. Zayas, J. Sonnenschein and D. Vaman, JHEP **0506**, 046 (2005); T. Sakai and S. Sugimoto, Prog. Theor. Phys. **113**, 843 (2005); *ibid* **114**, 1083 (2006); A. Paredes and P. Talavera, Nucl. Phys. B **713**, 438 (2005); J. Erdmenger, N. Evans and J. Grosse, arXiv:hep-th/0605241; A. V. Ramallo, arXiv:hep-th/0605261.
14. S. J. Brodsky and G. F. de Teramond, Phys. Rev. Lett. **96**, 201601 (2006).
15. S. J. Brodsky, arXiv:hep-ph/0412101; these proceedings.
16. P. Breitenlohner and D. Z. Freedman, Annals Phys. **144**, 249 (1982).
17. S. Eidelman *et al.* [Particle Data Group], Phys. Lett. B **592** (2004) 1.
18. A. A. Migdal, Annals Phys. **109**, 365 (1977).
19. J. Erlich, G. D. Kribs and I. Low, Phys. Rev. D **73**, 096001 (2006); A. V. Radyushkin, arXiv:hep-ph/0605116; O. Cata, arXiv:hep-ph/0605251.
20. S. J. Brodsky and G. F. de Teramond, in preparation.
21. M. Diehl, arXiv:hep-ph/0510221; A. F. Sill *et al.*, Phys. Rev. D **48**, 29 (1993); P. N. Kirk *et al.*, Phys. Rev. D **8** (1973) 63.

PROPERTIES OF HADRONS FROM D4/D8-BRANE SYSTEM*

TADAKATSU SAKAI

Department of Mathematical Sciences, Faculty of Sciences, Ibaraki University, Bunkyo 2-1-1, Mito, 310-8512 Japan
E-mail: tsakai@mx.ibaraki.ac.jp

SHIGEKI SUGIMOTO

Department of Physics, Nagoya University, Nagoya, Aichi 464-8602, Japan
E-mail: sugimoto@eken.phys.nagonya-u.ac.jp

We construct four dimensional QCD by using D4-branes and D8-branes in type IIA string theory and analyze it within the supergravity approximation. Replacing the D4-branes with the corresponding supergravity solution, a holographic five dimensional description of QCD is obtained. We show that various aspects of low energy (large N) QCD can be understood from this holographic description.

Keywords: QCD, D-brane, Supergravity

1. introduction

Since Maldacena proposed the duality between string theory in AdS space and conformal field theory [2], there have been various attempts to extend this idea to more general situations. (See for example Ref. 3 for a review.) A natural question one would ask is whether we can analyze the realistic QCD using this technique.

A key step toward this direction was given by Witten in Ref. 4, in which a supergravity dual of the four dimensional pure Yang-Mills theory was proposed. The basic idea is to use D4-branes compactified on a supersymmetry breaking circle to realize four dimensional Yang-Mills theory and replace the D4-branes with the corresponding supergravity solution to

---

*Talk given by S. Sugimoto at Continuous Advances in QCD 2006, William I. Fine Theoretical Physics Institute, Minnesota, May 11-14, 2006

obtain a holographic dual description. Though it is in general quite difficult to obtain a rigorous result in the duality of such non-supersymmetric theory, a lot of non-trivial evidence for the duality has been found and it encourages us to further investigate along this line toward the analysis of QCD via supergravity or superstring theory.

In Ref. 1, we proposed a way to extend this model to obtain a holographic description of QCD with fundamental quarks. Our strategy is to use the above D4-brane background that represents pure Yang-Mills theory and add probe D8-branes to add fundamental quarks to the system.[a] One of the advantages of our model is that the chiral $U(N_f)_R \times U(N_f)_L$ symmetry [b] in QCD with N_f massless flavors is manifestly realized. Thanks to this property, we can argue spontaneous chiral symmetry breaking and the appearance of the associated Nambu-Goldstone bosons which are interpreted as pions. Moreover, massive (axial-)vector mesons are also found in the meson spectrum. Some of the masses and couplings of these mesons are calculated and they are compared with experimental data. We should, however, keep in mind that the classical analysis in the supergravity side can only be trusted if the number of color and the 't Hooft coupling are large enough, which might sound still a bit far from the realistic situation. Nevertheless, as we will see, our model seems to nicely catch various features of QCD. The aim of this article is to explain the basic idea of the model and summarize the main results given in Ref. 1.

2. QCD & D4/D8 system

We construct $U(N_c)$ QCD with N_f massless flavors by using N_c D4-banes and N_f D8-$\overline{\text{D8}}$ pairs in type IIA string theory. D4-branes are extended along x^μ $(\mu = 0, \ldots, 3)$ directions and wrapped on an S^1 parametrized by $x^4 = \tau$. The period of the parameter τ is written as $2\pi M_{\text{KK}}^{-1}$, where M_{KK} gives the mass scale of massive Kaluza-Klein modes. Following Ref. 4, we impose the anti-periodic boundary condition along this S^1 to all the fermions in the system. The D8-branes and $\overline{\text{D8}}$-branes are placed at antipodal points on the S^1 and extended along all the other directions as depicted in left side of Fig. 1. Since we are going to consider the near horizon limit of the supergravity solution corresponding to the D4-branes in the next section, we

[a] The same idea was used in another interesting model proposed by Kruczenski et al. in Ref. 5, in which they used probe D6-branes to add quarks.

[b] The $U(1)_A$ subgroup of this chiral symmetry is actually anomalous. However, the effect of this anomaly is negligible in the large N_c limit.

are interested in the strings attached on the D4-branes, that is, 4-4 strings, 4-8 strings and 4-$\overline{8}$ strings. As it is well-known, a $U(N_c)$ gauge field is created by the 4-4 strings. Since the supersymmetry is completely broken by the anti-periodic boundary condition for fermions along the S^1, the other unwanted adjoint fields created by the 4-4 strings are expected to become massive. On the other hand, the massless modes in the 4-8 strings and 4-$\overline{8}$ strings turn out to be N_f flavors of fermions that belong to the fundamental representation of the $U(N_c)$ gauge group. Furthermore, as discussed in Ref. 7, the chirality of the fermions created by 4-8 strings are opposite to those created by the 4-$\overline{8}$ strings. Therefore the $U(N_f)_{\rm D8} \times U(N_f)_{\overline{\rm D8}}$ gauge symmetry of the N_f D8-$\overline{\rm D8}$ pairs is interpreted as the $U(N_f)_L \times U(N_f)_R$ chiral symmetry of QCD. After all, we obtain four dimensional $U(N_c)$ QCD with N_f massless flavors with manifest $U(N_f)_L \times U(N_f)_R$ chiral symmetry realized on the D4-brane world-volume.

3. Holographic description of QCD

Let us move to the supergravity description of the D4/D8/$\overline{\rm D8}$ system considered in the previous section. Here we use so-called probe approximation[6] and replace the D4-branes with the corresponding supergravity solution given in Ref. 4. The D8 and $\overline{\rm D8}$-branes are treated as probe branes embedded in this curved background and all the backreaction to the background caused by these probe branes are neglected. This approximation can be justified when $N_c \gg N_f \sim \mathcal{O}(1)$. The metric of this background is given as

$$ds^2 \propto \frac{4}{9} M_{\rm KK}^2\, u^{3/2}(\eta_{\mu\nu}dx^\mu dx^\nu + f(u)d\tau^2) + u^{-3/2}\left(\frac{du^2}{f(u)} + u^2 d\Omega_4^2\right) \quad (1)$$

where u (≥ 1) represents the radial direction transverse to the D4-brane and $f(u) \equiv 1 - u^{-3}$. $d\Omega_4^2$ is the line element of a unit S^4 surrounding the D4-brane. From this metric, we see that the radius of the S^1 parametrized by τ shrinks to zero at $u \to 1$. Though the metric looks singular at $u \to 1$, the geometry is actually everywhere smooth and the topology of the space-time is $\mathbb{R}^{1,3} \times \mathbb{R}^2 \times S^4$, where $\mathbb{R}^{1,3}$ is the Minkowski space parametrized by x^μ ($\mu = 0, \ldots, 3$) and $\mathbb{R}^2$ is the (u, τ) plane.

This implies that the D8-brane and $\overline{\rm D8}$-brane must be connected at $u = 1$ as depicted in the right side of Fig. 1. Therefore, we have only one connected component of the D8-brane in this background and the gauge group $U(N_f)_{\rm D8} \times U(N_f)_{\overline{\rm D8}}$ is now broken to the diagonal subgroup $U(N_f)_V$. This phenomena is interpreted as the chiral symmetry breaking in QCD.

In this model, the closed strings are interpreted as glueballs and the

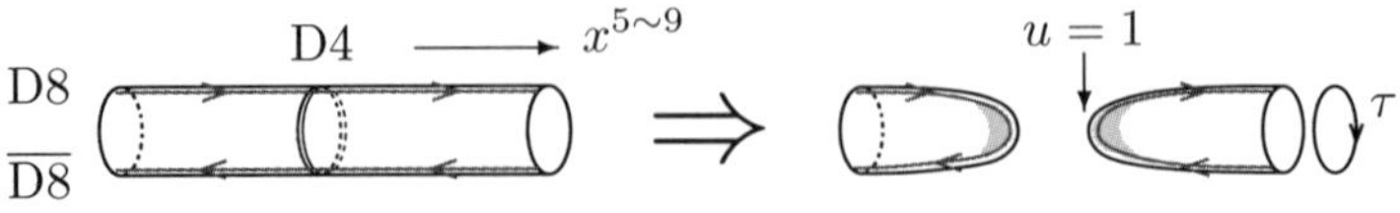

Fig. 1. Chiral symmetry breaking

D4-branes wrapped on the S^4 are interpreted as baryons. These objects are studied by many people around 1998 (See Ref. 3.). In the following, we consider the open strings attached on the probe D8-branes. Since they carry flavor indices, they are interpreted as mesons.

The effective theory of the open strings on the D8-branes is a 9 dimensional $U(N_f)$ gauge theory. In this article, we only consider fields that are invariant under $SO(5)$ symmetry that acts as rotation of the S^4, and ignore all the higher Kaluza-Klein modes. Then, the effective theory is reduced to a five dimensional gauge theory. By inserting the supergravity background to the effective action of the D8-brane and integrating over the S^4, we obtain the 5 dimensional effective action [c]

$$S_{\text{D8}} \simeq \kappa \int d^4x\, dz \,\text{Tr}\left(\frac{1}{2}K^{-1/3}F_{\mu\nu}^2 + M_{\text{KK}}^2 K F_{\mu z}^2\right) + \frac{N_c}{24\pi^2}\int_5 \omega_5(A) \ , \quad (2)$$

where $\kappa = \frac{\lambda N_c}{108\pi^3}$ is a constant, $\omega_5(A)$ is the Chern-Simons 5-form, z is the fifth coordinate related to u by $u^3 = 1+z^2$ and $K(z) = 1+z^2$. The claim is that this 5 dimensional $U(N_f)$ Yang-Mills - Chern-Simons theory is considered as the effective theory of mesons. It is quite interesting to note that this 5 dimensional description of mesons is closely related to 5 dimensional phenomenological models proposed in Refs. 8.

4. 5 dim gauge theory ⇒ 4 dim meson theory

Let us next explain how to extract four dimensional physics from the 5 dimensional action (2). First we expand the gauge field (A_μ, A_z) using some complete sets $\{\psi_n(z)\}_{n\geq 1}$ and $\{\phi_n(z)\}_{n\geq 0}$ as

$$A_\mu(x^\mu, z) = \sum_{n\geq 1} B_\mu^{(n)}(x^\mu)\psi_n(z) \ , \quad A_z(x^\mu, z) = \sum_{n\geq 0} \varphi^{(n)}(x^\mu)\phi_n(z) \ . \quad (3)$$

These complete sets are chosen so that the kinetic and mass terms for the four dimensional fields $B_\mu^{(n)}$ and $\varphi^{(n)}$ become diagonal. We choose $\{\psi_n\}_{n\geq 1}$ as eigen functions satisfying

$$-K^{1/3}\partial_z(K\partial_z\psi_n) = \lambda_n\psi_n \ , \quad \kappa\int dz\, K^{-1/3}\psi_n\psi_m = \delta_{nm} \ . \quad (4)$$

[c] There is also a scalar field on the D8-brane, but here we will omit it for simplicity.

Here λ_n are the eigen values. Then we choose $\{\phi_n\}_{n\geq 1}$ as $\phi_n = \partial_z \psi_n$ which satisfy the ortho-normal condition

$$\kappa \int dz\, K \phi_n \phi_m = \lambda_n \delta_{nm} \ . \tag{5}$$

An important point here is that there is one more normalizable mode

$$\phi_0(z) = \frac{c}{K(z)} \ , \tag{6}$$

where the normalization constant c is chosen as $c = M_{\rm KK}^{-1}(\kappa\pi)^{-1/2}$.

Inserting the expansion (3) into the action (2) we obtain

$$S_{D8} \sim \int d^4x \, {\rm Tr} \left[\partial_\mu \varphi^{(0)\,2} + \sum_{n\geq 1} \left(\frac{1}{2} F_{\mu\nu}^{(n)\,2} + \lambda_n M_{\rm KK}^2 \left(B_\mu^{(n)} - \partial_\mu \varphi^{(n)} \right)^2 \right) \right] \\ + (\text{interaction terms}) \ , \tag{7}$$

where $F_{\mu\nu}^{(n)} = \partial_\mu B_\nu^{(n)} - \partial_\nu B_\mu^{(n)}$. From this we see that $\varphi^{(n)}$ with $n \geq 1$ are eaten by $B_\mu^{(n)}$, which become massive vector fields. On the other hand, $\varphi^{(0)}$ does not have a partner vector field '$B_\mu^{(0)}$' in the expansion (3) and remain as a massless scalar field. We interpret $\varphi^{(0)}$ as the pion field, the lightest vector meson $B_\mu^{(0)}$ as the ρ meson, the second lightest vector meson $B_\mu^{(1)}$ as the a_1 meson and so on. The spin, parity and charge conjugation parity of π, ρ, a_1 mesons are consistent with this interpretation. Actually, from the fact that ϕ_0 and ψ_{2k-1} are even functions while ψ_{2k} are odd function, it can be shown that $\varphi^{(0)}$ is a pseudo-scalar meson, $B_\mu^{(2k-1)}$ are vector mesons and $B_\mu^{(2k)}$ are axial-vector mesons for $k = 1, 2, \cdots$. In contrast to usual construction of effective theory of mesons, in which each meson field is introduced independently, various mesons π, ρ, a_1, $\cdots$ are now elegantly unified in the five dimensional gauge field (A_μ, A_z) in our description.

As we have seen in (7), the mass of the n-th vector meson is given by $m_n^2 = \lambda_n M_{\rm KK}^2$ where λ_n is the eigen value of the eigen equation (4). It is tempting to compute the eigen values numerically and compare the results with the observed meson table [9]. Of course, we should not be too serious about the agreement since the approximation is still very crude. For example, we have assumed that $N_c \gg N_f \sim \mathcal{O}(1)$, for which the realistic values are $N_c = 3$, $N_f = 2$.[d] We also know that our model deviates from QCD above the energy scale around $M_{\rm KK}$. Therefore, our numerical results may only be useful to see if our model is too bad or not.

[d] Here we only consider up and down quarks whose masses are reasonably small compared to $\Lambda_{\rm QCD}$.

Here is the result:

	ρ	a_1	ρ'	(a_1')	ρ''
exp.(MeV)	776	1230	1465	(1640)	1720
our model	[776]	1189	1607	2023	2435

where we have fixed the value of $M_{\rm KK}$ to fit the rho meson mass.

We can also extract the interaction terms in (7). Here we skip all the details and show the results.

coupling	our model	fitting m_ρ and f_π	experiment
f_π	$1.13 \cdot \kappa^{1/2} M_{\rm KK}$	[92.4 MeV]	92.4 MeV
L_1	$0.0785 \cdot \kappa$	0.584×10^{-3}	$(0.1 \sim 0.7) \times 10^{-3}$
L_2	$0.157 \cdot \kappa$	1.17×10^{-3}	$(1.1 \sim 1.7) \times 10^{-3}$
L_3	$-0.471 \cdot \kappa$	-3.51×10^{-3}	$-(2.4 \sim 4.6) \times 10^{-3}$
L_9	$1.17 \cdot \kappa$	8.74×10^{-3}	$(6.2 \sim 7.6) \times 10^{-3}$
L_{10}	$-1.17 \cdot \kappa$	-8.74×10^{-3}	$-(4.8 \sim 6.3) \times 10^{-3}$
$g_{\rho\pi\pi}$	$0.415 \cdot \kappa^{-1/2}$	4.81	5.99
g_ρ	$2.11 \cdot \kappa^{1/2} M_{\rm KK}^2$	0.164 GeV^2	0.121 GeV^2
$g_{a_1\rho\pi}$	$0.421 \cdot \kappa^{-1/2} M_{\rm KK}$	4.63 GeV	$2.8 \sim 4.2$ GeV

The middle column of this table is the values obtained by fixing $M_{\rm KK}$ and κ to fit the rho meson mass and pion decay constant f_π. Though the agreement of our numerical results with the experimental data is not extremely good, we think it is much better than expected.

5. Other topics

We close this article by listing some other topics in our paper [1] that are omitted here.

Baryon A baryon is constructed as a D4-brane wrapped on the S^4 as proposed in Ref. 10. This wrapped D4-brane is realized as an instanton in the five dimensional Yang-Mills theory (2) and shown to be equivalent to a soliton (Skyrmion) in the Skyrme model [11].

Chiral anomaly and WZW term The chiral anomaly and the WZW term in QCD are reproduced from the CS-term in (2).

$U(1)_A$ anomaly and η' meson mass The $U(1)_A$ anomaly can also be understood in the supergravity description. Taking this anomaly into account, the mass of the η' meson is estimated and shown to satisfy the Witten-Veneziano formula [12].

Vector meson dominance We can show, in a certain gauge choice, the external photon field interacts with mesons only through vector meson exchange. This shows that the complete vector meson dominance is realized.

ω meson decay The Feynman diagrams relevant to $\omega \to \pi^0\gamma$ and $\omega \to \pi^0\pi^+\pi^-$ turn out to be the same as those proposed in Gell-Mann - Sharp - Wagner model [13].

Acknowledgments

S.S would like to thank the organizers for a stimulating CAQCD workshop and their kind hospitality.

References

1. T. Sakai and S. Sugimoto, Prog. Theor. Phys. **113**, 843 (2005); Prog. Theor. Phys. **114**, 1083 (2005).
2. J. M. Maldacena, Adv. Theor. Math. Phys. **2**, 231 (1998) [Int. J. Theor. Phys. **38**, 1113 (1999)].
3. O. Aharony, S. S. Gubser, J. M. Maldacena, H. Ooguri and Y. Oz, Phys. Rept. **323**, 183 (2000).
4. E. Witten, Adv. Theor. Math. Phys. **2**, 505 (1998).
5. M. Kruczenski, D. Mateos, R. C. Myers and D. J. Winters, JHEP **0405**, 041 (2004). J. L. F. Barbon, C. Hoyos, D. Mateos and R. C. Myers, JHEP **0410**, 029 (2004).
6. A. Karch and E. Katz, JHEP **0206**, 043 (2002).
7. S. Sugimoto and K. Takahashi, JHEP **0404**, 051 (2004).
8. D. T. Son and M. A. Stephanov, Phys. Rev. D **69**, 065020 (2004); J. Erlich, E. Katz, D. T. Son and M. A. Stephanov, Phys. Rev. Lett. **95** (2005) 261602; L. Da Rold and A. Pomarol, Nucl. Phys. B **721** (2005) 79. J. Hirn and V. Sanz, JHEP **0512**, 030 (2005).
9. S. Eidelman *et al.* [Particle Data Group Collaboration], Phys. Lett. B **592** (2004) 1.
10. E. Witten, JHEP **9807**, 006 (1998).
11. T. H. R. Skyrme, Proc. R. Soc. London A **260** (1961), 127; Proc. R. Soc. London A **262** (1961), 237; Nucl. Phys. **31** (1962), 556.
12. E. Witten, Nucl. Phys. B **156**, 269 (1979); G. Veneziano, Nucl. Phys. B **159**, 213 (1979).
13. M. Gell-Mann, D. Sharp and W. G. Wagner, Phys. Rev. Lett. **8**, 261 (1962).

ENERGY IN AdS

BAYRAM TEKIN

Department of Physics, Middle East Technical University,
Ankara, 06531 , Turkey
E-mail: btekin@metu.edu.tr

We construct conserved charges in curved backgrounds, specifically in the asymptotically AdS spacetime. As is well-known, the definition of energy in gravitating theories is a rather delicate issue. In this paper, which is a brief summary of our recent work, using the background Killing vectors, we define energy (and angular momenta) in asymptotically AdS spacetimes that are solutions to generic higher curvature gravity models as well as Einstein's gravity.

Keywords: Conserved charges, AdS, Gravity

1. Introduction

How one defines conserved charges, such as energy and angular momentum, in generic curved backgrounds is an unsettled issue. First of all, there is no meaningful (namely diffeomorphism invariant) local definition. One can give a definition that could work for the total energy of the whole spacetime. But even that is a matter of dispute for various spacetimes. For the asymptotically flat spaces, the celebrated Arnowitt-Deser-Misner (ADM) [1] energy does the job. But when one considers other spacetimes, such as asymptotically de-Sitter or Anti-de-Sitter, complications arise and one can find plenty of energy definitions which often disagree. Probably the most interesting and urging question in this venue is : how is energy defined in *de Sitter* spacetimes, since our Universe seems to be best described by one. Unfortunately, we are not aware of a satisfactory answer for this question. The cosmological horizon complicates the definition of global charges. It is ironic that while the Universe seems to be described by a de Sitter space, current theories (such as string theory an AdS/CFT dictionary) prefer Anti-de-Sitter spacetimes. What we shall do here also will work in the asymptotically AdS and flat cases but not in the dS case. But our exercise is not extremely futile: As we shall see, if we consider de Sitter spacetimes for which localized matter (such as a black hole) does not significantly change the location of the cosmological horizon, our formulation gives a meaningful energy expression.

2. Charges for generic gravity models

2.1. *Conserved Charges*

We first look at how conserved charges arise in a generic gravity theory coupled to a covariantly conserved bounded matter source $\tau_{\mu\nu}$ [2–4] . Consider the following D dimensional equations of motion which either comes from a proper Lagrangian or is endowed with the Bianchi identities and covariant conservation of the matter source:

$$\Phi_{\mu\nu}(g, R, \nabla R, R^2, ...) = \kappa \tau_{\mu\nu}, \tag{1}$$

where $\Phi_{\mu\nu}$ is a complicated tensor of a local, but otherwise arbitrary, gravity action and κ is an effective coupling constant.

Now we will decompose our metric into the sum of two parts: $g_{\mu\nu} = \bar{g}_{\mu\nu} + h_{\mu\nu}$, where $\bar{g}_{\mu\nu}$ solves (1) for $\tau_{\mu\nu} = 0$ and a deviation part $h_{\mu\nu}$ that vanishes sufficiently rapidly at infinity and is not necessarily small everywhere. [One can also work in the first order vielbein spin-connection formulation, which is necessary whenever fermionic fields are to be taken into account. Such a computation was carried out recently [5]]

Separating the field equations (1) into a part linear in $h_{\mu\nu}$ and collecting all other non-linear terms and the matter source $\tau_{\mu\nu}$ in $T_{\mu\nu}$ that constitute the total source, one obtains

$$\mathcal{O}(\bar{g})_{\mu\nu\alpha\beta} h^{\alpha\beta} = \kappa T_{\mu\nu}, \tag{2}$$

where $\Phi_{\mu\nu}(\bar{g}, \bar{R}, \bar{\nabla}\bar{R}, \bar{R}^2, ...) = 0$, by assumption; the operator $\mathcal{O}(\bar{g})$ depends only on the the background $\bar{g}_{\mu\nu}$. If the background spacetime admits a set of Killing vectors $\bar{\xi}_\mu^{(a)}$, $\bar{\nabla}_\mu \bar{\xi}_\nu^{(a)} + \bar{\nabla}_\nu \bar{\xi}_\mu^{(a)} = 0$, then the energy-momentum tensor can be used to construct the following conserved vector density current : $\bar{\nabla}_\mu(\sqrt{-\bar{g}} T^{\mu\nu} \bar{\xi}_\nu^a) \equiv \partial_\mu(\sqrt{-\bar{g}} T^{\mu\nu} \bar{\xi}_\nu^a) = 0$. Note that *the* crucial point here is that we are looking for ordinarily conserved charges and we can get this with the help of background Killing vectors. Therefore, up to a constant, the conserved Killing charges can be expressed as

$$Q^\mu(\bar{\xi}^a) = \int_{\mathcal{M}} d^{D-1}x \sqrt{-\bar{g}} T^{\mu\nu} \bar{\xi}_\nu^a = \int_\Sigma dS_i \mathcal{F}^{\mu i} \,. \tag{3}$$

Here $\mathcal{M}$ is a spatial (D–1) hypersurface and Σ is its $(D-2)$ dimensional boundary; $\mathcal{F}^{\mu i}$ is an antisymmetric tensor obtained from $\mathcal{O}(\bar{g})$, whose explicit form, depends on the theory.

Let us apply the outlined procedure in the most interesting case: The cosmological Einstein theory.

We define all terms of second and higher order in $h_{\mu\nu}$ and the matter source $\tau_{\mu\nu}$ to be the gravitational energy-momentum tensor and write $\mathcal{G}^L_{\mu\nu}$ as

$$\mathcal{G}^L_{\mu\nu} \equiv R^L_{\mu\nu} - \frac{1}{2}\bar{g}_{\mu\nu}R^L - \frac{2\Lambda}{D-2}h_{\mu\nu} \equiv \kappa T_{\mu\nu} \,. \tag{4}$$

As can be explicitly checked, the left hand side of this equation obeys the background Bianchi identity

In order to write the spatial volume integrals as surface integrals, we need to carry out the linearization of the relevant tensors.

2.2. *Stokes Theorem*

Recall that there are two facets of a proper conserved charge definition: First, identification of the "Gauss law", whose existence is guaranteed by gauge invariance; second, choice of the proper vacuum, possessing sufficient Killing symmetries with respect to which global, background gauge-invariant, generators can be defined; these will always appear as surface integrals in the asymptotic vacuum [4].

In converting the volume integrals to surface integrals for the cosmological Einstein theory described above, one should follow a route which will be convenient in the higher curvature cases. The details are described in [3] which we do not reproduce here for space limitations. After some computation, the Killing charges become

$$\begin{aligned} Q^\mu(\bar{\xi}) = \frac{1}{4\Omega_{(D-2)}G_D}\oint_\Sigma dS_i(&-\bar{\xi}_\nu\bar{\nabla}^i h^{\mu\nu} - \bar{\xi}^i\bar{\nabla}^\mu h + \bar{\xi}^i\bar{\nabla}_\nu h^{\mu\nu} + \bar{\xi}_\nu\bar{\nabla}^\mu h^{i\nu} \\ &+\bar{\xi}^\mu\bar{\nabla}^i h - \bar{\xi}^\mu\bar{\nabla}_\nu h^{i\nu} + h^{\mu\nu}\bar{\nabla}^i\bar{\xi}_\nu + h\bar{\nabla}^\mu\bar{\xi}^i - h^{i\nu}\bar{\nabla}^\mu\bar{\xi}_\nu) \,. \end{aligned} \tag{5}$$

Here $h = \bar{g}^{\mu\nu}h_{\mu\nu}$, and $dS_i \equiv \sqrt{-det\bar{g}}\ d\Omega_i$ where i ranges over $(1,2,....,D-2)$; the charge is normalized by dividing with the (D-dimensional) Newton's constant G_D and the solid angle Ω_{D-2}. Note that even though the integral is to be evaluated at the $(D-2)$ dimensional spatial boundary of the spacetime, one should use the determinant of the full metric in the integral (which is a reminder that this integral was converted from a volume integral.)

3. The energy of Schwarzschild (Anti)de-sitter solutions

We can now evaluate the energy of Schwarzschild-de Sitter (SdS) solutions. In static coordinates, the line element of D-dimensional SdS reads

$$ds^2 = -\left(1 - (\frac{r_0}{r})^{D-3} - \frac{r^2}{l^2}\right) dt^2 + \left(1 - (\frac{r_0}{r})^{D-3} - \frac{r^2}{l^2}\right)^{-1} dr^2 + r^2 d\Omega^2_{D-2},$$

where $l^2 \equiv (D-2)(D-1)/2\Lambda > 0$. The background ($r_0 = 0$) Killing vector is $\bar{\xi}^\mu = (-1, \mathbf{0})$, which is timelike everywhere for AdS ($l^2 < 0$), but remains timelike for dS ($l^2 > 0$) only inside the cosmological horizon: $\bar{g}_{\mu\nu}\bar{\xi}^\mu\bar{\xi}^\nu = -(1 - \frac{r^2}{l^2})$.

The $D = 4$ Case : To see how far we can go with the de-Sitter case, let us concentrate on $D = 4$ first and calculate the surface integral not at $r \to \infty$, but at some finite distance r from the origin; this will not be gauge-invariant, since energy is to be measured only at infinity. Nevertheless, for dS space (which has a horizon that keeps us from going smoothly to infinity), let us first keep r finite as an intermediate step. The integral becomes

$$E(r) = \frac{r_0}{2G} \frac{(1 - \frac{r^2}{l^2})}{(1 - \frac{r_0}{r} - \frac{r^2}{l^2})}. \tag{6}$$

For AdS, take $r \to \infty$ and we get the usual mass $\frac{r_0}{2G}$. For dS, we can only consider small r_0 limit, which corresponds to localized matter that does not change the location of the background horizon: in which case we also get $\frac{r_0}{2G}$ [3]. For $D = 3$, one needs to be careful in AdS space. Global AdS space becomes a negative energy bound state [6]. Below, among the rotating solutions, we shall allude to this case.

Energy For D Dimensions : We get the energy in D-dimensions as

$$E = \frac{(D-2)}{4G_D} r_0^{D-3}. \tag{7}$$

Here r_0 can be arbitrarily large in the AdS case but must be small in dS.

4. Conserved charges of higher D Kerr-AdS spacetimes

Rotating black holes solutions of D dimensional cosmological Einstein theory have recently been found by Gibbons *et. al* [7] Let us now show the mass and angular momenta of these metrics, whose explicit forms are to be found in [7] or in [8].[The latter also has the details of the computation that lead to the following expressions.]

$$E_D = \frac{m}{\Xi} \sum_{i=1}^{\frac{D-1-\epsilon}{2}} \left\{ \frac{1}{\Xi_i} - (1-\epsilon)(\frac{1}{2}) \right\}. \tag{8}$$

where

$$\Xi \equiv \prod_{i=1}^{\frac{D-1-\epsilon}{2}} (1 + \Lambda a_i^2), \quad \Xi_i \equiv 1 + \Lambda a_i^2. \tag{9}$$

$\epsilon = 1$ for even dimensions and vanishes for odd dimensions. This expression reduces to the standard limits as $a_i \to 0$ and $\Lambda \to 0$, Consider a given, say that i^{th} component, *i.e.*, the Killing vector is $\xi^{\mu}_{(i)} = (0, ..., 0, 1_i, 0, ..)$. Then the corresponding Killing charge (that is the angular momentum) becomes

$$J_i = \frac{m a_i}{\Xi \Xi_i}. \tag{10}$$

Let us now consider the $D = 3$ BTZ black hole which differs from its higher dimensional counterparts in one very important aspect: for it, AdS is not the correct-vacuum-background [9]. Our formula gives the usual answers $E = M$, $J = a$; . BTZ black hole also solves the topologically massive gravity [10] equations. In that case, its energy and angular momentum get corrections [11]

$$E = M - \frac{\Lambda a}{\mu}, \qquad J = a - \frac{M}{\mu}, \tag{11}$$

5. Negative mass solitons

We can apply our formula to certain negative mass solitons which are allowed in cosmological spacetimes. (Of course these solitons necessarily require a new statement of ‘positive mass“ in gravity).

The *AdS* Soliton

In [5] we computed the mass of the AdS soliton found by Horowitz and Myers [12] and found

$$E = -\frac{V_{D-3}\,\pi}{(D-1)\,\Omega_{D-2}\,G_D} \frac{r_0^{D-2}}{\ell^{D-2}}, \tag{12}$$

where V_{D-3} is the volume of the compact dimensions.

Eguchi-Hanson Solitons Recently [13] quite an interesting soliton solution appeared whose energy with the help of our formula was computed [5] to be

$$E = -\frac{(4\pi)^{(D-1)/2}\, a^{D-1}}{p\,\ell^2\,(D-1)^{(D-1)/2}\,\Omega_{D-2}\,G_D}\,. \tag{13}$$

6. Higher curvature gravity models

As we mentioned before, we can extend our discussion to higher order gravity models which often appear as some sort of low energy 'quantum gravity' models. Here let us consider the following quadratic model

$$I = \int d^D x\sqrt{-g}\{\frac{1}{\kappa}R + \alpha R^2 + \beta R_{\mu\nu}^2 + \gamma(R_{\mu\nu\rho\sigma}^2 - 4R_{\mu\nu}^2 + R^2)\}. \tag{14}$$

In $D = 4$, the Gauss-Bonnet part (that is the γ term) is a surface integral which does not contribute to the conserved charges as well as equations of motion. Here $\kappa = 2\Omega_{D-2}G_D$, where G_D is the D-dimensional Newton's constant [3]. We have not introduced an explicit cosmological constant (which is easy to put), but even in this case, AdS is a solution to the equations of motion. The 'effective' cosmological constant is

$$-\frac{1}{2\Lambda\kappa} = \frac{(D-4)}{(D-2)^2}(D\alpha + \beta) + \gamma\frac{(D-4)(D-3)}{(D-2)(D-1)}\,, \tag{15}$$

For this model, after a lengthy computation [3], one gets the charges as

$$\begin{aligned} Q^\mu(\bar\xi) = \{-\frac{1}{\kappa} + \frac{8\Lambda}{(D-2)^2}(\alpha D + \beta)\} \int d^{D-1}x\sqrt{-\bar g}\,\bar\xi_\nu\, \mathcal{G}_L^{\mu\nu} \\ +(2\alpha+\beta)\int dS_i\sqrt{-\bar g}\{\bar\xi^\mu\bar\nabla^i R_L - \bar\xi^i\bar\nabla^\mu R_L + R_L\bar\nabla^\mu\bar\xi^i\} \\ +\beta\int dS_i\sqrt{-\bar g}\{\bar\xi_\nu\bar\nabla^i\mathcal{G}_L^{\mu\nu} - \bar\xi_\nu\bar\nabla^\mu\mathcal{G}_L^{i\nu} - \mathcal{G}_L^{\mu\nu}\bar\nabla^i\bar\xi_\nu + \mathcal{G}_L^{i\nu}\bar\nabla^\mu\bar\xi_\nu\}. \end{aligned}$$

The last two lines vanish leaving a simple expression which exactly is the charge in cosmological Einstein's theory up to a constant. But observe that this constant may vanish for certain theories;

Our construction is based on the existence of background Killing vectors. For the background, we have chosen either constant curvature (AdS) or flat spacetimes. In our formalism, the background always has zero charge. We would like to stress that this brief review does not present new material but simply collect some of the work we have done recently. The interested reader is referred to the original material cited in our references.

7. Acknowledgments

This work is partially supported by the "Young Investigator Fellowship" of the Turkish Academy of Sciences (TÜBA) and by the TÜBİTAK Kariyer Grant 104T177. The author thanks S. Deser, Ö.Sarıoğlu, H. Cebeci, S.Ölmez and I. Kanik for collaboration on this subject.

References

1. R. Arnowitt, S. Deser and C. Misner, Phys. Rev. **116**, 1322 (1959); **117**, 1595 (1960); in *Gravitation: an introduction to current research*, ed. L. Witten (Wiley, New York, 1962);
2. L. F. Abbott and S. Deser, "Stability Of Gravity With A Cosmological Constant," Nucl. Phys. B **195**, 76 (1982).
3. S. Deser and B. Tekin, "Energy in generic higher curvature gravity theories," Phys. Rev. D **67**, 084009 (2003).
4. S. Deser and B. Tekin, "Gravitational energy in quadratic curvature gravities," Phys. Rev. Lett. **89**, 101101 (2002).
5. H. Cebeci, Ö. Sarıoğlu and B. Tekin, "Negative mass solitons in gravity," Phys. Rev. D **73**, 064020 (2006)
6. S. Ölmez, Ö. Sarıoğlu and B. Tekin, "Mass and angular momentum of asymptotically AdS or flat solutions in the topologically massive gravity," *Class. Quant. Grav.* **22**, 4355 (2005).
7. G. W. Gibbons, H. Lu, D. N. Page and C. N. Pope, "The general Kerr-de Sitter metrics in all dimensions," J. Geom. Phys. **53**, 49 (2005).
8. S. Deser, İ. Kanık and B. Tekin, "Conserved Charges of Higher D Kerr-AdS Spacetimes", Class. Quan. Grav. **22**, 3383 (2005).
9. M. Banados, C. Teitelboim and J. Zanelli, "The Black hole in three-dimensional space-time," Phys. Rev. Lett. **69**, 1849 (1992)
10. S. Deser, R. Jackiw and S. Templeton, "Topologically massive gauge theories," Annals Phys. **140**, 372 (1982)
11. S. Deser and B. Tekin, "Energy in topologically massive gravity," Class. Quant. Grav. **20**, L259 (2003).
12. G. T. Horowitz and R. C. Myers, "The AdS/CFT correspondence and a new positive energy conjecture for general relativity," Phys. Rev. D **59**, 026005 (1999)
13. R. Clarkson and R. B. Mann, *Phys. Rev. Lett.* **96** , 051104 (2006); "Eguchi-Hanson solitons in odd dimensions,"

SECTION 3
HEAVY QUARK PHYSICS

Conveners P. Colangelo, T. Mannel

SYSTEMS OF TWO HEAVY QUARKS WITH EFFECTIVE FIELD THEORIES

N. BRAMBILLA

Dipartimento di Fisica dell'Università di Milano and INFN
Milano, Italy
E-mail: nora.brambilla@mi.infn.it

I discuss results and applications of QCD nonrelativistic effective field theories for systems with two heavy quarks.

Keywords: QCD, Effective Field Theories, Heavy Quarks

1. Introduction

Systems made by to heavy quarks play an important role in several high energy experiments. The diversity, quantity and accuracy of the data currently being collected is impressive and includes: data on quarkonium formation from BES at BEPC, E835 at Fermilab, KEDR (upgraded) at VEPP-4M, and CLEO-III, CLEO-c; clean samples of charmonia produced in B-decays, in photon-photon fusion and in initial state radiation, from the B-meson factory experiments, BaBar at SLAC and Belle at KEK, including the unexpected observation of large amounts of associated $(c\bar{c})(c\bar{c})$ production; the CDF and D0 experiments at Fermilab measuring heavy quarkonia production from gluon-gluon fusion in $p\bar{p}$ annihilations at 2 TeV; the Selex experiment at Fermilab with the preliminary observation of possible candidates for doubly charmed baryons; ZEUS and H1, at DESY, studying charmonia production in photon-gluon fusion; PHENIX and STAR, at RHIC, and NA60, at CERN, studying charmonia production, and suppression, in heavy-ion collisions. In the near future, even larger data samples are expected from the BES-III upgraded experiment, while the B factories and the Fermilab Tevatron will continue to supply valuable data for several years. Later on, new facilities will become operational (LHC at CERN, Panda at GSI, hopefully a Super-B factory, a Linear Collider, etc.) offering fantastic challenges and opportunities in this field. A comprehensive review of the experimental and theoretical status of heavy quarkonium physics may be found in the Cern Yellow Report prepared by the Quarkonium Working Group [1]. See also the talks at the last QWG meeting at BNL (cf. http://www.qwg.to.infn.it/).

On the theory side, systems made by two heavy quarks are a rather unique laboratory. They are characterized by the existence of a hierarchy of energy scales in correspondence of which one can define a hierarchy of nonrelativistic effective field theries (EFT), each EFT has less degrees of freedom left dynamical and is simpler. Some of these physical scales are large and may be treated in perturbation theory. The occurrence of these two facts makes two heavy quark systems accessible in QCD. In particular the factorization of high and low energy scales realized in the EFTs allows us to study low energy QCD effects in a systematic and under control way. Ultimately the simplest EFT defined at the ultrasoft energy, pNRQCD, allows us to study the Quantum Mechanics of a non-Abelian field theory.

2. Scales and EFTs

The description of hadrons containing two heavy quarks is a rather challenging problem, which adds to the complications of the bound state in field theory those coming from a nonperturbative QCD low-energy dynamics. A simplification is provided by the nonrelativistic nature suggested by the large mass of the heavy quarks and manifest in the spectrum pattern. Systems made by two heavy quarks are thus characterized by three energy scales, hierarchically ordered by the quark velocity $v \ll 1$: the mass m (hard scale), the momentum transfer mv (soft scale), which is proportional to the inverse of the typical size of the system r, and the binding energy mv^2 (ultrasoft scale), which is proportional to the inverse of the typical time of the system. In bottomonium $v^2 \sim 0.1$, in charmonium $v^2 \sim 0.3$. In perturbation theory $v \sim \alpha_s$. Feynman diagrams will get contributions from all momentum regions associated with these scales. Since these momentum regions depend on α_s, each Feynman diagram contributes to a given observable with a series in α_s and a non trivial counting. For energy scales close to $\Lambda_{\rm QCD}$, the scale at which nonperturbative effects become dominant, perturbation theory breaks down and one has to rely on nonperturbative methods. Regardless of this, the non-relativistic hierarchy $m \gg mv \gg mv^2$ will persist also below the $\Lambda_{\rm QCD}$ threshold.

The wide span of energy scales involved makes also a lattice calculation in full QCD extremely challenging. We may, however, take advantage of the existence of a hierarchy of scales by substituting QCD with simpler but equivalent EFTs. A hierarchy of EFTs may be constructed by systematically integrating out modes associated to energy scales not relevant for the two quark system. Such integration is made in a matching procedure that enforces the equivalence between QCD and the EFT at a given order of the expansion in v and achieves a factorization between the high energy and the low energy contributions. By integrating out the hard modes one obtains Nonrelativistic QCD. In such EFT soft and ultrasoft scales are left dynamical and still their entanglement complicates calculation and power counting. We will focus here on the simplest EFT you can write down for two heavy quark systems, where only ultrasoft degrees of freedom remain dynamical. This is potential NRQCD [a].

3. pNRQCD

pNRQCD [5–7] is the EFT for two heavy quark systems that follows from NRQCD by integrating out the soft scale. Here the role of the potential and the quantum mechanical nature of the problem are realized in the fact that the Schrödinger equation appears as zero problem for two quark states. We may distinguish two situations: 1) weakly coupled pNRQCD when $mv \gg \Lambda_{\rm QCD}$, where the matching from NRQCD to pNRQCD may be performed in perturbation theory; 2) strongly coupled pNRQCD when $mv \sim \Lambda_{\rm QCD}$, where the matching has to be nonperturbative. Recalling that $r^{-1} \sim mv$, these two situations correspond to systems with inverse typical radius smaller than or of the same order as $\Lambda_{\rm QCD}$.

3.1. *Weakly coupled pNRQCD*

The effective degrees of freedom are: low energy $Q\bar{Q}$ states (that can be decomposed into a singlet and an octet field under colour transformations) with energy of order $\Lambda_{\rm QCD}, mv^2$ and momentum $\mathbf{p}$ of order mv, plus ultrasoft gluons with energy and momentum of order $\Lambda_{\rm QCD}, mv^2$. All the gluon fields are multipole expanded (i.e. expanded in r). The Lagrangian is then given by terms of the type

$$\frac{c_k(m,\mu)}{m^k} \times V_n(r\mu', r\mu) \times O_n(\mu', mv^2, \Lambda_{\rm QCD})\, r^n. \tag{1}$$

where the potential matching coefficients V_n encode the non-analytic behaviour in r. At leading order in the multipole expansion, the singlet sector of the Lagrangian gives rise to equations of motion of the Schrödinger

[a] for an alternative and equivalent EFT (in the case in which $\Lambda_{\rm QCD}$ is the smallest scale) see [4].

type. Each term in the pNRQCD Lagrangian has a definite power counting; retardation (or non-potential) effects start at the NLO in the multipole expansion and are systematically encoded in the theory, they are typically related to nonperturbative effects [6,7]. Resummation of large logs (typically logs of the ratio of energy and momentum scales) can be obtained using the renormalization group (RG) adapted to the case of correlated scales [12]; Poincaré invariance is not lost, but shows up in some exact relations among the matching coefficients [9]. The renormalon subtraction may be implemented systematically. [7]

Applications of weakly coupled pNRQCD

QCD Singlet Static potential. The singlet and octet potentials are well defined matching coefficients to be calculated in the perturbative matching. In [11] a determination of the singlet potential at three loops leading log has been obtained and correspondingly also a determination of α_V showing how this quantity starts to depend on the infrared behaviour of the theory at three loops. The perturbative calculation of the static potential at (almost) three loops and with the RG improvement has been compared to the lattice calculation of the potential and found in good agreement up to about 0.25 fm [17].

b and c masses.

Heavy quarkonium is one of the most suitable system to extract a precise determination of the mass of the heavy quarks b and c. Perturbative determinations of the $\Upsilon(1S)$ and J/ψ masses have been used to extract the b and c masses. The main uncertainty in these determinations comes from nonperturbative contributions (local and nonlocal condensates [6]) together with possible effects due to subleading renormalons [7]. A recent analysis performed by the QWG [1] and based on all the previous determinations indicates an error of about 50 MeV both for the bottom (1% error) and in the charm (4% error) mass.

Perturbative quarkonium spectrum.

B_c mass. Table 1 shows some recent determinations of the B_c mass in perturbation theory at NNLO accuracy compared with a recent lattice study [16] and the value of the CDF experimental B_c mass. This would support the assumption that nonperturbative contributions to the quarkonium ground state are of the same magnitude as NNLO or even NNNLO corrections, which would be consistent with a $mv^2 \gtrsim \Lambda_{\rm QCD}$ power counting. **Hyperfine splittings.** $c\bar{c}$, $b\bar{b}$, B_c ground state hyperfine splittings have been recently calculated at NLL in [19]. The prediction for η_b mass is $M(\eta_b) = 9421 \pm 10\,({\rm th})\,^{+9}_{-8}\,(\delta\alpha_{\rm s})$ MeV. The logs resummation seems to be important. If the experimental error in future measurements of $M(\eta_b)$ will not exceed few Mev, the bottomonium hyperfine separation will become a competitive source of $\alpha_s(M_Z)$ with an estimated accuracy of ± 0.003.

Radiative transitions (M1). A theory of M1 transitions in heavy quarkonium has been recently formulated using pNRQCD [20]. This may shed some light on recent CLEO results on radiative M1 transitions in the η_b search that have ruled out several models. No large anomalous quarkonium magnetic moment is generated.

Seminclusive radiative decays of $\Upsilon(1S)$. In [22] the end-point region of the photon spectrum in semi-inclusive radiative decays of heavy quarkonium has been discussed using Soft-Collinear Effective Theory and pNRQCD. Including the octet contributions a good understanding of the experimental data is obtained.

Gluelump spectrum. In pNRQCD [6,21] the full structure of the gluelump spectrum has been studied, obtaining model independent predictions on the shape, the pattern, the degeneracy and the multiplet structure of the hybrid static energies for small $Q\bar{Q}$ distances that well match and interpret the existing lattice data.

Properties of baryons made of two or three heavy quarks. Recently the SELEX experiment has detected first signals from three-body bound states made of two heavy quarks and a light one. The two heavy quark part of such systems may be treated in pNRQCD and a precise predication for the hyperfine interaction may be obtained [23].

B_c mass (MeV)				
[18] (expt)	[16] (lattice)	[15] (NNLO)	[13] (NNLO)	[14] (NNLO)
$6287 \pm 4.8 \pm 1.1$	$6304 \pm 12^{+12}_{-0}$	6326(29)	6324(22)	6307(17)

3.2. *Strongly coupled pNRQCD*

In this case the matching to pNRQCD is nonperturbative [25]. In the situation where the other degrees of freedom (like those associated with heavy-light meson pair threshold production and heavy hybrids) develop a mass gap of order $\Lambda_{\rm QCD}$, the quarkonium singlet field S remains as the only low energy dynamical degree of freedom in the pNRQCD Lagrangian (if no ultrasoft pions are considered), which reads [25,26,6,7]:

$$L_{\rm pNRQCD} = {\rm Tr}\left\{{\rm S}^\dagger \left(i\partial_0 - \frac{\mathbf{p}^2}{2m} - V_S(r)\right){\rm S}\right\}. \tag{2}$$

In this regime we recover the quark potential singlet model from pNRQCD. The matching potential V_S (static and relativistic corrections) is nonperturbative: the real part controls the spectrum and the imaginary part controls the inclusive decays. The potential is calculated in the nonperturbative matching procedure between NRQCD and pNRQCD [25,7]. Advantages of this approach include: factorization of hard (in the NRQCD matching coefficients) and soft scales (contained in Wilson loops or nonlocal gluon correlators); the low energy objects being only glue dependent, confinement investigations, on the lattice and in QCD vacuum models become feasible [24]; the existence of a power counting indicating leading and subleading terms in quantum-mechanical perturbation theory; the quantum mechanical divergences (like the ones coming from iterations of spin delta potentials) are absorbed by NRQCD matching coefficients. The potentials evaluated on the lattice once used in the Schrödinger equation produce the spectrum. The calculations involve only QCD parameters (at some scale and in some scheme).

Applications of strongly coupled pNRQCD

Nonperturbative potentials and Spectrum. Recently the multilevel algorithm has been applied to the lattice evaluation of field strength insertion inside the Wilson loop average, producing very precise data for the spin dependent potentials and a first evaluation of the nonperturbative potential at order $1/m$ [27]. This is the first step towards a precise determination of the nonperturbative matching potentials on the lattice.

Decays. The inclusive quarkonium decay widths in pNRQCD can be factorized with respect to the wave function (or its derivatives) calculated in zero, which is suggestive of the early potential models results: $\Gamma(\mathrm{H} \to \mathrm{LH}) = F(\alpha_s, \Lambda_{\mathrm{QCD}}) \cdot |\psi(0)|^2$. Similar expressions hold for the electromagnetic decays. However, the coefficient F depends here both on α_s and Λ_{QCD}. In particular all NRQCD matrix elements, including the octet ones, can be expressed through pNRQCD as products of universal nonperturbative factors by the squares of the quarkonium wave functions (or derivatives of it) at the origin. The nonperturbative factors are typically integral of nonlocal electric or magnetic correlators and thus depends on the glue but not on the quarkonium state [26]. Typically F contains both the NRQCD matching coefficients at the hard scale m and the nonperturbative correlators at the low energy scale Λ_{QCD}. The nonperturbative correlators, being state independent, are in a smaller number than the nonperturbative NRQCD matrix elements and thus the predictive power is increased in going from NRQCD to pNRQCD. Thus, having fixed the nonperturbative parameters on charmonium decays, new model-independent QCD predictions can be obtained for the bottomonium decay widths [26].

References

1. N. Brambilla *et al.*, CERN Yellow Report, CERN-2005-005, arXiv:hep-ph/0412158.
2. W. E. Caswell and G. P. Lepage, Phys. Lett. B **167** (1986) 437.
3. G. T. Bodwin, E. Braaten and G. P. Lepage, Phys. Rev. D **51** (1995) 1125
4. A. V. Manohar and I. W. Stewart, Phys. Rev. D **62**, 014033 (2000); A. H. Hoang and I. W. Stewart, Phys. Rev. D **67**, 114020 (2003).
5. A. Pineda and J. Soto, Nucl. Phys. Proc. Suppl. **64**, 428 (1998);
6. N. Brambilla, A. Pineda, J. Soto and A. Vairo, Nucl. Phys. B **566**, 275 (2000).
7. N. Brambilla, A. Pineda, J. Soto and A. Vairo, Rev. Mod. Phys. **77**, 1423 (2005)
8. A. Gray et al. arXiv:hep-lat/0507013.
9. N. Brambilla, D. Gromes and A. Vairo, Phys. Lett. B **576**, 314 (2003); Phys. Rev. D **64**, 076010 (2001);
10. N. Brambilla, A. Pineda, J. Soto and A. Vairo, Phys. Lett. **B470**, 215 (1999); B. A. Kniehl and A. A. Penin, Nucl. Phys. B **563**, 200 (1999).
11. N. Brambilla et al., Phys. Rev. D **60**, 091502 (1999).
12. A. Pineda, Phys. Rev. D **65**, 074007 (2002); A. Pineda and J. Soto, Phys. Lett. B **495**, 323 (2000).
13. N. Brambilla, Y. Sumino and A. Vairo, Phys. Lett. B **513**, 381 (2001).
14. N. Brambilla, Y. Sumino and A. Vairo, Phys. Rev. D **65**, 034001 (2002).
15. N. Brambilla and A. Vairo, Phys. Rev. D **62**, 094019 (2000).
16. I. F. Allison et al. arXiv:hep-lat/0411027.
17. A. Pineda, J. Phys. G **29**, 371 (2003); Y. Sumino, arXiv:hep-ph/0505034.
18. D. Acosta *et al.* [CDF Collaboration], arXiv:hep-ex/0505076.
19. B. A. Kniehl, A. A. Penin, A. Pineda, V. A. Smirnov and M. Steinhauser, Phys. Rev. Lett. **92**, 242001 (2004); A. A. Penin, A. Pineda, V. A. Smirnov and M. Steinhauser, Phys. Lett. B **593**, 124 (2004).
20. N. Brambilla, Y. Jia and A. Vairo, Phys. Rev. D **73**, 054005 (2006)
21. G. S. Bali and A. Pineda, Phys. Rev. D **69**, 094001 (2004).
22. X. Garcia i Tormo and J. Soto, Phys. Rev. D **72**, 054014 (2005). Phys. Rev. Lett. **96**, 111801 (2006)
23. N. Brambilla, A. Vairo and T. Rosch, Phys. Rev. D **72**, 034021 (2005); S. Fleming and T. Mehen, arXiv:hep-ph/0509313.
24. N. Brambilla and A. Vairo, arXiv:hep-ph/9904330;
25. A. Pineda and A. Vairo, Phys. Rev. D **63**, 054007 (2001) N. Brambilla et al. Phys. Rev. D **63**, 014023 (2001); Phys. Lett. B **580**, 60 (2004).
26. N. Brambilla, D. Eiras, A. Pineda, J. Soto and A. Vairo, Phys. Rev. **D67**, 034018 (2003); Phys. Rev. Lett. **88**, 012003 (2002).
27. Y. Koma, M. Koma and H. Wittig, arXiv:hep-lat/0607009. PoS **LAT2005**, 216 (2006)
28. N. Brambilla, P. Consoli and G. M. Prosperi, Phys. Rev. D **50**, 5878 (1994);

CONSTRAINING UNIVERSAL EXTRA DIMENSIONS THROUGH B DECAYS

F. DE FAZIO

Istituto Nazionale di Fisica Nucleare, Sezione di Bari,
Via Orabona 4, I-70126 Bari Italy
E-mail: fulvia.defazio@ba.infn.it

We analyze the exclusive rare $B \to K^{(*)}\ell^+\ell^-$, $B \to K^{(*)}\nu\bar{\nu}$ and $B \to K^*\gamma$ decays in the Applequist-Cheng-Dobrescu model, an extension of the Standard Model in presence of universal extra dimensions. In the case of a single universal extra dimension, we study the dependence of several observables on the compactification parameter $1/R$, and discuss whether the hadronic uncertainty due to the form factors obscures or not such a dependence. We find that, using present data, it is possible in many cases to put a sensible lower bound to $1/R$, the most stringent one coming from $B \to K^*\gamma$.

Keywords: Rare B decays; Universal Extra Dimensions.

1. Introduction

Rare B decays induced by $b \to s$ transition play a peculiar role in searching for new Physics, being induced at loop level and hence suppressed in the Standard Model (SM)[1]. They can also be useful to constrain extra dimensional scenarios[2]. This is the case of the Appelquist-Cheng-Dobrescu (ACD) model[3] in which universal extra dimensions are considered, which means that all the fields are allowed to propagate in all available dimensions. In the case of a single extra dimension compactified on a circle of radius R, Tevatron run I data allow to put the bound $1/R \geq 300$ GeV. To be more general, we analyze a broader range $1/R \geq 200$ GeV.

In Refs. 4, 5 the effective hamiltonian relative to inclusive $b \to s$ decays was computed within the ACD model. In this paper, we summarize the results obtained in Ref. 6 for exclusive $b \to s$-induced modes. In this case, the uncertainty in the form factors must be considered, since it can overshadow the sensitivity to the compactification parameter $1/R$. Indeed we find that computing the branching ratios of $B \to K^{(*)}\ell^+\ell^-$ and the forward-backward lepton asymmetry in $B \to K^*\ell^+\ell^-$ for a representative set of form factors, a bound can be put. We also study the modes $B \to K^{(*)}\nu\bar{\nu}$, for which no signal has been observed, so far, and the $BR(B \to K^*\gamma)$ versus $1/R$, which allows to establish the most stringent bound on $1/R$.

2. The ACD model with a single UED

The ACD model[3] consists in the minimal extension of the SM in $4+\delta$ dimensions; we consider $\delta = 1$. The fifth dimension $x_5 = y$ is compactified to the orbifold S^1/Z_2, i.e. on a circle of radius R and runs from 0 to $2\pi R$ with $y = 0, y = \pi R$ fixed points of the orbifold. Hence a field $F(x, y)$ (x denoting the usual 3+1 coordinates) would be a periodic function of y, and it could be expressed as $F(x,y) = \sum_{n=-\infty}^{n=+\infty} F_n(x) e^{i n \cdot y/R}$. If F is a massless boson field, the KK modes F_n obey the equation $\left(\partial^\mu \partial_\mu + n^2/R^2\right) F_n(x) = 0$, $\mu = 0, 1, 2, 3$ so that, apart the zero mode, they behave in four dimensions as massive particles with $m_n^2 = (n/R)^2$. Under the parity transformation $P_5 : y \to -y$ fields having a correspondent in the 4-d SM should be even, so that their zero mode in the expansion is interpreted as the ordinary SM field. On the other hand, fields having no SM partner should be odd, so that they do not have zero modes.

Important features of the ACD model are: i) there is a single additional free parameter with respect to the SM, the compactification radius R; ii) conservation of KK parity, with the consequence that there is no tree-level contribution of KK modes in low energy processes (at scales $\mu \ll 1/R$) and no production of single KK excitation in ordinary particle interactions. A detailed description of this model is provided in Ref. 4.

3. Decays $B \to K^{(*)}\ell^+\ell^-$

In the Standard Model the effective $\Delta B = -1$, $\Delta S = 1$ Hamiltonian governing the transition $b \to s\ell^+\ell^-$ is $H_W = 4\frac{G_F}{\sqrt{2}} V_{tb} V_{ts}^* \sum_{i=1}^{10} C_i(\mu) O_i(\mu)$. G_F is the Fermi constant and V_{ij} are elements of the Cabibbo-Kobayashi-Maskawa mixing matrix; we neglect terms proportional to $V_{ub}V_{us}^*$. O_1, O_2 are current-current operators, $O_3, ..., O_6$ QCD penguins, O_7, O_8 magnetic penguins, O_9, O_{10} semileptonic electroweak penguins. We do not consider the contribution to $B \to K^{(*)}\ell^+\ell^-$ with the lepton pair coming from $c\bar{c}$ resonances, mainly due to O_1, O_2. We also neglect QCD penguins whose coefficients are very small compared to the others. Therefore, in the case of the modes $B \to K^{(*)}\ell^+\ell^-$, the relevant operators are: $O_7 = \frac{e}{16\pi^2} m_b (\bar{s}_{L\alpha} \sigma^{\mu\nu} b_{R\alpha}) F_{\mu\nu}$, $O_9 = \frac{e^2}{16\pi^2} (\bar{s}_{L\alpha}\gamma^\mu b_{L\alpha})\ \bar{\ell}\gamma_\mu \ell$, $O_{10} =$

$\frac{e^2}{16\pi^2}(\bar{s}_{L\alpha}\gamma^\mu b_{L\alpha})\ \bar{\ell}\gamma_\mu\gamma_5\ell$. Their coefficients have been computed at NNLO in the Standard Model[7] and at NLO for the ACD model[4,5]: we use these results in our study. No new operators are found in ACD, while the coefficients are modified because particles not present in SM can contribute as intermediate states in loop diagrams. As a consequence, they are expressed in terms of functions $F(x_t, 1/R)$, $x_t = m_t^2/M_W^2$, generalizing the corresponding SM functions $F_0(x_t)$ according to $F(x_t, 1/R) = F_0(x_t) + \sum_{n=1}^{\infty} F_n(x_t, x_n)$, where $x_n = m_n^2/M_W^2$ and $m_n = n/R$[4,5]. For large values of $1/R$ the SM phenomenology should be recovered, since the new states, being more and more massive, decouple from the low-energy theory.

The exclusive $B \to K^{(*)}\ell^+\ell^-$ modes involve the matrix elements of the operators in the effective hamiltonian between the B and K or K^* mesons, for which we use the standard parametrization in terms of form factors:

$$< K(p')|\bar{s}\gamma_\mu b|B(p) >= (p+p')_\mu F_1(q^2) + \frac{M_B^2 - M_K^2}{q^2} q_\mu \left(F_0(q^2) - F_1(q^2)\right) \ ;$$

$$< K(p')|\bar{s}\ i\ \sigma_{\mu\nu}q^\nu b|B(p) >= \Big[(p+p')_\mu q^2 - (M_B^2 - M_K^2)q_\mu\Big] \frac{F_T(q^2)}{M_B + M_K} \ ;$$

$$< K^*(p',\epsilon)|\bar{s}\gamma_\mu(1-\gamma_5)b|B(p) >= \epsilon_{\mu\nu\alpha\beta}\epsilon^{*\nu}p^\alpha p'^\beta \frac{2V(q^2)}{M_B + M_{K^*}}$$
$$- i\Big[\epsilon^*_\mu(M_B + M_{K^*})A_1(q^2) - (\epsilon^*\cdot q)(p+p')_\mu \frac{A_2(q^2)}{(M_B + M_{K^*})}$$
$$- (\epsilon^*\cdot q)\frac{2M_{K^*}}{q^2}\left(A_3(q^2) - A_0(q^2)\right)q_\mu\Big] \ ;$$

$$< K^*(p',\epsilon)|\bar{s}\sigma_{\mu\nu}q^\nu \frac{(1+\gamma_5)}{2} b|B(p) >= i\epsilon_{\mu\nu\alpha\beta}\epsilon^{*\nu}p^\alpha p'^\beta\, 2\, T_1(q^2) +$$
$$+ \Big[\epsilon^*_\mu(M_B^2 - M_{K^*}^2) - (\epsilon^*\cdot q)(p+p')_\mu\Big]\ T_2(q^2)$$
$$+ (\epsilon^*\cdot q)\left[q_\mu - \frac{q^2}{M_B^2 - M_{K^*}^2}(p+p')_\mu\right]\ T_3(q^2)\ ,$$

where $q = p - p'$, $A_3(q^2) = \frac{M_B + M_{K^*}}{2M_{K^*}}A_1(q^2) - \frac{M_B - M_{K^*}}{2M_{K^*}}A_2(q^2)$ with the conditions $F_1(0) = F_2(0)$, $A_3(0) = A_0(0)$, $T_1(0) = T_2(0)$.

We use two sets of form factors: the first one (set A) obtained by three-point QCD sum rules based on the short-distance expansion[8]; the second one (set B) obtained by light-cone QCD sum rules[9]. For both sets we include in the numerical analysis the errors on the parameters.

In Fig. 1 we plot, for the two sets of form factors, the branching fractions relative to $B \to K^{(*)}\ell^+\ell^-$ versus $1/R$ and compare them with the experimental data provided by BaBar[10,11] and Belle[12,13]:

$$\begin{aligned} BR(B \to K\ell^+\ell^-) &= (5.50 \pm^{0.75}_{0.70} \pm 0.27 \pm 0.02) \times 10^{-7}\ (Belle) \\ &= (3.4 \pm 0.7 \pm 0.3) \times 10^{-7}\ (BaBar) \\ BR(B \to K^*\ell^+\ell^-) &= (16.5 \pm^{2.3}_{2.2} \pm 0.9 \pm 0.4) \times 10^{-7}\ (Belle) \\ &= (7.8 \pm^{1.9}_{1.7} \pm 1.2) \times 10^{-7}\ (BaBar). \end{aligned} \quad (1)$$

Set B excludes $1/R \leq 200$ GeV. Improved data will resolve the discrepancy between the experiments and increase the lower bound for $1/R$.

In the case of $B \to K^*\ell^+\ell^-$ the investigation of the forward-backward asymmetry $\mathcal{A}_{fb}$ in the dilepton angular distribution may also reveal effects beyond the SM. In particular, in SM, due to the opposite sign of the coefficients C_7 and C_9, $\mathcal{A}_{fb}$ has a zero the position of which is almost independent of the model for the form factors[14]. Let θ_ℓ be the angle between the ℓ^+ direction and the B direction in the rest frame of the lepton pair (we consider massless leptons). We define:

$$\mathcal{A}_{fb}(q^2) = \frac{\displaystyle\int_0^1 \frac{d^2\Gamma}{dq^2 dcos\theta_\ell} dcos\theta_\ell - \int_{-1}^0 \frac{d^2\Gamma}{dq^2 dcos\theta_\ell} dcos\theta_\ell}{\displaystyle\int_0^1 \frac{d^2\Gamma}{dq^2 dcos\theta_\ell} dcos\theta_\ell + \int_{-1}^0 \frac{d^2\Gamma}{dq^2 dcos\theta_\ell} dcos\theta_\ell} . \quad (2)$$

We show in Fig. 2 the predictions for the SM, $1/R = 250$ GeV and $1/R = 200$ GeV. The zero of $\mathcal{A}_{fb}$ is sensitive to the compactification parameter, so that its experimental determination would constrain $1/R$. At present, the analysis performed by Belle Collaboration indicates that the relative sign of C_9 and C_7 is negative, confirming that $\mathcal{A}_{fb}$ should have a zero[15].

4. The decays $B \to K^{(*)}\nu\bar{\nu}$

In the SM the effective Hamiltonian governing $b \to s\nu\bar{\nu}$ induced decays is

$$\mathcal{H}_{eff} = \frac{G_F}{\sqrt{2}} \frac{\alpha}{2\pi \sin^2(\theta_W)} V_{ts}V_{tb}^*\, \eta_X X(x_t)\, \bar{b}\gamma^\mu(1-\gamma_5)s\, \bar{\nu}\gamma_\mu(1-\gamma_5)\nu \quad (3)$$

obtained from Z^0 penguin and box diagrams dominated by the intermediate top quark. In (3) θ_W is the Weinberg angle. We put to unity the QCD factor η_X[17–19].The function X was computed in Refs. 16, 17, 18 in the SM and in the ACD model in Refs. 4, 5.

$B \to K^{(*)}\nu\bar{\nu}$ decays have been studied within the SM[20,21], while in Ref. 6 the dependence of $BR(B \to K\nu\bar{\nu})$ and $BR(B \to K^*\nu\bar{\nu})$ on $1/R$

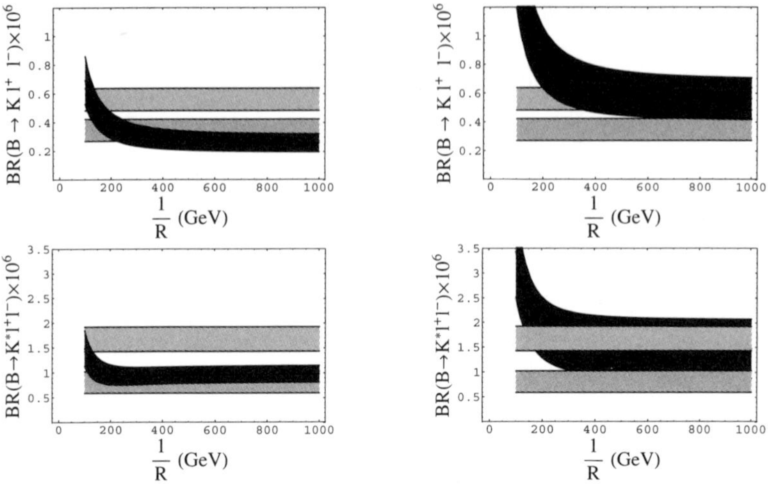

Fig. 1. $BR(B \to K\ell^+\ell^-)$ (upper figures) and $BR(B \to K^*\ell^+\ell^-)$ (lower figures) versus $1/R$ using set A (left) and B (right) of form factors. The two horizontal regions refer to BaBar (lower band) and Belle (upper band) data.

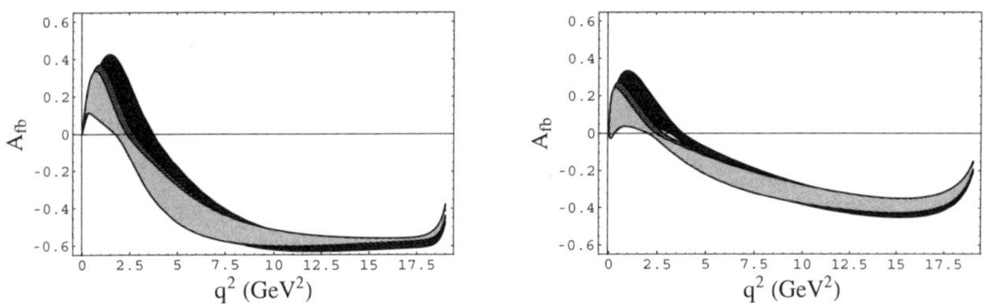

Fig. 2. Forward-backward lepton asymmetry in $B \to K^*\ell^+\ell^-$ versus $1/R$ using set A (left) and B (right). The dark (blue) bands correspond to the SM results, the intermediate (red) band to $1/R = 250$ GeV, the light (yellow) one to $1/R = 200$ GeV.

has been derived. However, only an experimental upper bound exists for $B \to K\nu\bar{\nu}$: $BR(B^- \to K^-\nu\bar{\nu}) < 3.6 \times 10^{-5}$ (90% CL) [22], $BR(B^- \to K^-\nu\bar{\nu}) < 5.2 \times 10^{-5}$ (90% CL)[23], furthermore the $1/R$ dependence turns out to be too mild for distinguishing values above $1/R \geq 200$ GeV.

5. The decay $B \to K^*\gamma$

The transition $b \to s\gamma$ is described by the operator O_7. The most recent measurements for the exclusive branching fractions are[24,25]:

$$
\begin{aligned}
BR(B^0 \to K^{*0}\gamma) &= (4.01 \pm 0.21 \pm 0.17) \times 10^{-5} \;\; (Belle) \\
&= (3.92 \pm 0.20 \pm 0.24) \times 10^{-5} \;\; (BaBar) \\
BR(B^- \to K^{*-}\gamma) &= (4.25 \pm 0.31 \pm 0.24) \times 10^{-5} \;\; (Belle) \\
&= (3.87 \pm 0.28 \pm 0.26) \times 10^{-5} \;\; (BaBar)
\end{aligned}
$$

In Fig. 3 the branching ratio computed in the ACD model is plotted versus $1/R$: the sensitivity to the this parameter is evident; a lower bound of $1/R \geq 250$ GeV can be put adopting set A, and a stronger bound of $1/R \geq 400$ GeV using set B, which is the most stringent bound that can be currently put on this parameter from the B decay modes we have considered.

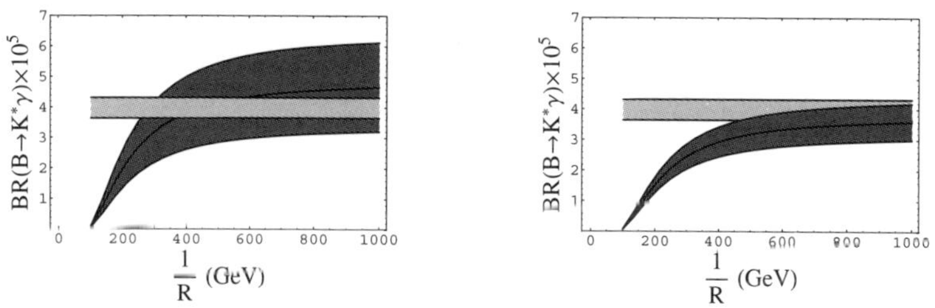

Fig. 3. $BR(B \to K^*\gamma)$ versus $1/R$ using set A (left) and B (right) of form factors . The horizontal band corresponds to the experimental result.

6. Conclusions and perspectives

We have shown how the predictions for $B \to K^{(*)}\ell^+\ell^-$, $B \to K^{(*)}\nu\bar{\nu}$, $B \to K^*\gamma$ decays are modified within the ACD scenario. The constraints on $1/R$ are slightly model dependent, being different using different sets of form factors. Nevertheless, various distributions, together with the lepton forward-backward asymmetry in $B \to K^*\ell^+\ell^-$ are very promising in order to constrain $1/R$, the most stringent lower bound coming from $B \to K^*\gamma$. Improvements in the experimental data, expected in the near future, will allow to establish more stringent constraints for the compactification radius.

Acknowledgments

I warmly thank the organizers of the workshop for their kind hospitality. I am grateful to P. Colangelo, R. Ferrandes and T.N. Pham for collaboration on the analyses discussed above and I acknowledge partial support from the EC Contract No. HPRN-CT-2002-00311 (EURIDICE).

References

1. For a review see T. Hurth, Rev. Mod. Phys. **75**, 1159 (2003) and Refs. therein.
2. K. Agashe, N. G. Deshpande and G. H. Wu, Phys. Lett. B **514**, 309 (2001).
3. T. Appelquist, H. C. Cheng and B. A. Dobrescu, Phys. Rev. D **64**, 035002 (2001).
4. A. J. Buras, M. Spranger and A. Weiler, Nucl. Phys. B **660**, 225 (2003).
5. A. J. Buras *et al.*, Nucl. Phys. B **678**, 455 (2004).
6. P. Colangelo, F. De Fazio, R. Ferrandes and T. N. Pham, Phys. Rev. D **73**, 115006 (2006).
7. C. Bobeth *et al.*, Nucl. Phys. B **574**, 291 (2000); H. H. Asatrian *et al.*, Phys. Lett. B **507**, 162 (2001); Phys. Rev. D **65**, 074004 (2002); Phys. Rev. D **66**, 034009 (2002); H. M. Asatrian *et al.*, Phys. Rev. D **66**, 094013 (2002); A. Ghinculov *et al.*, Nucl. Phys. B **648**, 254 (2003); A. Ghinculov *et al.*, Nucl. Phys. B **685**, 351 (2004); C. Bobeth *et al.*, JHEP **0404**, 071 (2004).
8. P. Colangelo *et al.*, Phys. Rev. D **53**, 3672 (1996) [Erratum-ibid. D **57**, 3186 (1998)].
9. P. Ball and R. Zwicky, Phys. Rev. D **71**, 014015 (2005); Phys. Rev. D **71**, 014029 (2005).
10. B. Aubert *et al.* [BaBar Collaboration], Phys. Rev. Lett. **93**, 081802 (2004).
11. B. Aubert *et al.* [BaBar Collaboration], arXiv:hep-ex/0507005.
12. M. Iwasaki *et al.* [Belle Collaboration], Phys. Rev. D **72**, 092005 (2005).
13. K. Abe *et al.* [Belle Collaboration], arXiv:hep-ex/0410006.
14. G. Burdman, Phys. Rev. D **57**, 4254 (1998).
15. A. Ishikawa *et al.*, Phys. Rev. Lett. **96**, 251801 (2006) [arXiv:hep-ex/0603018].
16. T. Inami and C. S. Lim, Prog. Theor. Phys. **65**, 297 (1981) [Erratum-ibid. **65**, 1772 (1981)].
17. G. Buchalla and A. J. Buras, Nucl. Phys. B **400**, 225 (1993).
18. G. Buchalla *et al.*, Rev. Mod. Phys. **68**, 1125 (1996).
19. G. Buchalla and A. J. Buras, Nucl. Phys. B **548**, 309 (1999).
20. P. Colangelo *et al.*, Phys. Lett. B **395**, 339 (1997).
21. G. Buchalla, G. Hiller and G. Isidori, Phys. Rev. D **63**, 014015 (2001).
22. K. Abe *et al.* [Belle Collaboration], arXiv:hep-ex/0507034.
23. B. Aubert *et al.* [BaBar Collaboration], Phys. Rev. Lett. **94**, 101801 (2005).
24. M. Nakao *et al.* [Belle Collaboration], Phys. Rev. D **69**, 112001 (2004).
25. B. Aubert *et al.* [BaBar Collaboration], Phys. Rev. D **70**, 112006 (2004).

LIGHT-CONE SUM RULES WITH B-MESON DISTRIBUTION AMPLITUDES

ALEXANDER KHODJAMIRIAN

Theoretische Physik 1, Fachbereich Physik, Universität Siegen, D-57068 Siegen, Germany
E-mail: khodjam@hep.physik.uni-siegen.de

I report on the work done in collaboration with Thomas Mannel and Nils Offen. We obtain new light-cone sum rules for heavy-to-light hadronic form factors using the correlation functions with B-meson distribution amplitudes .

1. Introduction

The method of light-cone sum rules (LCSR) suggested in [1] is well designed for calculating various hadronic form factors in QCD. The most important applications of this method to B physics are [2-5] the heavy-to-light form factors at small momentum transfer (at large recoil of the light hadron), the region which is currently not accessible to lattice QCD.

The standard LCSR derivation begins with the correlation function, e.g., for the phenomenologically interesting $B \to \pi$ form factor one defines the following vacuum-to-pion amplitude:

$$F_\lambda(q,p) = i \int d^4x e^{iqx} \langle \pi(p) \mid T\{\bar{u}(x)\gamma_\lambda b(x), \bar{b}(0) i\gamma_5 d(0)\} \mid 0\rangle \qquad (1)$$

with an on-shell pion state ($p^2 = m_\pi^2$), and with the B meson interpolated by a heavy-light quark current with a large virtuality $(p+q)^2 \ll m_b^2$. The operator product expansion (OPE) near the light-cone, $x^2 \sim 0$, is accessible at the momentum transfer squared $q^2 \ll m_b^2$, far from the threshold. The long-distance dynamics in the correlation function is described by a set of pion, (or kaon, ρ,K^*, etc.) meson distribution amplitudes (DA's) of low twists. The OPE result for (1) which is currently known with $O(\alpha_s)$ and twist-four accuracy, is combined with the hadronic dispersion relation and quark-hadron duality, hence the LCSR method has many common features with the original QCD sum rules [6]. The main uncertainties in LCSR for

$f^+_{B\pi}(q^2)$ and for other heavy-to-light form factors originate from the limited accuracy of the DA parameters. An additional "systematic" uncertainty is due to the quark-hadron duality approximation in the B-meson channel. Hence, in order to assess the accuracy of LCSR , it is desirable to calculate the form factors in some independent way, using different input and assumptions.

2. LCSR with B-meson distribution amplitudes

A different method to obtain LCSR for the $B \to \pi$ form factor was recently suggested by us in [7] and, independently, in the framework of SCET in [8]. In this approach the correlation function is "inverted " with respect to (1), that is, now the pion is interpolated with an appropriate light-quark current, while B meson is on-shell:

$$F^{(B)}_{\mu\nu}(q,p) = i \int d^4x \; e^{ip\cdot x} \langle 0|T\left\{\bar{d}(x)\gamma_\mu\gamma_5 u(x), \bar{u}(0)\gamma_\nu b(0)\right\}|\bar{B}(v)\rangle\,, \qquad (2)$$

where $\bar{u}\gamma_\nu b$ is the heavy-to-light weak current and $\bar{d}\gamma_\mu\gamma_5 u$ is the interpolating axial-vector current for the pion. The external momenta of the currents are q and p respectively; $v = (p+q)/m_B$ is the velocity of the B meson, $(p+q)^2 = m_B^2$. Changing the flavour content and/or spin-parity of the light-quark current in (2) one can easily access other light mesons and form factors (e.g. $B \to K, \rho, \omega, K^*$, etc).

Transforming the correlation function to HQET, it is possible to demonstrate [9], using the same line of arguments as for the product of two light-quark currents (see e.g., [10]), that at $q^2 \ll m_b^2$, and $p^2 < 0$, $|p^2| \gg \Lambda^2_{QCD}$, the u-quark in (2) propagates near the light-cone, $x^2 \leq 1/|p^2|$. This allows one to use the light-cone expansion of the quark propagator. In the approximation adopted here, we include the free-quark propagator and the one-gluon emission part. The OPE result for the correlation function (2) has a schematical form (omitting Lorentz indices):

$$\begin{aligned} F^{(B)}(q,p) &= i \int d^4x e^{iqx} \Bigg\{ \left[S_0(x^2,\mu) + \alpha_s S_1(x^2,\mu)\right] \otimes \langle 0 \mid \bar{d}(x)\Gamma b(0) \mid \bar{B}(v)\rangle|_\mu \\ &+ \int_0^1 du\; \tilde{S}(x^2, m_b^2, \mu, u) \otimes \langle 0 \mid \bar{d}(x)G(ux)\tilde{\Gamma} b(0)\} \mid \bar{B}(v)\rangle|_\mu \Bigg\}, \qquad (3) \end{aligned}$$

where S_0 , S_1, $\tilde{S}$ are calculable hard-scattering amplitudes convoluted with the long-distance matrix elements at a large separation scale μ . In the above expansion, S_0 and $\tilde{S}$ represent the quark-antiquark and quark-antiquark-gluon parts of the correlation function, respectively. The $O(\alpha_s)$ correction indicated by S_1 is neglected here.

The universal matrix elements of nonlocal operators on the light-cone entering the expansion (3) are expressed via DA's of B meson introduced [11] in the framework of HQET. The two-particle matrix element

$$\langle 0|\bar{d}_\alpha(x)[x,0]b_\beta(0)|\bar{B}(v)\rangle$$
$$= -\frac{if_B m_B}{4}\int\limits_0^\infty d\omega e^{-i\omega v\cdot x}\left[(1+\not{v})\left\{\phi_+^B(\omega)-\frac{\phi_+^B(\omega)-\phi_-^B(\omega)}{2v\cdot x}\not{x}\right\}\gamma_5\right]_{\beta\alpha} \quad (4)$$

(where $[x,0]$ is the path-ordered gauge factor) contains two DA's $\phi_\pm^B(\omega)$. The variable ω is the plus component of the light spectator-quark momentum in B. The same DA's are also used in factorization formulae for exclusive B-decays (see e.g., [12,13]).

For the three-particle DA's we employ the definition [14]:

$$\langle 0|\bar{d}_\alpha(x)[x,ux])G_{\lambda\rho}(ux)[ux,0]b_\beta(0)|\bar{B}(v)\rangle = \frac{f_B m_B}{4}\int\limits_0^\infty d\omega\int\limits_0^\infty d\xi\, e^{-i(\omega+u\xi)v\cdot x}$$
$$\times\left[(1+\not{v})\left\{(v_\lambda\gamma_\rho-v_\rho\gamma_\lambda)\Big(\Psi_A(\omega,\xi)-\Psi_V(\omega,\xi)\Big)-i\sigma_{\lambda\rho}\Psi_V(\omega,\xi)\right.\right.$$
$$\left.\left.-\left(\frac{x_\lambda v_\rho-x_\rho v_\lambda}{v\cdot x}\right)X_A(\omega,\xi)+\left(\frac{x_\lambda\gamma_\rho-x_\rho\gamma_\lambda}{v\cdot x}\right)Y_A(\omega,\xi)\right\}\right]_{\beta\alpha} \quad (5)$$

To obtain the sum rule the correlation function (2) is represented in a form of hadronic dispersion relation in the pion channel. Isolating the relevant invariant amplitude: $F_{\mu\nu}^{(B)} = p_\mu p_\nu F^{(B)}(p^2,q^2)+...$, we obtain

$$F^{(B)}(p^2,q^2) = \frac{2if_\pi f_{B\pi}^+(q^2)}{m_\pi^2-p^2}+\int\limits_{s_h}^\infty ds\,\frac{\rho^h(s,q^2)}{s-p^2}\,, \quad (6)$$

where the residue of the ground-state pion pole term contains the product of the pion decay constant and the $B\to\pi$ form factor defined in the standard way. The contributions of excited and continuum states with the pion quantum numbers are encoded by the spectral density ρ^h. The integral over ρ^h is, as usual, estimated with the help of (semi-local) quark-hadron duality.

One calculates the imaginary part of the OPE result (3), approximating:

$$\int_{s_h}^{\infty} ds\, \frac{\rho^h(s,q^2)}{s-p^2} = \int_{s_0^\pi}^{\infty} ds \frac{\mathrm{Im}F^{(B)}(s,q^2)}{s-p^2}\,. \tag{7}$$

The channel-dependent effective threshold s_0^π is determined from the two-point QCD sum rule [6] for f_π, similarly s_0^ρ in the LCSR for $B \to \rho$ form factors is fixed from the sum rule for f_ρ. Furthermore, calculating the $B \to K, K^*$ form factors we take into account $SU(3)$-flavour violation effects in terms of the s-quark mass and the differences between $s_0^{K(K^*)}$ and $s_0^{\pi(\rho)}$.

The OPE result (3), where the B meson DA's are substituted, is equated to the dispersion relation (6). After subtracting the contribution of excited and continuum states given by (7), one finally performs the Borel transformation in the variable $p^2 \to M^2$.

Let me present the resulting LCSR [7,8] for $f^+_{B\pi}(q^2)$, where the part containing two-particle DA's has a simple expression at $q^2 = 0$ ($m_\pi = 0$) depending only on $\phi_-^B(\omega)$:

$$f^+_{B\pi}(0) = \frac{f_B}{f_\pi\, m_B} \int_0^{s_0^\pi} ds\, e^{-s/M^2} \phi_-^B(s/m_B) + ... \tag{8}$$

Another example is the new LCSR for the $B \to \rho$ form factor $V(q^2)$ (of the $\bar{u}\gamma_\nu b$ current), where only $\phi_+^B(\omega)$ contributes:

$$V^{B\to\rho}(0) = \frac{f_B(m_B + m_\rho)}{2\, f_\rho\, m_B}\, e^{m_\rho^2/M^2} \int_0^{s_0^\rho} ds\, e^{-s/M^2} \phi_+^B(s/m_B) + ... \tag{9}$$

The three-particle DA corrections indicated by the ellipses in the above equations are suppressed by powers of $M^2 \sim 1$ GeV^2, the characteristic separation scale in these sum rules. They have more complicated expressions [9] in terms of four DA's defined in (5).

LCSR for all other $B \to P, V$ form factors ($P = \pi, K$, $V = \rho, K^*$) have been obtained [9] in the same way. The set of B-meson two- and three-particle DA's represent a uniform input for all these LCSR.

3. Modelling B-meson DA's

In [11,15] QCD sum rules in HQET were used to predict the shape and normalization parameters of the two-particle DA's $\phi_\pm^B(\omega)$. The idea is to use a special correlation function with two $\bar{q}\Gamma h_v$ currents, where h_v is the effective heavy-quark field. One current is local, and in the other one the

h_v and light-antiquark fields have a light-cone separation. The ground B-state contribution to this correlation function contains the product of the B decay constant and the nonlocal heavy-to-light matrix element (4). An appropriate choice of the Dirac-structure Γ allows one to isolate a certain DA. Important is that the leading-order perturbative contribution to this correlation function described by a simple loop diagram yields the dependence on ω. Matching the loop contribution to the B-meson term via quark-hadron duality, one reproduces the ω-behaviour of the DA , e.g. , $\phi_+^B(\omega) \sim \omega$ and $\phi_-^B(\omega) \sim const$ at $\omega \to 0$. We use the same method to reproduce the dependence of three-particle DA's on the two variables ω and ξ, the plus components of the light-quark and gluon momenta, respectively. The correlation function used in this case contains two quark-antiquark-gluon currents $\bar{q}G\Gamma h_v$. The perturbative loop approximation for this correlation function with an appropriate choice of Γ allows one to "read-off" the shape of different DA's at small ω, ξ. Combining this shape with an exponential fall-off at $\omega, \xi \to \infty$ we obtain the following model DA's:

$$\Psi_A(\omega,\xi) = \Psi_V(\omega,\xi) = \frac{\lambda_E^2}{6\omega_0^4}\,\xi^2 e^{-(\omega+\xi)/\omega_0}\,,$$

$$X_A(\omega,\xi) = \frac{\lambda_E^2}{6\omega_0^4}\,\xi(2\omega - \xi)\, e^{-(\omega+\xi)/\omega_0}\,,$$

$$Y_A(\omega,\xi) = -\frac{\lambda_E^2}{24\omega_0^4}\,\xi(7\omega_0 - 13\omega + 3\xi) e^{-(\omega+\xi)/\omega_0}\,. \tag{10}$$

They have to be combined with the two-particle DA's suggested in [11]:

$$\phi_+^B(\omega) = \frac{\omega}{\omega_0^2}\, e^{-\frac{\omega}{\omega_0}}\,,\quad \phi_-^B(\omega) = \frac{1}{\omega_0}\, e^{-\frac{\omega}{\omega_0}}\,. \tag{11}$$

The most important parameter is the inverse moment $\lambda_B = \int_0^\infty \phi_+^B(\omega)/\omega$. In the exponential model $\lambda_B = \omega_0$ and $\phi_-^B(0) = 1/\lambda_B$. Furthermore, we have checked that the set of DA's (11) and (10) satisfy the constraints derived from QCD equations of motion [11,12,14] if

$$\omega_0 = 2/3\bar{\Lambda},\quad \lambda_E^2 = 3/2\omega_0^2 = 2/3\bar{\Lambda}^2\,, \tag{12}$$

where $\bar{\Lambda} = m_B - m_b$ is the usual HQET parameter. The HQET sum rule calculation [15] in NLO predicts

$$\lambda_B = 460 \pm 110 \text{ MeV}\,, \tag{13}$$

yielding $\bar{\Lambda} = 690 \pm 165$ MeV or $m_b = 4.60 \pm 0.16$ GeV, within the usual interval of the b-quark pole mass. Also the estimate of λ_E^2 obtained in [11] is not far from the interval following from (12) and (13), indicating that our

form factor	this work	LCSR with light-meson DA's	ref.
$f^+_{B\pi}(0)$	0.267	0.258± 0.03	4
$f^+_{BK}(0)$	0.331	0.301± 0.041	4
$f^T_{B\pi}(0)$	0.228	0.253± 0.028	4
$f^T_{BK}(0)$	0.290	0.328± 0.04	4
$V^{B\rho}(0)$	0.339	0.323± 0.029	5
$V^{BK^*}(0)$	0.395	0.411± 0.033	5
$A_1^{B\rho}(0)$	0.253	0.242± 0.024	5
$A_1^{BK^*}(0)$	0.299	0.292± 0.028	5
$A_2^{B\rho}(0)$	0.211	0.221±0.023	5
$A_2^{BK^*}(0)$	0.257	0.259 ± 0.027	5
$T_1^{B\rho}(0)$	0.288	0.267 ± 0.021	5
$T_1^{BK^*}(0)$	0.338	0.333±0.028	5

model represents a reasonable and self-consistent approximation. Further improvements of three-particle DA's demand going beyond the perturbative loop approximation in HQET sum rules, as it was done in [15] for $\phi_+^B(\omega)$.

4. Estimates of form factors from the new sum rules

With the B-meson DA's determined by (10) -(13) we calculate various $B \to P, V$ form factors, taking $f_B = 180 \pm 30$ MeV (from two-point sum rules with $O(\alpha_s)$ accuracy). For light-meson channels we use the experimentally measured decay constants $f_\pi = 131$ MeV, $f_K = 160$ MeV, $f_\rho = 209$ MeV, $f_K^* = 217$ MeV and the duality thresholds $s_0^\pi = 0.7$ GeV^2, $s_0^K = 1.05$ GeV^2, $s_0^\rho = 1.6$ GeV^2, $s_0^{K^*} = 1.7$ GeV^2, determined from the respective two-point sum rules, as well as the interval $m_s(2 \text{ GeV}) = 105 \pm 10$ MeV. The appropriate range of Borel parameter $M^2 = 1.0 \pm 0.5$ GeV^2 controls the power suppressed corrections, in particular the magnitude of three-particle corrections in LCSR is at the level of 10-15%. The numerical results for the form factors at $q^2 = 0$ calculated at the central values of the input parameters are collected in the Table, compared with the results of "standard" LCSR with light-meson DA's. The details of the numerical analysis and related uncertainties, as well as the predictions at $q^2 > 0$ will be presented elsewhere [9]. Let me only mention that LCSR form factors are most sensitive to the choice of λ_B. In particular, the agreement between the predictions of the two types of LCSR displayed in the Table is lost if one shifts the inverse moment beyond the interval (13).

Concluding, we obtained a set of new LCSR relating various $B \to P, V$

transition form factors with the universal two- and three-particle DA's of B-meson. For the three-particle DA's we suggest a realistic model based on HQET sum rules. The fact that new LCSR agree with the form factors obtained from the "standard" LCSR with light-meson DA's is an important cross-check of both methods which use completely different inputs. Further improvements are possible in the new LCSR, including more accurate models of B meson DA's, radiative $O(\alpha_s)$ corrections (taking into account the nontrivial renormalization effects [16]) and elaborating on the issue of quark-hadron duality.

This work is supported by the Deutsche Forschungsgemeinschaft under the contract DFG KH205/1-1.

References

1. I. I. Balitsky, V. M. Braun and A. V. Kolesnichenko, Nucl. Phys. **B312** (1989) 509; V. M. Braun and I. E. Halperin, Z. Phys. C **44** (1989) 157; V. L. Chernyak and I. R. Zhitnitsky, Nucl. Phys. **B345** (1990) 137.
2. V. M. Belyaev, A. Khodjamirian and R. Ruckl, Z. Phys. C **60** (1993) 349; V. M. Belyaev, V. M. Braun, A. Khodjamirian and R. Ruckl, Phys. Rev. D **51** (1995) 6177; A. Khodjamirian, R. Ruckl, S. Weinzierl and O. I. Yakovlev, Phys. Lett. B **410** (1997) 275; E. Bagan, P. Ball and V. M. Braun, Phys. Lett. B **417** (1998) 154; P. Ball and R. Zwicky, JHEP **0110** (2001) 019.
3. P. Ball and V. M. Braun, Phys. Rev. D **55** (1997) 5561; Phys. Rev. D **58** (1998) 094016.
4. P. Ball and R. Zwicky, Phys. Rev. D **71** (2005) 014015.
5. P. Ball and R. Zwicky, Phys. Rev. D **71** (2005) 014029.
6. M. A. Shifman, A. I. Vainshtein and V. I. Zakharov, Nucl. Phys. B **147**, 385, 448 (1979).
7. A. Khodjamirian, T. Mannel and N. Offen, Phys. Lett. B **620** (2005) 52
8. F. De Fazio, T. Feldmann and T. Hurth, Nucl. Phys. B **733**, 1 (2006)
9. A. Khodjamirian, T. Mannel and N. Offen, preprint SI-HEP-2006-03, in preparation.
10. P. Colangelo and A. Khodjamirian, arXiv:hep-ph/0010175.
11. A. G. Grozin and M. Neubert, Phys. Rev. D **55** (1997) 272.
12. M. Beneke and T. Feldmann, Nucl. Phys. B **592** (2001) 3.
13. M. Beneke, G. Buchalla, M. Neubert and C. T. Sachrajda, Nucl. Phys. B **591** (2000) 313.
14. H. Kawamura, J. Kodaira, C. F. Qiao and K. Tanaka, Phys. Lett. B **523**, 111 (2001) [Erratum-ibid. B **536**, 344 (2002)].
15. V. M. Braun, D. Y. Ivanov and G. P. Korchemsky, Phys. Rev. D **69** (2004) 034014.
16. B. O. Lange and M. Neubert, Phys. Rev. Lett. **91** (2003) 102001.

THE CHARM QUARK
AS A MASSIVE COLLINEAR QUARK

THOMAS MANNEL

(Work in collaboration with Heike Boos, Thorsten Feldmann and Ben Pecjak)

Theoretische Physik I, Siegen University
57068 Siegen, Germany

We argue that in certain kinematic regions the charm quarks originating from decays of B hadrons may be treated in soft collinear effective theory, including a mass term. The mass of the charm quark is at the jet scale, i.e. $m_c^2 \sim \Lambda_{\rm QCD} m_b$, which is a perturbative scale. We investigate the phenomenological consequences of this power counting, e.g. the possibility to extract V_{ub}/V_{cb}.

1. Introduction

The expansion in inverse powers of the heavy-quark mass has become the standard tool in heavy-quark physics. The fact that $m_b \gg \Lambda_{\rm QCD}$ allows us to formulate this expansion in terms of effective field theories, which yield a model independent approach to many heavy-quark processes.

Heavy quark effective theory (HQET) or the heavy quark expension (HQE) describes systems in which the light degrees of freedom are entirely related to the soft scale $\Lambda_{\rm QCD}$. In this case a standard effective field theory may be formulated in terms of a local operator product expansion (OPE). HQET and HQE are the basis of the modern precision determinations of the CKM matrix element V_{cb} which is known currently with an accuracy of about 2%.

However, two-body non-leptonic decays as well as heavy-to-light form factors in certain kinematic regions involve light degrees of freedom where the large scale m_b is still relevant, since the light quarks and gluons carry a large energy of the order m_b in the rest frame of the decaying heavy hadron. It has been recently established that such a situation can as well be described in terms of an effective field theory, the so called soft collinear effective theory (SCET) [1,2]. The properties of SCET are more subtle than the ones of HQET and HQE; for example SCET is a field theory involving non-localities on the light cone.

While SCET is usually formulated for massless light quarks, it is trivial to introduce a mass term into SCET. The main question is the power counting of the mass appearing in the Lagrangian. For the light quarks the mass is clearly of the order Λ_{QCD}, while the charm quark is usually treated as a heavy quark.

However, it is well known that the charm quark is not particularly heavy, which means that there are examples where an expansion in $1/m_c$ does not yield useful results, e.g. when looking at the lifetimes of charmed hadrons. As an alternative on may treat the charm quark as a massive collinear quark, using SCET and the power counting $m_c^2 \sim m_b\Lambda_{QCD}$, i.e. the charm quark mass is an intermediate scale corresponding to the (still perturbative) jet scale [3,4]. In two recent papers [5,6] written in collaboration with Heike Boos, Thorsten Feldmann and Ben Pecjak we investigate the impact of such an approach to inclusive semileptonic decays in the endpoint region. Since we describe this process in the framework of SCET, we end up at leading order with the shape function known from semileptonic $b \to u$ decays or from radiative $b \to s$ decays and at subleading level only one additional shape functions is needed compared to the $b \to u$ case [7].

We calculate the hadronic tensor at tree-level, including $\mathcal{O}(\Lambda_{QCD}/m_b)$ power corrections and $\mathcal{O}(\alpha_s)$ corrections and show that it factorizes into a convolution of jet and shape functions. We identify a certain kinematical variable whose differential decay rate is directly proportional to the universal leading order shape function, and speculate as to whether information on this function can be extracted from a study of $b \to c$ decay. This variable is a generalization of the P_+ variable suggested in [8], which has been reconsidered in the framework of SCET in [9,10].

In the next section we discuss our power counting and construct the SCET Lagrangian for a massive collinear quark and give the corresponding transition currents. In Section 3 we use this version of SCET to calculate the hadronic tensor for semi-leptonic $b \to c$ transitions. QCD corrections to order α_s are considered in section 4, and in section 5 we sumamrize our results and discuss their impact on the determination of CKM matrix elements.

2. SCET lagrangian and transition currents

In order to calculated the decay distributions for inclusive $\bar{B} \to X_c \ell \bar{\nu}_\ell$ we start from the hadronic tensor expressed in terms of the forward scattering amplitude

$$W^{\mu\nu} = \frac{1}{\pi}\mathrm{Im}\langle \bar{B}(v)|T^{\mu\nu}|\bar{B}(v)\rangle, \tag{1}$$

where

$$T^{\mu\nu} = i\int d^4x e^{-iq\cdot x}\mathrm{T}\{J^{\dagger\mu}(x)J^{\nu}(0)\}, \quad J^{\mu} = \bar{c}\gamma^{\mu}(1-\gamma_5)b \qquad (2)$$

We write the hadronic tensor in terms of five scalar structure functions[11]

$$\begin{aligned} W_{\mu\nu} = & W_1\,(p_\mu v_\nu + v_\mu p_\nu - g_{\mu\nu} vp - i\epsilon_{\mu\nu\alpha\beta}\, p^\alpha v^\beta) \\ & - W_2\, g_{\mu\nu} + W_3\, v_\mu v_\nu + W_4\,(p_\mu v_\nu + v_\mu p_\nu) + W_5\, p_\mu p_\nu, \end{aligned} \qquad (3)$$

where the independent vectors are chosen to be v, the velocity of the $\bar{B}$ meson, and $p \equiv m_b v - q$ with m_b the b quark pole mass and q the momentum of the outgoing lepton pair.

The hadronic tensor is expressed in the effective theory as a double series in the perturbative coupling constant and a small parameter λ, which we define through the relations

$$\frac{\Lambda_{\rm QCD}}{m_b} \sim \lambda^2 \quad \text{and} \quad \frac{m_c^2}{m_b^2} \sim \lambda^2. \qquad (4)$$

This choice correlates the two scales $m_c/m_b \sim \sqrt{\Lambda_{\rm QCD}/m_b}$. The problem contains three widely separated scales, $m_b^2 \gg m_b^2\lambda^2 \gg m_b^2\lambda^4$, provided that the jet momentum p satisfies $p^2 - m_c^2 \sim m_b^2\lambda^2$. We will refer to these scales as the hard, jet, and soft scale respectively. We expect that the hadronic tensor can be factorized into a convolution of hard, jet and soft (shape) functions, corresponding to these three scales. This is similar to the massless case investigated in [1,12].

The B meson is described in terms of HQET and hence light degrees of freedom in the B meson are soft. The factorization formula for the hadronic tensor looks schematically like

$$\langle \bar{B}|T_i^{\rm eff}|\bar{B}\rangle = \langle \bar{B}|\bar{h}_v[\text{soft fields}]h_v|\bar{B}\rangle \otimes \langle 0|[\text{collinear fields}]|0\rangle \equiv S_i \otimes J_i\,, \quad (5)$$

where the $\otimes$ stands for a convolution. The vacuum matrix element of the collinear fields defines a set of jet functions J_i, the relevant scale of which is λm_b. The tree-level jet functions are given in terms of non-interacting propagators of the collinear fields, since we assume the jet scale to be still a perturbative scale. Calculating these jet functions removes the collinear degrees of freedom and defines the second step of the necessary two-step matching $QCD \to$ SCET $\to$ HQET. The matrix element of the soft fields between the $\bar{B}$ meson states is calculated in HQET and defines a set of non-perturbative shape functions S_i. The differential decay distributions are then written in terms of the W_i, each of which factorizes into a convolution of jet and shape functions.

The derivation of the SCET Lagrangian including a mass term proceeds along the same lines as in the massless case. One obtains

$$\mathcal{L}_{\rm QCD} = \bar{\psi}(i\not{D} - m_q)\psi$$
$$= \bar{\xi}\left((in_-D) + (i\not{D}_\perp - m_q)\frac{1}{in_+D}(i\not{D}_\perp + m_q)\right)\frac{\not{n}_+}{2}\xi \qquad (6)$$

where ξ is the massive collinear field. As in the massless case this expression is still completely equivalent to the full QCD Lagrangian; the expansion according to the SCET power counting proceeds in the same way as in the massless case through e.g. a multipole expansion [2].

Likewise the currents are expanded according to the SCET power counting. For $b \to c$ transitions, this matching takes the form

$$(\bar{c}\Gamma b)_{\rm QCD} \to e^{-im_b vx}\left[J^{(0)} + J^{(1)} + J^{(2)} + \ldots\right], \qquad (7)$$

where c becomes the collinear charm-quark field ξ. In general, the $J^{(i)}$ are convolutions of a short distance Wilson coefficient with operators built out of SCET and HQET fields. For the massless case the $J^{(i)}$ are well known[7]; in addition one finds three mass dependent terms

$$J_m^{(1)} = m_c\,\bar{\xi}\frac{\not{n}_+}{2}\frac{1}{-in_+\overleftarrow{D}_c}W_c\Gamma h_v, \quad J_{m1}^{(2)} = m_c\,\bar{\xi}\frac{\not{n}_+}{2}\frac{1}{-in_+\overleftarrow{D}_c}W_c\Gamma x_\perp D_{\perp s}h_v,$$
$$J_{m2}^{(2)} = m_c\,\bar{\xi}\frac{\not{n}_+}{2}\frac{1}{in_+\overleftarrow{D}_c}\Gamma\frac{\not{n}_-}{2m_b}[i\not{D}_{\perp c}W_c]h_v. \qquad (8)$$

where the arrow indicates that the derivative acts to the left. Furthermore, h_v is the static b quark field and W_c is a collinear Wilson line appearing also in the massless case.

3. Tree Level Results

It is well known that the standard HQE fails in the endpoint region for heavy to light decays, which is e.g. the region of lepton energies close to $E_\ell \sim m_b/2$. However, also on the case of a $b \to c$ the endpoint region calculated in HQE exhibits unphysical spikes, which can be attributed to terms of the form

$$\frac{d\Gamma}{dy} \sim \Theta(1-y-\rho)\left[2 + \frac{\lambda_1}{(m_b(1-y))^2}\left(\frac{\rho}{1-y}\right)^2\left\{3 - 4\frac{\rho}{1-y}\right\}\right], \qquad (9)$$

where $y = 2E_\ell/m_b$, $\rho = m_c^2/m_b^2$ and λ_1 is the kinetic energy parameter. Note that these spikes become δ functions and its derivatives in the limit $m_c \to 0$.

We argue in [5] that in this region the same shape function becomes relevant as in the massless case. In fact, this can be easily seen at tree level by considering the propagator of the collinear charm quark which becomes

$$S_c(p+k) = \frac{\not{n}_-}{2}\frac{i}{n_-(p+k)-\frac{m_c^2}{n_+p}} = \frac{\not{n}_-}{2}\frac{i}{u+n_-k}\,,\quad u=(n_-p)-\frac{m_c^2}{(n_+p)} \qquad (10)$$

where we have kept the charm quark mass according to the power counting advertised above. The form of this expression suggests to introduce a new variable u which takes the role of the hadronic light cone momentum $n_-P = P_+$ suggested in [8] for the massless case. In particular the non-perturbative shape functions depend in our case on this variable u, and the tree level result to subleading order in the SCET expansion is

$$\frac{1}{\Gamma_c}\frac{d\Gamma}{du} = \left(1-\frac{14}{3}\frac{u}{m_b}-8\frac{m_c^2}{m_b^2}\right)S(u)+\frac{s(u)}{2m_b}-4\frac{m_c^2}{m_b^2}t_1(u) \qquad (11)$$
$$+\frac{1}{3m_b}[t(u)+u_a(u)-5u_s(u)]$$

which can be compared to the massless result given in [10]. We note that in the massive case one additional nonperturbative function t_1 appears. Furthermore, the term $-8m_c^2/m_b^2$ is a remnant of the well known partonic phase space function, expanded to leading order in the SCET expansion.

4. Radiative corrections

QCD radiative corrections at the hard scale have to be the same as the ones in the massless case, which has been considerd in [10] The only change which appears is at intermediate scales since now additional dependences on the charm mass of the jet functions appear. To compute these radiative corrections we have to evaluate the SCET diagrams shown in fig. 1.

The calculation of these diagrams is straightforward, and the resulting jet function can be expressed in terms of $*$-distributions [10] as

$$J(u,n_+p) = \delta(u)+\frac{C_F\alpha_s}{4\pi}\Bigg\{ \qquad (12)$$
$$(7-\pi^2)\delta(u)-3\left(\frac{1}{u}\right)_*^{[\mu^2/n_+p]}+\left(\frac{\ln(u\,n_+p/\mu^2)}{u}\right)_*^{[\mu^2/n_+p]}$$
$$+\Theta(u)\left(\frac{u}{(u+m_c^2/n_+p)^2}-\frac{4}{u}\ln\left(1+\frac{u\,n_+p}{m_c^2}\right)\right)$$
$$+\left(1+\frac{2\pi^2}{3}\right)\delta(u)-\left(\frac{1}{u}\right)_*^{[m_c^2/n_+p]}+4\left(\frac{\ln(u\,n_+p/m_c^2)}{u}\right)_*^{[m_c^2/n_+p]}\Bigg\}$$

The first two lines are identical to the massless case, while the last two lines vanish as $m_c \to 0$.

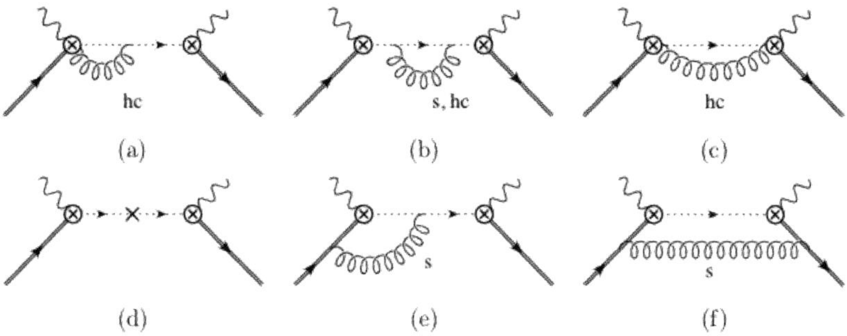

Fig. 1. Feynman diagrams in SCET to order α_s for $b \to c\ell\bar{\nu}$

5. Discussion of the results

We first switch to hadronic variables which means that the partonic momentum of the final state is replaced by the corresponding momentum of the hadrons. To the accuracy we are working this corresponds to a shift in both P_+ used in the massless case and the variable u suggested here:

$$U = u + \bar{\Lambda} \quad \leftrightarrow \quad P_+ = p_+ + \bar{\Lambda}$$

In order to compare $b \to u$ and $b \to c$ we consider a partially integrated spectrum

$$F_c(\Delta) = \frac{1}{\Gamma_c} \int_0^\Delta dU \, \frac{d\Gamma_c}{dU} = \frac{\Gamma_c(U < \Delta)}{\Gamma_c} = F_u(\Delta) + F_m(\Delta) \qquad (13)$$

which decomposes in a mass-independent term and a mass dependent one.

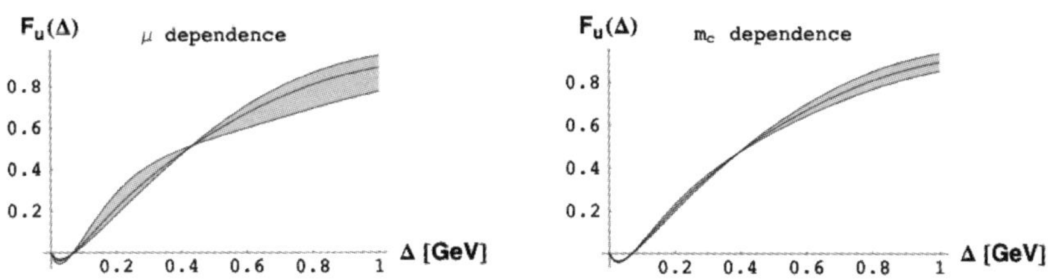

Fig. 2. Partially integrated rate. The left plot shows the factorization scale dependence, this right one shows the m_c dependence.

In the same way one may define a partially integrated spectrum for the massless case which is defined as

$$F_u(\Delta) = \int_0^{\Delta} dP_+ \frac{d\Gamma_u}{dP_+} \tag{14}$$

The relation between the case of $B \to X_u \ell \bar{\nu}$ and $B \to X_c \ell \bar{\nu}$ may now be expressed in terms of a calculable weight function

$$\int_0^{\Delta} dP_+ \frac{d\Gamma_u}{dP_+} = \frac{|V_{ub}|^2}{|V_{cb}|^2} \int_0^{\Delta} dU W(\Delta, U) \frac{d\Gamma_c}{dU} \tag{15}$$

which in principle yields a handle on the ratio V_{ub}/V_{cb}.

However, the use of this method is limited. One one hand, the window for the upper limit Δ is quite narrow. It has to be in the shape function region, but it should also be large enough to include not only the D and the D^* resonances. On the other hand, there are sizable uncertainties from different sources. As shown in fig. 2, there is an uncertainty due to the dependence on the factorization scale which can roughly be estimated as $10 - 15\%$. The charm mass dependence also shown in fig. 2 is rather mild and contributes less than 10% to the uncertainties. Subleading contributions are difficult to estimate due to out ignorance of the subleading shape functions, but we expect them to be sizable. Finally one may wonder about duality in this case, since the semileptonic $b \to c$ transitions are saturated to more than 70% by the two D meson ground states. This point needs further investigation.

References

1. C. W. Bauer, D. Pirjol and I. W. Stewart, Phys. Rev. D **65**, 054022 (2002).
2. M. Beneke, A. P. Chapovsky, M. Diehl and T. Feldmann, Nucl. Phys. B **643**, 431 (2002).
3. T. Mannel and M. Neubert, Phys. Rev. D **50**, 2037 (1994).
4. T. Mannel and F. J. Tackmann, Phys. Rev. D **71**, 034017 (2005).
5. H. Boos, T. Feldmann, T. Mannel and B. D. Pecjak, Phys. Rev. D **73**, 036003 (2006).
6. H. Boos, T. Feldmann, T. Mannel and B. D. Pecjak, JHEP **0605**, 056 (2006).
7. M. Beneke, F. Campanario, T. Mannel and B. D. Pecjak, JHEP **0506**, 071 (2005).
8. T. Mannel and S. Recksiegel, Phys. Rev. D **60**, 114040 (1999).
9. S. W. Bosch, B. O. Lange, M. Neubert and G. Paz, Nucl. Phys. B **699**, 335 (2004).
10. S. W. Bosch, B. O. Lange, M. Neubert and G. Paz, Phys. Rev. Lett. **93**, 221801 (2004).
11. F. De Fazio and M. Neubert, JHEP **9906**, 017 (1999).
12. G. P. Korchemsky and G. Sterman, Phys. Lett. B **340** (1994) 96.

HEAVY MESON MOLECULES IN EFFECTIVE FIELD THEORY

MOHAMMAD T. ALFIKY, FABRIZIO GABBIANI, ALEXEY A. PETROV

Department of Physics and Astronomy, Wayne State University,
Detroit, Michigan, 48201, USA
Presenter. E-mail: apetrov@wayne.edu
www.physics.wayne.edu/~apetrov/

We consider the implications from the possibility that the recently observed state $X(3872)$ is a meson-antimeson molecule. We write an effective Lagrangian consistent with the heavy-quark and chiral symmetries needed to describe $X(3872)$. We explore the consequences of the assumption that $X(3872)$ is a molecular bound state of D^{*0} and $\overline{D}^0$ mesons for the existence of bound states in the $D^0\overline{D}^0$ and $D^{*0}\overline{D}^{*0}$

1. Introduction

The unusual properties of $X(3872)$ state, recently discovered in the decay $X(3872) \to J/\psi\pi^+\pi^-$, invited some speculations regarding its possible non-$c\bar{c}$ nature [1,2]. Since its mass lies tantalizingly close to the $D^{*0}\overline{D}^0$ threshold of 3871.3 MeV, it is tempting to interpret $X(3872)$ as a $D^{*0}\overline{D}^0$ molecule with $J^{PC} = 1^{++}$ quantum numbers[3,4]. Such molecular states can be studied using techniques of effective field theories (EFT). This study is possible due to the multitude of scales present in QCD. The extreme smallness of the binding energy, $E_b = (m_{D^0} + m_{D^{0*}}) - M_X = -0.6 \pm 1.1$ MeV, suggests that this state can play the role of the "deuteron" [3] in meson-antimeson interactions. This fact allows us to use methods similar to those developed for the description of the deuteron[5,6], with the added benefit of heavy-quark symmetry. A suitable effective Lagrangian describing such a system contains only heavy-meson degrees of freedom with interactions approximated by local four-boson terms constrained only by the symmetries of the theory. While the predictive power of this approach is somewhat limited, several model-independent statements can be made. For instance, possible existence of a molecular state in $D^{*0}\overline{D}^0$ channel does not imply a molecular state in $D^{*0}\overline{D^{*}}^0$ or $D^0\overline{D}^0$ channels.

The tiny binding energy of this molecular state introduces an energy scale which is much smaller than the mass of the lightest particle, the pion, whose exchange can provide binding. This fact presents a problem with a straightforward EFT analysis of this bound state, which can be illustrated in the following example. Consider a low-energy s-wave scattering amplitude $A(k)$ of two heavy mesons with momentum k (ignore spin for a moment),

$$A(k) = \frac{4\pi}{m_D} \frac{1}{k \cot \delta - ik} = \frac{4\pi}{m_D} \frac{1}{-1/a + (r_0/2)\, k^2 + ... - ik}, \tag{1}$$

where a is a scattering length, which is related to the binding energy E_b of a meson-antimeson bound state as $a \sim (E_b)^{-1/2}$. Naturally, $a \sim 1/m$, where m is the mass of exchanged particle that provides binding. Scattering amplitude of Eq. (1) can be obtained from an effective Lagrangian expressed in terms of only heavy-meson degrees of freedom as a power series in momentum k,

$$A(k) = \frac{4\pi}{m_D} \sum_n C_n k^n = -\frac{4\pi a}{m_D} \left[1 - iak + \frac{a r_0 - 2a^2}{2} k^2 + ... \right], \tag{2}$$

where C_n are the coefficients of that effective Lagrangian. A problem with a simple application of EFT is apparent, as $a \to \infty$ for $E_b \to 0$, making the series convergence in Eq. (2) problematic. Indeed, $a \simeq 0.032$ MeV^{-1} for X(3872), which is much larger than the inverse masses of possible exchange particles, $1/m_\pi \simeq 7.1 \times 10^{-3}$ MeV^{-1}, $1/m_\rho \simeq 1.3 \times 10^{-3}$ MeV^{-1}, etc. This implies that all-order resummation of $(ak)^n$ series is required. In EFT language this would imply resummation of a class of "bubble" graphs, whose vertices are defined by C_n.

2. The effective Lagrangian

The general effective Lagrangian required for description of $D^{*0}\overline{D}^0$ molecular state and consistent with heavy-quark spin and chiral symmetries can be written as[7]

$$\mathcal{L} = \mathcal{L}_1 + \mathcal{L}_2, \tag{3}$$

where the two-body piece $\mathcal{L}_2$ describes the interactions between heavy meson degrees of freedom. The one-body piece $\mathcal{L}_1$ describes strong interactions of the heavy mesons P and P^* ($P = B, D$) containing one heavy quark Q and is well known[8]:

$$\mathcal{L}_1 = -\mathrm{Tr}\left[\overline{H}^{(Q)} \left(iv \cdot D + \frac{D^2}{2m_P} \right) H^{(Q)} \right] + \frac{\lambda_2}{m_P} \mathrm{Tr}\left[\overline{H}^{(Q)} \sigma^{\mu\nu} H^{(Q)} \sigma_{\mu\nu} \right] + \cdots \tag{4}$$

where the ellipsis denotes higher-order terms in chiral expansion, or describing pion-H interactions and antimeson degrees of freedom $H_a^{(\overline{Q})}$ and $H_a^{(\overline{Q})\dagger}$. A superfield describing the doublet of pseudoscalar heavy-meson fields $P_a = \left(P^0, P^+\right)$ and their vector counterparts with $v \cdot P_a^{*(Q)} = 0$, is defined as $H_a^{(Q)} = (1/2)(1+ \not{v}) \left[P_{a\mu}^{*(Q)} \gamma^\mu - P_a^{(Q)} \gamma_5\right]$. The second term in Eq. (4) accounts for the $P - P^*$ mass difference $\Delta \equiv m_{P*} - m_P = -2\lambda_2/m_P$. The two-body piece is[7]

$$\mathcal{L}_4 = -\frac{C_1}{4} \mathrm{Tr}\left[\overline{H}^{(Q)} H^{(Q)} \gamma_\mu\right] \mathrm{Tr}\left[H^{(\overline{Q})} \overline{H}^{(\overline{Q})} \gamma^\mu\right] - \frac{C_2}{4} \mathrm{Tr}\left[\overline{H}^{(Q)} H^{(Q)} \gamma_\mu \gamma_5\right] \mathrm{Tr}\left[H^{(\overline{Q})} \overline{H}^{(\overline{Q})} \gamma^\mu \gamma_5\right]. \quad (5)$$

Heavy-quark spin symmetry implies that the same Lagrangian governs the four-boson interactions of *all* $P_a^{(*)} = D^{(*)}$ states. Indeed, not all of these states are bound. Here we shall concentrate on $X(3872)$, which we assume to be a bound state of two *neutral* bosons, $P_a \equiv P^0 \equiv D^3$. Evaluating the traces yields for the $D\overline{D^*}$ sector

$$\mathcal{L}_{4,DD^*} = -C_1 D^{(c)\dagger} D^{(c)} D_\mu^{*(\bar{c})\dagger} D^{*(\bar{c})\mu} - C_1 D_\mu^{*(c)\dagger} D^{*(c)\mu} D^{(\bar{c})\dagger} D^{(\bar{c})} + C_2 D^{(c)\dagger} D_\mu^{*(c)} D^{*(\bar{c})\dagger\mu} D^{(\bar{c})} + C_2 D_\mu^{*(c)\dagger} D^{(c)} D^{(\bar{c})\dagger} D^{*(\bar{c})\mu} + \ldots \quad (6)$$

As we show later, the resulting binding energy depends on a *linear combination* of C_1 and C_2. Similarly, one obtains the component Lagrangian governing the interactions of D and $\overline{D}$,

$$\mathcal{L}_{4,DD} = C_1 D^{(c)\dagger} D^{(c)} D^{(\bar{c})\dagger} D^{(\bar{c})}. \quad (7)$$

Clearly, one cannot relate the existence of the bound state in the $D\overline{D^*}$ and DD channels, as the properties of the latter will depend only on C_1.

3. Properties of bound states

The lowest-energy bound state of D and $\overline{D^*}$ is an eigenstate of charge conjugation,

$$|X_\pm\rangle = \frac{1}{\sqrt{2}}\left[|D^* \overline{D}\rangle \pm |D \overline{D}^*\rangle\right]. \quad (8)$$

To find the bound-state energy of $X(3872)$ with $J^{PC} = 1^{++}$, we shall look for a pole of the transition amplitude $T_{++} = \langle X_+ | T | X_+ \rangle$. Defining DD^*-DD^* transition amplitudes,

$$T_{11} = \langle D^* \overline{D} | T | D^* \overline{D} \rangle, \quad T_{12} = \langle D^* \overline{D} | T | D \overline{D}^* \rangle,$$
$$T_{21} = \langle D \overline{D}^* | T | D^* \overline{D} \rangle, \quad T_{22} = \langle D \overline{D}^* | T | D \overline{D}^* \rangle, \quad (9)$$

we also have to include a "bubble" resummation of loop contributions, as existence of a bound state is related to a breakdown of perturbative expansion[5]. These amplitudes satisfy a system of Lippmann-Schwinger equations[7]. In an algebraic matrix form,

$$\begin{pmatrix} T_{11} \\ T_{12} \\ T_{21} \\ T_{22} \end{pmatrix} = \begin{pmatrix} -C_1 \\ C_2 \\ C_2 \\ -C_1 \end{pmatrix} + i\widetilde{A} \begin{pmatrix} -C_1 & C_2 & 0 & 0 \\ C_2 & -C_1 & 0 & 0 \\ 0 & 0 & -C_1 & C_2 \\ 0 & 0 & C_2 & -C_1 \end{pmatrix} \begin{pmatrix} T_{11} \\ T_{12} \\ T_{21} \\ T_{22} \end{pmatrix}. \tag{10}$$

The solution of Eq. (10) produces the T_{++} amplitude,

$$T_{++} = \frac{1}{2}\left(T_{11} + T_{12} + T_{21} + T_{22}\right) = \frac{\lambda}{1 - i\lambda\widetilde{A}}, \tag{11}$$

where $\lambda = C_2 - C_1$ and $\widetilde{A}$ is a (divergent) integral

$$\begin{aligned} \widetilde{A} &= \frac{i}{4} 2\mu_{DD^*} \int \frac{d^3q}{(2\pi)^3} \frac{1}{\vec{q}^{\,2} - 2\mu_{DD^*}(E-\Delta) - i\epsilon} \\ &= -\frac{1}{8\pi}\mu_{DD^*}|\vec{p}|\sqrt{1 - \frac{2\mu_{DD^*}\Delta}{\vec{p}^{\,2}}}. \end{aligned} \tag{12}$$

Here $E = \vec{p}^2/2\mu_{DD^*}$, with μ_{DD^*} being the reduced mass of the DD^* system. The divergence of the integral of Eq. (12) is removed by renormalization. We chose to define a renormalized λ_R within the MS subtraction scheme in dimensional regularization, which does not introduce any new dimension-full scales into the problem. In this scheme the integral $\widetilde{A}$ is finite, which corresponds to an implicit subtraction of power divergences in Eq. (12). This implies for the transition amplitude

$$T_{++} = \frac{\lambda_R}{1 + (i/8\pi)\lambda_R\,\mu_{DD^*}|\vec{p}|\sqrt{1 - 2\mu_{DD^*}\Delta/\vec{p}^{\,2}}}. \tag{13}$$

The position of the pole of the molecular state on the energy scale can be read off Eq. (13),

$$E_{\text{Pole}} = \frac{32\pi^2}{\lambda_R^2 \mu_{DD^*}^3} - \Delta. \tag{14}$$

Recalling the definition of binding energy E_b and that $m_{D^*} = m_D + \Delta$, we infer

$$E_b = \frac{32\pi^2}{\lambda_R^2 \mu_{DD^*}^3}. \tag{15}$$

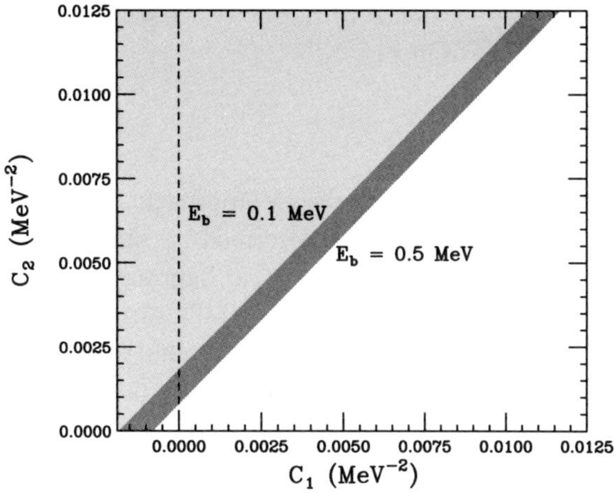

Fig. 1. The coupling constant C_2 is plotted vs. C_1. The lightly shaded area shows the region of parameter space allowed by postulating the existence of a $J^{PC} = 1^{++}$ bound state with $E_b = 0.1$ MeV, while excluding the orthogonal bound state with C = −1. The darker area becomes allowed in addition if we assume $E_b = 0.5$ MeV.

Assuming $E_b = 0.5$ MeV, which is one sigma below the central value [2], and the experimental values for the masses [9], we obtain $\lambda_R \simeq 8.4\times 10^{-4}$ MeV^{-2}.

Similar considerations apply to $D^0\overline{D}^0$ state, in which case the starting point is the Lagrangian term in Eq. (7). Since it involves only a single term, the calculations are actually easier and involve only one Lippmann-Schwinger equation. The resulting binding energy is then[7]

$$E_b = \frac{256\pi^2}{C_{1R}^2 m_D^3}. \tag{16}$$

Examining Eq. (16) we immediately notice that the existence of a bound state in the $D^*\overline{D}$ channel does not dictate the properties of a possible bound state in the $D^0\overline{D}^0$ or $B^0\overline{B}^0$ channels, since C_1 and C_2 are generally not related to each other.

If we assume that the orthogonal state with $J^{PC} = 1^{+-}$ is not bound, which is consistent with all the existing experimental observations, we can place some separate constraints on the renormalized values of C_1 and C_2. The amplitude orthogonal to T_{++},

$$T_{--} = \langle X_-|T|X_-\rangle = \frac{1}{2}\left(T_{11} - T_{12} + T_{21} - T_{22}\right) = \frac{\lambda'_R}{1 - i\lambda'_R \widetilde{A}_R}, \tag{17}$$

with $\lambda'_R = -C_1 - C_2$, does not have a pole that corresponds to a bound state if $C_1 + C_2 > 0$. The exclusion of the $C = -1$ state together with the assumption of the existence of the $C = +1$ state limits the (C_1, C_2) parameter space as shown in Fig. 1.

4. Conclusions

We introduced an effective field theory approach in the analysis of the likely molecular state $X(3872)$. We described its binding interaction with contact terms in a heavy-quark symmetric Lagrangian. The flexibility of this description allows us to ignore the details of the interaction and to concentrate on its effects, namely a shallow bound state and a large scattering length. We found that the existence of the bound state in the $D^*\overline{D}$ channel does not in general exclude a possibility of a bound state in the $D^0\overline{D}^0$ system, but does not require it. Future experimental studies of this state are interesting[10] and should provide lots of new information about properties of QCD bound states.

This work was supported in part by the U.S. Department of Energy under Contract DE-FG02-96ER41005. A.P. was also supported by the U.S. National Science Foundation CAREER Award PHY–0547794.

References

1. For a review of properties of X(3872), see E. S. Swanson, Phys. Rept. **429**, 243 (2006); C. Quigg, Nucl. Phys. Proc. Suppl. **142**, 87 (2005); S. L. Olsen [Belle Collaboration], Int. J. Mod. Phys. A **20**, 240 (2005). Possible interpretations of X(3872) also include a "hybrid charmonium" state: A. A. Petrov, proceeding of the 1st Meeting of the APS Topical Group on Hadronic Physics (GHP2004), Batavia, Illinois, 24-26 Oct 2004, J. Phys. Conf. Ser. **9**, 83 (2005); see also G. Chiladze, A. F. Falk, and A. A. Petrov, Phys. Rev. D **58**, 034013 (1998); F. E. Close and S. Godfrey, Phys. Lett. B **574**, 210 (2003); B. A. Li, Phys. Lett. B **605**, 306 (2005), a diquark-antidiquark state: L. Maiani, F. Piccinini, A. D. Polosa, and V. Riquer, Phys. Rev. D **71**, 014028 (2005), a glueball: K. K. Seth, Phys. Lett. B **612**, 1 (2005), and a cusp feature associated with the $D^0\overline{D}^0$ threshold: D. V. Bugg, Phys. Lett. B **598**, 8 (2004). See also T. Barnes and S. Godfrey, Phys. Rev. D **69**, 054008 (2004); E. J. Eichten, K. Lane, and C. Quigg, Phys. Rev. D **69**, 094019 (2004); T. Barnes, S. Godfrey, and E. S. Swanson, arXiv:hep-ph/0505002. Some older studies include M. B. Voloshin and L. B. Okun, JETP Lett. **23**, 333 (1976) [Pisma Zh. Eksp. Teor. Fiz. **23**, 369 (1976)]; A. De Rujula, H. Georgi, and S. L. Glashow, Phys. Rev. Lett. **38**, 317 (1977). S. Nussinov and D. P. Sidhu, Nuovo Cim. A **44**, 230 (1978); N. A. Törnqvist, Z. Phys. C **61**, 525 (1994); A. V. Manohar and M. B. Wise, Nucl. Phys. B **399**, 17 (1993).
2. S. K. Choi *et al.* [Belle Collaboration], Phys. Rev. Lett. **91**, 262001 (2003); B. Aubert *et al.* [BABAR Collaboration], Phys. Rev. D **71**, 071103 (2005);

D. Acosta *et al.* [CDF II Collaboration], Phys. Rev. Lett. **93**, 072001 (2004); V. M. Abazov *et al.* [D0 Collaboration], Phys. Rev. Lett. **93**, 162002 (2004).

3. N. A. Törnqvist, Phys. Lett. B **590**, 209 (2004); C. Y. Wong, Phys. Rev. C **69**, 055202 (2004). F. Close and P. Page, Phys. Lett. B **578**, 119 (2004); E. S. Swanson, Phys. Lett. B **588**, 189 (2004).
4. E. Braaten and M. Kusunoki, Phys. Rev. D **69**, 074005 (2004).
5. S. Weinberg, Nucl. Phys. B **363**, 3 (1991); Phys. Lett. B **251**, 288 (1990).
6. D. B. Kaplan, M. J. Savage and M. B. Wise, Nucl. Phys. B **534**, 329 (1998); *ibid* Phys. Lett. B **424**, 390 (1998).
7. M. T. AlFiky, F. Gabbiani and A. A. Petrov, Phys. Lett. B **640**, 238 (2006).
8. A. V. Manohar and M. B. Wise, "Heavy quark physics," Camb. Monogr. Part. Phys. Nucl. Phys. Cosmol. **10**, 1 (2000).
9. S. Eidelman et al., Phys. Lett. B592, 1 (2004).
10. M. B. Voloshin, Phys. Lett. B **604**, 69 (2004); E. Braaten and M. Kusunoki, Phys. Rev. D **71**, 074005 (2005); arXiv:hep-ph/0506087; E. Braaten, M. Kusunoki, and S. Nussinov, Phys. Rev. Lett. **93**, 162001 (2004); E. Braaten and M. Lu, arXiv:hep-ph/0606115; S. Pakvasa and M. Suzuki, Phys. Lett. B **579**, 67 (2004); J. L. Rosner, Phys. Rev. D **70**, 094023 (2004).

RECENT ADVANCES IN NRQCD

ANDRE HOANG and PEDRO RUIZ-FEMENIA

Max-Planck-Institut für Physik
(Werner-Heisenberg-Institut),
Föhringer Ring 6, 80335 Müenchen, Germany
E-mail: ahoang@mppmu.mpg.de, ruizfeme@mppmu.mpg.de

We discuss recent theoretical developments concerning the description of the production and decay of heavy quarks and colored scalars in the framework of non-relativistic QCD.

1. Threshold physics at the ILC

The e^+e^- center-of-mass (c.m.) energy in a linear collider (LC) can be very precisely monitored, allowing for an accurate exploration of the threshold regime. The top-quark mass can be determined from a measurement of $\sigma(e^+e^- \to Z^*, \gamma^* \to t\bar{t})$ line shape at a LC operating at c.m. energies around the $t\bar{t}$ threshold ($\sqrt{q^2} \sim 350$ GeV). The rise of the cross section with increasing c.m. energy is directly related to the mass of the top quark. Assuming a total integrated luminosity of 300 fb^{-1}, LC simulations of a threshold scan of the top-antitop total cross section have demonstrated that experimental uncertainties below 100 MeV for the top-quark mass determination can be obtained [1,2], even when beam effects, which lead to some smearing of the effective c.m. energy, are taken into account. If the normalization of the cross section line shape is well under control, it is possible to determine the strong coupling, the total top quark width and, if the Higgs boson is light, the top Yukawa coupling. In view of the accuracy obtainable at the LC the theoretical uncertainties for the total cross section should be lowered to a level of a few percent [3,4]. A precise knowledge of the top mass would also improve the analysis of electroweak precision observables and put indirect constraints on New Physics. It has been shown [5] that an accuracy of 100 MeV on the top quark mass would allow to perform stringent internal consistency checks of the SM and of some scenarios of Supersymmetry (SUSY).

Threshold studies would also be feasible for squarks at a next LC. Many models of SUSY predict that, due to large mixing, the lightest squark could correspond to one of the mass eigenstates of the third generation with $m_{\tilde{q}} < 500$ GeV, thus allowing for the production of stop pairs at a future e^+e^- LC operating below 1 TeV. A "threshold scan" of the total cross section line-shape at such a facility will yield precise measurements of the stop mass, lifetime and couplings [6], in close analogy to the program carried out in threshold studies for the top-antitop threshold [2].

2. Theoretical status of $t\bar{t}$ production

Close to threshold the top quark pairs are produced with small velocities $v \ll 1$ in the c.m. frame. Therefore the relevant physical scales governing the top-antitop dynamics are the top quark mass m_t, the relative three-momentum $\boldsymbol{p} \sim m_t v$ and the top quark non-relativistic kinetic energy $E \sim m_t v^2$. Since the ratios of the three scales can arise in matrix elements, the cross section cannot be calculated using the standard QCD expansion in the strong coupling α_s. The best known indication of the latter comes from the well-known "Coulomb singularity", which shows up as a singular $(\alpha_s/v)^n$ behaviour in the $v \to 0$ limit of the $t\bar{t}$ production amplitude at the n-loop order in perturbative QCD. The proper expansion scheme for the $t\bar{t}$ threshold region is a double expansion in both α_s and v, and one has to use the parametric counting $\alpha_s \sim v \ll 1$ to identify all effects contributing to a certain order of approximation. In this *fixed-order* expansion the leading order (LO) contributions correspond to terms in the total cross section proportional to $v(\alpha_s/v)^n$, $(n = 0, \ldots, \infty)$, next-to-leading (NLO) terms are proportional to $v(\alpha_s/v)^n \times [\alpha_s, v]$ and so on. At LO the total cross section is proportional to the absorptive part of the Green function of a Schrödinger equation containing the static QCD-potential [7]. Higher order corrections in this expansion are rigorously implemented employing the concept of non-relativistic effective quantum field theories, first proposed by Caswell and Lepage [8]. In this scheme, the original QCD Lagrangian is reformulated in terms of an effective non-relativistic Lagrangian called "non-relativistic Quantum Chromodynamics" (NRQCD) by using the hierarchy $m_t \gg \boldsymbol{p} \gg E$, which allows to separate short-distance physics at the "hard" scale of order the heavy quark mass from long-distance physics at the non-relativistic scales $\boldsymbol{p}$ and E. The hard-momentum effects are encoded as Wilson coefficients of the operators in the effective Lagrangian. Operators with increasing dimension are introduced in the effective theory to include the effects of higher orders in the non-relativistic expansion, but only a finite number is needed for a given precision. The NRQCD factorization properties were used in a number of NNLO calculations [9,10] of the total $t\bar{t}$ production cross section.

A common feature of the *fixed-order* calculations is that the running from the hard scale m down to the non-relativistic scales was not taken into account. At NNLO NRQCD matrix elements and Wilson coefficients involve logarithms of ratios of the hard scale and the non-relativistic scales, which in the case of top quark pair production close to threshold can be sizeable (for example $\alpha_s(m_t)\log[m_t/E] \simeq 0.8$ for $v \simeq 0.15$). Considering the parametric counting $\alpha_s \log v \sim 1$ introduces a modified expansion scheme for the size of the terms contributing to the total cross section, where the dominant contribution, proportional to $(\alpha_s/v)^n \sum_i (\alpha_s \log v)^i$, $(n = 0, \ldots \infty)$, is called leading-logarithmic order (LL). A number of different versions of NRQCD [11,12], each of which aiming (in principle) on applications in different physical situations, allow for renormalization improved calculations. The EFT vNRQCD ("velocity"NRQCD) [12,13] has been designed for predictions at the $t\bar{t}$ threshold. It treats the case $m_t \gg \boldsymbol{p} \gg E > \Lambda_{\rm QCD}$, i.e. all physical scales are perturbative, but also has the correlation $E_t = \boldsymbol{p}_t^2/m_t$ built in at the field theoretical level. QCD effects for the $t\bar{t}$ total cross section up to the NNLL order computed in this framework [4] significantly reduce the size and scale dependence to yield a 6% theoretical uncertainty.

The effective Lagrangian for the scalar version of vNRQCD, which describes the non-relativistic interaction between pairs of colored scalars and provides the needed ingredients for a summation of QCD effects at NLL order, has been given recently [14].

3. Finite width and electroweak effects

Up to now, no systematic and complete treatment of electroweak effects in the total cross section for top quark pair production has been achieved beyond the LO approximation. The large top width, being of the same order than the non-relativistic energy, is essential in the description of the $t\bar{t}$ threshold dynamics. It was shown [7] that in the non-relativistic limit the top-quark width can be consistently implemented by the replacement $E \to E + i\Gamma_t$ in the results for the total cross section for stable top quarks. Although this replacement rule can accomodate some of the NLO and NNLO electroweak corrections, a coherent treatment at the conceptual level requires the use of an extended NRQCD effective theory formalism.

In the NRQCD/vNRQCD framework electroweak corrections to the total cross section can be regarded as short-distance information to be encoded in the Wilson coefficients of the NRQCD Lagrangian and the NRQCD currents. As the particles involved in the electroweak corrections can be lighter than the top quark they can lead to nonzero imaginary parts in the matching conditions. These electroweak absorptive parts render the NRQCD Lagrangian non-hermitian, but the total cross section can still be obtained from the imaginary part of the $e^+e^- \to e^+e^-$ forward scattering amplitude by virtue of the optical theorem and the unitarity of the underlying theory, in complete analogy to the treatment of the inelastic processes in quantum mechanics where particle decay and absorption are implemented through potentials with complex coefficients. An important feature of the effective theory treatment of electroweak effects is that resonant and non-resonant contributions in NRQCD amplitudes can be systematically separated if the scaling relation $v \sim \alpha_s \sim \alpha^{1/2}$ is used. The latter is justified because numerically the top width is approximately equal to the typical top kinetic energy $\Gamma_t \sim m_t\alpha \sim E_{\rm kin} \sim m_t\alpha_s^2$. Including the non-resonant background diagrams leading to the same final states as those of top decay in the matching calculations is necessary in order to maintain gauge invariance.

In this approach the NNLL matching conditions accounting for the absorptive parts related to the $bW^+\bar{b}W^-$ final state were derived in [15] and shown to amount numerically to several percent. A very interesting new conceptual aspect of these corrections is that they have UV divergences that arise from the high energy behaviour of the $t\bar{t}$ effective theory phase space integration. The phase-space in the full theory is cut-off by the large top mass, but it extends to infinity in NRQCD where we have taken the limit $m_t \to \infty$ [a]. The divergences show up only when $\Gamma_t \neq 0$, as a finite width generates a distribution for the top invariant mass and thus allows for arbitrary large momenta in the non-relativistic phase space integration. Similar divergences had already been noted in the QCD NNLL relativistic corrections to the S-wave zero distance Green function if the unstable propagator was used [4,9] as well as in the leading order P-wave zero distance Green function which accounts for $t\bar{t}$ production through the Z-exchange [4]. The NNLL divergences renormalize $(e^+e^-)(e^+e^-)$ operators that contribute to the total cross section through the imaginary parts of their Wilson coefficients. The running induced in these Wilson coefficients by the divergences thus represents a NLL effect to the total cross section [15]

[a]If instead of using the optical theorem to obtain the total cross section from the forward scattering amplitude one integrates over the phase-space explicitly with cuts these UV divergences obviously do not arise

but their matching conditions at the hard scale are presently unknown.

Scalar particles are produced at leading order in the non-relativistic expansion in a P-wave state. In these systems the phase-space divergences constitute a more severe problem as they show up already in the leading order Green's function. A phenomenologically well motivated example is pair production of SUSY partners of the quarks at threshold, which has been studied at LO in several works [16] using a semi-phenomenological solution in order to deal with the phase-space UV divergences.

4. Non-relativistic currents with general quantum numbers

There are a number of issues concerning the consistent formulation of non-relativistic interpolating currents which describe color singlet heavy quark-antiquark and squark-antisquark pair production for general quantum numbers in $n = 3-2\epsilon$ dimensions . These involve in particular the generalization of spherical harmonics in n spatial dimensions and the role of evanescent operator structures (*i.e.* that vanish as $\epsilon \to 0$) for the description of the nonrelativistic spin.

Let us discuss first the spin singlet currents with arbitrary angular momentum L $({}^{2S+1}L_J = {}^1L_L)$. The generic structure of the production currents with total spin zero is $\psi^\dagger_{\boldsymbol{p}}(x)\,\Gamma(\boldsymbol{p})\,\tilde{\chi}^*_{-\boldsymbol{p}}(x)$, where $\Gamma(\boldsymbol{p})$ represents an arbitrary tensor depending on the c.m. momentum label $\boldsymbol{p}$ and $\tilde{\chi}^*_{-\boldsymbol{p}} = (i\sigma_2)\chi^*_{-\boldsymbol{p}}$. The interpolating currents associated to a definite angular momentum state L are related to irreducible representations of the tensor Γ with respect to the rotation group. The irreducible tensors are up to normalization just the spherical harmonics $Y_{LM}(n,\boldsymbol{p})$ of degree L, with $M = 1,\ldots,n_L$, that form an orthogonal basis of a n_L-dimensional space with $n_L = (2L+n-2)\,\frac{\Gamma(n+L-2)}{\Gamma(n-1)\Gamma(L+1)}$. A representation in terms of cartesian coordinates of the spherical harmonics of degree L is given by the totally symmetric and traceless tensors with L indices $T^{i_1\ldots i_L}(\boldsymbol{p})$, where the indices $i_1\ldots,i_L$ are cartesian coordinates:

$$T^{i_1\ldots i_L}(\boldsymbol{p}) = p^{i_1}\ldots p^{i_L} - \frac{\boldsymbol{p}^2}{2L+n-4}\left(\delta^{i_1 i_2}p^{i_3}\ldots p^{i_L} + \ldots\right) + \ldots\,, \quad (1)$$

and which satisfy the eigenvalue equation for the squared angular momentum operator $\boldsymbol{L}^2\, T^{i_1\ldots i_L}(\boldsymbol{p}) = L(L+n-2)\, T^{i_1\ldots i_L}(\boldsymbol{p})$. The currents with angular momenta S, P and D are *e.g.* relevant in the electromagnetic production of colored scalars from e^+e^- and $\gamma\gamma$ collisions. The use of the generalized currents built from (1) is mandatory to obtain consistent results in dimensional regularization in accordance with SO(n) rotational invariance. An instructive example regarding the computation of the nonrelativistic three-loop vacuum polarization diagram with two insertions of the Coulomb potential is explained in Ref. [17].

The interpolating currents describing the production of a fermion-antifermion pair in a spin triplet $S = 1$ state for arbitrary L (^3L_J) requires the treatment of Pauli $\boldsymbol{\sigma}$-matrices in n-dimensions. The $\boldsymbol{\sigma}$-matrices σ^i ($i = 1, \ldots, n$) are the generators of SO(n) rotations for spin 1/2 and satisfy the Euclidean Clifford algebra $\{\sigma^i, \sigma^j\} = 2\delta^{ij}$. As for the case of the γ-matrices [18] products of $\boldsymbol{\sigma}$-matrices in arbitrary number of dimensions cannot be reduced to a finite basis, but represent an infinite set of independent structures, which can be choosen as the antisymmetrized product of $\boldsymbol{\sigma}$ matrices: $\sigma^{i_1\cdots i_m} = \sigma^{[i_1}\sigma^{i_2}\cdots\sigma^{i_m]}$, $m = 0, 1, 2, \ldots$ For $m \leq 3$ the $\sigma^{i_1\cdots i_m}$ are related to physical spin operators, with eigenvalues of $\boldsymbol{S}^2$ equal to $(0, n-1, 2, n-3)$, which reduce to the known $n = 3$ values. The $m > 3$ operators are evanescent for $n \neq 3$ (although their spin eigenvalues are non-zero). S-wave currents in an arbitrary spin state thus have the form $\psi^\dagger_{\boldsymbol{p}}(x)\,\sigma^{i_1\cdots i_m}\,\tilde{\chi}^*_{-\boldsymbol{p}}(x)$, and can arise in important processes. The structure $\sigma^{i_1 i_2 i_3}$ for example arises in fermion pair production in $\gamma\gamma$ collisions, while the evanescent operator $\sigma^{i_1\cdots i_5}$ is present in the $\bar{f}f \to 3\gamma$ annihilation amplitude. Note that the differences between the two different singlet ($m = 0, 3$) and triplet ($m = 1, 2$) currents correspond to evanescent operators as well.

It is well known from subleading order computations based on the effective weak Hamiltonian that one needs to consistently account for the evanescent operator structures that arise in matrix elements of physical operators when being dressed with gluons. A renormalization scheme can be adopted such that a mixing of evanescent operators into physical ones does not arise [18]. Moreover it is also known [19] that modifications of the evanescent operator basis correspond to a change of the renormalization scheme. While this does not affect physical predictions, it does affect matrix elements, matching conditions and anomalous dimensions at nontrivial subleading order. Thus precise definitions of the schemes being used have to be given to render such intermediate results useful.

In the framework of the nonrelativistic EFT these properties still apply. However, using the velocity power counting in the EFT allows for even more specific statements. Concerning interactions throught potentials, transitions between the different S-wave currents built from $\sigma^{i_1\cdots i_m}$ cannot occur because the potentials are SO(n) scalars and the currents are inequivalent irreducible representations of SO(n). Even for currents with $L \neq 0$ and for the spin-dependent spin-orbit and tensor potentials (which is all we need to consider at NNLL order) one can show that transitions between currents containing $\sigma^{i_1 \cdots i_m}$ with a different number of indices cannot occur [17]. The same arguments apply to the exchange of soft gluons in vNRQCD. Concerning the exchange of ultrasoft gluons, transitions between currents containing $\sigma^{i_1 \cdots i_m}$ with a different number of indices can arise, but only if the interaction is spin-dependent. The dominant among these interactions corresponds to the operator $\psi_{\boldsymbol{p}}^\dagger \sigma^{ij} k^j \psi_{\boldsymbol{p}} A^i$ and can only contribute at N^4LL order, which is beyond the present need and technical capabilities.

So for the S-wave currents in n dimensions one can employ either one of the two spin singlet ($m = 0, 3$) or triplet currents ($m = 1, 2$) in the EFT and the difference corresponds to a change in the renormalization scheme. This means in particular that as long as the renormalization process is restricted to time-ordered products of the currents, one can freely use three-dimensional relations to reduce the basis of the physical currents. However, once the basis of the physical currents is fixed, one has to consistently apply the computational rules in n dimensions. Moreover, one can also conclude that currents containing evanescent $\sigma^{i_1 \cdots i_m}$ matrices ($m > 3$) can be safely dropped from the beginning as long as one does not need to account for spin-dependent ultrasoft gluon interactions.

Based on the later considerations it is straightforward to construct spin-triplet currents with arbitrary L (^3L_J). They can be obtained [17] by determining irreducible SO(n) representations from products of the tensors $T^{i_1 \ldots i_L}(\boldsymbol{p})$ describing angular momentum L and the spin-triplet $S = 1$ currents discussed previously. As for the case of the S-wave currents the physical basis for arbitrary spatial angular momentum is not unique due to the existence of evanescent operator structures. It is possible [17]. to construct currents with fully symmetric indices equal to the total angular momentum J by using the two spin triplet operators, σ^i and σ^{ij}. There are also currents having more than J indices, which transform according to more complicated patterns of SO(n). They become equivalent in $n = 3$ to the fully symmetric currents [17] and are also appropriate to describe production of ^3L_J states.

In this scheme, the NLL anomalous dimensions of the currents with arbitrary spin and angular momentum configurations, have been calculated [17]. The NLL running found for the currents shows a suppression $\propto 1/(2L+1)$, which suggests that the summation of logarithms of v for the production and annihilation rates of high angular momentum states is less significant.

Acknowledgments

I would like to thank the organizers of this workshop and their crew for the pleasant atmosphere during the conference.

References

1. M. Martinez and R. Miquel, Eur. Phys. J. C **27** (2003) 49 [arXiv:hep-ph/0207315].
2. A. H. Hoang *et al.*, Eur. Phys. J. directC **2** (2000) 1.
3. A. H. Hoang, talk at the *International Linear Collider Workshop*, Stanford, Califorina, USA, March 2005.
4. A. H. Hoang, A. V. Manohar, I. W. Stewart and T. Teubner, Phys. Rev. Lett. **86** (2001) 1951; A. H. Hoang, A. V. Manohar, I. W. Stewart and T. Teubner, Phys. Rev. D **65** (2002) 014014.
5. S. Heinemeyer, S. Kraml, W. Porod and G. Weiglein, JHEP **0309** (2003) 075.
6. H. Nowak, talk at the *ECFA Linear Collider Workshop*, Durham, UK, August 2004.
7. V. S. Fadin and V. A. Khoze, JETP Lett. **46** (1987) 525 [Pisma Zh. Eksp. Teor. Fiz. **46** (1987) 417]; V. S. Fadin and V. A. Khoze, Sov. J. Nucl. Phys. **48** (1988) 309 [Yad. Fiz. **48** (1988) 487].
8. W. E. Caswell and G. P. Lepage, Phys. Lett. B **167** (1986) 437.
9. A. H. Hoang and T. Teubner, Phys. Rev. D **60** (1999) 114027; A. H. Hoang and T. Teubner, Phys. Rev. D **58** (1998) 114023.
10. K. Melnikov and A. Yelkhovsky, Nucl. Phys. B **528** (1998) 59; O. I. Yakovlev, Phys. Lett. B **457** (1999) 170; T. Nagano, A. Ota and Y. Sumino, Phys. Rev. D **60** (1999) 114014; A. A. Penin and A. A. Pivovarov, Phys. Atom. Nucl. **64** (2001) 275 [Yad. Fiz. **64** (2001) 323]; M. Beneke, A. Signer and V. A. Smirnov, Phys. Lett. B **454** (1999) 137.
11. G. T. Bodwin, E. Braaten and G. P. Lepage, Phys. Rev. D **51**, 1125 (1995) [Erratum-ibid. D **55**, 5853 (1997)]; N. Brambilla, A. Pineda, J. Soto and A. Vairo, Nucl. Phys. B **566** (2000) 275; S. Fleming, I. Z. Rothstein and A. K. Leibovich, Phys. Rev. D **64**, 036002 (2001).
12. M. E. Luke, A. V. Manohar and I. Z. Rothstein, Phys. Rev. D **61**, 074025 (2000).
13. A. H. Hoang and I. W. Stewart, Phys. Rev. D **67**, 114020 (2003).
14. A. H. Hoang and P. Ruiz-Femenia, Phys. Rev. D **73** (2006) 014015.
15. A. H. Hoang and C. J. Reisser, Phys. Rev. D **71** (2005) 074022.
16. I. I. Y. Bigi, V. S. Fadin and V. A. Khoze, Nucl. Phys. B **377** (1992) 461; N. Fabiano, Eur. Phys. J. C **19** (2001) 547.
17. A A. H. Hoang and P. Ruiz-Femenia, "Renormalization of Heavy Pair Production Currents with General Quantum Numbers in Dimensionally Regularized NRQCD ," MPP-2006-48
18. M. J. Dugan and B. Grinstein, Phys. Lett. B **256**, 239 (1991).
19. S. Herrlich and U. Nierste, Nucl. Phys. B **455**, 39 (1995); K. G. Chetyrkin, M. Misiak and M. Munz, Nucl. Phys. B **520**, 279 (1998).

RADIATIVE TRANSITIONS AND THE QUARKONIUM MAGNETIC MOMENT

ANTONIO VAIRO

Dipartimento di Fisica dell'Università di Milano and INFN,
via Celoria 16, 20133 Milano, Italy
E-mail: antonio.vairo@mi.infn.it

I discuss heavy quarkonium radiative transitions and the related issue of the quarkonium magnetic moment inside effective field theories. Differences in set up and conclusions with respect to typical phenomenological approaches are outlined.

Keywords: Quarkonium, NRQCD, pNRQCD, radiative transitions

Heavy quarkonium radiative transitions have been studied for long time in the framework of phenomenological models[1–14] (for a recent review see[15]). In a typical set up[7], the starting point is a relativistic Hamiltonian whose dynamical fields are heavy quarks and photons; gluons are integrated out and replaced by a scalar and a vector interaction. The non-relativistic nature of heavy quarkonia is exploited by performing a Foldy–Wouthuysen transformation. Anomalous magnetic moments are added as if the system was made of two particles independently interacting with the external electromagnetic field.

Such an approach, despite phenomenological success, has an obvious limitation: its connection with QCD is unclear. More precisely, while it relies on a systematic (non-relativistic) expansion to describe the electromagnetic interaction of the heavy quarks, it does not to describe their strong interaction. As a result, the nature of the binding potential between the two heavy quarks and of the couplings of the heavy quarks with the electromagnetic field remains elusive. In particular, one may wonder if the heavy quarkonium anomalous magnetic moment gets contributions from the interaction of the heavy quarks in the bound state and if, in the non-relativistic expansion, the many couplings between the heavy quarks and the electromagnetic field, allowed by the symmetries of QCD, really organize as if they were generated by only two relativistic form factors (i.e. the scalar and vector interactions). Finally, the lack of a systematic expansion reflects in the fact that model-dependent determinations are affected by unknown uncertainties.

In the modern framework of non-relativistic effective field theories (EFTs)[16–20], one takes advantage of the small heavy quark velocity v to organize calculations of heavy quarkonium observables so that theoretical uncertainties can be evaluated and systematically reduced. Applied to radiative transitions, the EFT approach provides eventually definite answers to the above questions[21]. This is done through the following step.

First, one has to identify the relevant scales in the system. These are the heavy quark mass m, the typical momentum transfer in the bound state, which is of order mv, and the typical binding energy, which is of order mv^2. The momentum of the emitted photon, k_γ, is of order mv^2 for transitions between different radial levels (like magnetic-dipole hindered transitions) and usually of order mv^4 for transitions between levels with the same principal quantum number (like magnetic-dipole allowed transitions). Since the typical distance between the two heavy quarks is $r \sim 1/(mv)$, then $k_\gamma\, r \ll 1$ and the external electromagnetic field can be multipole expanded (at least for transitions between not too far away levels). We neglect virtual photons, whose contributions are suppressed by α. In QCD, it is also crucial to establish the size of these scales with respect to the typical hadronic scale $\Lambda_{\rm QCD}$. By definition of heavy quark: $m \gg \Lambda_{\rm QCD}$; however, $mv \gg \Lambda_{\rm QCD}$ holds possibly only for the lowest quarkonium resonances, while $mv \sim \Lambda_{\rm QCD}$ holds for the others. In the first case, called weakly coupled, the scale mv may be treated in perturbation theory (in this case $v \sim \alpha_{\rm s}$), in the last one, called strongly coupled, at the scale mv one cannot rely on an expansion in $\alpha_{\rm s}$ and non-perturbative techniques have to be used.

In order to construct an EFT suitable to describe heavy quarkonium transitions, we do not need to resolve scales larger than or of the same order as the soft scale mv. These scales may be integrated out from the EFT. The relevant degrees of freedom differ in the weakly and in the strongly coupled case. If $mv \gg \Lambda_{\rm QCD}$ these are singlet, S, and octet, O, quarkonium fields, gluons of energy and momentum mv^2 or $\Lambda_{\rm QCD}$ (sometimes also called ultrasoft) and photons. If $mv \sim \Lambda_{\rm QCD}$, at scales lower than mv colour confinement sets in and the relevant degrees of freedom are singlet quarkonium fields (all gluonic excitations between heavy quarks develop a mass gap of order $\Lambda_{\rm QCD}$ with respect to the lowest state and are integrated out) and photons. Fields scale in accordance to the lowest, still dynamical, energy scales, defining in this way the power counting of the EFT.

The EFT Lagrangian has the following form:

$$
\begin{aligned}
\mathcal{L} = & -\frac{1}{4}F^{\rm em}_{\mu\nu}F^{\mu\nu\,\rm em} + \int d^3r\, {\rm Tr}\left\{{\rm S}^\dagger\left(i\partial_0 - \frac{{\bf p}^2}{m} - V_s\right){\rm S}\right\} \\
& \left[-\frac{1}{4}F^a_{\mu\nu}F^{\mu\nu\,a} + \int d^3r\, {\rm Tr}\left\{{\rm O}^\dagger\left(iD_0 - \frac{{\bf p}^2}{m} - V_o\right){\rm O}\right\}\right. \\
& + \int d^3r\, V_A\, {\rm Tr}\left\{{\rm O}^\dagger {\bf r}\cdot g{\bf E}\,{\rm S} + {\rm S}^\dagger {\bf r}\cdot g{\bf E}\,{\rm O}\right\} \\
& \left. + \int d^3r\, V_B\, \frac{1}{2}{\rm Tr}\left\{{\rm O}^\dagger {\bf r}\cdot g{\bf E}\,{\rm O} + {\rm O}^\dagger{\rm O}{\bf r}\cdot g{\bf E}\right\} + \cdots\right]_{\text{weak coupling only}} \\
& + \mathcal{L}_\gamma, \qquad (1)
\end{aligned}
$$

where $\mathcal{L}_\gamma$ is the part of the Lagrangian that describes the interaction with the electromagnetic field. In the case of magnetic-dipole transitions, the relevant terms in $\mathcal{L}_\gamma$ up to relative order v^2 are:

$$
\begin{aligned}
\mathcal{L}_\gamma = & \int d^3r\, {\rm Tr}\left\{ V_A^{\rm em}\, {\rm S}^\dagger {\bf r}\cdot e{\bf E}^{\rm em}{\rm S} + \frac{1}{2m}\, V_1\, \left\{{\rm S}^\dagger, \boldsymbol{\sigma}\cdot e{\bf B}^{\rm em}\right\}{\rm S}\right. \\
& + \left[\frac{1}{2m}\, V_1\, \left\{{\rm O}^\dagger, \boldsymbol{\sigma}\cdot e{\bf B}^{\rm em}\right\}{\rm O}\right]_{\text{weak coupling only}} \\
& + \frac{1}{4m^2}\,\frac{V_2}{r}\, \left\{{\rm S}^\dagger, \boldsymbol{\sigma}\cdot[\hat{\bf r}\times(\hat{\bf r}\times e{\bf B}^{\rm em})]\right\}{\rm S} + \frac{1}{4m^2}\,\frac{V_3}{r}\, \left\{{\rm S}^\dagger, \boldsymbol{\sigma}\cdot e{\bf B}^{\rm em}\right\}{\rm S} \\
& \left. + \frac{1}{4m^3}\, V_4\, \left\{{\rm S}^\dagger, \boldsymbol{\sigma}\cdot e{\bf B}^{\rm em}\right\}\boldsymbol{\nabla}_r^2{\rm S} + \cdots\right\}, \qquad (2)
\end{aligned}
$$

where e is the electric charge of the heavy quark. The Wilson coefficients V_s, V_o, V_A, V_B, $V_A^{\rm em}$, V_i are in general functions of r and have to be determined by matching amplitudes calculated in the EFT with amplitudes calculated in QCD. The matching may be performed order by order in $\alpha_{\rm s}$ in the weak-coupling case, but has to be performed non-perturbatively in the strong-coupling one. In both situations, the matching can be done in two steps: the first one consists of integrating out hard modes of energy of the order of the heavy quark mass, the second one of integrating out soft modes of energy of the order of mv. The Wilson coefficients of the EFT will then have the factorized form: (hard)×(soft). For determinations of V_s, V_o, which are the singlet and octet potentials, and V_A, V_B we refer to[20] and references therein. The coefficients $V_A^{\rm em}$ and V_i have been studied in[21]. In the following, we summarize the main conclusions.

The coefficient V_1 may be interpreted as the heavy quarkonium magnetic moment. Its hard part is simply the sum of the heavy quark magnetic moments: $1 + 2\alpha_s/(3\pi) + \ldots$. Moreover, it can be shown that, in the $SU(3)_{\rm flavour}$ limit, to all orders in the strong-coupling constant, V_1 does not get soft contributions. The argument is somewhat similar to the factorization of the QCD corrections in $b \to u\, e^- \bar{\nu}_e$, which fixes the coefficient in the effective Lagrangian $\mathcal{L}_{\rm eff} = -4G_F/\sqrt{2}\, V_{ub}\, \bar{e}_L \gamma_\mu \nu_L\, \bar{u}_L \gamma^\mu b_L$ to be 1 to all orders in α_s (see, for example,[22]). This leads to the conclusion that the heavy quarkonium magnetic moment, if defined as the Wilson coefficient V_1, is just the sum of the heavy quark magnetic moments. Therefore, its anomalous part is small and positive and does not get any large low-energy and, in particular, non-perturbative contribution. This is consistent with a recent lattice determination[23]. Note that a large negative heavy quarkonium anomalous magnetic moment has been often advocated in potential models to accommodate the results with the data.

Reparametrization and Poincaré invariance[24,25] protect the coefficients V_2 and V_3. To all orders in perturbation theory and non-perturbatively it holds that $V_2 = r^2 V_s^{(0)\,\prime}/2$ ($V_s^{(0)}$ is the singlet static potential) and $V_3 = 0$. Since an effective scalar interaction would contribute to V_3, we conclude that such an interaction is not dynamically generated in QCD for magnetic-dipole transitions. Again, a scalar interaction is often employed in potential models.

The coefficient $V_A^{\rm em}$ is 1 up to possible corrections of order α_s^2. In general, the coefficients of the operators of order $1/m^3$ are not protected by any symmetry. V_4 is 1 plus $\mathcal{O}(\alpha_s)$ corrections. In the weak-coupling case, this is the only $1/m^3$ operator needed to describe magnetic-dipole transitions at relative order v^2. In the strong coupling regime, more terms are, in principle, necessary. Since $\alpha_s(1/r) \sim 1$ is no longer a suppression factor, more amplitudes shall contribute to the matching. These amplitudes will be encoded in the matching coefficients of the EFT in the form of static Wilson loop amplitudes with field strength insertions of the same kind as those that appear in the QCD potential at order $1/m^2$ [26]. Also, they may induce new operators in the EFT. A non-perturbative derivation of the EFT Lagrangian coupled to the electromagnetic field at order $1/m^3$ has not been worked out yet; note that to be of phenomenological impact such a calculation needs to be supplemented by lattice calculations of the relevant Wilson loop amplitudes. Since most of the potential model calculations rely on Lagrangians of the type of Eq. (2), we emphasize that, once cleaned up of the scalar interaction, they are suitable to describe the relativistic corrections for radiative transitions between weakly-coupled quarkonia ($mv^2 \gtrsim \Lambda_{\rm QCD}$) only. This is a rather severe constraint since most of the radiative transitions (e.g. magnetic dipole hindered, electric dipole, ...) involve higher excited quarkonium states, which may be difficult to accommodate as Coulombic bound states.

At present, allowed magnetic-dipole transitions between the quarkonium lowest states are those most reliably described in QCD. Moreover, it has been shown in[21] that for these processes, non-perturbative contributions mediated by quarkonium octet states and ultrasoft gluons cancel at relative order v^2. Therefore, at relative order v^2, transitions like $J/\psi \to \eta_c\,\gamma$ and $\Upsilon(1S) \to \eta_b\,\gamma$ are completely accessible in perturbation theory. An explicit calculation gives:

$$\Gamma(J/\psi \to \eta_c\,\gamma) = (1.5 \pm 1.0)\ \mathrm{keV}, \tag{3}$$

$$\Gamma(\Upsilon(1S) \to \eta_b\,\gamma) = \left(\frac{k_\gamma}{39\ \mathrm{MeV}}\right)^3 (2.50 \pm 0.25)\ \mathrm{eV}. \tag{4}$$

The error in Eq. (3) accounts for the large uncertainties coming from higher-order corrections; we recall that corrections of order $k_\gamma^3\,v^2/m^2$ affect the leading-order result by about 50%. Uncertainties may be reduced by higher-order calculations. The value given in Eq. (3) is consistent with the present experimental one[27]. This means that assuming the ground-state charmonium to be a weakly-coupled bound state leads to relativistic corrections to the transition width of the right sign and size. The decay width $\Gamma(\Upsilon(1S) \to \eta_b\,\gamma)$ depends on the η_b mass, which is unknown. The quoted value corresponds to a η_b mass of about 9421 MeV.

We conclude with few possible developments of the work discussed here. For magnetic-dipole transitions between higher resonances, the completion of the non-perturbative matching of the relevant operators in the EFT at order $1/m^3$ will be needed, possibly integrated by lattice calculations of the relevant Wilson-loop amplitudes. For magnetic-dipole hindered transitions of the type $\Upsilon(3S) \to \eta_b\gamma$, since the momentum of the emitted photon is comparable with the typical momentum transfer in the bound state, one cannot rely on the multipole expansion of the external electromagnetic field. In this case, one may, in principle, exploit the fact that the typical momentum transfer inside the η_b is much larger than that one inside the $\Upsilon(3S)$. A suitable treatment has not been developed yet. For electric-dipole transitions, much of the study still remains to be done. In the weak-coupling regime, octet contributions may not vanish; they can be worked out as in the case of the magnetic-dipole transitions. However, most of the electric-dipole transitions may need to be treated in a strong-coupling framework.

Acknowledgements

I thank Nora Brambilla and Yu Jia for an enjoyful collaboration on the work presented here. I acknowledge the financial support obtained inside the Italian MIUR program "incentivazione alla mobilità di studiosi stranieri e italiani residenti all'estero".

References

1. G. Feinberg and J. Sucher, Phys. Rev. Lett. **35**, 1740 (1975).
2. J. Sucher, Rept. Prog. Phys. **41**, 1781 (1978).
3. E. Eichten, K. Gottfried, T. Kinoshita, K. D. Lane and T. M. Yan, Phys. Rev. D **17**, 3090 (1978) [Erratum-ibid. D **21**, 313 (1980)]; E. Eichten, K. Gottfried, T. Kinoshita, K. D. Lane and T. M. Yan, Phys. Rev. D **21**, 203 (1980).
4. J. S. Kang and J. Sucher, Phys. Rev. D **18**, 2698 (1978).
5. K. J. Sebastian, Phys. Rev. D **26**, 2295 (1982).
6. G. Karl, S. Meshkov and J. L. Rosner, Phys. Rev. Lett. **45**, 215 (1980).
7. H. Grotch and K. J. Sebastian, Phys. Rev. D **25**, 2944 (1982).
8. P. Moxhay and J. L. Rosner, Phys. Rev. D **28**, 1132 (1983)
9. R. McClary and N. Byers, Phys. Rev. D **28**, 1692 (1983); V. Zambetakis and N. Byers, Phys. Rev. D **28**, 2908 (1983).
10. H. Grotch, D. A. Owen and K. J. Sebastian, Phys. Rev. D **30**, 1924 (1984).
11. Fayyazuddin and O. H. Mobarek, Phys. Rev. D **48**, 1220 (1993).
12. T. A. Lahde, Nucl. Phys. A **714**, 183 (2003) [arXiv:hep-ph/0208110].
13. D. Ebert, R. N. Faustov and V. O. Galkin, Phys. Rev. D **67**, 014027 (2003) [arXiv:hep-ph/0210381].
14. T. Barnes, S. Godfrey and E. S. Swanson, Phys. Rev. D **72**, 054026 (2005) [arXiv:hep-ph/0505002].
15. N. Brambilla *et al.*, *Heavy quarkonium physics*, CERN-2005-005, (CERN, Geneva, 2005) [arXiv:hep-ph/0412158].
16. W. E. Caswell and G. P. Lepage, Phys. Lett. B **167**, 437 (1986).
17. G. T. Bodwin, E. Braaten and G. P. Lepage, Phys. Rev. D **51**, 1125 (1995) [Erratum-ibid. D **55**, 5853 (1997)] [hep-ph/9407339].
18. A. Pineda and J. Soto, Nucl. Phys. Proc. Suppl. **64**, 428 (1998) [arXiv:hep-ph/9707481].
19. N. Brambilla, A. Pineda, J. Soto and A. Vairo, Nucl. Phys. B **566**, 275 (2000) [arXiv:hep-ph/9907240].
20. N. Brambilla, A. Pineda, J. Soto and A. Vairo, Rev. Mod. Phys. **77**, 1423 (2005) [arXiv:hep-ph/0410047].
21. N. Brambilla, Y. Jia and A. Vairo, Phys. Rev. D **73**, 054005 (2006) [arXiv:hep-ph/0512369].
22. M. Neubert, *Effective field theory and heavy quark physics*, arXiv:hep-ph/0512222.

23. J. J. Dudek, R. G. Edwards and D. G. Richards, Phys. Rev. D **73**, 074507 (2006) [arXiv:hep-ph/0601137].
24. A. V. Manohar, Phys. Rev. D **56**, 230 (1997) [hep-ph/9701294].
25. N. Brambilla, D. Gromes and A. Vairo, Phys. Lett. B **576**, 314 (2003) [hep-ph/0306107].
26. A. Pineda and A. Vairo, Phys. Rev. D **63**, 054007 (2001) [Erratum-ibid. D **64**, 039902 (2001)] [arXiv:hep-ph/0009145].
27. W. M. Yao *et al.* [Particle Data Group], J. Phys. G **33**, 1 (2006).

$B \to K^*\gamma$ VS $B \to \rho\gamma$ AND $|V_{td}/V_{ts}|$

PATRICIA BALL* and ROMAN ZWICKY†

IPPP, Department of Physics, University of Durham,
Durham DH1 3LE, UK
* *E-mail: patricia.ball@durham.ac.uk* † *roman.zwicky@durham.ac.uk*

We determine $|V_{td}/V_{ts}| = 0.179 \pm 0.022(\text{exp}) \pm 0.014(\text{th})$ from $B \to (K^*, \rho)\gamma$ as measured by the B-factories Babar and Belle.

1. Introduction

Decays like $B \to (K^*, \rho)\gamma$ are the exclusive counterparts of the famous flavour changing neutral current (FCNC) decays $b \to (s, d)\gamma$. It is well known that FCNC are absent in the Standard Model (SM) at the tree-level and further suppressed at the loop-level by the GIM-mechanism. Therefore they represent a sensitive probe for new physics (NP) and are widely studied in the literature. Moreover the weak transition involves the CKM-matrix elements $V_{t(d,s)}$.

The decays are described in the language of effective Hamiltonians $H_{\text{eff}} = \sum_i C_i(\mu)O_i(\mu)$ where the Wilson coefficients C_i are matched to the SM or some beyond SM framework at the electroweak scale. As a next step the renormalization group is employed to evolve the coefficients down to a scale μ where the actual decay of the B-meson takes place. This procedure has been performed in QCD to NLO [1] and the NNLO calculation are expected to be finished soon. As a final step the matrix elements of the operators O_i have to be taken, which is a notoriously difficult task due to bound-state effects. The matrix elements are parametrized in terms of form factors and it is the aim of this text to report on the reduction of the uncertainty of the relevant form factors in $B \to K^*\gamma$ and $B \to \rho\gamma$ [2].

On the experimental side the mode $B \to K^*\gamma$ has been measured as early as 1993 by CLEO. Its inclusive counterpart $b \to s\gamma$ is in very good agreement with NP calculations and gives stringent constraints on physics beyond the SM. The decay $B \to \rho\gamma$ has recently been measured by Belle [3] and very recently by Babar [4].

2. Reduced uncertainty in the ratio of branching ratios

The effective Hamiltonian for the $b \to D\gamma$ transition reads ($D = d, s$)

$$H_{\text{eff}}^{b\to D} = \frac{4G_F}{\sqrt{2}}\Big(\sum_{U=u,c} \lambda_U^{(D)} \Big[C_1 O_1^{U,(D)} + C_2 O_2^{U,D}\Big] + (-\lambda_t^{(D)}) \sum_{i=3..8} C_i O_i\Big)$$

where $O_{1,2}$ are current-current operators, $O_{3..6}$ penguin operators and $O_{7,8}$ are the electric and magnetic dipole operators. The most important ones are

$$O_7 \; = \frac{em_b}{32\pi^2}(\bar{D}\sigma_{\mu\nu}F^{\mu\nu}(1+\gamma_5)b) \quad , \tag{1}$$

$$O_2^{U,D} = \frac{1}{4}(\bar{D}\gamma_\mu(1-\gamma_5)U)(\bar{U}\gamma^\mu(1-\gamma_5)b) \quad , \tag{2}$$

where $U = u, c$ as above and the λ's are CKM factors, e.g. $\lambda_t^s = V_{\text{tb}}V_{\text{ts}}^*$. It has recently been shown in QCD factorization that at leading order in $1/m_b$ the following factorization formula applies [5]

$$\langle V\gamma|O_i|B\rangle = \left[T_1^{B\to V}(0)\, T_i^I + \int_0^1 d\xi\, du\, T_i^{II}(\xi,u)\, \phi_B(\xi)\, \phi_{V;\perp}(v)\right]\cdot\epsilon\,, \tag{3}$$

where ϵ is the photon polarisation 4-vector, O_i is one of the operators in the effective Hamiltonian, $T_1^{B\to V}$ is a $B \to V$ transition form factor, and ϕ_B, $\phi_{V;\perp}$ are leading-twist light-cone distribution amplitudes (DA) of the B meson and the vector meson V, respectively. These quantities are universal non-perturbative objects and describe the long-distance dynamics of the matrix elements, which is factorised from the perturbative short-distance interactions included in the hard-scattering kernels T_i^I and T_i^{II}. The amplitude reads

$$A(B\to(K^*,\rho)\gamma) = \Big[\frac{4G_F}{\sqrt{2}} V_{\text{tb}}V^*_{\text{t(s,d)}}\Big]\, a_7(K^*,\rho)\, \underbrace{\langle V\gamma|O_7|B\rangle}_{\sim T_1^{B\to V}(0)} + O(1/m_b)$$

with $a_7 = C_7 + O(\alpha_s)$. The most important $1/m_b$ corrections are due to the operator O_2 (2), which come with a numerically enhanced Wilson coefficient $C_2(m_b) \sim 3|C_7(m_b)|$. A further hierarchy is set by CKM factors, c.f. Tab. 1, which implies that for the $B \to K^*\gamma$ transition solely the current-current operator with a c-quark pair is numerically relevant. This contribution has been estimated a long time ago [6] and is considered to be under reasonable control. Moreover, assuming three generations, $|V_{ts}| = |V_{cb}|(1 + O(\lambda^2))$, and $|V_{cb}|$ is known with a precision of 2% [7]. Therefore, the major unknown quantity is the penguin form factor $T_1^{B\to K^*}(0)$. The value quoted by the

Table 1. CKM-hierarchy for operators $O_{2(1)}$. $\lambda \sim 0.22$ is a Wolfenstein-parameter.

λ_U^D	$U = u$	$U = c$
$\rho \ : D = d$	λ^3	λ^3
$K^* : D = s$	λ^4	λ^2

authors in [†,5] compares well to an update of the form factors from Light Cone Sum Rules (LCSR) [8]

$$T_1{}^{B\to K^*}(0)^\dagger = 0.28 \pm 0.02 \leftrightarrow T_1{}^{B\to K^*}_{\rm LCSR}(0) = 0.31 \pm 0.04$$

due to recent progress in the Kaon distribution amplitude to be discussed below. For the $B \to \rho\gamma$ transition things are more complicated because there also the current-current operators with an u-quark contribute which include annihilation contributions and an u-quark loop which has so far only been addressed by model dependent calculations, which makes the extraction of $|V_{\rm td}|$ by itself difficult.

On the other hand, in the ratio of branching ratios

$$\frac{\mathcal{B}(B \to \rho\gamma)}{\mathcal{B}(B \to K^*\gamma)} = \left|\frac{V_{td}}{V_{ts}}\right|^2 \left(\frac{m_B^2 - m_\rho^2}{m_B^2 - m_{K^*}^2}\right)^3 \underbrace{\left(\frac{T_1^\rho(0)}{T_1^{K^*}(0)}\right)^2}_{\equiv \xi^{-2}} \left|\frac{a_7^c(\rho)}{a_7^c(K^*)}\right|^2 [1+\Delta R] \quad , \tag{4}$$

the $1/m_b$-corrections ΔR are somewhat accidently CKM-suppressed. The parameters $|a_7^c(\rho\gamma)|$ and $|a_7^c(K^*\gamma)|$ are almost exactly equal, so we set $|a_7^c(\rho\gamma)/\ a_7^c(K^*\gamma)| = 1$. Non-factorizable contributions are unlikely to change this ratio significantly [6,9]. In order to compensate for relatively poor statistics an averaging is performed over isospin and we completely neglect the difference between the ρ and the ω meson. Then

$$\Delta R_{\pm,0} \simeq {\rm Re}\,(\delta a_\pm + \delta a_0) f_{\rm CKM} + \frac{1}{2}\left(|\delta a_\pm|^2 + |\delta a_0|^2\right) g_{\rm CKM} \quad ,$$

where $\delta a_{0,\pm} = a_7^u(\rho^{0,\pm}\gamma)/a_7^c(\rho^{0,\pm}\gamma) - 1$. The almost-equal sign indicates that other types of corrections are Cabibbo-suppressed and the subscript is a reminder of the isospin averaging. The CKM factors are given by

$$f_{\rm CKM} = \frac{R_b^2 - R_b\cos\gamma}{1 - 2R_b\cos\gamma + R_b^2} = 0.07 \pm 0.12$$

$$g_{\rm CKM} = \frac{R_b^2}{1 - 2R_b\cos\gamma + R_b^2} = 0.23 \pm 0.07 \quad ,$$

with γ being one angle of the unitarity triangle and $R_b \equiv (1 - \lambda^2/2)\frac{1}{\lambda}|V_{ub}/V_{cb}|$ one of its sides. The uncertainties are obtained by using $\gamma = (71 \pm 16)^\circ$, the value of R_b has only a minor impact as compared to the former.

2.1. *The form factor ratio* $T_1^{B\to K^*}/T_1^{B\to\rho}$

The penguin form factor is defined by the following matrix element

$$\langle V(p)|\bar{D}\sigma_{\mu\nu}q_\nu(1+\gamma_5)b|B(p_B)\rangle = i\epsilon_{\mu\nu\rho\sigma}\varepsilon^{*\nu}p_B^\rho p^\sigma 2T_1^{B\to V}(0) \quad (5)$$

at zero momentum transfer $q^2 = (p_B - p)^2 = 0$. In reference [8] this form factor was calculated using LCSR including radiative corrections up to next-to-leading twist. The uncertainty of the ratio

$$\xi \equiv \frac{T_1^{B\to K^*}(0)}{T_1^{B\to\rho}(0)} \quad , \quad (6)$$

we are aiming at, is considerably smaller than that of the individual form factors themselves as many uncertainties tend to cancel each other, e.g. the dependence on the b-quark mass and the normalization through the B-meson decay constant f_B. The form factor ratio ξ is basically a measure of $SU(3)$-breaking.

In reference [2] we have further improved this calculation w.r.t. [8] by including updated $SU(3)$-breaking in the twist-2 parameters, complete account of $SU(3)$-breaking in the twist 3 and 4 DA and we have tested the sensitivity of the ratio to different models of DA. It was found that solely the SU(3) breaking of the leading twist-2 DA has a major impact on the numerical value of ξ and we will therefore shortly discuss its status.

2.2. *SU(3)-breaking of distribution amplitudes*

The most important contribution to $T_1^{B\to V}$ comes from the leading-twist transverse DA

$$\langle 0|\bar{q}(z)\sigma_{\mu\nu}D(-z)|V(q,\lambda)\rangle = i(e_\mu^{(\lambda)}q_\nu - e_\nu^{(\lambda)}q_\mu)f_V^\perp(\mu)\int_0^1 du\, e^{i\xi q\cdot z}\phi_V^\perp(u),$$

and we refer to [10,11] for further details and references. The normalization of the transverse DA cannot be directly accessed by experiment unlike the longitudinal DA. We therefore rely on theoretical calculations. The updated QCD sum rule calculations for these quantities are [2]

$$f_\rho^\perp(1\,\text{GeV}) = (0.165 \pm 0.009)\,\text{GeV},\; f_{K^*}^\perp(1\,\text{GeV}) = (0.185 \pm 0.010)\,\text{GeV}. \quad (7)$$

The uncertainties of these quantities are rather large by themselves as compared to the longitudinal decay constants

$$f_{\rho}^{\parallel} = (0.209 \pm 0.002)\,\text{GeV}, \qquad f_{K^*}^{\parallel} = (0.217 \pm 0.005)\,\text{GeV}.$$

The uncertainties in (7) dominate the uncertainty in the final extraction of $|V_{\text{td}}/V_{\text{ts}}|$. Lattice QCD [12] has up to now only produced ratios of the longitudinal to transverse decay constants. The ratios of decay constants in Sum Rules [2] and Lattice-QCD calculations agree reasonably well. Nevertheless further efforts in this direction would be highly desirable.

The other important quantity for the tensor ratio (6) is the first Gegenbauer $a_1^{\perp}(V)$ moment in the conformal expansion of the DA,

$$\phi_{\perp}(u) = 6u\bar{u}\left(1 + \sum_{n\geq 1} a_n^{\perp}(V) C_n^{3/2}(2u-1)\right) \quad .$$

The first Gegenbauer moment a_1 is zero for particles with a definite G-parity as the ρ-meson. Moreover it is a measure for the average momentum of the strange quark as compared to the light quark in the Kaon and it is therefore expected to be positive on intuitive grounds. A positive value was found in [13] a long time ago, but when radiative corrections were evaluated and a sign mistake corrected, the value of a_1 became negative [14]. In the following year two papers appeared [15,16] which criticized the instability of the non-diagonal sum rule used in [13,14]. One paper used an exact operator relation [16] and the other one used stable diagonal sum rules [15]. Both analyses were completed in reference [10] and [11]. It is found that the diagonal sum rules are numerically superior to the operator method and the final values for a_1 are (at $\mu = 1\,\text{GeV}$) [10,11,17]

$$a_1(K) = 0.06 \pm 0.03, \quad a_1^{\parallel}(K^*) = 0.03 \pm 0.02, \quad a_1^{\perp}(K^*) = 0.04 \pm 0.03.$$

With these values we obtained the following result [2]

$$\begin{aligned}
\xi = \frac{T_1^{B\to K^*}(0)}{T_1^{B\to\rho}(0)} &= 1.17 \pm 0.08(f_{\rho,K^*}^{\perp}) \pm 0.03(a_1) \pm 0.02(a_2) \pm 0.02(\text{t-3,4}) \\
&\quad \pm 0.01(\text{sum-rule param., } m_b\text{, t-2 and -4 models}) \\
&= 1.17 \pm 0.09\,, \qquad (8)
\end{aligned}$$

where the total uncertainty of ± 0.09 is obtained by adding the individual terms in quadrature.

2.3. *Extraction of* $|V_{\rm td}/V_{\rm ts}|$

The Belle collaboration has measured the quantity

$$R_{\rm exp} \equiv \frac{\overline{\mathcal{B}}(B \to (\rho,\omega)\gamma)}{\overline{\mathcal{B}}(B \to K^*\gamma)},$$

where $\overline{\mathcal{B}}(B \to (\rho,\omega)\gamma)$ is defined as the CP-average of

$$\mathcal{B}(B \to (\rho,\omega)\gamma) = \frac{1}{2}\left\{\mathcal{B}(B^+ \to \rho^+\gamma) + \frac{\tau_{B^+}}{\tau_{B^0}}\left[\mathcal{B}(B^0 \to \rho^0\gamma) + \mathcal{B}(B^0 \to \omega\gamma)\right]\right\},$$

and $\overline{\mathcal{B}}(B \to K^*\gamma)$ is the isospin- and CP-averaged branching ratio of the $B \to K^*\gamma$ channels. HFAG averages Belle and Babar results into [18]

$$R_{\rm exp}^{\rm HFAG} = 0.024 \pm 0.006\,. \tag{9}$$

This has to be compared to the theoretical prediction

$$R_{\rm th} = \left|\frac{V_{td}}{V_{ts}}\right|^2 \left[0.75 \pm 0.11(\xi) \pm 0.02(a_7^{u,c}, \gamma, R_b)\right], \tag{10}$$

where in (2) the annihilation contributions have been evaluated in LCSR as well as QCD factorization and although the individual results differ slightly it has an absolutely negligible effect on the value and the uncertainty of the ratio due to the accidental CKM-supression [2]. Equating we extract

$$\left|\frac{V_{td}}{V_{ts}}\right|^{\rm HFAG}_{B\to V\gamma} = 0.179 \pm 0.022({\rm exp}) \pm 0.014({\rm th})\,, \tag{11}$$

which is our final result. It is interesting to compare this result to two other determinations of $|V_{\rm td}/V_{\rm ts}|$. First in the SM the ratio is related to other CKM-parameters by unitarity and allows an estimate from the CKM-fit

$$\left|\frac{V_{td}}{V_{ts}}\right|_{\rm SM} = \lambda(1 + R_b^2 - 2R_b\cos\gamma)^{1/2} = 0.216 \pm 0.029\,.$$

It is also interesting to note that for a typical LHCb uncertainty of the angle γ of about 4% the above uncertainty would drop to about ~ 0.007. Moreover this determination can be compared to that from $B_{(d,s)}$-oscillations which have been measured by CDF; they quote the following value [19]

$$\left|\frac{V_{td}}{V_{ts}}\right|_{{\rm CDF},\Delta {\rm M}_s} = 0.208^{+0.001}_{-0.002}({\rm exp})^{+0.008}_{-0.006}({\rm theo}) \quad . \tag{12}$$

3. Conclusions

We have calculated the ratio of tensor form factors determining the decays $B \to (K^*, \rho)\gamma$ as $\xi = 1.17 \pm 0.09$. Its uncertainty is dominated by the uncertainty of transversal decay constants. We then have calculated the isospin averaged ratio of branching ratios (10), including a preliminary estimate of power correction to be completed in [9]. From this we have obtained $|V_{\rm td}/V_{\rm ts}| = 0.179 \pm 0.022(\exp) \pm 0.014(\mathrm{th})$. This value is consistent, within uncertainties, with the value from the CKM-fit (12) and the value from $B_{(d,s)}$-oscillations (12) and therefore does not indicate signs of NP. It has to be mentioned that not only calculational uncertainties but also NP could cancel in the ratios considered. Therefore inspection of individual branching ratios is necessary and in particular an assessement of power corrections becomes indispensable [9].

Acknowledgements

R.Z. is grateful to the organizers of CAQCD06 for hospitality and the opportunity to present this work.

References

1. K. G. Chetyrkin, M. Misiak and M. Münz, Phys. Lett. B **400** (1997) 206
2. P. Ball and R. Zwicky, JHEP **0604** (2006) 046
3. K. Abe *et al.*, Phys. Rev. Lett. **96** (2006) 221601
4. B. Aubert, arXiv:hep-ex/0607099.
5. M. Beneke, T. Feldmann and D. Seidel, Eur. Phys. J. C **41** (2005) 173
   S. W. Bosch and G. Buchalla, JHEP **0501** (2005) 035
6. A. Khodjamirian *et al.*, Phys. Lett. B **402** (1997) 167
7. O. Buchmüller and H. Flächer, Phys. Rev. D **73**, 073008 (2006)
8. P. Ball and R. Zwicky, Phys. Rev. D **71** (2005) 014029
9. P. Ball and R. Zwicky, in preparation
10. P. Ball and R. Zwicky, Phys. Lett. **B 633** (2006) 289
11. P. Ball and R. Zwicky, JHEP **0602**, 034 (2006)
12. D. Becirevic *et al.*, JHEP **0305**, 007 (2003)
    V. M. Braun *et al.*, Phys. Rev. D **68** (2003) 054501
13. V. L. Chernyak, A. R. Zhitnitsky and I. R. Zhitnitsky, Nucl. Phys. B **204** (1982) 477 [Erratum-ibid. B **214** (1983) 547];
14. P. Ball and M. Boglione, Phys. Rev. D **68**, 094006 (2003)
15. A. Khodjamirian, T. Mannel and M. Melcher, Phys. Rev. D **70** (2004) 094002
16. V. M. Braun and A. Lenz, Phys. Rev. D **70**, 074020 (2004)
17. P. Ball, V. M. Braun and A. Lenz, JHEP **0605** (2006) 004
18. E. Barberio *et al.* [HFAG], arXiv:hep-ex/0603003;
19. A. Abulencia [CDF - Run II Collaboration], arXiv:hep-ex/0606027.

SECTION 4
HIGH TEMPERATURE/DENSITY PHYSICS

Convener E. Shuryak

CHARM AND CHARMONIUM IN THE QUARK-GLUON PLASMA

R. RAPP*, D. CABRERA and H. VAN HEES

Cyclotron Institute and Physics Department, Texas A&M University,
College Station, Texas 77843-3366, U.S.A.
** E-mail: rapp@comp.tamu.edu*

In the first part of the talk, we briefly review the problem of parton-energy loss and thermalization at the Relativistic Heavy-Ion Collider and discuss how heavy quarks (charm and bottom) can help to resolve the existing experimental and theoretical puzzles. The second part of the talk is devoted to the properties of heavy quarkonia in the (strongly interacting) Quark-Gluon Plasma (sQGP) and their consequences for observables in heavy-ion collisions.

Keywords: Quark-Gluon Plasma, Heavy Quarks and Quarkonia, Ultrarelativistic Heavy-Ion Collisions

1. Introduction

Experiments at the Relativistic Heavy-Ion Collider (RHIC) suggest that the matter created in semi-/central Au-Au collisions at $\sqrt{s_{NN}}$=200GeV constitutes an equilibrated, strongly interacting medium at high energy density, well above the critical one at the expected phase boundary, $\epsilon_c \simeq 1\,\mathrm{GeV/fm}^3$ (cf. the experimental assessment papers[1,2] and references therein). This conclusion is essentially based on three evidences: (i) at low transverse momenta, $p_T \lesssim$2-3 GeV (comprising ~99% of the produced hadrons), ideal relativistic hydrodynamics describes well the single-particle spectra and their azimuthal asymmetry, $v_2(p_T)$; the underlying collective expansion follows from kinetic equilibration times of around $\tau_0 \simeq 0.5\,\mathrm{fm}/c$, implying rapid (local) thermalization of the system at initial energy densities of about $\epsilon_0 \simeq 30\,\mathrm{GeV/fm}^3$. (ii) at high momenta, $p_T \gtrsim$5-6 GeV, the production of hadrons is suppressed by a factor of 4-5 relative to elementary p-p (or d-Au) collisions. This has been interpreted as energy loss of energetic partons traveling through an almost opaque high gluon-density medium. (iii) at intermediate momenta, $p_T \simeq$2-6GeV, surprisingly large baryon-to-meson ratios, as well as an empirical scaling of the hadron v_2 according to its constituent quark content, has been interpreted as quark coalescence as the prevalent hadronization mechanism in this regime, and thus as partonic degrees of freedom being active. The microscopic mechanisms underlying thermalization and energy loss, however, are not well understood yet. In fact, the jet quenching of light partons is so strong that high-p_T particle emission appears to be mostly limited to the surface of the fireball, rendering a precise determination of the corresponding transport coefficient (and maximal energy density) difficult. While earlier approaches assumed the dominance of radiative energy loss via medium-induced gluon radiation (which prevails in the limit of high jet energies) cf. [3], it has been realized subsequently that elastic ($2 \leftrightarrow 2$) scattering processes cannot be neglected at currently accessible $p_T \leq 10$-$15\,\mathrm{GeV}$ at RHIC. Moreover, none of the evidences (i)-(iii) is directly connected to the fundamental properties distinguishing the Quark-Gluon Plasma (QGP) from hadronic matter, namely the deconfinement of color charges and the restoration of the spontaneously broken chiral symmetry (which is believed to generate most of the visible mass in the universe).

Heavy-quark (HQ) observables are expected to provide new insights into the aforementioned problems (with the exception of chiral symmetry restoration). On the one hand, charm (c) and bottom (b) quarks, due to their relatively large mass, should suffer less energy loss and undergo delayed thermalization when traversing the QGP, and thus be more sensitive to the interactions with thermal partons than light quarks and gluons. Therefore, it came as a surprise when the measurement of single-electron ($e^{\pm}$) spectra associated with the semileptonic decay of open-charm (and -bottom) hadrons showed a factor 4-5 suppression[4,5], very comparable to what has been found for pions. In addition, the observed values for the $e^{\pm}$ elliptic flow, $v_2^e(p_T)$, reach up to 10% at $p_T \simeq 2\,\mathrm{GeV}$, indicating the build-up of substantial *early* collectivity of c quarks. These observations reinforced the implementation of elastic energy-loss processes into the theoretical description of the spectra[6–10]. On the other hand, HQ bound states (quarkonia), due to their small size/large binding, are suitable probes of a surrounding medium of sufficiently high density. E.g., for a typical (ground-state) charmonium of size r=0.25 fm, the relevant parton density at which significant modifications are to be expected is $n \approx 3/(4\pi r^3) \approx 10$-$20\,\mathrm{fm}^{-3}$, which for an ideal QGP (with N_f=2.5 massless flavors) translates into a temperature of $T \approx 270\,\mathrm{MeV} \approx 1.5 T_c$. Indeed, recent lattice QCD (lQCD) calculations indicate that J/ψ and η_c states survive in a QGP up to $\sim 2T_c$. The theoretical challenge is then to disentangle and quantify (a) medium modifications of the binding potential (e.g., color screening), (b) parton-induced dissociation reactions, and (c) in-medium changes of the HQ mass which affects both the bound-state mass and its decay threshold, as well as to identify and establish connections to observables in heavy-ion collisions.

In the first part of the talk (Sec. 2), we evaluate HQ diffusion and thermalization in the QGP employing both perturbative and *nonperturbative* elastic interactions. This problem is particularly suited to a Fokker-Planck approach for Brownian motion in a thermal background[11], which allows to describe both the quasi-thermal and kinetic regime, and its transition. A Langevin simulation for c and b quarks in an expanding QGP at RHIC is supplemented with coalescence (and fragmentation) at T_c, thus implementing all of the three main features (i)-(iii) of the initial RHIC data. In the second part of the talk (Sec. 3), we discuss in-medium properties of heavy quarkonia, including information from lQCD. We outline a T-matrix approach in which HQ potentials from lQCD are used as input to simultaneously describe bound and scattering states. This enables a comprehensive evaluation of euclidean correlation functions, which in turn can be checked against rather accurate results from lQCD. We briefly discuss how in-medium quarkonium properties reflect themselves in observables in ultrarelativistic heavy-ion collisions. Sec. 4 contains our conclusions.

2. Open charm and bottom in the QGP

Recent calculations of hadronic spectral functions in a QGP, both within lQCD and lQCD-based potential models, suggest that resonance (or bound) states persist up to $2T_c$, for both heavy- ($Q\bar{Q}$) and light-quark ($q\bar{q}$) systems. We conjectured[6] that this also holds for heavy-light systems, and therefore could lead to significantly faster thermalization for c and b quarks as compared to perturbative QCD (pQCD) processes. Even at T=350 MeV (which roughly corresponds to initial temperatures at RHIC) and for α_s=0.4, elastic pQCD scattering, which is dominated by t-channel gluon exchange, results in a thermal relaxation time for c quarks above 10 fm/c, well above a the typical QGP lifetime of $\tau_{\rm QGP}\sim$5 fm/c. Recent data on single-electron spectra associated with semileptonic heavy-meson decays have corroborated the need for nonperturbative HQ interactions in the QGP. As shown in Fig. 1, perturbative energy-loss calculations appear insufficient to describe the strong suppression in the p_T-spectra (left panel, using both elastic and radiative interactions) and the large $v_2(p_T)$ (right panel, using radiative energy loss with a substantially upscaled transport coefficient). While the

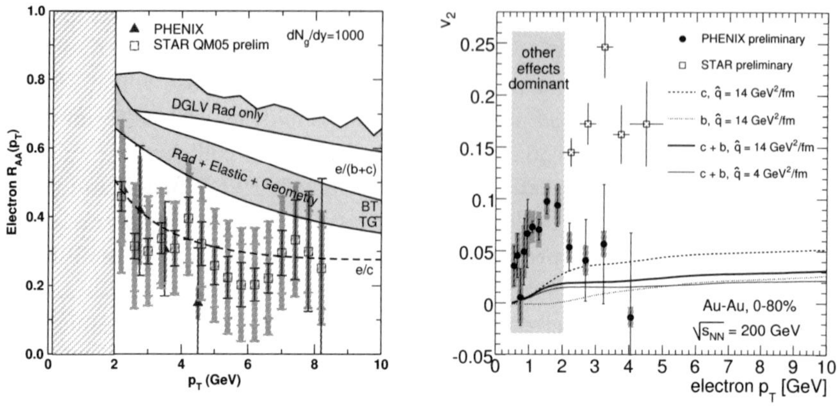

Fig. 1. Perturbative QCD energy-loss calculations for $e^{\pm}$ spectra from semileptonic heavy-meson decays. Left panel: nuclear suppression factor[10], right panel: elliptic flow[12].

plots indicate a (lower) applicability limit of the pQCD energy-loss approach of $p_T^e \simeq 2\,\text{GeV}$, we will argue below that nonperturbative effects may be relevant (or even dominant) up to at least $p_T^e \simeq 5\,\text{GeV}$. In particular, radiative energy-loss calculations typically do not include the backward reactions (detailed balance) which are essential for building up collective behavior of the heavy quarks in the expanding thermal medium. As mentioned above, in this context a Brownian-motion approach is well suited to address the thermalization of the heavy quarks[6,7,11,13]. Upon expanding the Boltzmann equation in small momentum transfers, one arrives at a Fokker-Planck equation for the HQ distribution function, f,

$$\frac{\partial f}{\partial t} = \gamma \frac{\partial (pf)}{\partial p} + D_p \frac{\partial^2 f}{\partial p^2} \ , \tag{1}$$

with (momentum) drag (γ) and diffusion (D_p) constants (related via $T = D_p/\gamma M_Q$). The latter are calculated from corresponding matrix elements for HQ scattering off light partons. As mentioned above, our main ingredient here are resonance-mediated elastic interactions in s- and u-channel (cf. left panel of Fig. 2) which we model with an effective Lagrangian according to[6]

$$\mathcal{L} = -G\, \bar{Q}\, \Gamma\, \frac{1+ \not{v}}{2}\, \Phi\, q + \text{h.c} \ ; \tag{2}$$

Φ is the resonance field representing "D" and "B" mesons, which we dress by evaluating the corresponding one-loop (Q-$\bar{q}$) selfenergy. Parameters of the model are the resonance mass (which we fix at 0.5 GeV above the Q-$\bar{q}$ threshold) and coupling constant, G, which determines the resonance width. HQ and chiral symmetry (above T_c) imply degeneracy of the pseudo-/scalar

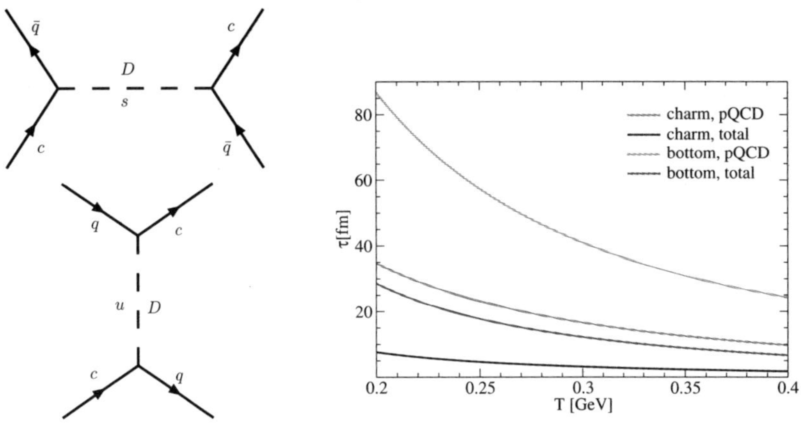

Fig. 2. Elastic HQ-scattering processes via D-(B-)meson like resonances (left panel) and resulting thermalization times compared to pQCD results (right panel)[6].

and axial-/vector channels represented by the Dirac matrices, Γ. The temperature dependence of the resulting thermal relaxation times, $\tau_Q = \gamma^{-1}$, for c- and b-quarks shows a factor of $\sim$3 reduction compared to calculations where only pQCD elastic scattering is accounted for (right panel of Fig. 2). This is very significant for c-quarks as τ_c is now comparable to (or even below) the duration of the QGP phase at RHIC ($\tau_{\rm QGP}\sim 5\,{\rm fm}/c$), whereas $\tau_b > \tau_{\rm QGP}$ still, due to the large b-quark mass.

The T- and p-dependent drag and diffusion coefficients have been implemented into a relativistic Langevin simulation[6] using an (elliptic) thermal fireball expansion for (semi-) central Au-Au collisions at RHIC as a background medium, where the bulk flow is adjusted to hydrodynamic simulations in accordance with experiment. At the end of the QGP phase (which terminates in a mixed phase at T_c=180 MeV), the c- and b-quark output distributions from the Langevin simulation are subjected to hadronization into D- and B-mesons using the quark-coalescence model of Ref. [14]. Since the probability for coalescence is proportional to the light-quark distribution functions, it preferentially occurs at lower p_T; "left-over" heavy quarks are hadronized via δ-function fragmentation. The extra contribution to momentum and v_2 from the light quarks *increases both* the $R_{AA}(p_T)$ and $v_2(p_T)$ of the heavy-meson spectra, relative to a scheme with fragmentation only. Rescattering in the hadronic phase has been neglected. The resulting D- and B-meson spectra are decayed semileptonically resulting in a nuclear suppression factor and elliptic flow which compare reasonably well with recent RHIC data[4,5,15] up to $p_T \simeq 5$ GeV (Fig. 3). At higher momenta, radiative

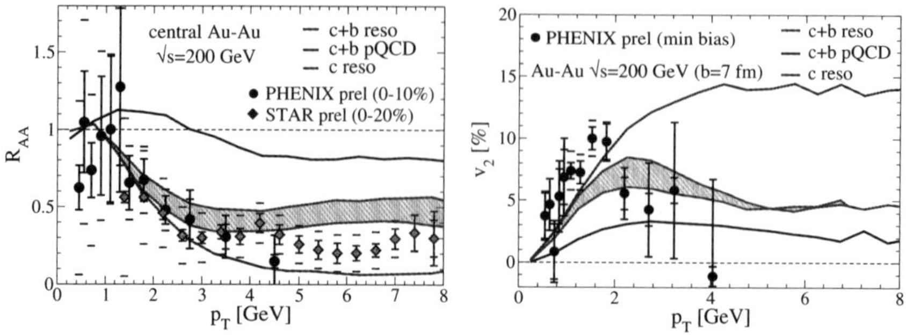

Fig. 3. Calculations for single-electron spectra arising from semileptonic heavy-meson decays. Left panel: Nuclear suppression factor right panel: elliptic flow.

energy loss (not included here) is expected to contribute significantly[a].

Another issue in interpreting the semileptonic $e^{\pm}$ spectra concerns the relative contributions of charm and bottom. Since bottom quarks are much less affected by rescattering effects, their contribution reduces the effects in both R_{AA} and v_2 appreciably. Following Refs. [9,17], we have determined the initial charm spectra by reproducing D- and D^*-spectra in d-Au collisions[18], calculated the pertinent $e^{\pm}$ spectra and assigned the missing yield relative to the data[19] to bottom contributions. In this way, the crossing between charm and bottom contribution in p-p collisions occurs at $p_T \simeq 5$ GeV. Due to the strong quenching of the c-quark spectra in the QGP this crossing is shifted down to $p_T \simeq 2.5$-3 GeV in Au-Au collisions. For charm contributions only, our nonperturbative rescattering mechanism results in a pertinent electron R_{AA} as low as 0.1 and an elliptic flow as large as 14%. Thus, the severeness of the "$e^{\pm}$ puzzle" partially hinges on the baseline spectra in more elementary systems (p-p and d-Au). Obviously, a direct measurement of D-mesons in Au-Au would be very valuable.

3. Heavy quarkonia in the QGP

3.1. *Potential models, spectral functions and lattice QCD*

Heavy quarkonia (c-$\bar{c}$ and b-$\bar{b}$ bound states) have long been identified as suitable objects to quantitatively investigate the properties of QCD, including color confinement[20]. A particularly appealing feature is the applicability of potential models as a nonrelativistic effective-theory approximation to

[a]Note that in a chemically equilibrated QGP, the effects of induced gluon radiation are presumably smaller than in a pure gluon plasma as assumed in Refs. [10,12]; an estimate of (some) 3-body scattering diagrams has been performed in Ref. [16].

QCD (cf. Ref. [21] for a recent review). In the vacuum, the heavy-quarkonium spectrum is indeed reasonably well described by the "Cornell potential", consisting of a Coulomb plus linear confining term. Subsequently it was realized that, when immersing quarkonia into a medium of deconfined color charges, Debye screening will reduce the binding and eventually lead to the dissolution of the bound state[22,23]; already slightly above T_c, a substantial (factor ~2-3) reduction in charmonium and bottomonium binding energies, ε_B, has been found, together with the possibility that the (S-wave) ground states ($\eta_{c,b}$, J/ψ, $\Upsilon(1S)$) survive (well) above T_c. Modern lQCD (-based) calculations qualitatively support this picture: On the one hand, directly extracted charmonium spectral functions (in quenched approximation) indicate resonance peaks up to $\sim 2T_c$[24,25]. On the other hand, (both quenched and unquenched) lQCD results for the (color-singlet) free energy,

$$F_{Q\bar{Q}}(r;T) = U_{Q\bar{Q}}(r;T) - TS_{Q\bar{Q}}(r;T) \ , \tag{3}$$

have been implemented into potential models within a Schrödinger[26–29] or Lippmann-Schwinger equation[30]. There is an ongoing debate as to which quantity ($F_{Q\bar{Q}}$, the internal energy $U_{Q\bar{Q}}$, or a linear combination thereof[31]) is the most appropriate one to use as a potential. Maximal binding is obtained with $U_{Q\bar{Q}}$, and even in this case, the ground-state charmonium (bottomonium) binding energies are reduced to $\varepsilon_B \simeq$0.1-0.2 GeV (0.3-0.6 GeV) at $T \simeq 1.5\ T_c$, compared to ~0.6 GeV (1.1 GeV) in vacuum. However, a quantitative description of in-medium modifications of quarkonia requires at least two additional components, i.e., the effects of parton-induced dissociation reactions and of in-medium HQ masses. The objective is thus to construct an approach that (i) is consistent with finite-T lQCD and (ii) allows for reliable applications to observables in heavy-ion collisions. Only then the original idea[22] of using quarkonia as a tool to identify (and characterize) confinement (or at least color screening) may be realized.

The interplay of medium effects becomes more transparent in terms of the charmonium propagator, schematically written as (nonrelativistic)[b]

$$D_\Psi(\omega;T) = \left[\omega - (2m_c^* - \varepsilon_B) + \mathrm{i}\Gamma_\Psi\right]^{-1} \ . \tag{4}$$

It shows that screening (affecting the binding energy and thus the real part of the inverse propagator), and parton-induced dissociation (governing the width, Γ_Ψ, of the spectral function, Im D_Ψ) are not mutually exclusive but have to be taken into account simultaneously[32].

[b]Writing Eq. (4) in this form implies that the Q-$\bar{Q}$ potential is strong enough to generate a pole in the scattering amplitude. A microscopic (T-matrix) approach is discussed below.

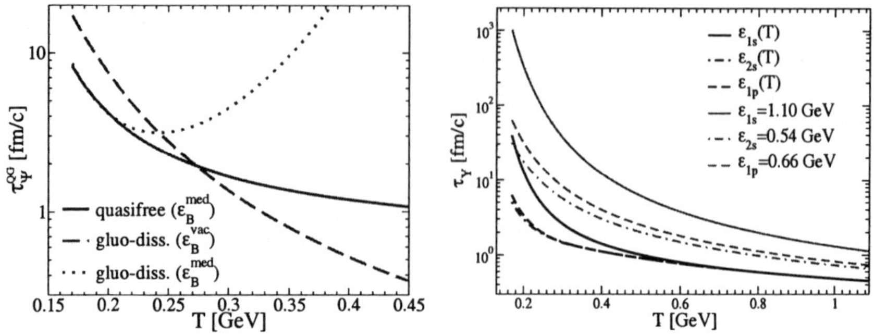

Fig. 4. Lifetimes of J/ψ (left panel, gluo-dissociation vs. quasifree dissociation mechanisms[34]) and Y states (right panel, quasifree dissociation[35]) in the QGP.

The inelastic width (or dissociation rate) can be expressed as

$$\Gamma_\Psi = (\tau_\Psi)^{-1} = \int \frac{d^3k}{(2\pi)^3} \, f^p(\omega_k, T) \, v_{\rm rel} \, \sigma_\Psi^{\rm diss}(s) \tag{5}$$

in terms of (thermal) parton-distribution functions, f^p (p=$q, \bar{q}, g$), and the parton-induced break-up cross section, $\sigma_\Psi^{\rm diss}$. The latter has first been evaluated for gluo-dissociation[33], $g + \Psi \to c + \bar{c}$, using Coulomb wave-functions for the quarkonia, leading to a T-dependent J/ψ lifetime as shown in the left panel of Fig. 4. For the vacuum J/ψ binding energy, ε_B=0.63 GeV, the lifetime is well behaved showing a rather steep decrease with T (dashed line). However, if an in-medium binding energy is used as following from screened (or lQCD) potentials, the lifetime is initially reduced (dotted line), but if the binding energy drops below ~0.1 GeV, the phase space in the gluo-dissociation reaction strongly shrinks leading to an unphysical increase of the J/ψ lifetime with T. This artifact is even more pronounced for the less bound excited states (χ_c, ψ'). This problem has been remedied by introducing the "quasifree" dissociation process[34], $p + \Psi \to p + c + \bar{c}$, where a parton knocks out the Q (or $\bar{Q}$) from the bound state. With reasonable values for the strong coupling constant ($\alpha_s \simeq 0.25$) the resulting charmonium widths are in the ~0.1 GeV range for T~1.5 T_c (as relevant for RHIC) with a well-behaved T-dependence. The order of magnitude of the inelastic width is easily reproduced using a simplified estimate of Eq. (5) according to $\Gamma_\Psi \approx n_p(T) \sigma_{\rm diss} v_{\rm rel}$; with a parton density $n_p \simeq 10\,{\rm fm}^{-3}$ (at $T \approx 0.25$ GeV), a cross section of 1 mb and $v_{\rm rel}$=1/2 one finds $\Gamma_\Psi \approx 0.1$ GeV.

When applied to bottomonia, the same arguments apply[35]; the right panel of Fig. 4 illustrates the impact of in-medium binding energies on bottomonium dissociation: in the RHIC temperature regime, screening may lead to up to a factor of ~20 (!) reduction in the ground-state $\Upsilon(1S)$ life-

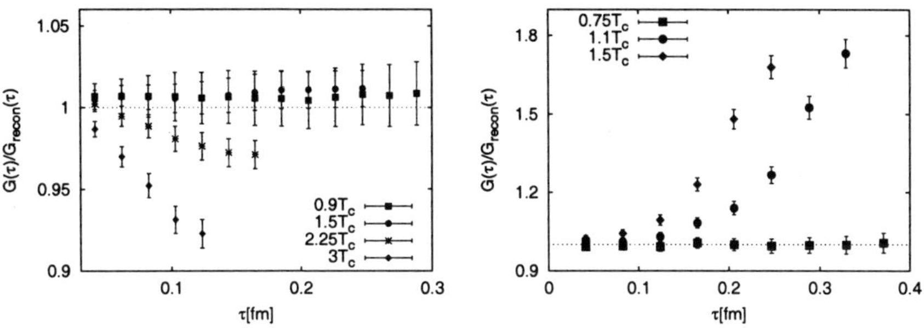

Fig. 5. Finite-temperature euclidean correlation functions computed in lQCD in the pseudoscalar (η_c, left panel) and scalar (χ_c, right panel) channels[25].

time, bringing it down to ~2-5fm/c at T=0.3-0.4 GeV. This has important consequences for bottomonium suppression at RHIC (as elaborated in Sec. 3.2 below), rendering it a sensitive probe of color screening[35].

A key issue for model descriptions of quarkonium spectral functions in the QGP is the implementation of constraints from lQCD. The standard object computed on the lattice is (the thermal expectation value of) a hadronic current-current correlation function in euclidean time, τ (and at zero 3-momentum), in quantum-number channel, α[36],

$$G_\alpha(\tau;T) = \langle j_\alpha(\tau) j_\alpha^\dagger(0)\rangle \ . \tag{6}$$

The connection to the spectral function, σ_α, is given by a convolution with a temperature kernel as

$$G_\alpha(\tau;T) = \int_0^\infty d\omega \ \sigma_\alpha(\omega;T) \ \frac{\cosh[\omega(\tau - 1/2T)]}{\sinh[\omega/2T]} \ . \tag{7}$$

While euclidean correlators can be computed with good accuracy (Fig. 5), their limited temporal extent at finite T, $0\leq\tau\leq 1/T$, renders the extraction of (Minkowski) spectral functions more difficult. On the contrary, model calculations of spectral functions are readily integrated via Eq. (7) and compared to the rather precise euclidean correlators from lQCD [37,29]. To better exhibit the T-dependence induced by the spectral function, both lattice and model results for $G_\alpha(\tau;T)$ are commonly normalized to the so-called "reconstructed" correlator, which follows from Eq. (7) by inserting a free spectral function, $\sigma_\alpha(\omega, T=0)$.

Eq. (6) requires the knowledge of the quarkonium spectral function at all (positive) energies. A schematic decomposition consists of a bound-state part and a continuum,

$$\sigma_\alpha(\omega) = \sum_i 2M_i F_i^2 \delta(\omega^2 - M_i^2) + \frac{3}{8\pi^2}\,\omega^2 \, f_{\rm cont}(\omega, E_{\rm thr})\Theta(\omega - E_{\rm thr}) \ , \tag{8}$$

where, for fixed α, i runs over all bound states with masses M_i and couplings F_i, and the threshold energy ($E_{\rm thr}$) and function $f_{\rm cont}$ characterize the onset and plateau of the continuum (usually taken from pQCD).

In Ref. [29], T-dependent bound-state masses and couplings have been evaluated by solving a Schrödinger equation with in-medium potentials (screened Cornell or lQCD internal energy, $U_{Q\bar{Q}}$); the threshold energy has been inferred from the asymptotic ($r \to \infty$) value of the respective potential, $E_{\rm thr}^{\rm med}(T) = 2m_c + V(r \to \infty; T)$ (which monotonously decreases for $T > T_c$). For the scalar (χ_c) channel it has been found that, despite a rather quickly dissolving bound-state contribution, the correlator increases significantly (consistent with the lattice result in Fig. 5, right panel), due to the decreasing continuum threshold. The latter also leads to an increasing pseudoscalar correlator, which is not favored by lQCD (Fig. 5, left panel).

In Ref. [38] the separation of bound-state and continuum parts is improved by employing a T-matrix approach[39,30] for the Q-$\bar{Q}$ interaction,

$$T_\alpha(E;q',q) = V_\alpha(q',q) - \frac{2}{\pi}\int_0^\infty dk\, k^2\, V_\alpha(q',k)\, G_{\bar{Q}Q}(E;k)\, T_\alpha(E;k,q) \; . \quad (9)$$

For V_α the Fourier transformed and partial-wave expanded lQCD internal energy has been used. Spin-spin (hyperfine) interactions are neglected implying degeneracy of states with fixed angular momentum (S-wave: η_c and J/ψ, P-wave: $\chi_{c0,1,2}$). $G_{\bar{Q}Q}(E;k)$ denotes the intermediate 2-particle propagator including quark selfenergies. The correlation and spectral functions follow from closing the external legs with 3-momentum integrations as

$$G_\alpha(E) = \int G_{\bar{Q}Q} + \int G_{\bar{Q}Q} \int T_\alpha\, G_{\bar{Q}Q} \; , \quad \sigma_\alpha(E) = a\ {\rm Im} G_\alpha(E) \; , \qquad (10)$$

where the coefficient a depends on the channel α. For a fixed c-quark mass of m_c=1.7 GeV, the resulting charmonium spectral functions confirm that bound states are supported in the S-wave up to $\sim 3T_c$ while dissolved in the P-wave below 1.5 T_c, see Fig. 6. In both cases, however, one finds a large (nonperturbative) enhancement of strength in the threshold region. The corresponding euclidean correlators are displayed in Fig. 7, normalized to a reconstructed one using the vacuum bound-state spectrum and a perturbative continuum with onset at the free $D\bar{D}$ threshold, $E_{\rm thr}^{\rm vac}=2m_D\simeq 3.75$ GeV. The χ_c correlator (right panel) is well above one, both due to a lower in-medium threshold, $E_{\rm thr}^{\rm med}=2m_c=3.4$ GeV, and the nonperturbative rescattering enhancement. While this is qualitatively consistent with lQCD, the temperature dependence is opposite, indicative for a further threshold decrease with temperature. In the η_c channel (left panel), the significant reduction in binding energy (due to color screening) reduces the correlator

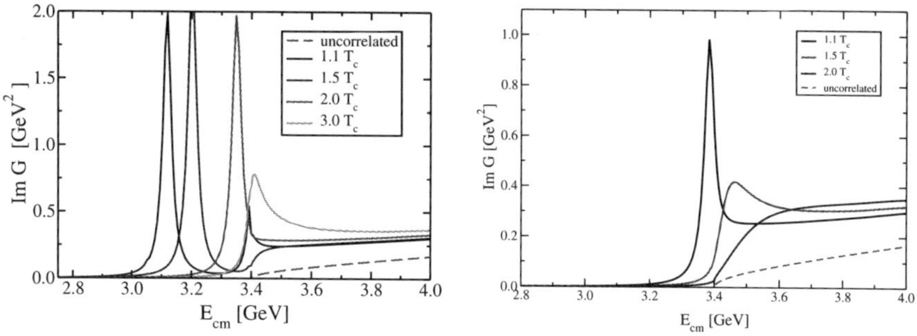

Fig. 6. S- (left panel) and P-wave (right panel) charmonium-spectral functions from the T-matrix approach with an lQCD-internal energy as potential[38].

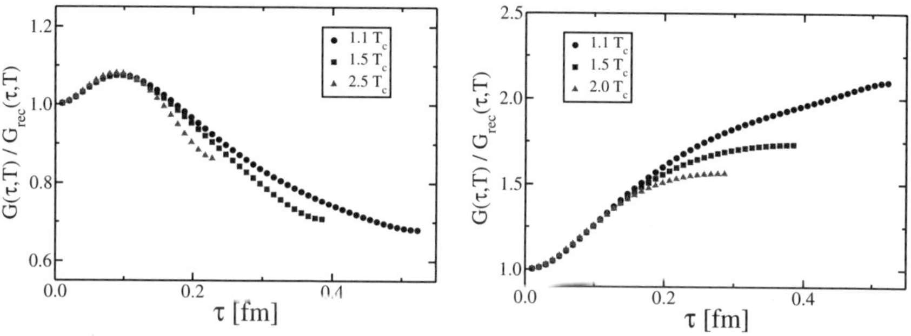

Fig. 7. Normalized euclidean correlation function for S- (left panel) and P-wave (right panel) charmonium states within a T-matrix approach using lQCD-internal energies[38].

at large τ, qualitatively in line with lQCD. In fact, the latter shows less reduction, leaving room for a T-dependent reduction of $E_{\rm thr}^{\rm med}$ as well.

The scattering equation (9) is furthermore well suited to study finite-width effects. E.g., when dressing the c-quarks with an imaginary self-energy corresponding to Γ_c=50 MeV, inducing a charmonium width of $\Gamma_\Psi \simeq 100$ MeV (as in the left panel of Fig. 4), the euclidean correlators vary by no more than 2%. This indicates that there is rather little sensitivity to phenomenologically relevant values for the quarkonium widths.

3.2. *Quarkonium phenomenology in heavy-ion collisions*

The formation of thermalized matter above T_c in ultrarelativistic heavy-ion collisions (recall the discussion in the introduction) provides the basis for describing the production systematics of quarkonia in terms of their in-medium properties as discussed in the previous section. The suppression of the initially produced number of quarkonia, N_Ψ^i, may be schematically

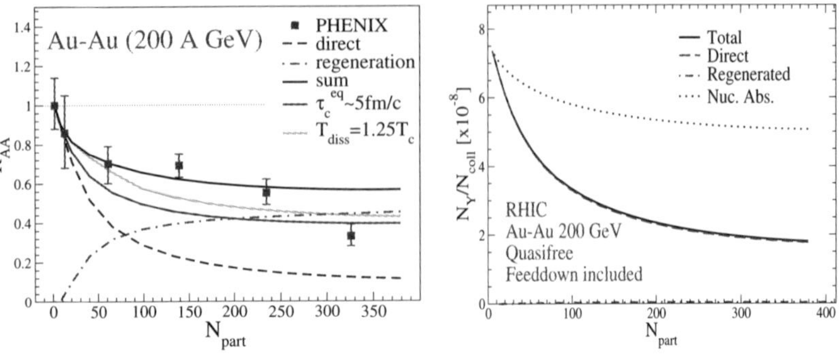

Fig. 8. Centrality dependence of the nuclear modification factor for J/ψ (left panel)[40,46] and Υ (right panel)[35] at RHIC.

written as proceeding through 3 stages,

$$S_{\rm tot} = \exp(-\sigma_{\rm nuc}^{\rm abs}\ \varrho_N\ L)\ \exp(-\Gamma_{\rm QGP}\ \tau_{\rm QGP})\ \exp(-\Gamma_{\rm HG}\ \tau_{\rm HG}). \qquad (11)$$

With the suppression in the hadron gas (HG) believed to be small, and the ("pre-equilibrium") nuclear absorption (cross section) inferred from p-A experiments, the suppression factor directly probes the dissociation rate of Ψ's in the QGP. The observed J/ψ-suppression pattern at SPS can indeed be well described using the quasifree dissociation process with in-medium binding energies consistent with lQCD-potential models[34,40].

At RHIC energy, copious production of $c\bar{c}$ pairs ($N_{c\bar{c}}\simeq 20$ in central $\sqrt{s}$=200 AGeV Au-Au[41], compared to 0.2 at SPS) opens the possibility of secondary charmonium production[42,34]. This is, after all, required by detailed balance in the dissociation reaction, $\Psi + g \rightleftharpoons c + \bar{c} + X$, provided the Ψ state still exists at the given temperature[40,43]. If, in addition, the c-quarks are close to thermal equilibrium, one can apply the rate equation,

$$\frac{dN_\Psi}{dt} = -\Gamma_\Psi [N_\psi - N_\psi^{eq}(T)] \quad , \qquad (12)$$

for the time evolution of the J/ψ number. Pertinent predictions (upper solid line in the left panel of Fig. 8) agree reasonably well with current PHENIX data[44] suggesting that the regeneration component becomes substantial in central Au-Au. Corrections due to incomplete c-quark thermalization, as well as a lower dissociation (and thus formation) temperature, reduce the regeneration (lower solid curves in the left panel of Fig. 8). It has also been suggested[45] that available J/ψ data at SPS and RHIC are compatible with suppression of only χ_c and ψ' (which make up ~40% of the inclusive J/ψ yield in p-p), with the direct J/ψ's being unaffected. However, if at

RHIC the J/ψ width is of order 0.1 GeV (cf. previous section) and the QGP lifetime at least 2fm/c, one finds a QGP suppression factor of $S_{\rm QGP}\simeq0.4$.

Finally, turning to Υ production, the increase of the Υ dissociation rate due to color screening (recall right panel of Fig. 4) could lead to a Υ yield in central Au-Au (right panel of Fig. 8) which is more suppressed for the J/ψ (left panel), and thus provide a rather striking QGP signature[35].

4. Conclusions

Charm and bottom hadrons are valuable sensors of the sQGP. In the open-flavor sector, the inadequacy of pQCD to account for the suppression and collectivity of current $e^{\pm}$ spectra at RHIC may be a rather direct indication of nonperturbative rescattering processes above T_c. We have elaborated these in a concrete example using (elastic) resonance interactions. The latter will have to be (i) evaluated more microscopically, e.g., using input interactions from lQCD, and (ii) combined with radiative energy loss.

In the quarkonium sector, steady progress is made in implementing first-principle information from finite-T lattice QCD into effective models suitable for tests in heavy-ion reactions. T-matrix approaches incorporate effects of color-screening, parton-induced dissociation, and in medium masses and widths of heavy quarks (which, in turn, connect to the open-flavor sector). We have suggested an sQGP signature in terms of stronger suppression for Υ relative to J/ψ, which would be a direct proof of J/ψ regeneration and carries large sensitivity to color screening of bottomonia.

Acknowledgments

This work was supported in part by a fellowship of the Spanish M. E. C., a Feodor-Lynen fellowship of the A.-v.-Humboldt foundation, and a U.S. National Science Foundation CAREER Award under grant PHY-0449489.

References

1. K. Adcox *et al.* [PHENIX Collaboration], *Nucl. Phys.* **A757**, 184 (2005).
2. J. Adams *et al.* [STAR Collaboration], *Nucl. Phys.* **A757**, 102 (2005).
3. M. Gyulassy, I. Vitev, X.N. Wang and B.W. Zhang, arXiv:nucl-th/0302077.
4. S.S. Adler *et al.* [PHENIX Collaboration], *Phys. Rev. Lett.* **96**, 032301 (2006).
5. B.I. Abelev *et al.* [STAR Collaboration], arXiv:nucl-ex/0607012.
6. H. van Hees and R. Rapp, *Phys. Rev. C* **71**, 034907 (2005).
7. G.D. Moore and D. Teaney, *Phys. Rev. C* **71**, 064904 (2005).
8. M.G. Mustafa, *Phys. Rev. D* **72**, 014905 (2005).
9. H. van Hees, V. Greco and R. Rapp, *Phys. Rev. C* **73**, 034913 (2006).

10. S. Wicks *et al.*, arXiv:nucl-th/0512076.
11. B. Svetitsky *Phys. Rev. D* **37**, 2484 (1988).
12. N. Armesto *et al.*, *Phys. Lett. B* **637**, 362 (2006).
13. D. Pal and M. G. Mustafa, *Phys. Rev. C* **60**, 034905 (1999).
14. V. Greco, R. Rapp and C. M. Ko *Phys. Lett. B* **595**, 202 (2004).
15. S.S. Adler *et al.* [PHENIX Collaboration], *Phys. Rev. C* **72**, 024901 (2005); S. Sakai, arXiv:nucl-ex/0510027.
16. W. Liu and C. M. Ko, arXiv:nucl-th/0603004.
17. R. Rapp, V. Greco and H. van Hees, arXiv:hep-ph/0510050.
18. J. Adams *et al.* [STAR Collaboration], *Phys. Rev. Lett.* **94**, 062301 (2005).
19. A.A.P. Suaide *et al.* [STAR Collaboration] *J. Phys. G* **30**, S1179 (2004).
20. V.A. Novikov *et al.*, *Phys. Rept.* **41**, 1 (1978).
21. N. Brambilla *et al.*, arXiv:hep-ph/0412158.
22. T. Matsui and H. Satz, *Phys. Lett. B* **178**, 416 (1986).
23. F. Karsch, M.T. Mehr and H. Satz, *Z. Phys. C* **37**, 617 (1988).
24. M. Asakawa and T. Hatsuda, *Phys. Rev. Lett.* **92**, 012001 (2004).
25. S. Datta *et al.*, *Phys. Rev. D* **69**, 094507 (2004).
26. E.V. Shuryak and I. Zahed, *Phys. Rev. C* **70**, 021901 (2004).
27. C.Y. Wong, *Phys. Rev. C* **72**, 034906 (2005).
28. W.M. Alberico *et al.*, *Phys. Rev. D* **72**, 114011 (2005).
29. A. Mocsy and P. Petreczky, *Phys. Rev. D* **73**, 074007 (2006).
30. M. Mannarelli and R. Rapp, *Phys. Rev. C* **72**, 064905 (2005).
31. C.Y. Wong, arXiv:hep-ph/0606200.
32. G. Röpke, D. Blaschke and H. Schulz, *Phys. Rev. D* **38**, 3589 (1988).
33. E.V. Shuryak, *Sov. J. Nucl. Phys.* **28**, 408 (1978); G. Bhanot and M. Peskin, *Nucl. Phys.* **B156**, 391 (1979).
34. L. Grandchamp and R. Rapp, *Phys. Lett. B* **523**, 60 (2001); *Nucl. Phys.* **A709**, 415 (2002).
35. L. Grandchamp *et al.*, *Phys. Rev. C* **73**, 064906 (2006).
36. F. Karsch and E. Laermann, arXiv:hep-lat/0305025.
37. R. Rapp, *Eur. Phys. J. A* **18**, 459 (2003).
38. D. Cabrera and R. Rapp, in preparation (2006).
39. L.S. Celenza, B. Huang and C.M. Shakin, *Phys. Rev. C* **59**, 1030 (1999).
40. L. Grandchamp *et al.*, *Phys. Rev. Lett.* **92**, 212301 (2004).
41. S.S. Adler *et al.* [PHENIX Collaboration], *Phys. Rev. Lett.* **94**, 082301 (2005).
42. P. Braun-Munzinger and J. Stachel, *Phys. Lett. B* **490**, 196 (2000).
43. L. Yan, P. Zhuang and N. Xu, arXiv:nucl-th/0608010.
44. H. Pereira Da Costa *et al.* [PHENIX Collaboration], arXiv:nucl-ex/0510051.
45. F. Karsch, D. Kharzeev and H. Satz, *Phys. Lett. B* **637**, 75 (2006).
46. X. Zhao and R. Rapp, work in progress (2006).

CONFINEMENT-DECONFINEMENT PHASE TRANSITION AND FRACTIONAL INSTANTON QUARKS IN DENSE MATTER

ARIEL R. ZHITNITSKY

Department of Physics and Astronomy, University of British Columbia, Vancouver, Canada

We present arguments suggesting that large size overlapping instantons is the driving mechanism of the confinement-deconfinement phase transition at nonzero chemical potential μ. The arguments are based on the picture that instantons at very large chemical potential in the weak coupling regime are localized configurations with small sizes $\rho \sim \mu^{-1}$. At the same time, the same instantons at smaller chemical potential in the strong coupling regime are well represented by the so-called instanton-quarks with fractional topological charge $1/N_c$. We estimate the critical chemical potential $\mu_c(T)$ where transition between these two regimes takes place. We identify this transition with confinement- deconfinement phase transition.

1. Introduction

This talk is based on a number of original results[1]-[3] obtained with different collaborators at different times.

Instantons were discovered 30 years ago [4]. However, their role in QCD_4 remains unclear even today due to the divergence of the instanton density for large size instantons. The development of the instanton liquid model (ILM) [5,6] has encountered successes (chiral symmetry breaking, resolution of the $U(1)$ problem, etc) and failures (confinement could not be described by well separated and localized lumps with integer topological charges). Therefore, it is fair to say that at present, the widely accepted viewpoint is that the ILM can explain many experimental data (such as hadron masses, widths, correlation functions, decay couplings, etc), with one, but crucial exception: confinement. There are many arguments against the ILM approach, see e.g. [7], there are many arguments supporting it [6].

In this talk we present new arguments supporting the idea that the instanton-quarks [8] along with instantons[4] are the relevant quasiparticles in the strong coupling regime. In this case, many problems formulated in [7] are naturally resolved as both phenomena, confinement and chiral symmetry breaking are originated from the same vacuum configurations, instantons, which may have arbitrary scales: the finite size localized lumps with integer topological charges, as well as set of N_c fractionally $1/N_c$ -charged correlated objects sitting at arbitrary large distances from each other. In this picture when fractionally charged $1/N_c$ constituents propagate far away from each other, the confinement could be a natural consequence of a dynamics of these well correlated objects. More importantly, we make some very specific predictions which can be tested with traditional Monte Carlo techniques, by studying QCD at nonzero isospin chemical potential.

2. Instantons at large μ

At low energy and large chemical potential, the η' is light and described by the Lagrangian derived in [2]:

$$L_\varphi = f^2(\mu)[(\partial_0\varphi)^2 - u^2(\partial_i\varphi)^2] + a(\mu,T)\mu^2\Delta^2\cos(\varphi-\theta), \tag{1}$$

where f is the decay constant and $u^2=1/3$ is the velocity. We define baryon and isospin chemical potentials as $\mu_{B,I}=(\mu_u \pm \mu_d)/2$. The nonperturbative potential $V_{inst} = -a(\mu,T)\mu^2\Delta^2\cos(\varphi-\theta)$ can be explicitly calculated[2] and is strongly suppressed at large chemical potential, $a \ll 1$. The approach presented above is valid as long as the φ field is lighter than the gap, $\sim 2\Delta$, the mass of the other mesons in the system that is

$$a(\mu,T) \leq 8f^2(\mu)/\mu^2. \tag{2}$$

This is exactly the vicinity where the Debye screening scale and the inverse gap become of the same order of magnitude [2], and therefore, where the instanton expansion breaks down. For reasons which will be clear soon, we want to represent the Sine-Gordon (SG) partition function (1) in the equivalent dual Coulomb Gas (CG) representation [2],

$$Z = \sum_{M_\pm=0}^{\infty} \frac{(\lambda/2)^M}{M_+!M_-!}\int d^4x_1\ldots d^4x_M e^{-i\theta\sum Q_a - \frac{1}{2f^2u}\sum Q_aQ_bG(x_a-x_b)}, \tag{3}$$

where $G(x_a-x_b)$ is $4d$ Green's function and $\lambda \equiv \frac{a\mu^2\Delta^2}{u}$. Physical interpretation of the dual CG representation (3) can be formulated as follows: **a)** Since $Q_{\rm net} \equiv \sum_a Q_a$ is the total charge and it appears in the action multiplied be the parameter θ, one concludes that $Q_{\rm net}$ *is* the total topological charge of a given configuration; **b)** Each charge Q_a in a given configuration should be identified with an integer topological charge well localized at the point x_a. This, by definition, corresponds to a small instanton positioned at x_a; **c)** While the starting low-energy effective Lagrangian contains only a colorless field φ we have ended up with a representation of the partition function in which objects carrying color (the instantons) can be studied; **d)** In particular, II and $I\bar{I}$ interactions (at very large distances) are exactly the same up to a sign, order g^0, and are Coulomb-like. This is in contrast with semiclassical expressions when II interaction is zero and $I\bar{I}$ interaction is order $1/g^2$; **e)** The very complicated picture of the bare II and $I\bar{I}$ interactions becomes very simple for dressed instantons/anti-instantons when all integrations over all possible sizes, color orientations and interactions with background fields are properly accounted for; **f)** As expected, the ensemble of small $\rho \sim 1/\mu$ instantons can not produce confinement. This is in accordance with the fact that there is no confinement at large μ.

3. Instantons at small μ

We want to repeat the same procedure that led to the CG representation in the confined phase at small μ to see if any traces from the instantons can be recovered. We keep only the diagonal elements of the chiral matrix $U = \exp\{i\mathrm{diag}(\phi_1, \ldots, \phi_{N_f})\}$ which are relevant in the description of the ground state. Singlet combination is defined as $\phi = \mathrm{Tr}\, U$. The effective Lagrangian for the ϕ is

$$L_\phi = f^2(\partial_\mu\phi)^2 + E\cos\left(\frac{\phi-\theta}{N_c}\right) + \sum_{a=1}^{N_f} m_a \cos\phi_a \tag{4}$$

A Sine-Gordon structure for the singlet combination corresponds to the following behavior of the $(2k)^{\mathrm{th}}$ derivative of the vacuum energy in pure gluodynamics [9],

$$\frac{\partial^{2k} E_{vac}(\theta)}{\partial\,\theta^{2k}}|_{\theta=0} \sim \int \prod_{i=1}^{2k} dx_i \langle Q(x_1)...Q(x_{2k})\rangle \sim (\frac{i}{N_c})^{2k}, \tag{5}$$

where $Q \equiv \frac{g^2}{32\pi^2} G_{\mu\nu}\widetilde{G}_{\mu\nu}$ is topological density. The same structure was also advocated in [10] from a different perspective. As in (3) the Sine-Gordon effective field theory (4) can be represented in terms of a classical statistical ensemble (CG representation) similar to (3) with the replacements $\lambda \to E,\ u \to 1$, more precisely,

$$Z = \sum_{M_\pm=0}^{\infty} \frac{(E/2)^M}{M_+!M_-!} \int d^4x_1 \ldots d^4x_M e^{-i\theta\sum_{a=0}^{M} Q_a - \frac{1}{2f^2}\sum Q_a Q_b G(x_a - x_b)}. \tag{6}$$

The fundamental difference in comparison with the previous case (3) is that while the total charge is integer, the individual charges are **fractional** $\pm 1/N_c$. This is a direct consequence of the θ/N_c dependence in the underlying effective Lagrangian (4) before integrating out ϕ fields. Physical Interpretation of the CG representation (6) of theory (4) is: **a)** As before, one can identify $Q_{\rm net} \equiv \sum_a Q_a$ with the total topological charge of the given configuration; **b)** Due to the 2π periodicity of the theory, only configurations which contain an integer topological number contribute to the partition function. Therefore, the number of particles for each given configuration Q_i with charges $\sim 1/N_c$ must be proportional to N_c; **c)** Therefore, the number of integrations over d^4x_i in CS representation exactly equals $4N_ck$, where k is integer. This number $4N_ck$ exactly corresponds to the number of zero modes in the k-instanton background. This is basis for the conjecture [3] that at low energies (large distances) the fractionally charged species, $Q_i = \pm 1/N_c$ are the **instanton-quarks** suspected long ago; **d)** For the gauge group, G the number of integrations would be equal to $4kC_2(G)$ where $C_2(G)$ is the quadratic Casimir of the gauge group (θ dependence in physical observables comes in the combination $\frac{\theta}{C_2(G)}$). This number $4kC_2(G)$ exactly corresponds to the number of zero modes in the k-instanton background for gauge group G; **e)** The CG representation corresponding to eq.(4) describes the confinement phase of the theory.

4. Conjecture

We thus conjecture that the confinement-deconfinement phase transition takes place at precisely the value where the dilute instanton calculation breaks down. At large μ the weakly interacting phase (CS) is realized. Instantons are well localized configurations with a typical size μ^{-1}. Color in CS phase is not confined. At low μ the strong interacting regime is realized and color is confined. Instantons are not well localized configurations, but rather are represented by N_c instanton quarks which can propagate far away from each other. The value of the critical chemical potential as a function of temperature, $\mu_c(T)$ is given by saturating the inequality (2).

Few remarks are in order. First, we can estimate the critical μ_c not only at $T = 0$, but also at $T \neq 0$ as long as the temperature is relatively small such that our approach is justified. Indeed, in the weak coupling regime the T dependence of the instanton density is determined by a simple insertion $\sim \exp[-(\frac{1}{3}(2N_c + N_f)\pi^2T^2)\rho^2]$ into the expression for the density. The temperature dependence also enters the expression for $\Delta(T)$. As long as $\Delta(T)$ does not vanish and we are in CS phase, our calculations are justified, and the critical $\mu_c(T)$ can be estimated as a function of T at small T. As

	$N_c = 3, N_f = 2$	$N_c = 3, N_f = 3$	$N_c = N_f = 2$
μ_{Bc}/Λ	2.3	1.4	3.5
μ_{Ic}/Λ	2.6	1.5	3.5

a final remark: while we expect that the instanton density suffers from large uncertainties at $\mu \sim \mu_c$, the numerical results for $\mu_c(T)$ are not very sensitive to these uncertainties due to the extraction of the large power from the instanton density, $\Lambda_{QCD}/\mu_c \sim \sqrt[b]{a(\mu_c(T), T)}$, $b \sim 11/3N_c - 2/3N_f$.

5. Concluding comments

Main Results

The main leitmotiv of this talk is based on the conjecture that the confinement-deconfinement phase transition at nonzero chemical potential and small temperature is driven by instantons. The instantons qualitatively change the shapes at the transition: they small well-localized objects at large $\mu \gg \mu_c$; they become arbitrary large, strongly overlapped configurations at small $\mu \ll \mu_c$ in which case description in terms of the instanton quarks become appropriate. While the instanton quarks can be arbitrary far away from each other, they keep the information about their origin; they are correlated. Therefore, instanton quarks form not a random, but rather, the coherent large size configurations. Furthermore we make a quantitative prediction for the critical value of the chemical potential where this transition between two descriptions takes place: $\mu_c \sim 3\Lambda_{\text{QCD}}$ at $T = 0$. This prediction can be readily tested on the lattice at nonzero isospin chemical potential.

Future Directions

There are well established lattice method which allow to introduce isospin chemical potential into the system, see e.g. [11]. Independently, there are well- established lattice methods which allow to measure the topological charge density distribution, see e.g. [7,12,13]. We claim that the topological charge density distribution measured as a function of μ_I will experience sharp changes at the same critical value $\mu_I = \mu_c(T)$ where the phase transition (or rapid crossover) occurs. Indeed, the changes in the topological charge density distribution are expected due to the fundamental differences in θ dependence in two different regimes. We identify these changes with confinement-deconfinement transition based on the arguments presented above.

Relation to Other Studies

i). As we already mentioned, at the intuitive level there seems to be a close relation between instanton quarks and the "periodic instanton" [14,15,12]. Indeed, in these papers it has been shown that the large size instantons and monopoles are intimately connected and instantons have the internal structure resembling the instanton-quarks. Also, it has been shown that the constituents carry the magnetic charges. More than that, it has been also argued that large size instantons likely were missing in the lattice simulations, which is consistent with the picture advocated in the present work. Unfortunately, one should not expect to be able to account for large instantons using semiclassical technique to bring this intuitive correspondence onto the quantitative level. However, such a mapping may help us to understand the relation between pictures advocated by 't Hooft and Mandelstam [16] on one hand and picture where instanton-quarks are the key players, on the other hand.

ii). There is an interesting recent development in lattice computations which in principle would allow to study the topological charge fluctuations in QCD vacuum without any assumptions or guidance based on some specific models for QCD vacuum configurations[13]. Our remark here that the picture based on the instanton quarks advocated here is consistent with these recent lattice results[13]. Indeed, the most profound finding of ref.[13] is demonstration that the topological density distribution in QCD has " inherently global " structure. It is definitely consistent with our picture when the point like instanton quarks can be far away from each other, but still keep the correlation at arbitrary large distances.

Also, the dual picture of our CG representation (describing the instanton quarks) is nothing but the effective chiral lagrangian for Goldstone fields, see eq.(4). This " obvious" connection between confinement and chiral symmetry breaking phenomenon in our framework is consistent with speculation of ref.[13] that the corresponding long distance correlations might be associated with long range propagation of Goldstone fields.

It is too early to say whether ref.[13] finds precisely the features we have been advocating to exist for quite a while[3], but the results of ref.[13] look very exciting and promising to us.

iii). As the final remark: the θ parameter played a key role in all discussions presented above. However, the region of $\mu_c(T)$ where transition is expected to occur (see Table 1) is not very sensitive to value of θ. Indeed, the θ dependence in physical observable comes with extra suppression $\sim m_q$ which is very small factor. This is exactly the reason why all results for $\mu_c(T)$ are quoted for $\theta = 0$. This is definitely not the case when transition from normal to superfluid phase is considered as a function of baryon chemical potential at $N_c = 2$, or as a function of isotopical chemical potential at $N_c \geq 3$. In these cases the transitions are happening at $\mu \sim m_\pi(\theta)$ where very nontrivial dependence $\mu_c(T)$ on θ is expected[17].

References

1. D. Toublan and A. R. Zhitnitsky, Phys. Rev. D **73**, 034009 (2006).
2. D. T. Son, M. A. Stephanov, and A. R. Zhitnitsky, Phys. Lett. **B510**, 167 (2001).
3. S. Jaimungal and A.R. Zhitnitsky, [hep-ph / 9904377], [hep-ph / 9905540], unpublished.
4. A. A. Belavin, A. M. Polyakov, A. S. Shvarts and Y. S. Tyupkin, Phys. Lett. B **59**, 85 (1975).
5. E. V. Shuryak, Nucl. Phys. B **203**, 93 (1982); D. Diakonov and V. Y. Petrov, Nucl. Phys. B **245**, 259 (1984);
6. T. Schafer and E. V. Shuryak, Rev. Mod. Phys. **70**, 323 (1998).
7. I. Horvath, N. Isgur, J. McCune and H. B. Thacker, Phys. Rev. D **65**, 014502 (2002); I. Horvath *et al.*, Phys. Rev. D **66**, 034501 (2002)
8. A. Belavin et al, Phys. Lett. **83B** (1979) 317.
9. G. Veneziano, Nucl. Phys. **B159** (1979) 213.
10. I. Halperin and A.R. Zhitnitsky,*Phys. Rev. Lett.* **81**, 4071 (1998); *Phys. Rev.* **D58**, 054016 (1998).
11. J. B. Kogut and D. K. Sinclair, Phys. Rev. D **66**, 034505 (2002); Phys. Rev. D **70**, 094501 (2004).
12. C. Gattringer, Phys. Rev. D **67**, 034507 (2003); C. Gattringer and S. Schaefer, Nucl. Phys. B **654**, 30 (2003).
13. I. Horvath *et al.* Phys. Rev. D **67**, 011501 (2003); Phys. Rev. D **68**, 114505 (2003); Phys. Lett. B **612**, 21 (2005); I. Horvath, Nucl. Phys. B **710**, 464 (2005) [Erratum-ibid. B **714**, 175 (2005)]
14. T.C. Kraan and P. van Baal, Nucl. Phys. **B533** (1998) 627; Phys. Lett.**B428** (1998) 268, Phys. Lett.**B435** (1998) 389; M.G.Perez, T.G.Kovacs, P. van Baal, Phys. Lett.**B472** (2000) 295; K.Lee and P.Yi, Phys. Rev. **D56** (1997) 3711; K.Lee, Phys. Lett.**B426** (1998) 323.
15. D. Diakonov, N. Gromov, V. Petrov and S. Slizovskiy, Phys. Rev. D **70**, 036003 (2004); D. Diakonov and N. Gromov, Phys. Rev. D **72**, 025003 (2005)
16. G't Hooft, in "Recent Developments in Gauge Theories" Charges 1979, Plenum Press, NY 1980; Nucl. Phys. **B190** (1981) 455; S. Mandelstam, *Phys. Rep.* **23** (1976) 245.
17. M. A. Metlitski and A. R. Zhitnitsky, Nucl. Phys. B **731**, 309 (2005); Phys. Lett. B **633**, 721 (2006).

SURPRISES FOR QCD AT NONZERO CHEMICAL POTENTIAL

K. SPLITTORFF* and J.J.M. VERBAARSCHOT**

Niels Bohr Intitute,
Blegdamsvej 17, Copenhagen
* *Email: split@nbi.dk*
** *On leave from Stony Brook University, Stony Brook, NY 11794.*
Email: jacobus.verbaarschot@stonybrook.edu

In this lecture we compare different QCD-like partition functions with bosonic quarks and fermionic quarks at nonzero chemical potential. Although it is not a surprise that the ground state properties of a fermionic quantum system and a bosonic quantum system are completely different, the behavior of partition functions with bosonic quarks does not follow our naive expectation. Among other surprises, we find that the partition function with one bosonic quark only exists at nonzero chemical potential if a conjugate bosonic quark and a conjugate fermionic quark are added to the partition function.

Keywords: QCD at nonzero chemical potential, bosonic quarks

1. Introduction

The QCD phase diagram in the chemical potential temperature plane has far reaching phenomenological implications ranging from heavy ion collisions to the interior of neutron stars. Unfortunately, first principle lattice simulations are only possible at zero chemical potential, and our knowledge of the phase diagram mainly relies on model calculations (see for example 1, 2). Physically, we know that at zero temperature the baryon density is zero below a chemical potential equal to $m_N/3$. Therefore, in the thermodynamic limit, the QCD free energy and its derivatives, such as for example the chiral condensate, do not depend on the chemical potential for $\mu < m_N/3$. Since the Dirac operator depends on the chemical potential this requires miraculous cancellations in the microscopic theory a problem that was coined[3] as the *The Silver Blaze Problem.* This problem becomes particularly manifest in terms of the eigenvalues of the Dirac operator which are distributed homogeneously in a strip[4] with a width that increases a function

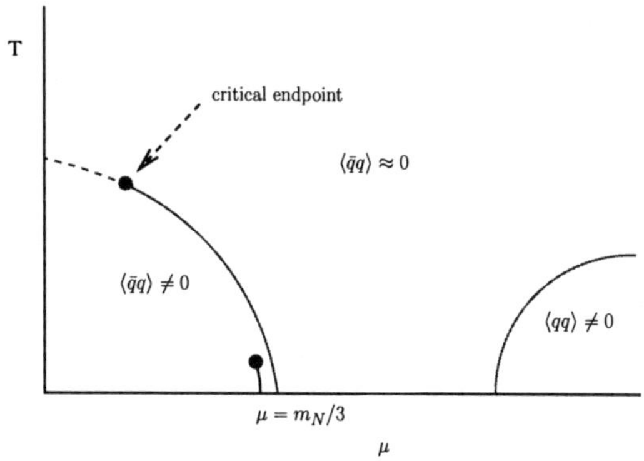

Fig. 1. Possible phases of the QCD partition function at nonzero temperature and chemical potential.

of μ. In this lecture we will mainly focus on the zero temperature axis of the phase diagram. To better understand the effect of the baryon chemical potential in QCD, we will consider four different partition functions listed in Table 1 which were discussed in 5 and 6. Although, the first partition

Theory	Number of Charged Goldstone Modes for $\mu < \mu_c$	Critical Chemical Potential
$\langle \det(D + \mu\gamma_0 + m) \rangle$	0	$\mu_c = \frac{1}{3} m_N$
$\langle \lvert \det(D + \mu\gamma_0 + m) \rvert^2 \rangle$	2	$\mu_c = \frac{1}{2} m_\pi$
$\langle \frac{1}{\det(D+\mu\gamma_0+m)} \rangle$	4	$\mu_c = \frac{1}{2} m_\pi$
$\langle \frac{1}{\lvert \det(D+\mu\gamma_0+m) \rvert^2} \rangle$	na	$\mu_c = 0$

Table 1. Summary of properties of low energy QCD at nonzero chemical potential and zero temperature. These partition functions will be denoted by $Z^{N_f=1}$, $Z_{n=1}$, $Z^{N_f=-1}$, $Z_{n=-1}$, in this order.

function is physically the most relevant, the other partition functions have important applications. Because lattice QCD simulations of full QCD at nonzero chemical potential are not possible, one sometimes uses the phase quenched approximation where

$$\langle \det^2(D + \mu\gamma_0 + m) \rangle \to \langle |\det(D + \mu\gamma_0 + m)|^2 \rangle, \tag{1}$$

which can be interpreted as a partition function of quarks and conjugate quarks[7]. Then Goldstone bosons made out of quarks and conjugate anti-quarks have nonzero baryon number resulting in a critical chemical potential of $m_\pi/2$ instead of $m_N/3$.

The bosonic partition function occurs in the formula for the quenched spectral density in the microscopic domain of QCD which is given by[15]

$$\rho^{\rm quen}(z, \mu) = \frac{|z|^2}{2} Z_{n=1}(z, \mu) Z_{n=-1}(z, \mu). \tag{2}$$

In a future publication[8] we will consider the expectation value of the phase of the fermion determinant given by

$$\langle e^{2i\theta} \rangle = \left\langle \frac{\det(D + \mu\gamma_0 + m)}{\det(-D + \mu\gamma_0 + m^*)} \right\rangle. \tag{3}$$

This partition function is not among the above list, but based on our insights from the bosonic partition function, we will be able to predict the its phase diagram.

2. Gauge Invariance and the Phases of QCD at $\mu \neq 0$

The principle that underlies the independence of the free energy on the chemical potential is gauge invariance [9,10]. The Dirac operator can be written as

$$D + \mu\gamma_0 + m = e^{-\mu\tau}(D + m)e^{\mu\tau}, \tag{4}$$

which implies that the μ-dependence can be transformed into the boundary conditions. A μ-independent free energy is possible in a phase that is not sensitive to the boundary conditions. This is the case for $\mu < \mu_c$ when the quarks do not loop around the torus in the time direction.

Although the partition function $Z_{n=-1}$ is naively gauge invariant it turns out that the regulator of the partition function breaks gauge invariance so that a μ-independent phase cannot exist. The need for regularization is best seen by writing the partition function in terms of eigenvalues[11]

$$Z_{n=-1} = \int_{\mathbb{C}/C_m(\epsilon)} \prod_k d^2 z_k \frac{\rho(\{z_k\})}{\prod_k (z_k^2 - m^2)(z^{*2}_k - m^2)}, \tag{5}$$

where $\mathbb{C}/C_m(\epsilon)$ is the complex plane except two small spheres with radius ϵ around $\pm m$. Because of the complex conjugated pole the integral diverges as $\log \epsilon$. Instead of this regularization we prefer to regularize the partition function as[15] (also known as hermitization[12])

$$Z_{n=-1} = \left\langle \det{}^{-1} \begin{pmatrix} \epsilon & D + \mu\gamma_0 + m \\ -D + \mu\gamma_0 + m^* & \epsilon \end{pmatrix} \right\rangle . \tag{6}$$

Since the matrix inside the determinant is Hermitian, this partition function can be written as a convergent bosonic integral. However, ϵ breaks gauge invariance, and for $\epsilon \neq 0$ it is not possible to gauge away the chemical potential. We thus find that $\mu_c = 0$ in this case.

Let us now consider the partition function with one bosonic flavor. This partition function *cannot* be written as a convergent bosonic integral and therefore *cannot* be interpreted in terms of bosonic quarks only. The correct interpretation is to rewrite this partition function as[6]

$$Z^{N_f=-1} = \left\langle \frac{\det^*(D + \mu\gamma_0 + m)}{\det(D + \mu\gamma_0 + m)\det^*(D + \mu\gamma_0 + m)} \right\rangle . \tag{7}$$

and regulate the denominator as in (6). However, contrary to the case of a pair of conjugate bosonic quarks, this partition function does not diverge for $\epsilon \to 0$, and it is possible to gauge away the chemical potential. In this case the free energy will be μ-independent in the thermodynamic limit below the lightest particle with nonzero baryon number which is a Goldstone boson made out of a bosonic quark and a conjugate bosonic anti-quark.

3. Low Energy Limit of QCD

The low-energy limit of the partition functions in Table 1 uniquely follows from chiral symmetry and gauge invariance. In Table 2 and Fig. 2 we compare the bosonic and fermionic (see 7, 10, 13, 14, 15) partition functions with a pair of conjugate flavors.

For $N_f = -1$ the partition function is given by the ratio in eq. (7). In this case the partition function is finite for vanishing regulator and the gauge symmetry (4) is not obstructed. Therefore, we have a μ-independent phase for $\mu < \mu_c$. In this phase the chiral condensate is given by

$$\begin{aligned} \langle \bar{\psi}\psi \rangle^{N_f=-1} &= \frac{1}{V} \left(\Sigma_k \frac{1}{z_k + m} + \Sigma_k \frac{1}{z_k + m} - \Sigma_k \frac{1}{z_k + m} \right) \quad \text{for} \quad \mu < \mu_c \\ &= 2\Sigma - \Sigma = \Sigma. \end{aligned} \tag{8}$$

For $\mu > \mu_c$ the bosonic contribution to the chiral condensate rotates into a pion condensate (see Fig. 3) so that for $\mu \gg \mu_c$ only the fermionic contribution remains:

$$\langle \bar{\psi}\psi \rangle = -\Sigma \quad \text{for} \quad \mu \gg \mu_c. \tag{9}$$

	$\langle \lvert \det(D+\mu\gamma_0+m)\rvert^2 \rangle$	$\langle \lvert \det(D+\mu\gamma_0+m)\rvert^{-2} \rangle$
	no regularization	regularization
Goldstone Manifold[a]	$U \in U(2)$	$Q \in Gl(2)/U(2)$
Chiral Lagrangian	$\mathcal{L}^{\rm kin} = \frac{F^2}{4}{\rm Tr}\nabla_\mu U \nabla_\mu U^\dagger$	$\mathcal{L}^{\rm kin} = \frac{F^2}{4}{\rm Tr}\nabla_\mu Q \nabla_\mu Q^{-1}$
Covariant Derivative	$\nabla_0 U = \partial_0 U + \mu[U, B]$	$\nabla_0 Q = \partial_0 Q + \mu\{Q, B\}$
μ_c	$\mu_c = \frac{m_\pi}{2}$	$\mu_c = 0$

Table 2. Comparison of the $n = 1$ and the $n = -1$ partition function.

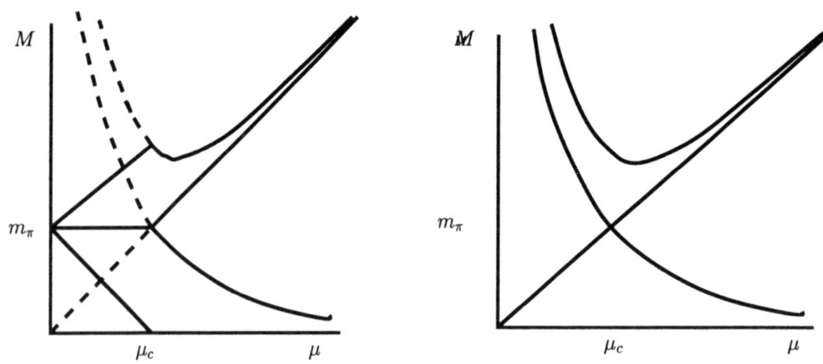

Fig. 2. Goldstone spectrum for $Z_{n=1}$ (left) and $Z_{n=-1}$ (right)

4. Chiral Symmetry Breaking at $\mu \neq 0$ and the Dirac Spectrum

In the domain where the kinetic term of the chiral Lagrangian factorizes from the partition function, i.e. for $\mu \ll 1/L$ and $|z| \ll 1/L^2$, the quenched spectral density satisfies the relation (2). This offers the possibility to test our results for $Z_{n=-1}$ by means of lattice QCD simulations. Calculations both with staggered fermions and overlap fermions show an impressive

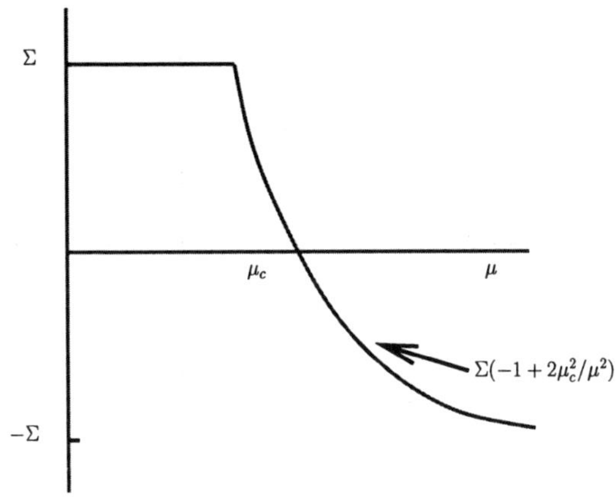

Fig. 3. The μ-dependence of the chiral condensate for $N_f = -1$

agreement[16] with (2).

In the thermodynamic limit the eigenvalues are distributed homogeneously inside the strip

$$|\text{Re}(z)| < \frac{2\mu^2 F^2}{\Sigma}. \tag{10}$$

Therefore, inside this strip, the quenched chiral condensate goes to zero linearly. The behavior of the chiral condensate for full QCD is quite different. In that case the chiral condensate remains nonzero for $m \to 0$. On the other hand, the eigenvalues still spread out in the complex plane. To explain[5] this so called "Silver Blaze Problem" we introduce the "spectral density"

$$\rho^{\text{full}}(z,\mu) = \frac{\langle \det(D+\mu\gamma_0+m)\sum_k \delta^2(z-z_k)\rangle}{\langle \det(D+\mu\gamma_0+m)\rangle}. \tag{11}$$

Because of the phase of the fermion determinant $\rho^{\text{full}}(z,\mu)$ is in general complex. Its microscopic limit is known analytically[17] and can be decomposed as

$$\rho^{\text{full}}(z,\mu) = \rho^{\text{quen}}(z,\mu) + \rho^{\text{osc}}(z,\mu), \tag{12}$$

where ρ^{osc} is complex with oscillations with a period of $O(1/V)$ and an amplitude that diverges exponentially with the volume. For $V \to \infty$ it vanishes outside a region with $m < |\text{Re}(z)| < \frac{8}{3}\mu^2 F^2/\Sigma - \frac{m}{3}$. The chiral condensate follows the same decomposition

$$\Sigma^{\text{full}} = \Sigma^{\text{quen}} + \Sigma^{\text{osc}}. \tag{13}$$

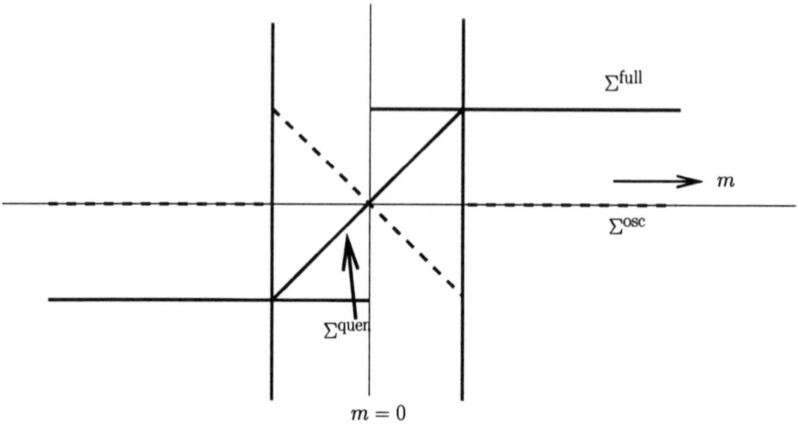

Fig. 4. The chiral condensate in quenched QCD (Σ^{quen}) and in full QCD (Σ^{full}) as a function of the mass. The support of the Dirac spectrum is in between the vertical lines.

In Fig. 4 we show the behavior of the different contributions in the thermodynamic limit. This shows that a nonzero chiral condensate for $m \to 0$ is due to the oscillatory contribution to the spectral density[5]. Therefore these oscillations solve the Silver Blaze Problem.

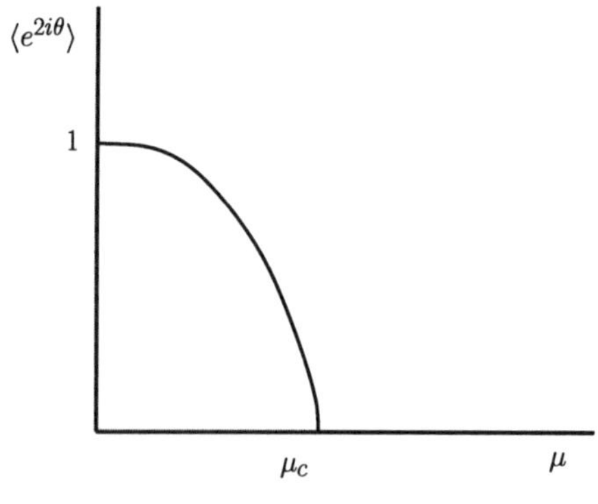

Fig. 5. Average phase of the fermion determinant as a function of μ.

5. Conclusions

The behavior of bosonic partition functions at nonzero chemical potential is quite different from what could be expected naively. This surprising behavior can be understood from gauging the chemical potential into the boundary conditions. In particular, this shows that the partition function with a pair of conjugate bosonic quarks has no μ-independent phase. The free energy of the theory with one bosonic quark, on the other hand is μ-independent for $\mu < m_\pi/2$. However, this partition function only exists as a partition function of a pair of conjugate bosonic quarks and a fermionic quark with the same mass. Finally, an analysis along the lines of this paper[8] shows that the expectation value of the phase of the fermion determinant, eq. (3), behaves as in Fig. 5.

Acknowledgments. This work was supported by DOE Grant No. DE-FG-88ER40388. KS was supported by the Carslberg Foundation.

References

1. M. A. Halasz, A. D. Jackson, R. E. Shrock, M. A. Stephanov and J. J. M. Verbaarschot, Phys. Rev. D **58**, 096007 (1998).
2. J. Berges and K. Rajagopal, Nucl. Phys. B **538**, 215 (1999).
3. T. D. Cohen, Phys. Rev. Lett. **91**, 222001 (2003).
4. I. Barbour, N. E. Behilil, E. Dagotto, F. Karsch, A. Moreo, M. Stone and H. W. Wyld, Nucl. Phys. B **275**, 296 (1986).
5. J. C. Osborn, K. Splittorff and J. J. M. Verbaarschot, Phys. Rev. Lett. **94**, 202001 (2005)
6. K. Splittorff and J. J. M. Verbaarschot, Nucl. Phys. B (in press) [arXiv:hep-th/0605143].
7. M. Stephanov, Phys. Rev. Lett. **76**, 4472 (1996).
8. K. Splittorff and J.J.M. Verbaarschot, in preparation.
9. J. Gasser and H. Leutwyler, Ann. Phys. **158**, 142 (1984); Nucl. Phys. B **250**, 465 (1985); H. Leutwyler, Ann. Phys. **235**, 165 (1994).
10. J.B. Kogut, M.A. Stephanov, and D. Toublan, Phys. Lett. B **464**, 183 (1999).
11. G. Akemann, J. C. Osborn, K. Splittorff and J. J. M. Verbaarschot, Nucl. Phys. B **712**, 287 (2005).
12. K.B. Efctov, Phys. Rev. Lett. **79**, 491 (1997); Phys. Rev. **B56**, 9630 (1997); J. Feinberg and A. Zee, Nucl. Phys. **B504**, 579 (1997); R. A. Janik, M. A. Nowak, G. Papp, J. Wambach and I. Zahed, Phys. Rev. E **55**, 4100 (1997); R. A. Janik, M. A. Nowak, G. Papp and I. Zahed, Nucl. Phys. B **501**, 603 (1997).
13. D. Toublan and J. J. M. Verbaarschot, Int. J. Mod. Phys. B **15**, 1404 (2001).
14. J.B. Kogut, M.A. Stephanov, D. Toublan, J.J.M. Verbaarschot, and A. Zhitnitsky, Nucl. Phys. B **582**, 477 (2000); K. Splittorff, D. T. Son, and M. A. Stephanov, Phys. Rev. D **64**, 016003 (2001); D. T. Son and M. A. Stephanov, Phys. Rev. Lett. **86**, 592 (2001).
15. K. Splittorff and J. J. M. Verbaarschot, Nucl. Phys. B **683**, 467 (2004).
16. J. C. Osborn and T. Wettig, PoS **LAT2005**, 200 (2005); J. Bloch and T. Wettig, arXiv:hep-lat/0604020.
17. J. C. Osborn, Phys. Rev. Lett. **93**, 222001 (2004).

STABILITY CONDITIONS IN GAPLESS SUPERCONDUCTORS

E. GUBANKOVA

Center for Theoretical Physics, Massachusetts Institute of Technology, Cambridge, MA 02139, USA
E-mail: elena1@mit.edu

Gapless superconductivity can arise when pairing occurs between fermion species with different Fermi surface sizes, provided there is a sufficiently large mismatch between Fermi surfaces and/or at sufficiently large coupling constant. In gapless states, secondary Fermi surfaces appear where quasiparticle excitation energy vanishes. This work focuses on homogeneous and isotropic superfluids in the s-wave channel, with either zero (conventional superconductor), one, or two spherical Fermi surfaces. The stability conditions for these candidate phases are analyzed. It is found that gapless states with one Fermi surface are stable in the BEC region, while gapless states with two Fermi surfaces are unstable in all parameter space. The results can be applied to ultracold fermionic atom systems.

Keywords: Unconventional Superconductivity; Fermi Surface; Topology.

1. Introduction

This work focuses on stability conditions and their possible relations to different Fermi surface topologies in a superconductor with unequal number densities of fermions (or unequal chemical potentials)[1]. The physics of paired fermion systems with unequal densities of the two fermion species is of interest in the study of two physical systems: (1) Ultracold atomic fermionic gases, where one can freely choose populations in two hyperfine states of the fermionic atom. Experimental work is currently being conducted with these systems. (2) Quark matter in the interior of neutron stars, which is believed to be a color superconductor. There, the mismatch in quark Fermi surfaces is driven by differences in quark masses and electric and color neutrality conditions. In both cases, the ground state of such a system is the subject of current debate. The study of fermionic superfluids with imbalanced populations is a novel subject, because until recently, experimental evidence for this situation was elusive. In a superconductor, an imbalance in population between spin-up and spin-down electrons can be created by a magnetic field. However, due to the Meissner effect, the magnetic field is either completely shielded from the superconductor bulk or enters in the form of quantized flux lines or vortices.

Cold fermionic atomic gases provide new possibilities in the experimental exploration of fermionic superfluids with unequal mixtures of fermions. In cold atoms, Feshbach resonance permits tuning of the interaction from (1) attraction and a resulting superfluid of loosely bound pairs (the BCS regime at $\zeta > n^{-1/3}$); to (2) repulsion and a resulting Bose-Einstein condensate of tightly bound molecules (the BEC regime at $\zeta < n^{-1/3}$), where ζ is the size of a pair and $n^{-1/3}$ is the interparticle separation (or, more precisely, the mean free path). With equal mixtures, the BEC–BCS crossover is smooth, with no phase transition. With asymmetric densities, one or more phase transitions are expected, and a more complex phase diagram may result.

In the weak coupling regime, a BCS superfluid remains stable as long as the difference in chemical potential is small compared to the pairing gap, $\delta\mu < \Delta$; the gap prevents the excess unpaired atoms from entering the superfluid state. By either increasing the mismatch or reducing the binding energy (and hence decreasing the gap), a quantum phase transition from the superfluid to normal state takes place, and superflidity ceases. The point at which the phase transition occurs is known as the Clogston limit, which can be estimated as $\delta\mu \sim \Delta \sim \mu \exp(-1/g) \ll \mu$, where μ is the Fermi energy. Thus only an exponentially small population imbalance is allowed in weak coupling. In strong coupling, as one approaches Feshbach resonance, the situation is quite different. On the BEC side, superfluidity with imbalanced population is robust over a wide range of parameter space. Surprisingly, on the BCS side, macroscopic imbalance is also possible in a superfluid state, possibly due to the formation of a gapless superfluid at strong coupling. This gapless superfluid incorporates large numbers of unpaired fermions which reside at the secondary Fermi surfaces. Here, we consider stability conditions for gapless states with different Fermi surface topologies.

2. Definitions for the screening masses and susceptibility

We consider two species of nonrelativistic fermions, $\psi = (\psi_1, \psi_2)$ with the same mass but with different chemical potentials, $\mu = \mathrm{diag}(\mu_1, \mu_2)$. There is an attractive interaction only between different species of fermions, $g\left(\psi^\dagger \sigma_2 \psi^*\right)\left(\psi^T \sigma_2 \psi\right)$. The order parameter is $\Phi^\dagger = \Delta\sigma_2$ where $\Delta = 2g <$

$\psi^T \sigma_2 \psi >$, it defines pairing in the singlet channel, spin 0, $[2] \times [2] = [1] + [3]$. The order parameter breaks the original group associated with conservation of particle numbers of each species down to the diagonal subgroup associated with conservation of difference in particle numbers, $U(1)_{\alpha_1} \times U(1)_{\alpha_2} \rightarrow U(1)_{\alpha_1 - \alpha_2}$ which is invariant under simultaneous rotation $\alpha_1 = -\alpha_2$. Total number of particles is not conserved. Formally, one can express this pattern of symmetry breaking by gauging the theory. We introduce two external gauge fields, and couple each species of fermions to its gauge field, ψ_1 couples to A_1 with g_1 and ψ_2 couples to A_2 with g_2, that is reflected by the generators of the gauge group, $T_1 = \mathrm{diag}(1, 0)$ and $T_2 = \mathrm{diag}(0, 1)$. According to the symmetry breaking pattern, gauge fields mix in a superconductor, rotated sum of fields $\tilde{A}_1 = A_1 + A_2$ is screened with non-zero Meissner mass via Andersen-Higgs mechanism, $m_M^2 \neq 0$ corresponding to $U(1)_{\alpha_1 + \alpha_2}$, and rotated difference of fields $\tilde{A}_2 = A_1 - A_2$ propagates in a superconductor unscreened, Meissner mass is zero $m_M^2 = 0$ corresponding to $U(1)_{\alpha_1 - \alpha_2}$. There is a different mixing of gauge fields in case of Debye masses. In QCD, with color and electromagnetic gauge groups, $SU(3)_c \times U(1)_{EM}$, mixing of gauge fields is more complicated.

We treat four-fermion interaction on a mean field level. We integrate over the fermions, obtain fermion determinant,

$$\mathcal{Z} = \int \mathcal{D}A \exp\left[S_A + \frac{|\Delta|^2}{4g} - \frac{1}{2}\mathrm{Tr}\ln(\mathcal{S}^{-1} + \mathcal{A})\right]. \tag{1}$$

We use the Nambu-Gorkov formalism with particle-hole basis, $\Psi = (\psi, \psi^*)$. Inverse fermion propagator and gauge field are 4×4 matrices in this basis, including the Nambu-Gorkov and two fermion species indices,

$$\mathcal{S}^{-1} \equiv \begin{pmatrix} [G_0^+]^{-1} & \Phi^- \\ \Phi^+ & [G_0^-]^{-1} \end{pmatrix}, \tag{2}$$

where the inverse free fermion propagators are $[G_0^{\pm}]^{-1} = i\partial_t \pm \frac{\nabla^2}{2m} \pm \mu$, and we have abbreviated for the gauge fields $\mathcal{A} = \mathrm{diag}(A^+, A^-)$ with $A^{\pm} = \pm\Gamma_a A_a^0 \mp \frac{\Gamma_a^2}{2m}\mathbf{A}_a^2 - \frac{i\Gamma_a}{2m}(\nabla \cdot \mathbf{A}_a + \mathbf{A}_a \cdot \nabla)$ and $\Gamma_a = g_a T_a$. Performing derivative expansion and collecting terms quadratic in the gauge field, we produce fermion loops, Π_{ab}^{00}, Π_{ab}^{i0}, Π_{ab}^{0i}, Π_{ab}^{ij}. Debye mass in one-loop is defined by the temporal component of polarization operator, Meissner mass is given by the spatial component, $m_{D,ab}^2 \equiv -\lim_{\mathbf{p}\to 0} \Pi_{ab}^{00}(0, \mathbf{p})$ and $m_{M,ab}^2 \equiv \frac{1}{2}\lim_{\mathbf{p}\to 0}(\delta_{ij} - \hat{p}_i\hat{p}_j)\Pi_{ab}^{ij}(0, \mathbf{p})$, where $\hat{p}_i \equiv p_i/p$. Screening masses are given[1]

$$m_{D,ab}^2 = -\lim_{P\to 0} \frac{1}{2}\frac{T}{V}\sum_K \mathrm{Tr}[S(K)\Gamma_a^- S(K-P)\Gamma_b^-],$$

$$m_{M,ab}^2 = \frac{1}{2m}\lim_{P\to 0}\frac{T}{V}\sum_K \Bigg(\delta_{ab}\mathrm{Tr}[S(K)\bar{\Gamma}_a^2] + \frac{k^2}{2m}[1-(\hat{p}\cdot\hat{k})^2]\,\mathrm{Tr}[S(K)\Gamma_a^+ S(K-P)\Gamma_b^+]\Bigg), \quad (3)$$

where S is the fermion propagator and we have introduced the following matrices in Nambu-Gorkov space, $\Gamma_a^\pm = \mathrm{diag}(\Gamma_a, \pm\Gamma_a)$ and $\bar{\Gamma}_a^2 = \mathrm{diag}(\Gamma_a^2, -\Gamma_a^2)$. Physically, Debye mass to all loops, including ladder diagrams, is equivalent to compressibility of a system, and Meissner mass can be associated with the density of superconducting fermions. These 2×2 matrices in two-fermion space shall be evaluated in the following section in order to obtain stability conditions for gapless superconductors.

One may derive the pressure from the partition function in (1) using the Cornwall-Jackiw-Tomboulis formalism. The pressure is the negative of the effective potential at its stationary point (i.e., with the propagators determined to extremize the effective potential). The fermionic part of the pressure is $p = \frac{1}{2}\frac{T}{V}\mathrm{Tr}\ln\mathcal{S}^{-1} + \frac{1}{2}\frac{T}{V}\mathrm{Tr}[\mathcal{S}_0^{-1}\mathcal{S} - 1] + \Gamma_2[\mathcal{S}]$, where $\mathcal{S}_0 = \mathrm{diag}(G_0^+, G_0^-)$ is the tree-level fermion propagator in Nambu-Gorkov space and $\Gamma_2[\mathcal{S}]$ is the sum of all two-particle irreducible diagrams. The number densities is defined $n_a = \partial p/\partial\mu_a$. The number susceptibility χ is defined as the derivative of the number density with respect to the chemical potential (at constant volume and temperature). Using the expression for the pressure, we obtain[1]

$$\chi_{ab} = \frac{\partial n_a}{\partial \mu_b} = -\frac{1}{2g_a g_b}\frac{T}{V}\sum_K \mathrm{Tr}[\Gamma_a^- \mathcal{S}(K)\Gamma_b^- \mathcal{S}(K)] - \frac{1}{2g_a}\frac{T}{V}\sum_K \mathrm{Tr}\left[\Gamma_a^- \mathcal{S}(K)\frac{\partial\Sigma(K)}{\partial\mu_b}\mathcal{S}(K)\right], \quad (4)$$

where Σ is the fermion self-energy, $\mathcal{S}^{-1} = \mathcal{S}_0^{-1} + \Sigma$. The first term on the right-hand side of this equation is given by the one-loop result for the electric screening mass, cf. Eq. (3). For the second term, we assume that the self-energy Σ depends on μ only through the gap, then we obtain[1]

$$\chi_{ab} = \frac{m_{D,ab}^2}{g_a g_b} + \frac{\partial n_a}{\partial\Delta}\frac{\partial\Delta}{\partial\mu_b}. \quad (5)$$

In general, the self-energy Σ contains terms of any number of fermion loops. Consequently, the number susceptibility contains terms of arbitrary many fermion loops too, corresponding to the exact Debye mass including all possible perturbative insertions. Remarkably, the free fermion result for χ, i.e. $\Sigma = 0$, gives the one-loop result for m_D^2. Susceptibility is equivalent to compressibility of a system. We shall use Eq. (5) in the following section to compute the number susceptibility. As this equation shows, it goes beyond the one-loop result for the electric screening mass.

3. Results

We consider three cases distinguished by how many zeros quasiparticle dispersion has, $\varepsilon_k^- = \sqrt{(k^2/2m - \bar{\mu})^2 + \Delta^2} - \delta\mu$, where μ is the average chemical potential and $\delta\mu$ is the difference between potentials, Fig. 1. In this case, number of zeros is equivalent to number of the Fermi surfaces. No zeros corresponds to the fully gapped state. In case of one Fermi surface, momenta outside the Fermi surface contribute to the pairing, while the excess of fermions resides inside the Fermi ball. In case of two Fermi surfaces, the excess of fermions resides between the two Fermi surfaces in momentum space. As was noticed by Son, apart from number of zeros, dispersion can have two different characteristic behaviors, distinguished by the position of the minimum. Minimum is located at nonzero momentum for positive $\bar{\mu}$, and corresponds to BCS, and minimum shifts to $p = 0$ for negative μ, and corresponds to BEC. This behavior manifests itself in stability conditions. We depict different topologies which are distinguished by the number of Fermi surfaces on the phase diagram in dimensionless average chemical potential $\bar{\mu}/\Delta$ and difference in chemical potentials $\delta\mu/\Delta$, the gap is the energy scale, in Fig. 2 regions between the solid lines. We have fully gapped F_0, and gapless states with one F_1 and two F_2 Fermi surfaces. At small mismatch, there is a fully gapped state F_0, which is at positive $\bar{\mu}$ and $\delta\mu < \Delta$, BCS, and at small and negative $\bar{\mu}$, $\bar{\mu} < -\sqrt{\delta\mu^2 - \Delta^2}$, BEC. Increasing mismatch, when mismatch is at least larger than the gap, there is a gapless state with one Fermi surface F_1 when μ is around the Feshbach resonance, $-\sqrt{\delta\mu^2 - \Delta^2} < \bar{\mu} < \sqrt{\delta\mu^2 - \Delta^2}$, and gapless state with two Fermi surfaces F_2 for positive μ restricted from below, $\sqrt{\delta\mu^2 - \Delta^2} < \bar{\mu}$. We therefore expect that F_1 exists at strong coupling, while F_2 probably exists only at weak coupling.

The question we are solving here is, "What is the ground state in a degenerate Fermi system with asymmetric number densities of fermions?". We avoid solving for the ground state explicitly. Instead, we check stabil-

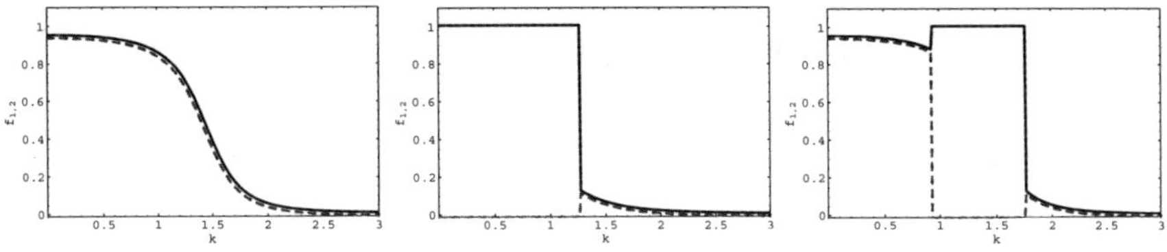

Fig. 1. Schematic plot of possible quasiparticle occupation numbers (in arbitrary units).

ity criteria, positive definite eigenvalues of screening masses and number susceptibility. If stability conditions are satisfied, homogeneous superconducting gapless state is indeed the ground state in this parameter range. If stability conditions are not satisfied, the alternative state may be realized as the ground state. These include LOFF, spatially separated mixtures, normal (not superconducting) states.

We analyze stability conditions, Eq. (3) and Eq. (5), in all parameter space $(\bar{\mu}, \delta\mu)$, and depict stable/unstable regions together with topology regions, Fig. 2 left panel. Debye mass e.v. are positive in all parameter space, hence Debye mass does not impose a constraint. All entries for the Meissner mass matrix are the same, it is trivial to diagonalize. We obtain[1] $m_M^2 = 0$ corresponding to the unbroken sector $U(1)_{\alpha_1-\alpha_2}$, and $m_M^2 = 2L$ corresponding to the broken group $U(1)_{\alpha_1+\alpha_2}$, where $L = \tilde{I} - \frac{\rho_+^{3/2}+\rho_-^{3/2}}{2\eta\sqrt{\eta^2-1}}$, which defines the dashed-dotted (blue online) curve in Fig. 2 left panel. It renders all states between the dashed-dotted curve and the solid vertical line unstable. There is a strip left in gapless superconductor state with two Fermi surfaces F_2 which is stable, gray area. F_0 and F_1 are stable with respect to m_M^2 everywhere. In magnetic sector, mixing does not depend on chemical potentials and it is defined by the pattern of symmetry breaking. In electric sector, mixing depends on chemical potentials, e.g. in QCD mixing depends on $\delta\mu$. Mixing in electric and magnetic sectors is the same only for the fully gapped case.

Analyzing the number susceptibility matrix, Eq. (5), we obtain[1] the expression defining the sign of e.v., which is very similar in structure to that for the Meissner mass, see expression for L, $R = I - \frac{\rho_+^{1/2}+\rho_-^{1/2}}{2\eta\sqrt{\eta^2-1}}$. Here, we defined $I \equiv I_\rho(0,\infty) - I_\rho(\sqrt{\rho_-}, \sqrt{\rho_+})$ and $\tilde{I} \equiv \tilde{I}_\rho(0,\infty) - \tilde{I}_\rho(\sqrt{\rho_-}, \sqrt{\rho_+})$ with elliptic integrals $I_\rho(a,b) \equiv \int_a^b dx\, x^2/[(x^2-\rho)^2+1]^{3/2}$ and $\tilde{I}_\rho(a,b) \equiv \int_a^b dx\, x^4/[(x^2-\rho)^2+1]^{3/2}$, and $\rho_\pm = \rho \pm \sqrt{\eta^2-1}$ are zeros of $\varepsilon_k^- = 0$

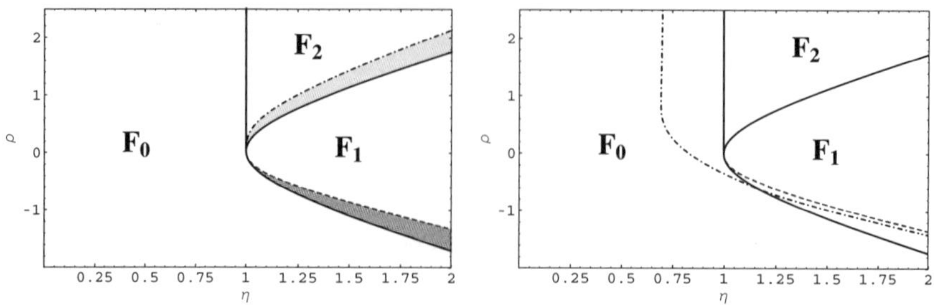

Fig. 2. Different topologies of the effective Fermi surfaces, and stability conditions.

with $\rho = \bar{\mu}/\Delta$, $\eta = \delta\mu/\Delta$. In expression for R one should put $\rho_- = 0$ when applied to F_1 region. It defines the dashed (red online) line, and it renders all states between dashed and solid vertical lines unstable. Thus all F_2 states are unstable,and there is a strip in F_1 which is stable, dark gray area. F_0 is stable everywhere. Stable states with respect to χ correspond to local maximum of pressure.

We obtained stable regions which are local maxima of pressure, Fig. 2 left panel. Now we consider global maxima, Fig. 2 right panel. For this we compare pressure of the superconducting and normal states. Superconducting states which pressure is higher than that of the normal state are stable, $\Delta p = p_s - p_n > 0$. The dashed-dotted (blue on line) line is $\Delta p = 0$, it renders all states above and to the right of it unstable. All unstable regions with respect to the screening masses and number susceptibility are subset of unstable region with respect to the pressure. In a weak coupling, the vertical dashed-dotted line reproduces the known Clogston limit, $\delta\mu = \Delta/\sqrt{2}$, above which BCS is unstable. The global stability line cuts through the stable strip of F_1 state, below is a stable superconducting state, above is a metastable state. Both lines coincide at large mismatches.

Currently, experiments are being performed with unequal mixtures of fermions to map superfluid regions as a function of population imbalance, interaction strength and temperature. The experimental signature of superfluidity is the existence of vortices, which prove phase coherence in a sample.

References

1. E. Gubankova, A. Schmitt, F. Wilczek, cond-mat/0603603; accepted for publication in Phys. Rev. B; and references therein.

RECENT RESULTS IN COLOR SUPERCONDUCTIVITY

G NARDULLI

Department of Physics, University of Bari and INFN-Bari
Via E. Orabona 4, 70126, Bari, Italy
E-mail: giuseppe.nardulli@ba.infn.it

Some recent results in color superconductivity are reviewed, In particular I stress some recent advances in the extension of the Larkin-Ovchinnikov-Fulde-Ferrell phase of QCD to the case of three flavors.

Keywords: Color Superconductivity; High Density QCD; LOFF Phase.

1. Introduction

In Quantum-Chromo Dynamics (QCD) the color antisymmetric channel between quarks is attractive. Therefore at high baryonic density and small temperatures the Cooper's theorem predicts color superconductivity [1–8]. For very large values of the baryonic chemical potential the energetically favored state is the Color-Flavor-Locking (CFL) phase [4], characterized by a homogenous spin 0 diquark condensate, antisymmetric in both color and flavor. This can be proved rigorously in QCD, albeit the proof holds for values of the baryonic chemical potential μ so high to be of no phenomenological interest. One expects however the dominance of the CFL phase also for high, but not extremely high μ, as can be seen by the use of phenomenological Nambu- Jona Lasinio hamiltonians. At intermediate densities the situation is more involved because the quark Fermi surfaces are unpaired, due to the strange quark mass ($M_s \neq 0$) and the differences $\delta\mu$ in the quark chemical potentials, induced by β equilibrium. Several ground states have been considered for this kinematic regime. For example in the 2SC phase [2], one takes the limit $M_s \to \infty$. This phase is probably of limited phenomenological interest since the condition $M_s \gg \mu$ can be hardly satisfied. More interesting appear the phases g2SC [9] and gCFL [10,11]. Being gapless these phases have interesting phenomenological consequences in the astrophysical domain [12]. However these gapless phases cannot realize the true vacuum even though their free energies are smaller than in the normal phase. This is so because they are chromo-magnetic instable, as for both models some of the gluon Meissner masses are imaginary (for g2SC see [13], for gCFL see [14] and also [15,16]).

This short review is devoted to a different color superconductive phase, the Larkin-Ovchinnikov-Fulde-Ferrell (LOFF) phase (for the original papers see [17,18], for applications to QCD see Refs. [19–25] and the review [26]). The relevance of this phase is based on the circumstance that, for appropriate values of $\delta\mu$, pairing with non-vanishing total momentum: $\mathbf{p_1}+\mathbf{p_2} = 2\mathbf{q} \neq 0$ is favored. The phase is characterized by a gap parameter not uniform in space that, in the simplest case, has the form of a single plane wave: $\Delta(\mathbf{r}) = \Delta\exp(2i\mathbf{q}\cdot\mathbf{r})$. Recently an extension of the original analysis to the case of three flavors has been obtained and the aim of this talk is to review it. The review is organized as follows. We discuss in Section 2 the results of the LOFF phase in the Ginzburg-Landau approximation and under the hypothesis of gap parameters having the space modulation of a single plane wave [25]. In Section 3 we present some recent results showing that this phase is chromomagnetic stable [27]. Both studies were performed in the high density limit. A recent analysis of higher order corrections to this approximation can be found in Ref. [28].

2. LOFF Phase of QCD with three flavors in the Ginzburg Landau approximation

We shall describe in this section some results obtained in the Ginzburg Landau (GL) approximation [25]. Within the limits of this approximation they agree with the outcome of a slightly different study, which, although based on assumptions valid in the GL approximation, does not perform an expansion in the gap parameters [29]. We shall limit our presentation to the case of a single plane wave (more recently an analysis, also based on the GL limit, has appeared, showing that the presence of more plane waves is energetically favored [30]).

The free energy per unit volume Ω in the GL limit is

$$\Omega = \Omega_n + \sum_{I=1}^{3}\left(\frac{\alpha_I}{2}\Delta_I^2 + \frac{\beta_I}{4}\Delta_I^4 + \sum_{J\neq I}\frac{\beta_{IJ}}{4}\Delta_I^2\Delta_J^2\right) + O(\Delta^6) \quad (1)$$

where Ω_n refers to the normal phase

$$\Omega_n = -\frac{3}{12\pi^2}\left(\mu_u^4+\mu_d^4+\mu_s^4\right) - \frac{\mu_e^4}{12\pi^2}\,, \quad (2)$$

and one has assumed for the condensate the pairing ansatz

$$< \psi_{i\alpha}\, C\, \gamma_5\, \psi_{\beta j} > \propto \sum_{I=1}^{3} \Delta_I(\mathbf{r})\, \epsilon^{\alpha\beta I}\, \epsilon_{ijI} \tag{3}$$

with

$$\Delta_I(\mathbf{r}) = \Delta_I \exp\left(2i\mathbf{q_I}\cdot\mathbf{r}\right) \ . \tag{4}$$

In other words, for each inhomogeneous pairing a Fulde-Ferrell ansatz is assumed and $2\mathbf{q_I}$ represents the momentum of the Cooper pair. In (2) μ_j are chemical potentials for quarks or for the electron while the coefficients of (1) can be found in [25]. One works in the approximation of vanishing color chemical potentials $\mu_3 = \mu_8 = 0$, which is valid for the present case, see the discussion in [28]. $\beta-$equilibrium is imposed together with the electric neutrality condition

$$-\frac{\partial\Omega}{\partial\mu_e} = 0 \ . \tag{5}$$

These conditions, together with the gap equations, give, for each value of the strange quark mass, the electron chemical potential μ_e and the gap parameters Δ_I. Moreover one determines $\mathbf{q_I}$ by solving, together with the gap equation and Eq. (5), also:

$$0 = \frac{\partial\Omega}{\partial q_I} = \Delta_I \frac{\partial\alpha_I}{\partial q_I} + \Delta_I \sum_{J=1}^{3} \Delta_J^2 \frac{\partial\beta_{IJ}}{\partial q_I} \ , \qquad I = 1,2,3 \ . \tag{6}$$

The condition (5) gives

$$\mu_e \approx \frac{M_s^2}{4\mu} \ , \tag{7}$$

which is valid up to terms of the order $(1/\mu)$. This result is identical to the free fermion case, which was expected since one works near the transition point between the LOFF and the normal phase. It follows that

$$\delta\mu_{du} = \delta\mu_{us} \equiv \delta\mu \tag{8}$$

and

$$\delta\mu_{ds} = 2\delta\mu \ . \tag{9}$$

To evaluate (6), it is sufficient to work at the $\mathcal{O}(\Delta^2)$, which leads to $q = 1.1997|\delta\mu|$. As to the orientation of $\mathbf{q_j}$, the results obtained in Ref. [25] indicate that the favored solution has $\Delta_1 = 0$ and therefore $\mathbf{q_1} = 0$. Furthermore $\mathbf{q_2} = \mathbf{q_3}$ and $\Delta_2 = \Delta_3$. These results are consequences of the GL limit. In fact, as shown by Eqns. (8) and (9), the surface separation of d and s quarks is larger, which implies that Δ_1 pairing is disfavored. On the other hand the surface separations of d-u and u-s quarks are equal, which implies that Δ_2 and Δ_3 must be almost equal. Finally $\mathbf{q_2}$, $\mathbf{q_3}$ are parallel because the pairing region on the $u-$quark surface at the GL point is formed by two distinct rings (in the northern and southern emisphere respectively), while for antiparallel $\mathbf{q_2}$, $\mathbf{q_3}$ the two rings overlap, which reduces the phase space available for pairing.

The results one obtains can be summarized as follows. Leaving the strange quark mass as a parameter (though a complete calculation would need its determination, considering also the possibility of condensation in the $q\bar{q}$ channel) the free energy can be evaluated as a function of the variable M_s^2/μ. One finds an enlargement of the range where color superconductivity is allowed, with the LOFF state having free energy lower than the normal and the gCFL phases. For example, for $\mu = 500$ MeV, and with the coupling fixed by the value $\Delta_0 = 25$ MeV of the homogeneous gap for two flavors, at about $M_s^2/\mu = 150$ MeV the LOFF phase has a free energy lower than the normal one. This point corresponds to a second order transition. Then the LOFF state is energetically favored till the point where it meets the gCFL line at about $M_s^2/\mu = 128$ MeV. This is a first order transition since the gaps are different in the two phases.

3. Stability of the LOFF Phase of QCD with three flavors

In [27] the gluon Meissner masses in the three flavor LOFF phase of QCD where computed using the High Density Effective Theory (HDET) [31] and the GL approximation. In the effective theory there are two vertices describing the coupling of gluons and quarks. Either the gluon couples to a quark and an antiquark (three-body vertex, coupling $\sim g$) or two quarks couple to two gluons (four-body vertex, coupling $\sim g^2$). At the order of g^2 the four-body coupling gives rise to the contribution $g^2\mu^2/(2\pi^2)$ identical for all the eight gluons; this result for the LOFF phase is identical to those of the normal or the CFL case as this term is independent of the gap parameters. The three-body coupling gives rise to the polarization tensor:

$$i\Pi_{ab}^{\mu\nu}(x,y) = -Tr[\,i\,S(x,y)\,i\,H_a^\mu\,i\,S(y,x)\,i\,H_b^\nu] \tag{10}$$

where the trace is over all the internal indexes; $S(x,y)$ is the quark propagator, and H_a^μ is the vertex matrix in the HDET formalism that can be found in [27]. In the Nambu-Gorkov (NG) formalism, as modified in the HDET, the quark propagator S has components S^{ij} $(i,\, j = 1,2)$, where each S^{ij} is a 9×9 matrix in the color-flavor space. At the fourth order in Δ one has:

$$S^{11} = S_0^{11} + S_0^{11}\,\boldsymbol{\Delta}\,\left[S_0^{22}\boldsymbol{\Delta}^\star\left(S_0^{11} + S_0^{11}\,\boldsymbol{\Delta}\,S_0^{22}\,\boldsymbol{\Delta}^\star\,S_0^{11}\right)\right], \quad (11)$$

$$S^{21} = S_0^{22}\,\boldsymbol{\Delta}^\star\left(S_0^{11} + S_0^{11}\boldsymbol{\Delta}\,S_0^{22}\boldsymbol{\Delta}^\star\,S_0^{11}\right), \quad (12)$$

where S_0^{ij} is the 18×18 matrix:

$$S_0 = \begin{pmatrix} [S_0^{11}]_{AB} & 0 \\ 0 & [S_0^{22}]_{AB} \end{pmatrix} = \delta_{AB}\begin{pmatrix} (p_0 - \xi + \bar{\mu}_A)^{-1} & 0 \\ 0 & (p_0 + \xi - \bar{\mu}_A)^{-1} \end{pmatrix}, \quad (13)$$

A, $B = 1, \cdots 9$ are indexes in the basis $A = (u_r, d_g, s_b, d_r, u_g, s_r, u_b, s_g, d_b)$, $\boldsymbol{\Delta}$ and $\boldsymbol{\Delta}^\star$ are 9×9 matrices containing the gap parameters Δ_I, p_0 is the energy, $\xi = |\mathbf{p}| - \mu$, $\bar{\mu}_A = \mu_A - \mu$. S^{12} and S^{22} are obtained by the changes $11 \leftrightarrow 22$ and $\boldsymbol{\Delta} \leftrightarrow \boldsymbol{\Delta}^\star$.

In the region where the use of the GL expansion is justified, i.e. 128 MeV $< M_s^2/\mu <$ 150 MeV all the squared gluon Meissner masses were found positive and therefore the LOFF phase of three flavor QCD is free from the chromo-magnetic instability.

Acknowledgments

It is a pleasure to thank R. Casalbuoni, M. Ciminale, R. Gatto, N. Ippolito and M. Ruggieri for their precious collaboration.

References

1. J. C. Collins and M. J. Perry, *Phys. Rev. Lett.* **34**, 1353 (1975); B. Barrois, *Nucl. Phys. B* **129**, 390 (1977); S. Frautschi, *Proc. workshop on hadronic matter at extreme density*, Erice 1978; D. Bailin and A. Love, *Phys. Rept.* **107**, 325 (1984).
2. M. G. Alford, K. Rajagopal and F. Wilczek, *Phys. Lett. B* **422**, 247 (1998) [arXiv:hep-ph/9711395].
3. R. Rapp, T. Schäfer, E. V. Shuryak and M. Velkovsky, *Phys. Rev. Lett.* **81**, 53 (1998) [arXiv:hep-ph/9711396].
4. M. G. Alford, K. Rajagopal and F. Wilczek, *Nucl. Phys. B* **537**, 443 (1999) [arXiv:hep-ph/9804403].
5. K. Rajagopal and F.Wilczek, in *Handbook of QCD*, M. Shifman ed. (World Scientific, Singapore, 2001) [arXiv:hep-ph/0011333];
6. M. G. Alford, *Ann. Rev. Nucl. Part. Sci.* **51**, 131 (2001) [arXiv:hep-ph/0102047].
7. T. Schafer, arXiv:hep-ph/0304281.
8. G.Nardulli, *Riv. Nuovo Cim.* **25N3**, 1 (2002) [arXiv:hep-ph/0202037].
9. I. Shovkovy and M. Huang, *Phys. Lett. B* **564**, 205 (2003) [arXiv:hep-ph/0302142].

10. M. Alford, C. Kouvaris and K. Rajagopal, *Phys. Rev. Lett.* **92**, 222001 (2004) [arXiv:hep-ph/0311286];
11. M. Alford, C. Kouvaris and K. Rajagopal, *Phys. Rev. D* **71**, 054009 (2005) [arXiv:hep-ph/0406137].
12. M. Alford, P. Jotwani, C. Kouvaris, J. Kundu and K. Rajagopal, *Phys. Rev. D* **71**, 114011 (2005) [arXiv:astro-ph/0411560].
13. M. Huang and I. A. Shovkovy, *Phys. Rev. D* **70**, 051501 (2004) [arXiv:hep-ph/0407049]; M. Huang and I. A. Shovkovy, *Phys. Rev. D* **70**, 094030 (2004) [arXiv:hep-ph/0408268].
14. R. Casalbuoni, R. Gatto, M. Mannarelli, G. Nardulli and M. Ruggieri, *Phys. Lett. B* **605**, 362 (2005) [Erratum-*ibid. B* **615**, 297 (2005)] [arXiv:hep-ph/0410401].
15. K. Fukushima, *Phys. Rev. D* **72**, 074002 (2005) [arXiv:hep-ph/0506080].
16. M. Alford and Q. h. Wang, *J. Phys. G* **31**, 719 (2005) [arXiv:hep-ph/0501078].
17. A. I. Larkin and Yu. N. Ovchinnikov, *Zh. Eksp. Teor. Fiz.* **47**, 1136 (1964) (*Sov. Phys. JETP* **20**, 762 (1965)).
18. P. Fulde and R. A. Ferrell, *Phys. Rev.* **135**, A550 (1964).
19. M. G. Alford, J. A. Bowers and K. Rajagopal, *Phys. Rev. D* **63**, 074016 (2001) [arXiv:hep-ph/0008208].
20. J. A. Bowers, J. Kundu, K. Rajagopal and E. Shuster, *Phys. Rev. D* **64**, 014024 (2001) [arXiv:hep-ph/0101067];
21. A. K. Leibovich, K. Rajagopal and E. Shuster, *Phys. Rev. D* **64**, 094005 (2001) [arXiv:hep-ph/0104073].
22. J. A. Bowers and K. Rajagopal, *Phys. Rev. D* **66**, 065002 (2002) [arXiv:hep-ph/0204079].
23. R. Casalbuoni, R. Gatto, M. Mannarelli and G. Nardulli, *Phys. Rev. D* **66**, 014006 (2002) [arXiv:hep-ph/0201059].
24. R. Casalbuoni, M. Ciminale, M. Mannarelli, G. Nardulli, M. Ruggieri and R. Gatto, *Phys. Rev. D* **70**, 054004 (2004) [arXiv:hep-ph/0404090].
25. R. Casalbuoni, R. Gatto, N. Ippolito, G. Nardulli and M. Ruggieri, *Phys. Lett. B* **627**, 89 (2005) [arXiv:hep-ph/0507247].
26. R. Casalbuoni and G. Nardulli, *Rev. Mod. Phys.* **76**, 263 (2004) [arXiv:hep-ph/0305069].
27. M. Ciminale, G. Nardulli, M. Ruggieri and R. Gatto, *Phys. Lett. B* **636**, 317 (2006) [arXiv:hep-ph/0602180].
28. R. Casalbuoni, M. Ciminale, R. Gatto, G. Nardulli and M. Ruggieri, arXiv:hep-ph/0606242.
29. M. Mannarelli, K. Rajagopal and R. Sharma, arXiv:hep-ph/0603076.
30. K. Rajagopal and R. Sharma, arXiv:hep-ph/0605316; K. Rajagopal and R. Sharma, arXiv:hep-ph/0606066.
31. D. K. Hong, *Phys. Lett. B* **473**, 118 (2000) [arXiv:hep-ph/9812510]; D. K. Hong, *Nucl. Phys. B* **582**, 451 (2000) [arXiv:hep-ph/9905523]; S. R. Beane, P. F. Bedaque and M. J. Savage, *Phys. Lett. B* **483**, 131 (2000) [arXiv:hep-ph/0002209]; R. Casalbuoni, R. Gatto and G. Nardulli, *Phys. Lett. B* **498**, 179 (2001) [Erratum-*ibid. B* **517**, 483 (2001)] [arXiv:hep-ph/0010321].

HEAVY QUARKONIA ABOVE DECONFINEMENT

Á. MÓCSY

RIKEN-BNL Research Center, Brookhaven National Laboratory
Upton, NY 11973, USA
E-mail: mocsy@bnl.gov

In this talk I summarize our current understanding of quarkonium states above deconfinement based on phenomenological and lattice QCD studies.

Keywords: Quarkonium. Deconfinement. Finite temperature QCD. Potential models.

Quarkonium as a Signal of Deconfinement. Quarkonium, the common name of a meson state made by a heavy quark Q and antiquark $\bar{Q}$ has been in the center of interest since the discovery of the J/ψ at BNL and SLAC in 1974. These heavy mesons allowed for a careful testing of Quantum Chromodynamics (QCD), and are essential diagnostic tools of the deconfinement transition in hot QCD.

The 20 year old prediction by Matsui and Satz[2], that the melting of heavy quark-antiquark bound states at the deconfinement temperature could be considered an unambiguous signal for deconfinement, has led to an intense line of studies. It was predicted[2] that color screening in a deconfined medium causes the dissolution of the J/ψ. The main idea behind this prediction is that in the deconfined phase of QCD, refered to as quark-gluon plasma, there is a screening of the color force between a heavy quark and antiquark, much like the known Debye-screening in QED. This screening is due to the light quarks, antiquarks and gluons present in the plasma. The range of screening is characterized by the screening length, which is dependent on temperature. When the screening length is smaller than the size of the boundstate then this will dissociate. In case of the J/ψ this would happen at T_c, the temperature of deconfinement. The melting of this state would in turn manifests in the suppression of the J/ψ peak in the dilepton spectrum, providing an experimentally detectable signal for deconfinement. Experiments colliding heavy ions at ultrarelativistic energies at SPS-CERN and RHIC-BNL have been looking for J/ψ suppression[3].

It is clear that understanding the modification of the properties of the J/ψ, and other quarkonium states in a hot medium is therefore crucial for understanding deconfinement. Until recently, theoretical studies, including the one the original prediction is based on, have been mostly phenomenological, using potential models as the basic tool[2,4,5]. Today, first principle numerical calculations of QCD carried out on the lattice are also available, and provide new information about quarkonia at high temperatures[7–9]. In what follows, I first discuss potential models and lattice results, then point out the agreements and inconsistencies between these.

Potential Models. The essence of potential models is to assume that the interaction between a heavy quark and antiquark is mediated by a potential. This assumption is feasible when the $Q-\bar{Q}$ interaction is instantaneous. This adiabatic approximation is applicable in the nonrelativistic situation, when the timescale associated with the relative motion of the Q and $\bar{Q}$ is much larger than that associated with the gluons. The bound state properties are determined by solving the Schrödinger equation with the potential. At zero temperature the phenomenologically introduced Cornell potential[10], made of a Coulomb and a linear part, was found to describe quarkonia spectroscopy rather well. It was understood much later, that the existence of the hierarchy of energy scales, $m \gg mv \gg mv^2$, in nonrelativistic systems of mass m and velocity v, allows for the construction of a sequence of effective field theories: nonrelativistic QCD (NRQCD) and potential NRQCD (pNRQCD)[11]. The Cornell potential can be obtained as the leading order approximation of pNRQCD. The matching between QCD and these effective theories has been done on a perturbative level, but by emphasizing the importance of non-perturbative effects the validity of the potential model approach has been early on challenged by[12].

Potential models have been used also at finite temperature in context of deconfinement. At finite temperature the form of the potential is not known. A sequence of effective field theories that might yield a potential have not yet been derived from hot QCD. Such a calculation is intrinsically more involved than that at zero temperature, due to the presence of additional scales determined by the temperature, T, gT, g^2T. So at finite T it is more difficult to establish a well-defined potential. It is thus questionable whether a temperature-dependent potential is adequate for understanding the properties of quarkonia in a hot medium.

Nevertheless, in lack of other theoretical means potential models have been widely used to investigate quarkonium states at finite temperature. A usual choice for the potential is a screened version of the Cornell potential with a temperature-dependent screening. The screening radius can be determined exactly in perturbative analysis, as the inverse of the Debye-mass $\sim gT$. Given the small size of the quarkonium states one can argue that at the relevant short distances perturbative calculations are acceptable. In finite temperature lattice QCD the potential between a static heavy quark and antiquark has been identified from Polyakov loop correlations measuring the free energy, from which the color screening has also been obtained. Solving the Schrödinger equation provides the temperature-dependence of the quarkonium properties, allowing to monitor at what temperature the screening radius drops below the radius of the bound state. The prediction is that the J/ψ disappears around $1.1T_c$[5]. Since the excited states have a larger radius than the ground state, a pattern of sequential suppression can be identified, meaning that higher excited states dissolve earlier. This sequential effect is predicted to be seen in the J/ψ suppression pattern as a function of the energy density[5,6] as determined in experiments[3].

Quarkonium from Lattice. A few years ago an alternative way to study the temperature dependence of quarkonia properties became available in the form of numerical simulations of QCD carried out on a lattice. Correlation functions of hadronic currents are reliably calculated in Euclidean time τ. The spectral representation of the correlator $G(\tau, T) = \int d\omega\sigma(\omega, T)K(\tau, \omega, T)$ allows for the reconstruction of the meson spectral function $\sigma(\omega, T)$ using the Maximum Entropy Method. Correlators can be conveniently used to determine the modification of the quarkonium properties in a hot medium at temperatures above T_c. To eliminate the trivial T-dependence of the kernel $K(\tau, \omega, T)$ the ratio of the correlator to the so-called reconstructed one, $G_{recon}(\tau, T) = \int d\omega\sigma(\omega, T = 0)K(\tau, \omega, T)$ is determined. Any deviation from one of the ratio G/G_{recon} indicates modification of the spectral function with temperature.

The first results[7,8] were obtained in quenched simulations, i.e. a purely gluonic background, in which the $Q - \bar{Q}$ is at rest with respect to the thermal medium. To the surprise of the community, the data contradicts what has been theoretically expected from the potential model calculations: The ratio of correlators for the ground state charmonium are technically flat even at temperatures well above T_c, indicating that the 1S J/ψ and η_c survive at least up to 1.5 T_c. The correlator of the 1P state increases for all $T > T_c$, suggesting that the χ_c dissolves by 1.16 T_c. The corresponding spectral functions not only reinforce these findings, but also indicate that the properties of the 1S states, the mass and within errors the amplitude do not change up to these temperatures[8]. Although the temperature-dependence of the ground state charmonium correlators was found to be small, a small difference between the behavior of the J/ψ and η_c correlators has been identified. Such a difference was a priori not expected. Since hyperfine splitting is not taken into account explicitly, the source for this must be elsewhere. The findings for bottomonium are similar[9]: no modification of the ground state, Υ and η_b, up to temperatures of $2.3T_c$, while dissolution of the χ_b already at $1.15T_c$. Such early melting of the χ_b is in contrast with expectations based on the screening argument. The size of this state is similar to that of the J/ψ, so similar dissolution temperatures were expected. Calculations performed in two-flavor QCD[13] sofar support the qualitative findings of the quenched results. Examining the behavior of quarkonium in motion with respect to the heat bath have also began. The first results[14] show modifications of the finite momenta charmonium when inserted in a deconfined gluonic medium.

Potential Models Revisited. After the appearance of the lattice data on quarkonium, potential models have been reconsidered, using different temperature dependent potentials[15–17]. The internal energy of a $Q-\bar{Q}$ pair as determined on the lattice[18] and identified as the potential[15] is used to study the possibility of strongly coupled Coulomb bound states. One should be aware though that in leading order perturbation theory, which is valid at high temperatures, the potential is equal to the free energy of the quark-antiquark pair. Beyond leading order there is an entropy contribution to the free energy which determines the internal energy. There is a sharp peak in the entropy near T_c[19] making the identification of the internal energy as potential conceptually difficult[20]. It is exactly this entropy contribution however that makes the internal energy a deeper potential than the free enregy, allowing thus for some of the quarkonium states to remain bound up to temperatures beyond T_c. Because of this, the lattice internal energy remains a popular choice for the potential[17]. A combination of the lattice internal and free energy has also been suggested as potential[16]. A common feature of these potentials is that they incorporate temperature-dependent screening and yield quarkonium dissociation temperatures in accordance with the above quoted numbers from the lattice.

Correlators and Spectral Functions. We performed the first phenomenological study of quarkonium correlators[21,23] using potential models and pointed out that although the agreement of dissociation temperatures from potential models with lattice data is necessary, it is not sufficient to claim understanding the disappearance of quarkonia in the quark-gluon plasma: Even though potential models with certain screened potentials can reproduce qualitative features of the lattice spectral function, such as the survival of the ground state and the melting of the excited states, the temperature dependence of the quarkonium correlators, especially in the S-channel is not reproduced. Furthermore, the properties of the states determined with these screened potentials do not seem to reproduce the results indicated by the lattice spectral functions.

As mentioned above, solving the Schrödinger equation yields the temperature-dependence of the quarkonium mass, size, and amplitude. It does not however give the spectral function and the correlator. This needs to be determined by different means. We considered two appraches. One is to design the spectral function as the sum of bound state/resonance contributions and the perturbative continuum above a threshold. This threshold is determined by the asymptotic value of the potential, and decreases with increasing temperature. To make direct comparison with the lattice data we calculate the ratio of correlators G/G_{recon}. The behavior for the χ_c correlator qualitatively agrees with what is seen on the lattice. But there is no agreement for the η_c correlator[21] . In the model calculations one can identify a more complex substructure in the η_c correlator: The reduction of the continuum threshold and that the amplitude of the states are distinguishable contributions[21] . The small difference between the J/ψ and η_c correlators detected on the lattice can also be seen in the model calculations. The difference is attributed to the transport contribution additional in the vector channel compared to the pseudoscalar channel[22]. The increase of the χ_b correlator is in accordance with lattice results, and is explained as the effect of the decreasing continuum threshold. All these conclusions are unchanged for the different potentials.

In another approach we perform a full non-relativistic calculation of the Green's function[23,24], whose imaginary part provides the quarkonium spectral function. The main advantage of this approach is that resonances and continuum are considered together. The behavior of the ratio of S-wave correlators obtained for the different lattice fitted potentials again are not flat, indicating that the spectral functions of the η_c, J/ψ and Υ are significantly different than at zero temperature. The spectral function of these states are severely modified in this approach[24], which is not supported by the lattice results.

Possible Improvements. Since none of the popular lattice fitted potentials lead to results in agreement with the lattice, it is reasonable to repeat the question whether such temperature-dependent screened potentials are the right way to describe modification of quarkonia properties with temperature? If not, then what is the mechanism responsible for the dissociation of quarkonia at high temperatures? If yes, then what is the potential that can reproduce the data for S and P charmonium and bottomonium states simultaneously?

As a first attempt to answer this question we consider a simple toy model[25]: Because no modification in the properties of the ground state charmonium compared to the zero temperature values has been observed up to well above T_c, keep these at their Particle Data Group values. Plus, since higher excited states seem to disappear near T_c, we "melt" these by removing them from the design spectral function at T_c. The main idea is to compensate for the melting of the higher excited states above T_c with the decrease of the only parameter, the continuum threshold. With such a simple model, that does not include temperature dependent screening we were able to recover the flatness of the η_c and the increase of the χ_c correlator.

Another attempt is to use a "screened" Cornell potential as input to determine the nonrelativistic Green's function. Here "screening" is a parameter that is related neither to Debye-screening, nor has anything to do with the free- and internal energies determined on the lattice. Our preliminary finding is that we can tune this parameter such that some qualitative agreement for the S-waves can be obtained. Determining the P-wave correlator is part of our currently ongoing research.

Final Remaks. We do not yet posses a comprehensive phenomenological tool that can explain consistently all the lattice observations on heavy quarkonium. It is not clear at the moment to what extent the modification of the quarkonium spectral functions can be understood in terms of the color Debye-screening picture. When the time-scale of screening is not short compared to the time-scale of the heavy quark motion, the role of gluo-dissociation in the medium modification of the spectral functions has to be understood. This is a topic of our ongoing investigations.

Acknowledgments. This presentation is based on work done in collaboration with P. Petreczky. I thank the Organizers for a successful workshop and J. Casalderrey-Solana, D. Kharzeev and H. Satz for helpful discussions.

References

1. N. Brambilla *et al.*, arXiv:hep-ph/0412158, and references therein.
2. T. Matsui and H. Satz, Phys. Lett. B **178**, 416 (1986).
3. P. Cortese *et al.* [NA50 Collaboration], J. Phys. G **31** (2005) S809; R. Arnaldi *et al.* [NA60 Collaboration], Eur. Phys. J. C **43** (2005) 167; M. Calderon de la Barca Sanchez, arXiv:nucl-ex/0606009; R. G. de Cassagnac [the PHENIX Collaboration], arXiv:nucl-ex/0608041.
4. F. Karsch, M. T. Mehr and H. Satz, Z. Phys. C **37**, 617 (1988).
5. S. Digal, P. Petreczky and H. Satz, Phys. Rev. D **64**, 094015 (2001) [arXiv:hep-ph/0106017].
6. F. Karsch, D. Kharzeev and H. Satz, Phys. Lett. B **637**, 75 (2006) [arXiv:hep-ph/0512239].
7. T. Umeda, K. Nomura and H. Matsufuru, Eur. Phys. J. C **39S1**, 9 (2005) [arXiv:hep-lat/0211003]; M. Asakawa and T. Hatsuda, Phys. Rev. Lett. **92**, 012001 (2004) [arXiv:hep-lat/0308034].
8. S.Datta et al, Nucl. Phys. Proc. Suppl. **119**, 487 (2003) [arXiv:hep-lat/0208012]; Phys. Rev. D **69**, 094507 (2004) [arXiv:hep-lat/0312037]; A. Jakovac et al, arXiv:hep-lat/0603005.
9. K. Petrov et al, PoS **LAT2005**, 153 (2006).
10. E. Eichten et al, Phys. Rev. Lett. **34**, 369 (1975) [Erratum-ibid. **36**, 1276 (1976)].
11. G. S. Bali, Phys. Rept. **343**, 1 (2001) [arXiv:hep-ph/0001312].
12. M. B. Voloshin, Nucl. Phys. B **154**, 365 (1979); H. Leutwyler, Phys. Lett. B **98**, 447 (1981).
13. G. Aarts et al, Nucl. Phys. Proc. Suppl. **153**, 296 (2006) [arXiv:hep-lat/0511028]; arXiv:hep-lat/0608009.
14. S. Datta et al, arXiv:hep-lat/0409147.
15. E. V. Shuryak and I. Zahed, Phys. Rev. C **70**, 021901 (2004) [arXiv:hep-ph/0307267]; Phys. Rev. D **70**, 054507 (2004) [arXiv:hep-ph/0403127].
16. C. Y. Wong, Phys. Rev. C **72**, 034906 (2005) [arXiv:hep-ph/0408020].
17. W.M.Alberico et al, Phys. Rev. D **72**, 114011 (2005) [arXiv:hep-ph/0507084]; M.Mannarelli and R.Rapp, arXiv:hep-ph/0509310; R.Rapp, D.Cabrera and H.van Hees, arXiv:nucl-th/0608033.
18. O. Kaczmarek et al, Nucl. Phys. Proc. Suppl. **129**, 560 (2004) [arXiv:hep-lat/0309121].
19. O. Kaczmarek and F. Zantow, Eur. Phys. J. C **43**, 63 (2005) [arXiv:hep-lat/0502011].
20. P. Petreczky, Eur. Phys. J. C **43**, 51 (2005) [arXiv:hep-lat/0502008].
21. Á.Mócsy and P.Petreczky, Eur. Phys. J. C **43**, 77 (2005) [arXiv:hep-ph/0411262]; Phys. Rev. D **73**, 074007 (2006) [arXiv:hep-ph/0512156].
22. P. Petreczky and D. Teaney, Phys. Rev. D **73**, 014508 (2006) [arXiv:hep-ph/0507318]; P. Petreczky et al, PoS **LAT2005**, 185 (2005) [arXiv:hep-lat/0510021].
23. Á. Mócsy and P. Petreczky, arXiv:hep-ph/0606053.
24. Á. Mócsy, P. Petreczky and J.Casalderrey-Solana, in preparation.
25. Á. Mócsy, arXiv:hep-ph/0606124.

θ-PARAMETER IN QCD-LIKE THEORIES AT FINITE DENSITY

M. A. METLITSKI

Department of Physics and Astronomy, University of British Columbia,
Vancouver, BC, Canada, V6T 1Z1
E-mail: mmetlits@physics.ubc.ca

In this talk I present some recent results[1,2] on the θ dependence of QCD-like theories at finite density. I discuss $N_c = 3$ QCD at finite isospin chemical potential, and $N_c = 2$ QCD at both finite baryon and isospin chemical potentials. The phase diagram of these theories is constructed. The validity of anomalous Ward identities is confirmed in the normal and superfluid phases. Careful attention is paid to physics at $\theta \sim \pi$. Finally, the behaviour of the gluon condensate at finite density is computed. The results are compared to recent lattice simulations.

1. Introduction

The θ-parameter of gauge theories has long attracted attention as it is a probe of the topological properties of the theory. In almost every context, from pure Yang-Mills theories to QCD the θ dependence of the theory is highly non-trivial and, frequently, non-analytic. However, if one introduces a massless quark into the theory the θ dependence disappears due to the existence of fermion zero-modes in topologically nontrivial sectors. If the quark has a finite mass, the chiral anomaly dictates that the θ-parameter can be incorporated into the phase of the quark mass matrix. Therefore, in theories with light quarks the θ dependence becomes a chiral property and can be reliably explored in the chiral Lagrangian framework. Hence, the θ dependence of the QCD phase diagram in vacuum in the chiral regime is by now fairly well understood[3–7].

The purpose of this work is to understand the θ-dependence of QCD in a non-trivial environment when a finite chemical potential is present. The main physical motivation for such a study is the attempt to understand the cosmological phase transition when θ, being non-zero and large at the very beginning of the phase transition, slowly relaxes to zero, as the axion resolution of the strong CP problem suggests. Of course, in real world, we are mostly interested in the effects of θ on matter at finite baryon density. However, analytical control over QCD is absent at moderate baryon density and appears only at asymptotically large baryon chemical potential, where one expects the color-superconducting state to be realized[8]. Nevertheless, one may resort to QCD-like theories, such as $N_c = 2$ QCD at finite baryon chemical potential μ_B[9] and $N_c = 3$ QCD at finite isospin chemical potential μ_I[10],[a] where analytical control is present, to gain some insight into real dense QCD.

The tool that gives us control over the above theories is again the chiral Lagrangian. There exists a well-known procedure for including the effects of finite chemical potential coupled to one of the chiral charges into the chiral Lagrangian[9,10]. As explained above, one can also incorporate the θ angle into the QCD chiral Lagrangian. We, thus, expect that we may adequately describe QCD at finite density and $\theta \neq 0$ in the effective chiral Lagrangian approach, as long as the chemical potentials μ_B, μ_I and the quark masses are much smaller than the QCD scale Λ_{QCD}.

Using the above approach, we obtain a wide range of information about the phase diagram of the two and three colour QCD in the μ, θ plane. We concentrate on the case of two quark flavours. We show that the transition to the superfluid phase occurs at μ equal to the θ dependent goldstone mass, $m_\pi(\theta)$. This implies that for fixed μ of order of the goldstone mass, the θ dependence of the theory becomes non-analytic. Two second order phase transitions, accompanied by a jump in the topological susceptibility, occur as θ relaxes from 2π to 0. We find that the θ dependence in the superfluid phase near $\theta = \pi$ is much smoother than in the normal phase. In particular, for $m_u = m_d$, we show that the first order phase transition across $\theta = \pi$ present in the normal phase, disappears in the superfluid phase.

We also discuss expectation values involving gluon degrees of freedom, namely the topological susceptibility and the gluon condensate, computing their dependence on the chemical potential.

We would like to point out that a large part of our results can be explicitly confirmed by lattice simulations. Indeed, both the $N_c = 2$ theory at finite μ_B or μ_I and $N_c = 3$ theory at finite μ_I have a non-negative Dirac determinant and, thus, at $\theta = 0$ are devoid of the sign problem that plagues simulations of $N_c = 3$ QCD at finite baryon chemical potential. Thus, the

---

[a]The reason that the baryon chemical potential can be analyzed in the $N_c = 2$, but not in the $N_c = 3$ case, is that the two colour theory has a larger chiral symmetry, which renders the lightest baryon (the diquark) a goldstone.

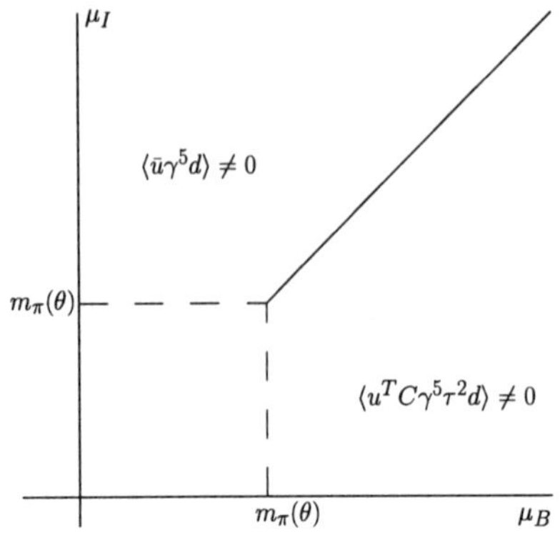

Fig. 1. The phase diagram of $N_c = 2$, $N_f = 2$ QCD at finite baryon and isospin chemical potentials.

dependence of the gluon condensate and the topological susceptibility on chemical potential at $\theta = 0$, can be explicitly checked on the lattice. I will show that our results are in qualitative agreement with recent lattice simulations[11,12].

2. Phase diagram

As noted in the introduction our principal tool is the effective Chiral Lagrangian. We refer the reader to papers [1,2] for the precise form of the Chiral Lagrangian in $N_c = 2$ and $N_c = 3$ theories at finite μ, θ. We concentrate on the two flavour case $N_f = 2$. It turns out that the behaviour of $N_c = 2$ and $N_c = 3$ theories at finite isospin chemical potential is identical, so we confine ourselves to a discussion of the $N_c = 2$ theory in what follows.

By finding the classical minimum of the effective chiral Lagrangian we compute the thermodynamic potential Ω and arrive at the the phase diagram of the $N_c = 2$ theory shown on Fig. 1. At $\theta = 0$, this phase diagram has been first constructed by [9]. There are three phases: the normal phase realized for $\mu_B < m_\pi$, $\mu_I < m_\pi$, the baryon phase $\mu_B > m_\pi$, $\mu_B > \mu_I$ and the isospin phase $\mu_I > m_\pi$, $\mu_I > \mu_B$. Here, m_π is the goldstone mass in vacuum. In the baryon phase the $U(1)_B$ symmetry is spontaneously broken by the diquark condensate $\langle u^T C\gamma^5\tau^2 d\rangle$ and in the isospin phase the $U(1)_I$ symmetry is spontaneously broken by the pion condensate $\langle \bar{u}\gamma^5 d\rangle$. The transitions from the normal to baryon and isospin phases are of second order and the baryon to isospin phase transition is of first order. The phase diagram is symmetric under the interchange $\mu_I \leftrightarrow \mu_B$ due to the enlarged symmetry of the $N_c = 2$ theory, so from here on, we don't explicitly specify whether we discuss dependence on μ_B or μ_I.

The only effect of the θ parameter on the phase diagram is to replace the critical chemical potential μ_c for transition into superfluid phases by the θ-dependent pseudo-goldstone mass,

$$\mu_c^2 = m_\pi^2(\theta) = \frac{m(\theta)|\langle\bar{\psi}\psi\rangle_0|}{4F^2} \tag{1}$$

Here, F is the pion decay constant and $\langle\bar{\psi}\psi\rangle_0$ is the chiral condensate in the limit $\mu_{B,I} = 0$, $\theta = 0$ and $m_u = m_d = m \to 0^+$. The pseudo-goldstone mass m_π acquires a dependence on θ through the effective quark mass parameter $m(\theta)$,

$$m(\theta) = \frac{1}{2}\left((m_u + m_d)^2 \cos^2(\theta/2) + (m_u - m_d)^2 \sin^2(\theta/2)\right)^{\frac{1}{2}} \tag{2}$$

For a quantitative description of the phase diagram in terms of chiral, pion and diquark condensates and densities, see the papers[1,2].

3. θ-dependence

It is instructive to look at the phase diagram in the μ-θ plane, as shown on Fig. 2 a). We see that due to the θ-dependence of the critical chemical potential for transition into the superfluid phase $\mu_c = m_\pi(\theta)$ the phase diagram in this plane is quite non-trivial. In particular, if we fix $m_\pi(\theta = \pi) < \mu < m_\pi(\theta = 0)$ the θ dependence becomes non-analytic: one experiences two normal to superfluid phase transitions of the second order as θ increases from 0 to 2π. One can check that the topological susceptibility $\chi = \frac{\partial^2\Omega}{\partial\theta^2}$ experiences a jump across these phase transitions.

Moreover, the critical chemical potential for transition into the superfluid phase is greatly reduced at $\theta = \pi$ compared to $\theta = 0$ if the quark masses m_u, m_d are close together:

$$\frac{\mu_c(\theta = \pi)}{\mu_c(\theta = 0)} = \frac{m_\pi(\theta = \pi)}{m_\pi(\theta = 0)} = \left(\frac{|m_u - m_d|}{m_u + m_d}\right)^{\frac{1}{2}} \tag{3}$$

In fact, eq. (3) suggests that the goldstone mass vanishes at $\theta = \pi$ for degenerate quark masses $m_u = m_d$. This phenomenon is actually an artifact of working to leading order in quark mass. When higher order mass terms in the chiral Lagrangian are taken into account, the goldstone mass at $\theta = \pi$ will be restored to a non-vanishing value[7] leading to the phase diagram of Fig. 2 b). In this case, one observes the so-called Dashen's phenomenon in the normal phase: a first order phase transition across $\theta = \pi$ characterized

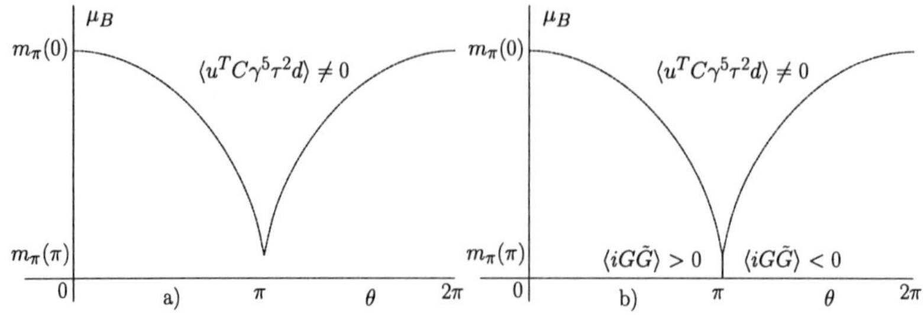

Fig. 2. Phase diagram of $N_c = 2, N_f = 2$ QCD at finite baryon chemical potential μ_B and θ angle: a) Non-degenerate case (m_u, m_d sufficiently far from each other), b) Degenerate limit $m_u = m_d$.

by spontaneous breaking of parity[3–7]. An order parameter for this phase transition is the topological charge density $\langle iG\tilde{G}\rangle$. We observe from Fig. 2 b) that the Dashen's transition disappears in the superfluid phase by splitting into two second order normal to superfluid phase transitions. Thus, the θ dependence is smooth across $\theta = \pi$ in the superfluid phase.

4. Topological susceptibility. Ward identities

As we know the θ-dependence of the thermodynamic potential, we may compute correlators of the topological charge density simply by differentiating with respect to θ. In particular, the topological susceptibility χ is given by,

$$\chi = \frac{\partial^2 \Omega}{\partial \theta^2} = \int d^4x \langle T \frac{g^2 G\tilde{G}}{32\pi^2}(x) \frac{g^2 G\tilde{G}}{32\pi^2}(0)\rangle_{conn} \tag{4}$$

At $\theta = 0$ and $m_u = m_d$, we find in the normal phase,

$$\chi(\mu) = -\frac{1}{4} m \langle \bar{\psi}\psi\rangle_0 \tag{5}$$

Clearly, χ is μ independent in the normal phase. In the superfluid phase,

$$\chi(\mu) = -\frac{1}{4}\frac{m_\pi^2}{\mu^2} m \langle \bar{\psi}\psi\rangle_0 \stackrel{!}{=} -\frac{1}{4} m \langle \bar{\psi}\psi\rangle(\mu) \tag{6}$$

where the chiral condensate in the superfluid phase is independently computed as,

$$\langle \bar{\psi}\psi\rangle(\mu) = \frac{\partial \Omega}{\partial m} = \frac{m_\pi^2}{\mu^2}\langle \bar{\psi}\psi\rangle_0 \tag{7}$$

Thus, we see that the topological susceptibility at $\theta = 0$ is continuous across the transition to the superfluid phase and falls off as $1/\mu^2$ for $\mu > m_\pi$.

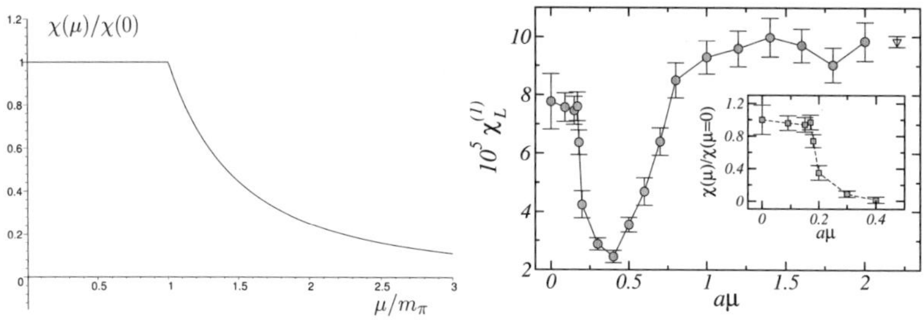

Fig. 3. Topological susceptibility in $N_c = 2$ QCD ($\theta = 0$): a)Chiral perturbation theory prediction. b)Lattice Simulation (see particularly the inset). Figure taken from [11].

Moreover, as can be seen from eqs. (5),(6) the topological susceptibility appears to be related to the μ-dependent chiral condensate both in the normal and superfluid phases. This is not a coincidence but a consequence of the anomalous Ward identity [5,13],

$$\chi = -\frac{1}{N_f^2}\langle\bar{\psi}M\psi\rangle + O(M^2) \tag{8}$$

This Ward identity follows from the chiral anomaly and, thus, should be independent of infra-red effects such as the chemical potential and/or temperature. Our explicit computations show that although the expectation values on both sides of the identity depend on the chemical potential, the identity is preserved.

We now compare our results with the recent lattice study[11] of the topological susceptibility at finite baryon chemical potential in the $N_c = 2$ theory. Fig. 3 displays the results of the lattice simulation[11] for $\chi(\mu)$ alongside with our prediction. As Fig. 3 b) shows, the topological susceptibility is flat before the phase transition and decreases after the phase transition, in qualitative agreement with our prediction 3 a). The subsequent increase of χ for $a\mu > 0.5$, as argued in [11], is likely an artifact of the saturation effects on the lattice and, thus, should be disregarded (so that only the data in the inset graph of Fig. 3 b) is physical).

We note that it would also be interesting to test the Ward identity (8) (and its dependence on μ) on the lattice. To do so one must have detailed lattice results for the topological susceptibility and the chiral condensate. Moreover, to neglect the $O(M^2)$ term in the expression (8) one must be in the chiral limit, $m_\pi^2 \ll m_{\eta'}^2$, which is difficult to achieve on the lattice.

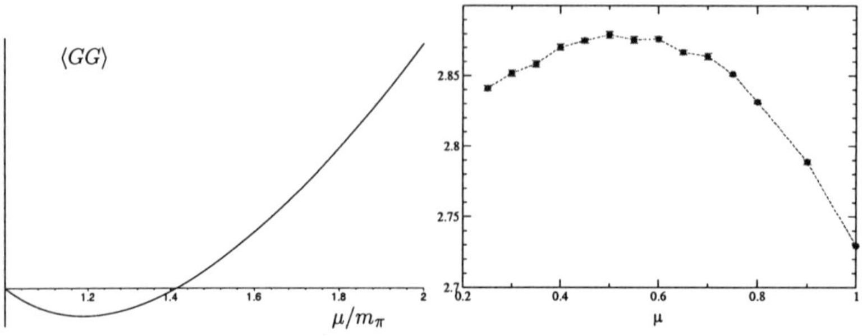

Fig. 4. a) Chiral perturbation theory prediction for the gluon condensate in $N_c = 2$ QCD, b) Lattice simulation of plaquette $\langle Tr\Box\rangle(\mu)$. Figure taken from [12].

5. Gluon condensate

It turns out that we can extract the gluon condensate $\langle\frac{-\beta(g)}{2g}G^a_{\mu\nu}G^a_{\mu\nu}\rangle$ from the thermodynamic potential Ω by using the expression for conformal anomaly of QCD (see [1,2] for details). We find,

$$\langle\frac{-\beta(g)}{2g}G^a_{\mu\nu}G^a_{\mu\nu}\rangle_\mu - \langle\frac{-\beta(g)}{2g}G^a_{\mu\nu}G^a_{\mu\nu}\rangle_{\mu=0} = 4F^2(\mu^2-m_\pi^2)\left(1-2\frac{m_\pi^2}{\mu^2}\right) \quad (9)$$

where eq. (9) holds in the superfluid phase (in the normal phase the gluon condensate does not depend on μ). The result (9) is curious as it implies that the gluon condensate displays non-monotonic behaviour, first continuously decreasing after the transition into the superfluid phase for $m_\pi < \mu < 2^{1/4}m_\pi$ and then increasing. The fact that the gluon condensate is increasing with μ for $m_\pi \ll \mu \ll \Lambda_{QCD}$ is counterintuitive, as we expect the chemical potential to suppress gluon degrees of freedom.

The gluon condensate can be extracted on the lattice from the trace of the plaquette operator. Indeed, taking the naive continuum limit,

$$\frac{1}{N_c}\langle Tr\Box\rangle \rightarrow 1 - \frac{a^4g^2}{48N_c}\langle G^a_{\mu\nu}G^a_{\mu\nu}\rangle \quad (10)$$

The plaquette in the $N_c = 2$ theory as a function of baryon chemical potential has been studied in a few recent lattice simulations[11,12] (see Fig. 4 b) for a plot of $\langle Tr\Box\rangle(\mu)$ taken from [12]). The simulation shows that the plaquette is increasing after the transition into the finite density phase (corresponding to decreasing gluon condensate) in qualitative agreement with our predictions. The subsequent decrease of the plaquette at larger μ (corresponding to increasing gluon condensate), although also predicted by our calculations, should be interpreted with care, as it may be due to unphysical saturation effects on the lattice.

6. Conclusion

In this talk I have presented the recent results on the interplay between the θ-parameter and chemical potential in $N_c = 2$ QCD. It would be interesting to verify these results by lattice simulations. As indicated, preliminary lattice investigations[11,12] are in qualitative agreement with our predictions. To obtain quantitative agreement better control of the chiral limit on the lattice and elimination of saturation effects is needed.

Our study is also a prelimenary step to understanding whether there is a topologically driven transition from the hadronic phase of QCD to the color-superconducting phase at $\mu \sim \Lambda_{QCD}$, whose existence has been argued in [14]. Unfortunately, as our chiral Lagrangian calculations are justified only for small chemical potentials $\mu \ll \Lambda_{QCD}$, the transition (if it exists) is outside the region where our approach is valid. Nevertheless, it is instructive to understand the θ-dependence in both the low and high density regimes before we approach the transition itself.

References

1. M. A. Metlitski and A. R. Zhitnitsky, Nucl. Phys. B **731**, 309 (2005).
2. M. A. Metlitski and A. R. Zhitnitsky, Phys. Lett. B **633**, 721 (2006).
3. R. Dashen, Phys. Rev. D**3**, 1879 (1971).
4. E. Witten, Ann. of Phys. **128**, 363 (1980).
5. P. Di Vecchia and G. Veneziano, Nucl. Phys. **B 171**, 253 (1980).
6. M. Creutz, Phys. Rev. D**52**, 2951 (1995); Phys. Rev. Lett. **92**, 162003 (2004); Phys. Rev. Lett. **92**, 201601 (2004).
7. A. V. Smilga, Phys. Rev. D **59**, 114021 (1999).
8. M. Alford, K. Rajagopal, and F. Wilczek, Phys. Lett. **B 422**, 247 (1998); R. Rapp, T. Schäfer, E. V. Shuryak, and M. Velkovsky, Phys. Rev. Lett. **81**, 53 (1998).
9. J.B. Kogut, M.A. Stephanov, and D. Toublan, Phys. Lett. **B 464**, 183-191 (1999); J.B. Kogut, M.A. Stephanov, D. Toublan, J.J.M. Verbaarschot, and A. Zhitnitsky, Nucl. Phys. **B 582**, 477-513 (2000); K. Splittorff, D. T. Son and M. A. Stephanov, Phys. Rev. D **64**, 016003 (2001).
10. D. T. Son and M. A. Stephanov, Phys. Rev. Lett. **86**, 592 (2001); Phys. Atom. Nucl. **64**, 834 (2001) [Yad. Fiz. **64**, 899 (2001)].
11. B. Alles, M. D'Elia and M. P. Lombardo, [arXiv:hep-lat/0602022].
12. S. Hands, S. Kim and J. I. Skullerud, [arXiv:hep-lat/0604004].
13. R.Crewther, Phys. Lett. **B70**, 349 (1977); E. Witten, Nucl. Phys. **B156**, 269 (1979); V. Novikov, M.A. Shifman, A.I. Vainshtein, and V.I. Zakharov, Nucl. Phys. **B166**, 493 (1980).
14. D. Toublan and A. R. Zhitnitsky, Phys. Rev. D **73**, 034009 (2006)

THE DECONFINING PHASE OF SU(2) YANG-MILLS THERMODYNAMICS

RALF HOFMANN

Universität Heidelberg
Institut für Theoretische Physik
Philosophenweg 16
69120 Heidelberg, Germany
E-mail: r.hofmann@thphys.uni-heidelberg.de
http://www.thphys.uni-heidelberg.de/ hofmann/

We discuss nonperturbative SU(2) Yang-Mills thermodynamics in the deconfining phase. The maximal resolution of trivial-topology fluctuations is set by coarse-grained, interacting calorons and anticalorons: The effective loop expansion is very efficient. Postulating that $\mathrm{SU(2)_{CMB}} \overset{\mathrm{today}}{=} \mathrm{U(1)}_Y$, a modification of thermalized, low-momentum photon propagation is predicted for temperatures a few times 2.7 K. Phenomenological implications are: magnetic-field induced dichroism and birefringence at a temperature of 4.2 K (PVLAS), stability of cold and dilute H1 clouds, and absence of low-l correlations in the TT CMB power spectrum.

Keywords: caloron, holonomy, coarse-graining, polarization tensor

1. A nonperturbative ground state at high temperature

In spite of the innocent-looking title I will point out that the deconfining phase of an SU(2) Yang-Mills theory relies, in an essential way, on interspersed nonperturbative delicacies. As a consequence, unexpected results emerge in low-temperature, low-momentum photon propagation.

Let me first sketch the phase diagram of an SU(2) (or SU(3)) Yang-Mills theory as derived in [1]: There are a deconfining, high-temperature phase, a preconfining, thin intermediate phase, and a low-temperature confining phase. In the former two phases excitations are partially and exclusively massive gauge bosons, respectively. In contrast to perturbative screening temperature-dependent masses are induced by topological field configurations upon a spatial coarse-graining: Masses appear at tree-level by Higgs mechanisms in the associated effective theories. In the confining phase excitations are spin-1/2 fermions with equidistant mass spectrum set by the Yang-Mills scale[a]. Accurate results are obtained by (i) the consideration of BPS saturated, topological configurations (calorons and anticalorons) and (ii) a self-consistent spatial coarse-graining. An essential singularity in the weight of (anti)calorons at zero coupling forbids them in weak-coupling expansions. This is at the heart of the magnetic-sector instability encountered in perturbation theory [9,10].

One writes a (unique) definition for the kernel $\mathcal{K}$ of a to-be-determined differential operator $\mathcal{D}$ in terms of the composite

$$\mathrm{tr}\,\frac{\lambda^a}{2}\,F_{\mu\nu}\,((\tau,0))\,\{(\tau,0),(\tau,\vec{x})\}\,F_{\mu\nu}\,((\tau,\vec{x}))\,\{(\tau,\vec{x}),(\tau,0)\} \qquad (1)$$

of fundamental field variables. In (1) $F_{\mu\nu}$ is the field strength and $\{(\tau,0),(\tau,\vec{x})\}$ denotes a fundamental Wilson line[b]. $\mathcal{K}$ contains the phase $\hat{\phi}^a$ of an emerging adjoint scalar field ϕ^a. Due to an indefinite spatial coarse-graining $\hat{\phi}^a$ depends, in a periodic way, only on euclidean time τ: No information on dimensional transmutation enters in $\hat{\phi}^a(\tau)\in\mathcal{K}$. Therefore, it suffices to evaluate the object in (1) on absolutely stable classical configurations and the Wilson lines are along straight lines[c]. But only BPS saturated configurations are absolutely stable[d]. In passing I mention that adjointly transforming *local* composites vanish on BPS saturated configurations. For $Q=0$ BPS saturated configurations are pure gauges, $F_{\mu\nu}\equiv 0$, and (1) vanishes identically. For $|Q|=1$ stable BPS saturated configurations are trivial-holonomy or Harrington-Shepard (HS) [11,12] (anti)calorons [e]. The integration over the independent[f] moduli of HS (anti)calorons (scale parameter ρ) must be subject to a flat measure since no scale exists which

---

[a]Except for a small range of temperatures above the critical temperature T_c the pressure is positive in the deconfining phase and reaches the Stefan-Boltzmann limit in a power-like way. While the total pressure is negative in the preconfining phase it is precisely zero at $T=0$.

[b]On the level of BPS saturated configurations, see below, no scale is available for a shift $0\to\vec{y}$. As a consequence, the definition in (1) is no restriction of generality.

[c]No scale determining a curvature of a spatial path is available.

[d]All other solutions have higher euclidean action: A departure from classical trajectories takes place by their decay into BPS saturated plus topologically trivial configurations.

[e]Each solution enters the definition (1) separately, and the sum over $Q=\pm 1$ is taken subsequently.

[f]The integral over global spatial or color rotations is contained in the spatial average because of the particular structure of the HS (anti)caloron. Nontrivial periodicity excludes the integration over time translations. The integration over space translations leaves $\mathcal{K}$ invariant because each shift is compensated for by an according parallel transport: This integration is already performed.

would set a 'spectral slope' for this dimensionless quantity. Also, there is no a priori cutoff for the spatial coarse-graining. It is easily checked by dimensional counting that both adding higher n-point functions of the field strength to (1) and BPS saturated configurations with $|Q| > 1$ are forbidden (dimensionful space and moduli integrations). Thus $\mathcal{K}$ is defined by integrating the $Q = \pm 1$ sum of the expression in (1) with the weight $\int d^3x \int \rho$.

In the radial (r) part of space integral a logarithmic divergence occurs for the magnetic-to-magnetic correlation of the field strength [1,3]. At the same time, the azimuthal angular integration yields zero. The former divergence can be regularized in a rotationally invariant way (dimensional regularization). This is not true for the latter zero: an apparent breaking of rotational symmetry is required for regularization. Namely, a defect (or surplus) angle needs to be defined with respect to a fixed direction in the azimuthal plane. Since distinct directions are connected by global gauge rotations no breaking of rotational symmetry is detected in a physical quantity. Thus the angular regularization is admissible.

Performing the integrals, undetermined normalizations appear for each contribution (caloron or anticaloron). Moreover, there are undetermined global phase shifts $\tau \rightarrow \tau + \tau_{C,A}$. The convergence towards $\mathcal{K} = \left\{ \hat{\phi}^a \middle| D\hat{\phi}^a \equiv \left[\partial_\tau^2 + \left(\frac{2\pi}{\beta}\right)^2\right] \hat{\phi}^a = 0;\ \text{fixed ang. reg.} \right\}$ is extremely fast. That is, with finite upper limits ρ_u and r_u in both the ρ- and the r-integration the τ-dependence of the results resembles the limiting behavior ($\rho_u = r_u = \infty$) within a small error already for ρ_u and r_u a few times $\beta \equiv 1/T$ [1,3]. This, however, makes the introduction of a finite cutoff $|\phi|^{-1}$ self-consistent: At fixed global gauge the infinite-volume coarse-graining, determining the τ-dependence of $\hat{\phi}^a$, is saturated on a finite ball of radius $\sim |\phi|^{-1}$.

How large is $|\phi|^{-1}$? Since a sufficiently large cutoff $|\phi|^{-1}$ saturates $\mathcal{K}$, since $D\hat{\phi}^a = 0$ is a linear equation, and since $|\phi|$ is τ-independent[g] we also have $D\phi = 0$. Moreover, since a (finite) coarse-graining over noninteracting, BPS saturated configurations implies the BPS saturation of the field ϕ we need to find an appropriate square root of $D\phi = 0$. Assuming the existence of a scale Λ, which together with β determines the scale $|\phi|$, the right-hand side of the BPS equation must not depend on β explicitly and must be analytic and linear in ϕ. The only consistent option (up to global gauge rotations) is $\partial_\tau \phi = \pm i \lambda_3 \Lambda^3 \phi^{-1}$ where $\phi^{-1} \equiv \frac{\phi}{|\phi|^2}$. Solutions

[g]Composed of coarse-grained, large quantum fluctuations $\Rightarrow$ no finite Matsubara frequencies.

are $\phi(\tau) = \sqrt{\frac{\Lambda^3\beta}{2\pi}}\,\lambda_1 \exp\left(\mp\frac{2\pi i}{\beta}\lambda_3(\tau-\tau_0)\right)$ where τ_0 is a physically irrelevant integration variable (global gauge rotation). A critical temperature $2\pi T_c = 11.65\,\Lambda$ exists, see [1]. Thus, expressing the cutoff $|\phi|^{-1} = \sqrt{\frac{2\pi}{\Lambda^3\beta}}$ in units of β, yields 6.32 at T_c; for $T > T_c$ this number grows as $(T/T_c)^{3/2}$. But for $\rho_u \sim r_u \geq 6.32$ the kernel $\mathcal{K}$ is practically that of the infinite-volume limit, see also [1,14,3].

Coarse-graining the $Q = 0$ sector alone, leaves the Yang-Mills action form-invariant[h]. One can shown that the field ϕ is inert: Quantum fluctuations of resolution $< |\phi|$ do not deform ϕ making it a background for the coarse-grained $Q = 0$-dynamics[1]. The gauge-invariant extension of the kinetic term tr $(\partial_\tau\phi)^2$ in the (gauge-dependent) action for the field ϕ alone is $\partial_\tau \to D_\tau$ (D_τ the adjoint covariant derivative): A unique effective action emerges. The equations of motion for the $Q = 0$ sector (subject to the coarse-grained $|Q| = 1$-background) possess a pure-gauge solution a_μ^{bg}: The ground-state energy density ρ^{gs} and pressure P^{gs} then are $\rho^{gs} = -P^{gs} = 4\pi\Lambda^3\,T$[i].

2. Constraints on resolution in the effective theory

Two color directions acquire mass (adjoint Higgs mechanism). In unitary gauge, $\phi = \lambda_3\,|\phi|\,,\ \ a_\mu^{bg} = 0$, one has $m_{1,2} \propto |\phi|$. Fixing the remaining U(1) by $\partial_i a_i^{a=3} = 0$ a given mode's momentum is physical. To distinguish between quantum and thermal fluctuations we work in the real-time formalism when integrating out gauge-field fluctuations in the effective theory.

Two classes of constraints emerge: (i) Only propagating modes of resolution $\Delta p \leq |\phi|$ need to be considered. (ii) Since coarse-graining generates (quasi)particle masses for $\Delta p \leq |\phi|$ we need assure that the exchange of unresolved massless particles contributing to an effective, local vertex does not involve momentum transfers larger than $|\phi|$. Condition (i) reads

$$|p^2 - m^2| \leq |\phi|^2 \quad \text{(massive mode)}\,, \qquad |p^2| \leq |\phi|^2 \quad \text{(massless mode)} \tag{2}$$

where $|\phi| = \sqrt{\frac{\Lambda^3}{2\pi T}}$. For a three-vertex (ii) is contained in (i) by momentum conservation. For a four-vertex condition (ii) distinguishes s, t, and u chan-

[h] By all-order perturbative renormalizability interaction effects are absorbed into redefinitions of the parameters of the bare action [8].

[i] A negative ground-state pressure is expected microscopically due to the dominating dynamics of small-holonomy calorons leading to finite life-time cycles of magnetic dipoles (a magnetic monopoles attracts its antimonopole, the pair annihilates, and is recreated)[12].

nels in the scattering process. Labelling the ingoing (outgoing) momenta by p_1 and p_2 (p_3 and $p_4 = p_1 + p_2 - p_3$), we have

$$|(p_1+p_2)^2| \leq |\phi|^2\,, \ (s) \ |(p_3-p_1)^2| \leq |\phi|^2 \ (t)\,, \ |(p_2-p_3)^2| \leq |\phi|^2 \ (u)\,. \quad (3)$$

Notice conditions (3) reduce to the first condition if one computes the one-loop tadpole contribution to the polarization tensor or the four-vertex induced two-loop contribution to a thermodynamical quantity[j]. The pressure was computed up to two loops in [1,13,6]: Two-loop corrections are smaller than (depending on temperature) $\sim 0.1\%$ of the one-loop result.

We expect that the contribution of N-particle irreducible (NPI) polarizations to the dressing of propagators vanishes for $N > N_{\rm max} < \infty$ since (2) and (3) then impose more independent conditions than there are independent loop-momentum components. It is instructive to analyze the two bubble diagrams in Fig. 1. While, due to (2) and (3), the two-dimensional region of integration for $|\vec{k_1}|$ and $|\vec{k_2}|$ in diagram (a) is non-compact the three-dimensional region of integration for $|\vec{k_1}|$, $|\vec{k_2}|$, and $|\vec{k_3}|$ is compact in diagram (b) [15]. In one-particle reducible diagrams so-called pinch-singularities

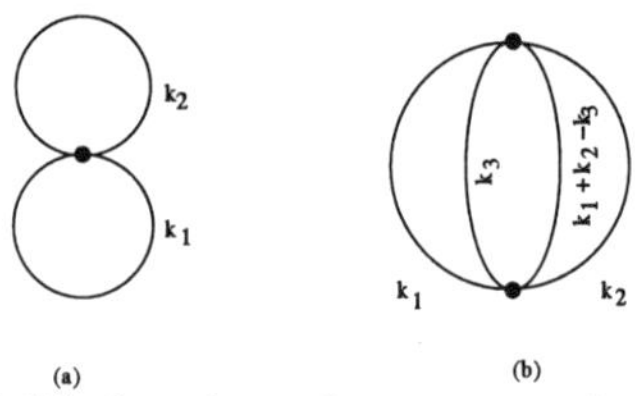

Fig. 1. (a) Two-loop and (b) three-loop diagram contributing to the pressure in the deconfining phase of SU(2) Yang-Mills thermodynamics. The solid lines refer to thermal massive-mode propagation.

arise in the real-time dressing of propagators (powers of delta functions). But a re-summation of 1PI polarizations modifies [k] the scalar part of the tree-level propagators by momentum dependent screening functions with finite imaginary parts. This makes powers of spectral functions well-defined. To summarize, the effective loop expansion should be given by infinite re-summations of a finite number of NPI polarizations [15]. Because the latter

---

[j]The t-channel condition is then trivially satisfied while the u-channel condition reduces to the s-channel condition by letting the loop momentum $k \rightarrow -k$ in $|(p-k)^2| \leq |\phi|^2$, see [1,13,6].

[k]To avoid a logical contradiction the 1PI polarizations are first computed in real time subject to the constraints (2) and (3). Subsequently, a continuation to imaginary time in the external momentum variable p^0 is performed. Then the re-summation is carried out, and finally the result is continued back to real time.

dramatically decrease with N radiative effects are reliably approximated at the two-loop level.

3. Application: $\mathrm{SU}(2)_{\mathrm{CMB}}$

A (falsifiable, see below) postulate emerges[6,1,4,5]: SU(2) Yang-Mills dynamics of scale $T_{\mathrm{CMB}} \sim \Lambda_{\mathrm{CMB}} = 1.77 \times 10^{-4}\,\mathrm{eV}$ (thus the name $\mathrm{SU}(2)_{\mathrm{CMB}}$) masquerades today as the $\mathrm{U}(1)_Y$ of the Standard Model (SM). Λ_{CMB} derives from the b.c. that light propagates in an unadulterated way today[l]. All low-temperature ($T \ll 0.5\,\mathrm{MeV}$) dynamics of the SM with momentum transfers considerably below the 'electroweak scale' $\sim 200\,\mathrm{GeV}$ is unaffected by this assignment if one distinguishes between propagating (the massless excitations of $\mathrm{SU}(2)_{\mathrm{CMB}}$) and interacting photons[m]. An exception takes place for T a few times $T_{\mathrm{CMB}} = 2.18 \times 10^{-4}\,\mathrm{eV} \sim 2.7\,\mathrm{K}$ [n]. Namely, the photon's dispersion law modifies, see [6]:

$$\omega^2(\vec{p}) = \vec{p}^2 \quad \longrightarrow \quad \omega^2(\vec{p},T) = \vec{p}^2 + G(\omega(\vec{p},T),\vec{p},T)\,. \tag{4}$$

At photon momenta $p > 0.2...0.3\,T$ small antiscreening takes place ($G < 0$ in Eq. (4)) which dies off exponentially with p. There is a power suppression of $|G|$ with increasing T. For p being a small fraction of T ($T \sim 5\,\mathrm{K} \to p \lesssim 0.2\,T$) antiscreening converts into screening ($G > 0$) which rapidly grows for decreasing p. The effect is negligible for photon-gas temperatures sufficiently above T_{CMB}, say $T > 80\,K$, and it is absent at $T = T_{\mathrm{CMB}}$, see [6]. A modification of the black-body spectrum emerges at low temperatures and low momenta [7]: At $T = 10\,\mathrm{K}$ the spectral intensity vanishes frequencies $0 < \omega \leq 0.1\,T$. For $0.1\,T \leq \omega \leq 0.25\,T$ there is excess of spectral power. This prediction tests the postulate $\mathrm{SU}(2)_{\mathrm{CMB}} \overset{\mathrm{today}}{=} \mathrm{U}(1)_Y$. The relative deviation to the U(1) black-body pressure peaks at $T \sim 2\,T_{\mathrm{CMB}}$ on the 10^{-3}-level coinciding with the strength of the CMB dipole[o]. The observation of cold, *old*, and dilute clouds of atomic hydrogen in between the spiral arms of our galaxy [17] hints to $\mathrm{SU}(2)_{\mathrm{CMB}} \overset{\mathrm{today}}{=} \mathrm{U}(1)_Y$ being true [6,7p]. The suppression of low-momentum photons could be the reason for the missing power in TT

[l]This is the case only for $T_{\mathrm{CMB}} = T_c$.

[m]Interaction with electroweak matter dynamically invokes the Weinberg angle by a rotation of the propagating to the interacting photon, for a discussion see [5].

[n]Due to Interactions with the massive excitations of $\mathrm{SU}(2)_{\mathrm{CMB}}$ referred to as $V^{\pm}$ in the following.

[o]We expect that besides the contribution due to the Doppler-effect [16] also a dynamical part to generate the CMB dipole.

[p]The forbidden wavelengths at brightness temperature $T = 5\,\mathrm{K}$ [17], range from 1.95 cm to 19.87 cm. This is comparable to the mean distance between H-atoms [17]: The dipole

CMB spectra at low l [18q]. Also there is the result of the PVLAS experiment [19r] To quantitatively investigate this the two one-loop diagrams for the photon polarization involving the full $V^{\pm}$-propagator in the external magnetic field must be calculated. We hope to tackle this task in the near future.

References

1. R. Hofmann, Int. J. Mod. Phys. A **20**, 4123 (2005).
2. R. Hofmann, Mod. Phys. Lett. A **21**, 999 (2006).
3. U. Herbst and R. Hofmann, hep-th/0411214.
4. R. Hofmann, PoS **JHW2005**, 021 (2006).
5. F. Giacosa and R. Hofmann, hep-th/0512184.
6. M. Schwarz, R. Hofmann, and F. Giacosa, hep-th/0603078.
7. M. Schwarz, R. Hofmann, and F. Giacosa, hep-ph/0603174.
8. G. 't Hooft, Nucl. Phys. B **33** (1971) 173. G. 't Hooft and M. J. G. Veltman, Nucl. Phys. B **44**, 189 (1972). G. 't Hooft, Int. J. Mod. Phys. A **20** (2005) 1336 [arXiv:hep-th/0405032].
9. A. D. Linde, Phys. Lett. B **96**, 289 (1980).
10. A. M. Polyakov, Phys. Lett. B *59*, 82 (1975).
11. B. J. Harrington and H. K. Shepard, Phys. Rev. D **17**, 105007 (1978).
12. D. Diakonov, N. Gromov, V. Petrov, and S. Slizovskiy, Phys. Rev. D **70**, 036003 (2004).
13. U. Herbst, R. Hofmann, and J. Rohrer, Acta Phys. Polon. B **36**, 881 (2005).
14. U. Herbst, Diploma thesis (Universität Heidelberg, 2005), hep-th/0506004.
15. R. Hofmann, to be published.
16. P. J. Peebles and D. T. Wilkinson, Phys. Rev.**17**, 2168 (1968).
17. L. B. G. Knee and C. M. Brunt, Nature **412**, 308 (2001).
18. C. Copi, D. Huterer, D. Schwarz, and G. Starkman, astro-ph/0605135.
19. E. Zavattini *et al.*, Phys. Rev. Lett. **96**, 110406 (2006).
20. E. Masso and J. Redondo, hep-ph/0606163.

---

force, which would cause all atoms to convert into H_2 molecules within a time two orders of magnitude lower than the inferred age of the cloud, is switched off.

[q]The suppression of 'messenger' photons weakens the correlation between temperature fluctuations at large angular separation in the sky.

[r]A dichroism induced by a 5 Tesla homogeneous magnet on linearly polarized laser light with the temperature of the aparatus being $\sim 4.2\,$K. When fitted to an axion model the inferred axion mass is about 1 meV with a too large coupling (contradicting solar bounds on axion-induced X-ray emission [20]). The observation making contact with $SU(2)_{CMB}$ is that at $T = 4.2\,$K one has $m_{V^\pm} = 0.7\,$meV: A valuecomparable to the axion mass. It is the mass of the propagating intermediary particle and which is a nearly model independent quantity in non-standard theories of photon-photon coupling. Thus $m_{V^\pm} = 0.7\,$meV is an encouraging observation. Moreover, the small photon-to-photon coupling would be explained by the smallness of a kinematically strongly constrained loop-propagation of $V^\pm$ excitations, see Sec. 2.

SECTION 5

SOLITONS IN GAUGE THEORIES

Convener D. Tong

FRACTIONAL STRINGS ON DOMAIN WALLS

R. AUZZI

William I. Fine Theoretical Physics Institute, University of Minnesota, Minneapolis, MN 55455, USA
E-mail: auzzi@physics.umn.edu

A version of softly broken $\mathcal{N} = 2$ supersymmetric quantum electrodynamics (SQED) with 2 flavors is considered. A Fayet–Iliopoulos parameter and a mass term β which breaks the extended supersymmetry down to $\mathcal{N} = 1$ are added. The bulk theory has two vacua; at $\beta = 0$ the BPS-saturated domain wall interpolating between them has a moduli space parameterized by a $U(1)$ phase σ which can be promoted to a scalar field in the effective low-energy theory on the wall world-volume. At small nonvanishing β this field gets a sine-Gordon potential. As a result, only two discrete degenerate BPS domain walls survive. We find an explicit solitonic solution for domain lines — string-like objects living on the surface of the domain wall which separate wall I from wall II. The domain line is seen as a BPS kink in the world-volume effective theory. The domain line carries a magnetic flux which is exactly 1/2 of the flux carried by the flux tube living in the bulk on each side of the wall.

Keywords: Solitons, Domain Walls

1. Introduction

In the last few years we witnessed extensive explorations of various $\mathcal{N} = 2$ models at weak coupling. Solitonic BPS-saturated objects of novel types were found and studied, such as non-Abelian flux tubes (strings), domain-wall junctions, domain-wall-string junctions (boojums), trapped monopoles and so on, for recent reviews see [1–3] (a complete bibliography is not attempted here).

Another example of a composite BPS soliton, a strings lying on the domain wall world-volume, is discussed here. These strings carry one half of the magnetic flux of the bulk strings (hence, the name "fractional"). From 3D perspective of the wall world-volume theory this strings resemble Polyakov's confining strings of (1+2)-dimensional compact electrodynamics [4].

We consider $\mathcal{N}=2$ supersymmetric quantum electrodynamics (SQED) with 2 flavors, the Fayet–Iliopoulos parameter ξ, and a mass term β which breaks the extended supersymmetry down to $\mathcal{N}=1$. The theory at $\beta=0$ has a moduli space of BPS walls which can be parameterized by a phase $0<\sigma<2\pi$. This modulus can be promoted to a massless scalar on the wall world-volume. The scalar σ can be dualized à la Polyakov [4] into a gauge vector on the wall world-volume. The field strength tensor is given by (see e.g. Ref. [7]):

$$F_{nm}^{(2+1)} = \frac{e_{2+1}^2}{2\pi}\,\epsilon_{nmk}\,\partial^k\sigma\,. \tag{1}$$

In the bulk the probe magnetic charges, monopoles, are confined since their magnetic flux gets squeezed into the Abrikosov–Nielsen–Olesen (ANO) flux tubes [5]. On the other hand inside the wall the magnetic flux can spread all over the brane, so that the probe magnetic charges on the wall world-volume are in the Coulomb phase.

Once we introduce a small β perturbation, the modulus σ ceases to be a massless field. We will consistently work in the first nontrivial order in β. A potential of the form

$$V = \frac{\beta^2\xi}{m}\cos^2\sigma + \mathcal{O}(\beta^3) \tag{2}$$

develops. The potential (2) lifts the moduli space leaving us with two isolated vacua, at $\sigma=\pi/2$ and $\sigma=3\pi/2$. Thus, we have two degenerate BPS domain walls to be referred to as walls of type I and type II. The domain line is a "wall on a wall", it divides the the wall into two domains – one in which we have the type I wall from another of type II wall. In terms of the effective world-volume description the domain line is a sine-Gordon kink in the effective world-volume theory.

Magnetic charges on the wall are confined by these domain lines; the magnetic flux carried by them is one half of the one carried by the the Abrikosov–Nielsen–Olesen flux tubes in the bulk.

The effective description of the world-volume physics found in this theory is different from the Chern-Simon description proposed for the domain-wall world-volume theory in $\mathcal{N}=1$ super-Yang–Mills in Ref. [8]. In particular, the Chern-Simon term is local only in terms of the dual gauge field description while, on the other hand, the sine-Gordon description is local only in terms of the phase field σ.

This work has been done in collaboration with A. Yung and M. Shifman. Many intermediate points are not included in this short presentation; Ref. [9] contains all the details.

2. Theretical set-up

The bulk theory which we start from has the gauge group $U(1)_G$, extended $\mathcal{N} = 2$ supersymmetry, and $N_f = 2$ hypermultiplets of matter $\{Q^A, \tilde{Q}_A\}$, $A = 1, 2$, with mass m and $-m$. A Fayet-Iliopolous term is added in the superpotential. A non-diagonal mass term β for the hypermultiplets is added too; the extended supersymmetry is broken by this term. The superpotential is:

$$\mathcal{W} = \frac{1}{\sqrt{2}}\mathcal{A}Q^B\tilde{Q}_B + m\left(Q^1\tilde{Q}_1 - Q^2\tilde{Q}_2\right) + \frac{\beta}{\sqrt{2}}\left(Q^1\tilde{Q}_2 - \tilde{Q}_1 Q^2\right) - \frac{1}{2\sqrt{2}}\xi\mathcal{A}. \tag{3}$$

The bosonic part of the action can be written as

$$S = \int d^4x \left\{ \frac{1}{4g^2}F_{\mu\nu}^2 + \frac{1}{g^2}|\partial_\mu a|^2 + \bar{\nabla}_\mu \bar{q}_A \nabla_\mu q^A + \bar{\nabla}_\mu \tilde{q}_A \nabla_\mu \bar{\tilde{q}}^A + V_D + V_F \right\}, \tag{4}$$

where ∇_μ is the covariant derivative and a, q and $\tilde{q}$ are the lowest components of the chiral superfields $\mathcal{A}$, Q and $\tilde{Q}$, respectively. The potential is the sum of the D and F terms,

$$V_D = \frac{g^2}{8}\left(|q^B|^2 - |\tilde{q}_B|^2\right)^2, \tag{5}$$

and

$$V_F = \frac{1}{2}\left|q^1(a + \sqrt{2}m) - \beta q^2\right|^2 + \frac{1}{2}\left|\tilde{q}_2(a - \sqrt{2}m) - \beta\tilde{q}_1\right|^2$$
$$+\frac{1}{2}\left|(a + \sqrt{2}m)\tilde{q}_1 + \beta\tilde{q}_2\right|^2 + \frac{1}{2}\left|(a - \sqrt{2}m)q^2 + \beta q^1\right|^2 + \frac{g^2}{2}\left|\tilde{q}_A q^A - \frac{\xi}{2}\right|^2. \tag{6}$$

It is convenient to introduce two parameters,

$$\Omega = \frac{1}{2}\sqrt{\xi + \frac{m\xi}{\sqrt{m^2 - \beta^2/2}}}, \qquad \omega = \frac{1}{2}\sqrt{-\xi + \frac{m\xi}{\sqrt{m^2 - \beta^2/2}}}. \tag{7}$$

Note that $\omega = 0$ and $\Omega = \sqrt{\frac{\xi}{2}}$ at $\beta = 0$. For $\beta < \sqrt{2}m$ the theory has two vacua. The first vacuum is

$$a = -\sqrt{2m^2 - \beta^2}, q^1 = \bar{\tilde{q}}_1 = \Omega, \quad q^2 = -\bar{\tilde{q}}_2 = \omega. \tag{8}$$

The second vacuum is

$$a = \sqrt{2m^2 - \beta^2}, q^1 = -\bar{\tilde{q}}_1 = \omega, \quad q^2 = \bar{\tilde{q}}_2 = \Omega. \tag{9}$$

We will work in the Sigma Model Limit ($\beta, m \ll g\sqrt{\xi}$). Under the above choice of parameters we can integrate out the scalar field a and the gauge field. The low-energy effective description is given by a sigma model with the target space on $q^A, \tilde{q}_A$ which is the Eguchi–Hanson manifold. It is possible to parameterize the target space of the Sigma Model with the radial coordinate $r \geq 0$ plus three angles θ, ψ and φ with

$$0 \leq r < \infty\,, \quad 0 \leq \theta \leq \pi\,, \quad 0 \leq \varphi \leq 2\pi\,, \quad 0 \leq \psi \leq 2\pi\,. \tag{10}$$

The explicit expression which relates the squark fields to the sigma model coordinates is:

$$\begin{aligned}
q^1 &= \frac{e^{i\varphi/2}}{2}\left(e^{i\psi/2} g(r)\cos\frac{\theta}{2} + e^{-i\psi/2} f(r)\sin\frac{\theta}{2}\right),\\
q^2 &= \frac{e^{-i\varphi/2}}{2}\left(e^{i\psi/2} g(r)\sin\frac{\theta}{2} - e^{-i\psi/2} f(r)\cos\frac{\theta}{2}\right),\\
\tilde{q}_1 &= \frac{e^{-i\varphi/2}}{2}\left(e^{-i\psi/2} g(r)\cos\frac{\theta}{2} - e^{i\psi/2} f(r)\sin\frac{\theta}{2}\right),\\
\tilde{q}_2 &= \frac{e^{i\varphi/2}}{2}\left(e^{-i\psi/2} g(r)\sin\frac{\theta}{2} + e^{i\psi/2} f(r)\cos\frac{\theta}{2}\right).
\end{aligned} \tag{11}$$

See Ref. [9] for more details, including the explicit form of the effective sigma model action.

3. The domain walls

In the gauge-theory approach the solution of the BPS equations are rather contrived. It turns out that the problem is much easier in the sigma model approach. If we plot the potential in different slices at r, θ constant, we find that there are always two minima, one at $\varphi = \psi = -\pi/2$ and the other at $\varphi = \psi = \pi/2$. Therefore, this is the appropriate *ansatz* we have to use in order to find two distinct wall solutions (Wall I at $\varphi = \psi = -\pi/2$ and Wall II at $\varphi = \psi = \pi/2$). Two profile functions $r(z)$ and $\eta(z)$ are introduced for the sigma-model coordinates r and θ. The Bogomolny completion of the energy [6] for a wall oriented in the $\hat{z}$ direction is:

$$\int d^3x\,\mathcal{H} = \int d^3x \left\{ \left(\frac{\partial_z r}{\sqrt[4]{4r^2+\xi^2}} - \frac{2mr - (\beta/\sqrt{2})\sqrt{4r^2+\xi^2}}{\sqrt[4]{4r^2+\xi^2}} \right)^2 \right.$$

$$+ \left(\frac{\sqrt[4]{4r^2+\xi^2}}{2} (\partial_z \eta) - \frac{\sqrt{m^2-\beta^2/2}\,\xi \sin\eta}{\sqrt[4]{4r^2+\xi^2}} \right)^2$$

$$\left. - \xi\sqrt{m^2-\beta^2/2}\,(\partial_z \cos\eta) \right\} . \tag{12}$$

The solution for $r(z)$ is just the constant value $r_0 = \frac{\beta\xi}{2\sqrt{2m^2-\beta^2}}$. The solution for θ is:

$$\eta(z) = 2\arctan\left[\exp\left((2m-\beta^2/m)z\right)\right] . \tag{13}$$

The tension is given by the total derivative term

$$T_{\mathrm{wall}} = 2\xi\sqrt{m^2-\beta^2/2} . \tag{14}$$

At $\beta = 0$ the domain walls were studied in Ref. [7]. There exist a compact $U(1)$ modulus $0 \le \sigma \le$ which parameterizes degenerates solutions (this modulus is due to an $U(1)$ flavor symmetry). At $\beta \neq 0$ this symmetry is explicitly broken, but a $\mathbb{Z}_2$ symmetry is still surviving. This symmetry acts trivially on the vacuum and exchanges the two domain wall solutions (see Ref. [9] for the details). The two BPS solutions found here correspond to $\sigma = -\pi/2, \pi/2$ of the moduli space of solutions found at $\beta = 0$.

4. The effective world-volume action

When $\beta \neq 0$ there are just two stable domain wall solutions interpolating between the two vacua of the bulk theory. At generic $\sigma \neq \pi/2$ or $3\pi/2$ we expect that the wall becomes unstable (perhaps, it is better to say, quasistable, at small β). From the standpoint of the world-volume theory this can be interpreted as an effective potential with two degenerate minima for the world-volume modulus σ. The tension of the unstable walls in the first nontrivial order in β has been calculated in Ref. [9]. The result is:

$$T_{\mathrm{wall}}(\sigma) = 2\xi m + \frac{\beta^2\xi}{2m}\cos(2\sigma) + \mathcal{O}(\beta^3) . \tag{15}$$

At small β the effective description for the domain wall physics is a supersymmetric sine-Gordon model with 2 supercharges (we leave aside the translational modulus and its superpartner). The bosonic part of the

Lagrangian (at $\mathcal{O}(1)$ in β for the kinetic part and $\mathcal{O}(\beta^2)$ for the potential) is

$$S = \int d^3x \left[\frac{\xi}{4m} (\partial_n \sigma)^2 - \frac{\beta^2 \xi}{m} \cos^2 \sigma \right]. \tag{16}$$

5. The wall junction

The sigma model which describes the world-volume physics has two vacua. We denote them as Vacuum I (the one at $\sigma = -\pi/2$, which corresponds to Wall I) and Vacuum II (the one at $\sigma = \pi/2$, which corresponds to Wall II).

We will build the domain line as a kink in the effective 2 + 1 Sine-Gordon theory. The Bogomolny completion of the energy functional [6] for a kink oriented in the $\hat{y}$ direction can be written as:

$$\begin{aligned} \mathcal{H} &= \int d^2x \left(\frac{\xi}{4m} (\partial_n \sigma)^2 + \frac{\beta^2 \xi}{m} \cos^2 \sigma \right) \\ &= \int d^2x \left\{ \frac{\xi}{2m} \left(\frac{1}{\sqrt{2}} (\partial_x \sigma) \mp \sqrt{2} \beta \cos \sigma \right)^2 \pm \frac{\xi \beta}{m} \partial(\sin \sigma) \right\}. \end{aligned} \tag{17}$$

There exist two distinct kink solutions interpolating between vacuum I and vacuum II:

$$\sigma = 2 \arctan \left(\exp(2\beta x) \right) - \frac{\pi}{2}, \tag{18}$$

(this solution has $\sigma(x \to -\infty) = -\pi/2$, $\sigma(x \to +\infty) = \pi/2$, let us call it a-string) and

$$\sigma = -2 \arctan \left(\exp(2\beta x) \right) - \frac{\pi}{2}, \tag{19}$$

(this solution has $\sigma(x \to -\infty) = -\pi/2$, $\sigma(x = +\infty) \to -3\pi/2$, let us call it b-string). The transverse size of these objects is of the order of $1/\beta$ and the tension is $T_{\rm dl} = \frac{2\beta\xi}{m}$.

Note that at this order in β the kink is BPS-saturated in the world-volume theory. We expect that the domain line saturates the $\{\frac{1}{2}, \frac{1}{2}\}$ central charge of the $(1 + 3)$-dimensional theory and that these solitons remain BPS-saturated to all orders in β. From the standpoint of $(1+3)$-dimensional theory the above domain lines are strings each of which carries the magnetic flux that is 1/2 of the magnetic flux of the flux tube living in the bulk (see Ref. [9] for the detailed proof).

The two different solutions we obtained — the domain lines a and b — correspond to two possible orientations of the magnetic flux in the y

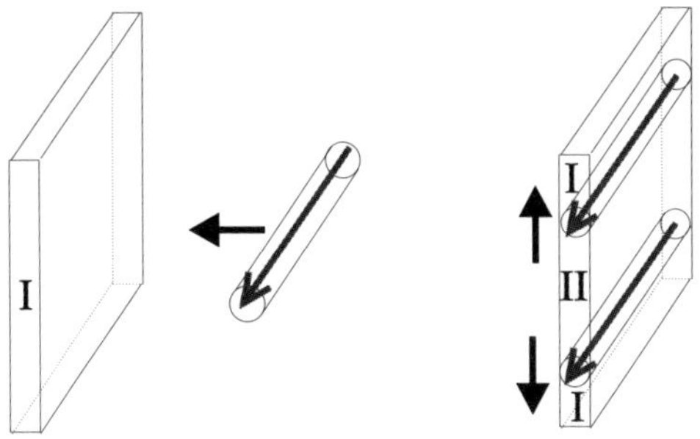

Fig. 1. In the bulk the string tension is of order ξ while on the wall of order $\xi\beta/m$. So for small β the vortex is attracted by the wall. On the wall world-volume the string is a composite Sine-Gordon double kink; two Sine-Gordon kinks repel each other and so a region of type II domain wall will appear in between.

direction. An isolated string can not be taken out of the wall, because it connects two different vacua of the wall world-volume theory. On the other hand, if we take a bound state of an a-line and a $\bar{b}$-line, this configuration interpolates, through one winding, between one and the the same vacuum of the world-volume theory. Indeed, this bound state carries the same magnetic flux as the flux tube living in the bulk, and, therefore, can be pulled out from the wall. If we take a bound state of an a- and $\bar{a}$-lines we get a topologically trivial configuration: a and $\bar{a}$ domain lines annihilate each other. Bulk strings are attracted by the domain wall (see Fig. 1). If they reach the wall surface, they decay in two component domain lines.

References

1. N. Sakai and D. Tong, *JHEP* **0503**, 019 (2005) [hep-th/0501207].
2. D. Tong, *TASI Lectures on Solitons,* hep-th/0509216.
3. M. Eto, Y. Isozumi, M. Nitta, K. Ohashi and N. Sakai, hep-th/0602170.
4. A. M. Polyakov, Nucl. Phys. B **120**, 429 (1977).
5. A. Abrikosov, Sov. Phys. JETP **32**, 1442 (1957) [Reprinted in *Solitons and Particles*, Eds. C. Rebbi and G. Soliani (World Scientific, Singapore, 1984), p. 356]; H. Nielsen and P. Olesen, Nucl. Phys. **B61**, 45 (1973) [Reprinted in *Solitons and Particles*, Eds. C. Rebbi and G. Soliani (World Scientific, Singapore, 1984), p. 365].
6. E. Bogomolny, Sov. J. Nucl. Phys. **24**, 449 (1976). [Reprinted in *Solitons and Particles*, Eds. C. Rebbi and G. Soliani (World Scientific, Singapore, 1984), p. 389].
7. M. Shifman and A. Yung, Phys. Rev. D **67** (2003) 125007 [hep-th/0212293].
8. B. S. Acharya and C. Vafa, hep-th/0103011.
9. R. Auzzi, M. Shifman and A. Yung, arXiv:hep-th/0606060.

MAGNETIC MONOPOLES IN HOT QCD

C. P. KORTHALS ALTES

Centre Physique Théorique au CNRS,
Luminy, 13288 Marseille Cedex, France
E-mail: ab_altes@cpt.univ-mrs.fr

In this talk we review how a dilute gas of magnetic monopoles in the adjoint describes the spatial k-Wilson loops. We formulate an effective theory from S_{MQCD} by integrating out dof's down to scales in between the magnetic screening mass and the string tension and relate the 3d pressure and the string tension. Lattice data are consistent with the gas being dilute for all temperatures.

Keywords: monopoles, QCD plasma, effective action.

1. Motivation and outline

The QCD plasma phase consists of colour electric quasi-particles, gluons, whose interactions get screened by the Debye effect. Their flux is not confined, like in a glueball.

The plasma is a nice theoretical laboratory to understand effects of quasi-particles on flux-loops. Flux-loops are either of the colour-electric variety (spatial 't Hooft loop) or of the magnetic variety (spatial Wilson loop).

Why do we mention "magnetic"? In QED plasmas there is no reason to suspect magnetic quasi-particles because long range static magnetic fields are possible, in contrast to the screening of long range electric fields. But in QCD there is indeed magnetic screening, as seen by lattice simulations. So there must be magnetic activity in the QCD plasma. That is corroborated by the behaviour of the spatial Wilson loop: through Stokes law it measures the flux of magnetic quasi-particles (MQP's or "monopoles")¿ Its thermal average behaves like an area law, as expected from a gas of free screened magnetic charges. In contrast to electric screening the magnetic screening is non-perturbative. The same is valid for the respective loops.

These magnetic charges are supposed to be in a representation of a global magnetic SU(N) group. They are non-perturbative objects. In order to produce a pressure proportional to $N^2 - 1$ (valid to all orders in perturbation theory) the monopoles be better in the adjoint representation. And this is enough to compute the tension for the Wilson loop in the totally antisymmetric representation built from k quark representations. Its highest weight Y_k controls the Stokes form for the loop. The adjoint representation has a multiplicity $2k(N-k)$ of non-zero charges $\pm 2\pi/g$ in $Y_k|_{adj}$. Every magnetic charge contributes the same irrespective of the sign, and independently. Hence the tension is proportional to this multiplicity and this is up to 1 or 2 percent reproduced by the lattice data. In the last section we set up an effective theory for the monopole field.

2. Dimensional reduction, and the magnetic sector

Dimensional reduction is valid at temperatures well above T_c. We will be brief and only mention the essentials. Notation is g for the coupling, and N for the number of colours.Potentials are NxN matrices.

For the purpose of computing magnetic loops it reduction is particularly suitable. This is because the original 4d QCD action reduces at high T to 3d Magnetostatic QCD, the 3d Yang Mills theory:

$$S_{QCD} \rightarrow S_{EQCD} \rightarrow S_{MQCD}, \tag{1}$$

$$S_{EQCD} = (\vec{D}A_0)^2 + m_D^2 A_0^2 + \lambda_E A_0^4 + F_{ij}^2 + ... \tag{2}$$

$$S_{MQCD} = F_{ij}^2 + ... \tag{3}$$

Both of the reduction steps are perturbative. The first integrates out the hard T modes, leaving only static modes. The resulting Electrostatic QCD Lagrangian is three dimensional and contains the A_0 potential as a massive excitation with the Debye mass $m_D^2 = (N/3)g^2T^2$, and of the 3d Yang-Mills action, with the typical magnetic scale g^2T. The second step is valid when the Debye mass is much larger than the magnetic scale g^2T. Then we can integrate perturbatively the A_0 potential, and what stays is the 3d Yang-Mills action S_{MQCD}. This action has no perturbative components left at the scale g^2T.

2.1. *Magnetic observables*

These are the screening of the force between two Dirac Z(N) monopoles [2] and the spatial Wilson loop. The first gives a temperature dependent screening mass $m_M(T)$:

$$V_M(r) = d\frac{\exp(-m_M r)}{r}. \tag{4}$$

It can be shown [2] that this mass equals the mass gap in a 4d SU(N) gauge theory with one periodic space dimension of period $1/T$. This relation is valid for all T, in particular $T = 0$. At very large T this mass can be simulated in 3d S_{MQCD}, a considerable simplification.

The same magnetic activity that causes screening can be monitored by the spatial Wilson loop, that measures the average colour magnetic flux. The loop is carrying a representation R built from a number k of quark representations. Then the tension σ_k is defined by:

$$\langle W_R(C)\rangle = \int DA_0 D\vec{A} \exp(ig\oint \vec{A}_R.d\vec{l}) \sim \exp(-\sigma_k A) \tag{5}$$

A the area of the minimal surface subtended by C.

In three and four dimensions σ_k only depends on the N-allity k [a]. So we need to discuss only one representation of N-allity k, and we will limit ourselves to the fully antisymmetric one, built from k boxes. Its highest weight Y_k, written as a traceless NxN matrix, equals:

$$Y_k = \frac{1}{N}\mathrm{diag}(k, k, ..., k, k-N, k-N, ..., k-N) \tag{6}$$

This highest weight generalizes hypercharge. The average of the loop in this representation admits a Stokes law:

$$\langle W_k(C)\rangle = \int DA_0 D\vec{A} D\Omega \exp\left(ig\int d\vec{S}.Tr(\vec{B} - \frac{i}{g}\vec{D}(A)Y_k^{\Omega} \times \vec{D}(A)Y_k^{\Omega})Y_k^{\Omega}\right). \tag{7}$$

The gauge transformation Ω acts on Y_k as on a frozen Higgs field: $Y_k^{\Omega} = \Omega Y_k \Omega^{\dagger}$. As such this integral will only converge when we admit a Higgs field that fluctuates around its frozen value [3], i.e. a Higgs phase with the unbroken components $SU(k) \times SU(N-k) \times U(1)$. It is known from simulations that these are the only stable broken phases [4]. In the gauge $\Omega = 1$ the commutator terms in $\vec{B}$ and in the second covariant derivative term do cancel. This property determines the Y_k uniquely. For the surviving Abelian expression Stokes theorem is true and a gauge invariant projection of the

[a] In three dimensions a proof exists [5]. It is based on the confining ground states being due to global magnetic Z(N) being broken. The unique tension σ_k is that of the domain wall separating two vacua differing by the phase $\exp(ik\frac{2\pi}{N})$. In 4d there is a consensus, based on circumstantial proof.

line integral results. For the relation of this line integral to the original Wilson loop see ref. [3].

Both magnetic screening and tension are at high T dominated by the 3d S_{MQCD}.

2.2. *Electric observables*

Electric observables are the familiar Debye screening mass m_D between two heavy quarks, and the mean flux as measured by the spatial 't Hooft loop.

The spatial 't Hooft loop, given by a macroscopic closed Z(N) magnetic vortex [1] can be written as a flux loop in the physical Hilbert space:

$$V_k(C) = \exp\left(i\frac{4\pi}{g}\int d\vec{S}.Tr\vec{E}Y_k\right). \tag{8}$$

The surface in the integral is subtended by C.

Note the similarity to the first term in eq.(7).

The average of the loop obeys an area law in the deconfined phase, due to the free colour charges in the plasma:

$$\langle V_k\rangle = \exp\left(-\rho_k A\right)) \tag{9}$$

3. Theoretical results compared to lattice data

The electric loop can be computed in perturbation theory through integration of the hard modes and Debye modes[6]. This has been done including order $O(g^3)$. Including effects of hard modes to order g^4 is desirable as will become evident below. The analytic results give Casimir scaling $k(N-k)$ for the contribution of the hard modes.

To get an insight in the analytic results one considers a quasi-particle gas of Debye screened longitudinal gluons with density n per gluon species in the octet (or adjoint for general N). One species should contribute to the tension proportional to its density n. To get the dimension for the tension right one should scale n by m_D. This result follows from Poissonian fluctuations in the density. Casimir scaling is now simply due to the multiplicity $2k(N-k)$ of gluons in the adjoint with non-zero Y_k charge: $\rho_k/m_D^2 \sim k(N-k)n/m_D^3 \sim 1/(g\sqrt{N})^3 >> 1$, the weak coupling plasma condition.

Fig.1 shows the data for various groups, with Casimir scaling divided out. Remarkably a universal curve results dow to near T_c! Doing the same for the theoretical result gives the curve. Including the hard corrections to g^4 for the coupling [7] brings the theoretical curve down. Hence the need for g^4 for the tension itself.

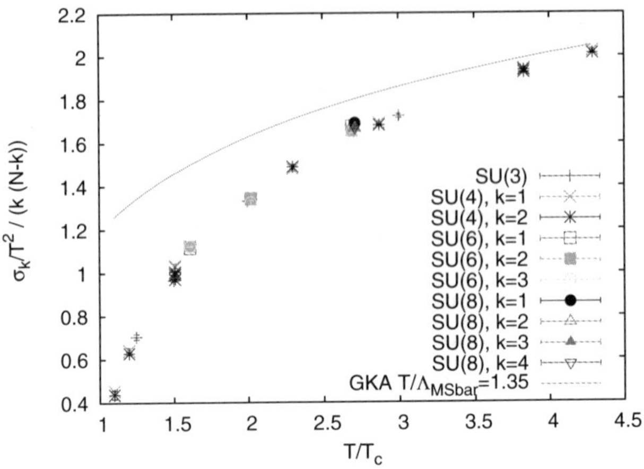

Fig.1: Reduced e-tension for $SU(N_c)$, $N_c \leq 8$, PdF et al.,hep-lat/0510081

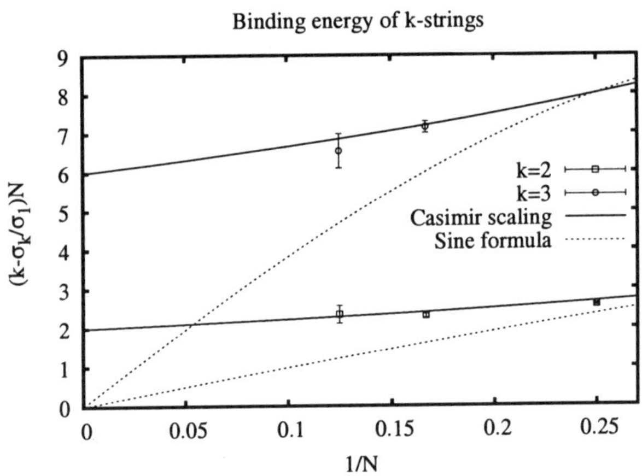

Fig.2: m- flux-tension for $SU(N_c)$, $N_c \leq 8$; Meyer, hep-lat/0412021)

Turning to the magnetic loop, we are faced with its dominant behaviour being due to the non-perturbative magnetic sector. For lack of analytic calculations we proposed [8] a magnetic quasi-particle model that should explain qualitatively the pressure. As the pressure scales like $N^2 - 1$ to all orders in perturbation theory an adjoint multiplet of dilute monopoles is needed. The screening, eq. 4, of the 3d gluons is thought to be so strong that bound states (monopoles) are formed with size on the order of the screening length m_M^{-1}. The diluteness assumption means that the density

n_M of the monopoles is much smaller than m_M^3, the inverse size of the monopoles.

With an adjoint the same multiplcity argument as for the electric loop is valid. So the k-tension of the Wilson loop has again Casimir scaling:

$$\sigma_k \sim k(N-k)n_M/m_M. \tag{10}$$

In particular: $\frac{\sigma_1}{m_M^2} \sim (N-1)\frac{n_M}{m_M^3}$. In striking contrast to the weak coupling plasma condition in the electric case, lattice data tell the left hand side of this relation is for all N a *small* number, ~ 0.05 [10], justifying a posteriori the diluteness assumption. So for finite N the diluteness $\delta = \frac{n_M}{m_M^3}$ is small. For large N the l.h.s. is $O(1)$, hence $n_M = O(1/N)$. In fig.2 are shown the 3d data, which are at most off by two percent from Casimir scaling.

4. Effective action for the monopoles

The raison d'être of the monopole model is that it gives the right group factor for pressure and Wilson loops. At the same time the diluteness is insured by the empirically small ratio σ_1/m_M^2. This warrants a more methodical approach.

The question is whether one can define an effective action starting from S_{MQCD}. The new action, S_{MQP}, follows by integrating out all scales smaller or equal to the size of the monopole, m_M. The monopole field M is a traceless hermitean field, transforming under a magnetic global SU(N) group: $M \rightarrow UMU^\dagger$.

The new action should be invariant under this group. This invariance was not there in the magnetostatic action. There are many examples of a symmetry appearing in an effective action and being broken by the higher order terms.

With S_{MQP} we should be able to compute the k-tensions and the pressure. When computing the pressure we will use the $\overline{MS}$ scheme. It reads:

$$S_{MQP} = \frac{g_3^6 N^3(N^2-1)}{(4\pi)^4}(ylog(\bar{\mu}/2g_3N)+c)+Tr[M(-\partial^2+m_0^2)M]+O(M^3). \tag{11}$$

The first term is the perturbative 4 loop contribution. The coefficient y is known and equals $\frac{43}{12} - \frac{157}{768}\pi^2$ [7].

$\bar{\mu}$ is a scale yet to be fixed by perturbative matching. The constant c and the mass term $m_0 = O(g_3^2 N)$ are unknown. Doing the integration over M gives then the pressure:

$$p_{\overline{MS}} = \frac{g_3^6 N^3(N^2-1)}{(4\pi)^4}((\frac{43}{12} - \frac{157}{768}\pi^2)log(\bar{\mu}/2g_3N) + B_G + O(\epsilon)). \tag{12}$$

where $B_G = c + \frac{64}{3}(\pi \frac{m_0}{g_3^2 N})^3$.

Computation of the tension σ_1 fixes m_0. Note that we did not show how to compute the tension without some periodic potential, which is present in the case that the weak coupling plasma condition applies, as in the classic paper by Polyakov [1]. A recent lattice determination [9] gives a result consistent with zero for B_G.

5. Conclusions

We argued that not only electric quasi-particles- the longitudinal gluons but also magnetic quasi-particles (with spin zero as well) explain nicely the flux loop data. These data do not admit dyonic quasi-particles, which would strongly correlate the loops.

What happens to the model when we go down in temperature?

The variation of the ratio σ_1/m_M^2 from very high T to T=0 is from about 0.05 to 0.09. So the gas can still be considered dilute at T=0. If so, T_c would the transition where the Bose gas becomes superfluid, and the magnetic symmetry gets spontaneously broken. Vortices in the superfluid become the locus of electric flux strings. Above T_c they are absent, so the Polyakov loop develops a VEV above T_c. One should have experimental signals for this superfluid phase.

I thank Maxim Chernodub, Mikko Laine and Harvey Meyer for discussions.

References

1. G. 't Hooft Nucl. Phys B138, 1 (1978); A. M. Polyakov, Nucl. Phys.B120, 429 (1977).
2. P. de Forcrand, C. P. Korthals Altes, O. Philipsen, to appear in Nucl.Phys.B, hep-ph/0510140 .
3. M. Diakonov, V. Petrov, Phys.Lett.B224, 131 (1989).
4. A. Rajantie, Eger 1997, Strong electroweak matter '97, hep-ph/9709368.
5. C. P. Korthals Altes, H.B. Meyer, in preparation.
6. For a review, see P. Giovannangeli, C.P. Korthals Altes, Nucl.Phys.B721:25-49,2005, hep-ph/0412322.
7. Y. Schroeder, hep-lat/03091; M. Laine, Y. Schroeder, JHEP 0503:067,2005.
8. P. Giovannangeli, C. P. Korthals Altes Nucl. Phys.B608, 203 (2001), hep-ph/0102022; C.P. Korthals Altes, H. B. Meyer, hep-ph/0509018.
9. F. di Renzo, M. Laine, V. Miccio, Y. Schroeder, C. Torrero, hep-ph/0605042.
10. M. Teper, Phys.Rev.D59:014512,1999; hep-lat/980400.

DOMAIN WALL JUNCTIONS IN $\mathcal{N} = 1$ SUPER YANG-MILLS AND QUANTUM HALL EDGES

A. RITZ

Department of Physics and Astronomy, University of Victoria
Victoria BC, V8P 5C2, Canada

We discuss various aspects of 1/4-BPS domain wall intersections in $\mathcal{N}=1$ super Yang-Mills, focussing on the allowed Kaluza-Klein mometum contribution to the mass in a dimensionally reduced system, which allows for a close relation to quantum-Hall edge states within the Acharya-Vafa construction of BPS walls in M-theory.

1. Introduction

$\mathcal{N}=1$ super Yang-Mills is the archctypal example of a confining supersymmetric gauge theory, and one that many suspcct realizes confinement in a manner very similar to QCD. The constraints of supersymmetry are not overwhelming in this case, and further input is required to fully understand the dynamics. In this regard, it is also expected that at large N a simplcr description cxists in terms of a weakly coupled string theory, but a precise realization of this is still lacking within the AdS/CFT correspondence. This is despite the existence of the Klebanov-Strassler and Maldacena-Nunez backgrounds which clearly describe closely related thcories.

Our knowledge of this theory is primarily limited to the vacuum structure and the spectrum and index of BPS states, namely domain walls. At large N, these domain walls have a tension $T_1 \sim \mathcal{O}(N)$, and thus can naturally play the role of D-branes for the SYM string [1]. However, the strings themselves are non-BPS and difficult to detect directly. Here, we would like to discuss the remaining BPS object – the 1/4-BPS 2-wall intersection – which is in some sense the closest tractable relative of the SYM string.

Relatively little is known about these objects in pure $\mathcal{N}=1$ SYM, primarily because they preserve only a single supercharge, which does not afford much in the way of nonrenormalization theorems and thus we are not at liberty to utilize convenient weak-coupliong deformations of the theory. A second rather generic problem in actually verifying the existence of a wall junction appears to be the lack of a local order parameter in SYM which can identify the distinct BPS walls. We will come back to this point again below. Finally, a third issue is that we do not expect the tension of wall intersections to scale with N (this may be seen by studying the wall configurations in SQCD as in [2]). This has the important consequence that there is presumably no D-brane interpretation for these intersections, which hinders any simple identification within a string dual.

Getting around the latter point will be the primary issue discussed here, and we will do so by making use of one of the interesting features of the central charge, $\mathcal{Z}_\mu$, supported by the junction. Since it is a Lorentz vector, is not algebraically independent of the momentum, i.e. in 3+1D,

$$\{Q_\alpha, \bar{Q}_{\dot{\alpha}}\} = 2(\gamma^\mu)_{\alpha\dot{\alpha}}(P_\mu + \mathcal{Z}_\mu) \ . \tag{1}$$

Orienting the junction in the x_3 direction, and ignoring the contribution from the walls, the BPS bound takes a somewhat unusual form [3,2]

$$T_j \geq P_3 + \mathcal{Z}_3 \geq \mathcal{Z}_3 \ , \tag{2}$$

where the second relation follows on noting that the allowed boost in the x_3 direction is 'chiral', namely in a basis with positive $\mathcal{Z}_3$ then P_3 is required to be strictly positive [3]. Note that the residual worldvolume supersymmetry pairs e.g. the left-moving bosonic mode with a single goldstino, realizing a chiral (0,1) algebra.

Since we are free to choose the momentum P_3 as we wish, we can artifically render $T_j \sim \mathcal{O}(N)$, by allowing P_3 to scale with N. To make this more precise, from now on we will take the worldvolume of the walls to be compactified on a T^2, and will wrap the junction around the contractible cycle, of length R. Taking $P_3 \sim 1/R \sim \mathcal{O}(N)$, the effective 'mass' of the junction in the dimensionally reduced theory is

$$M_j \sim \Lambda^2 R + \frac{1}{R} \sim \mathcal{O}(N). \tag{3}$$

Note that this trick is essentially just playing with the kinematics, and at large N suppresses any contribution from the intrinsic junction tension. However, exploring this kinematic feature leads to some interesting consequences as we outline below.

1.1. *SYM on* $\mathbb{R}^3 \times S^1$

If we compactify $\mathcal{N}=1$ SYM on $\mathbb{R}^3 \times S^1$, the 4th component of the gauge field $\phi^a \sim \int_{S^1} A_3^a$, or more precisely the corresponding Wilson line round the compact direction, can now Higgs the gauge group. If $R\Lambda \ll 1$, the Higgs vev is generically large and the theory can be studied at weak coupling.

Let us consider the SU(2) theory for simplicity. Generically ϕ then Higgses the gauge group to U(1), and the corresponding photon can be dualized to a compact scalar σ. Classically the massless sector of the reduced theory corresponds to a sigma model on T^2, coordinatized by $X = \phi + \tau\sigma - i\pi\tau$, where τ is the complexified coupling. The Kähler and superpotentials are

$$\mathcal{K}_{\rm cl} = \frac{1}{16\pi {\rm Im}\tau R} X\bar{X}, \qquad \mathcal{W}_{\rm inst} = 2\pi R\Lambda_0^3 \cosh X. \tag{4}$$

There are vacua at $X = 0, i\pi$ allowing for BPS walls interpolating between them. There are two inequivalent solutions that can easily be found explicitly, but it is important here only that the solutions are purely imaginary.

From the form of the central charge, the junction mass (in the absence of any momentum P_3) takes the form [4],

$$M_j = \frac{1}{16\pi {\rm Im}\tau R} \oint (i\bar{X} \overleftrightarrow{\partial_k} X) dx_k, \tag{5}$$

where the contour is a large rectangle in the plane transverse to the junction. It follows, since ${\rm Re}(X) = 0$ along the soliton trajectory, that $M_j = 0$. Indeed, since the Kahler potential is real, this somewhat surprising conclusion appears stable to quantum corrections.

The resolution of this puzzle is that to actually form a junction, the fields must pass through the region near $e^X = 0$ where the effective theory breaks down. The contribution to the tension that is missed in (5) is actually of $\mathcal{O}(1/R)$, which for $\Lambda R \ll 1$ is parametrically larger than the intrinsic strong coupling scale Λ. Assuming analyticity in R, It is natural to expand,

$$M_j \sim \frac{1}{R} + \Lambda + \Lambda^2 R + \cdots, \tag{6}$$

and interpret the first contribution as Kaluza-Klein momentum, the third as a 'winding' contributution from the intrinsic junction tension in 3+1D, while the second term is hard to interpret in 3+1D and may well be absent.

The fact that the first term in this expansion is zero in the case above, implies that the wall intersection is parametrically thin. The wall thickness in this regime is $\mathcal{O}(1/(R^2\Lambda^3))$ while, since $M_j \ll 1/R$, the junction thickness must be parametrically smaller and thus naturally invisible in the effective theory as the gradients required to describe it would be too large.

It is interesting to reformulate the above expansion as follows. We can generalize straightforwardly to an SU(N) gauge group, as the junction only involves two component walls which, up to rescaling, effectively reduces to the SU(2) case above. Let us then take $R \sim 1/(N\Lambda)$ which is consistent with the weak coupling requirement, and then we have

$$M_j \sim N\Lambda + \Lambda + \frac{\Lambda}{N} + \cdots, \tag{7}$$

where the winding contribution then becomes subleading at large N, while the leading term naturally suggests an interpretation in terms of D-branes, i.e. if we now allow for $P_3 \sim \mathcal{O}(N\Lambda)$, then we could interpret $M_j \sim \mathcal{O}(1/g_s)$.

2. Acharya-Vafa and threshold bound states

The most convenient string dual in which to address some of these questions is the realization by Acharya and Vafa [5] of BPS walls as M5-branes wrapped on a Lens space $S^3/\mathbb{Z}_N$, which itself forms part of a G_2-manifold rendering the required number of supersymmetries.

The IIA picture primarily considered in [5], involved reduction on the S^1 fibre of the Lens space, so that BPS walls reduced to D4 branes wrapped on a finite S^2 with N units of RR flux. The worldvolume description of this system for a single 1-wall is $\mathcal{N}=1$ Maxwell Chern-Simons theory in 2+1D. It is apparent that there are no light fields in this system with which to form a solitonic junction configuration, as the nontrivial index arises purely from quantum mechanical Chern-Simons modes. However, a very generic feature of this system is that the insertion of an 'edge' does introduce massless chiral fields, due to pure gauge modes of the Chern-Simons theory which become physical on a manifold with boundary. This is precisely the kinematics expected for the wall intersection, and thus we are led to identify the wall junction in this language as a quantum-Hall edge state. More precisely, this refers to the kinematic part associated with the introduction of P_3 which will be the momentum carried by the chiral edge modes. The 'intrinsic' component of the junction, which is important for forming the edge in the first place, can be taken as subleading at large N, which is presumably the regime in which this picture is relevant.

To compare this viewpoint with that of the previous section, we lift back to M-theory and reduce on another S^1 cycle – giving a D4 brane wrapping $S^3/\mathbb{Z}_N$. The BPS walls then correspond to the inequivalent flat U(1) connections on the D4 worldvolume, of which there are N. As described by Gopakumar and Vafa [6], geometrically one can think of the wall ground states as threshold bound states of a D4 with a 'twisted' fundamental string (twisted under $\mathbb{Z}_N$) corresponding to the flat U(1) connection.

In this context, a 2-wall junction corresponds to a situation where the flat connection on the D4 worldvolume undergoes a jump at a point in the

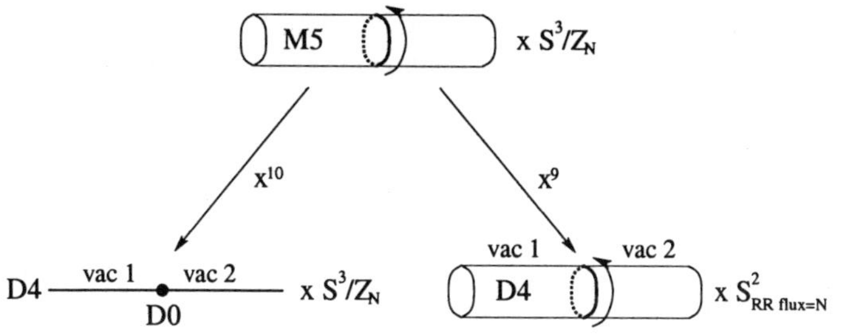

Fig. 1. Reduction of an M5 wrapped on $S^1_{10} \times S^3/\mathbb{Z}_N$ either on the 11th dimension x^{10}, or on the S^1 fibre of the Lens space x^9. In the former case one finds a D4 wrapped on $S^3/\mathbb{Z}_N$ and in the latter on an S^2 with N units of RR 2-form flux. The edge state, representing a BPS junction in the latter, then corresponds to a bound state of a D0 with the D4 wrapped on $S^3/\mathbb{Z}_N$ in the former.

remaining noncompact direction. On way to achieve this would be to break the twisted string on another D-brane. The options are D0's and D2's, the latter e.g. wrapping a 2-cycle in the Lens space. In fact, by considering the lift of the edge state configuration to M-theory and the subsequent reduction on a transverse S^1 (see Fig. 2), we can see that the edge state indeed maps precisely to a D0 brane bound to the D4 wrapped on $S^3/\mathbb{Z}_N$. Now, in flat space the D0-D4 system and its T-dual relations have been extensively studied, and interestingly enough a single D0 and D4 can form an exactly marginal bound state [7]. For present purposes, the best way to understand this state is simply to lift the system back up to M-theory, where we have an M5 wrapped on $S^3/\mathbb{Z}_N$ and the D0 becomes a wave along the M5 around the M-theory circle. There is a corresponding charge under the KK gauge field, and the state is BPS [8]. It seems from this point of view that the wrapping of the 5-brane around the Lens space will not disturb the existence of the bound state. However, care is needed here since part of the supersymmetry of the background is broken and this is often sufficient to remove any threshold bound states from the BPS spectrum. This picture is nonetheless remakably consistent with the large N limit of (7), with $M_j \sim M_{D0} \sim 1/g_s$ and provides a nice check of some of the kinematics of the Acharya-Vafa construction. Clearly, we can also identify the original compactified S^1 of radius R in the field theory with the M-theory circle.

3. Quantum-Hall edges and mirror symmetry

The above check of the tension formula at large N suggests we might go further and ask whether the picture of the junction as an edge state can also find some justification within field theory. The first point which resonates with this is the apparent lack of a local order parameter noted above. This is one of the characteristic features of the quantum-Hall system, going under the name of 'topological order'. We have also emphasized that the chiral nature of the edge excitation also follows naturally within this system.

To proceed, recall that since the wall worldvolume was compactified on a T^2 we really need to consider a configuration with a junction and an anti-junction. Thus far, we have assumed that the torus is very thin, dimensionally reducing the system on the contractible cycle, and have ignored the presence of the anti-junction. Let us now consider a thick torus, with the edge and anti-edge playing a role (see Fig. 2). Within the quantum-Hall system, this configuration was studied by Wen [9], who argued that its low energy description was a non-chiral 1+1D sine-Gordon theory,

$$\mathcal{L} = (\partial_\mu \phi)^2 - g\cos(\sqrt{N}\phi), \tag{8}$$

where the nonchiral field comprises the chiral and anti-chiral components from the edge and anti-edge, while the potential arises from tunneling between the edges. In this regime bulk modes are heavy and can thus be ignored, realizing an effective dimensional reduction on the *non*-contractible cycle of the torus. Note that because of topological order, the quantum-Hall vacua between the edges can be changed by moving a unit of magnetic flux inside the torus, via the creation and subsequent propagation and annihilation of a quasiparticle-antiquasiparticle pair around the contractible cycle of the torus. This is realized as a kink in the sine-Gordon theory.

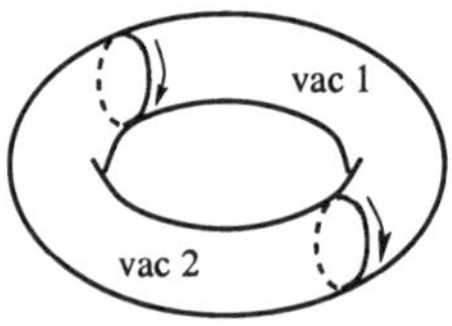

Fig. 2. The wall worldvolume compactified on T^2 with an edge–anti-edge configuration.

We will finish with some speculative remarks on a rather similar description of BPS junctions, obtained by starting from their description within SQCD with $N_f = N$, with $N = 2$ for simplicity. After compactification over the *non*-contractible cycle, the junctions remain 1+1-dimensional, and the wall worldvolume theory reduces to an $\mathcal{N}=2$ $\mathbb{C}P^1$ sigma model in 1+1D [2] where $R_{\mathbb{C}P^1} \sim 1/m$, with m the mass of the lightest additional flavor. As we take the mass parameters more disparate the geometry of the $\mathbb{C}P^1$ becomes more elongated, and it is natural to T-dualize over the small cycle. This is the mirror map of Hori and Vafa [10], and the dual is an $\mathcal{N}=2$ sine-Gordon theory, with superpotential $\mathcal{W} = \cosh Y$! Although it is difficult to make the relation between this description of the theory and Wen's description of the analogous quantum-Hall setup very precise, it is rather tantalizing.

4. Concluding remarks

As alluded to earlier, much of this discussion is based on using the kinematics of the superalgebra to study – by hand – junctions in a regime where a D-brane interpretation seems possible. Of course, its also of interest to seek a string-theoretic interpretation of the intrinsic junction tension, and we will end with the following observation. In the Acharya-Vafa construction, since we expect the tension not to scale with N, the only clear candidate appears to be an M2-brane wrapping the S^1 fibre of the lens space. The resulting effective IIA string carries the required Z_N charge and was identified by Acharya as the confining SYM string [11]. That a bound state of this confining string with the BPS wall might in fact realize the wall junction would be quite consistent with the related picture recently described in [12].

Acknowledgements: I'd like to thank the organizers for the invitation to this stimulating meeting, M. Shifman for initial collaboration on this work, and D. Tong for useful discussions. This work was supported in part by NSERC, Canada.

References

1. E. Witten, Nucl. Phys. B **507**, 658 (1997).
2. A. Ritz, M. Shifman and A. Vainshtein, Phys. Rev. D **66**, 065015 (2002); Phys. Rev. D **70**, 095003 (2004).
3. G. W. Gibbons and P. K. Townsend, Phys. Rev. Lett. **83**, 1727 (1999).
4. A. Gorsky and M. A. Shifman, Phys. Rev. D **61**, 085001 (2000).
5. B. S. Acharya and C. Vafa, arXiv:hep-th/0103011.
6. R. Gopakumar and C. Vafa, Adv. Theor. Math. Phys. **2**, 399 (1998).
7. S. Sethi and M. Stern, Nucl. Phys. B **578**, 163 (2000).
8. M. R. Douglas *et al.*, Nucl. Phys. B **485**, 85 (1997).
9. X. G. Wen, Phys. Rev. B **41**, 12838 (1990).
10. K. Hori and C. Vafa, arXiv:hep-th/0002222.
11. B. S. Acharya, arXiv:hep-th/0101206.
12. R. Auzzi, these proceedings, and R. Auzzi *et al.*, arXiv:hep-th/0606060.

MULTI-VORTICES WITH LARGE MAGNETIC FLUX

S. BOLOGNESI

The Niels Bohr Institute,
Blegdamsvej 17, DK-2100 Copenhagen O, Denmark
E-mail: bolognesi@nbi.dk

We study the Abrikosov-Nielsen-Olesen(ANO) vortices in the large magnetic flux limit. We find that their structure dramatically simplify and they can described by a domain wall compactified on a cylinder. We describe the passages that brought to this idea and the numerical proof. We also speculate on possible application to k-string in large N Yang-Mills theory.

1. Introduction

Consider a $U(1)$ gauge theory coupled to a charged scalar field q

$$\mathcal{L} = -\frac{1}{4e^2} F_{\mu\nu}F^{\mu\nu} - \frac{1}{2}|(\partial_\mu - iA_\mu)q|^2 - V(|q|) \ . \tag{1}$$

We consider a generic potential that has a minimum in the Higgs phase $|q| = q_0 \neq 0$. In this vacuum there is an ANO vortex [1] obtained by choosing an element n of the homotopy group $\pi_1(\mathbf{S}^1)$. The cylindrical symmetric ansatz for this soliton is $q = e^{in\theta}q(r)$ and $A_\theta = \frac{n}{r}A(r)$, where $q(r)$ and $A(r)$ are profile functions to be determined by the equation of motion, plus the boundary conditions $A, \chi(r \to 0) \to 0$ and $A, \chi(r \to \infty) \to 1$. In this talk we will study the large n limit of these vortices. We will see that a great simplification occurs.

2. Genesis of the idea

The idea of studying the large n limit started in a rather indirect way when, mainly motivated by supersymmetric models, we were facing the following problem: a $U(1)$ gauge theory that has two degenerate vacua, one in the Coulomb phase and the other in the Higgs phase (see (A) of Figure 1).[4] This theory admits a kink that interpolates between the two vacua. Consider the domain wall of tension $T_{\rm W}$ that interpolates between

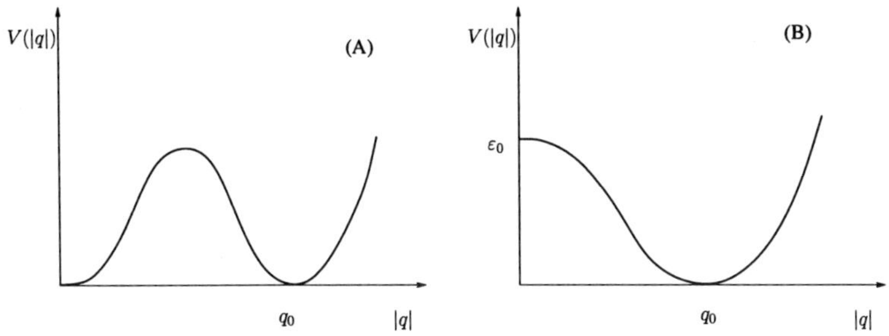

Fig. 1. A potential with two degenerate vacua. $q = 0$ is in the Coulomb phase while $|q| = q_0$ is in the Higgs phase.

these two vacua. We can build a flux tube rolling the wall into a cylinder of radius R, keeping the Coulomb phase inside the tube and turning on a magnetic flux inside as in Figure 2. We will call this configuration a *wall vortex*. Let us analyze the energy per unit of length of the tube. The

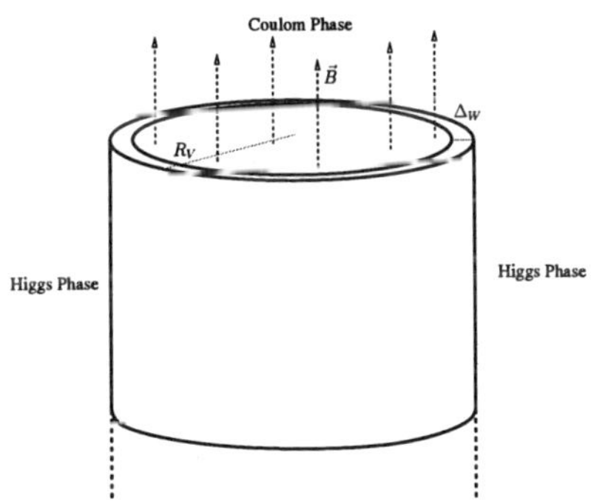

Fig. 2. The wall vortex. A wall of thickness Δ_W is compactified on a circle of radius R_V and stabilized by the magnetic field inside.

tension of the wall gives a contribution $T_W 2\pi R$ to the energy density. The magnetic flux is $\Phi_B = B\pi R^2$ where B is the magnetic field. Varying the radius R, the flux Φ_B remains constant, so the contribution of the flux to the energy density is $\Phi_B/2\pi R^2$. The magnetic flux is quantized in integer values $\Phi_B = 2\pi n$. The tension of the tube is the sum of two pieces, the one that comes from the wall and the other that comes from the flux:

$$T(R) = \frac{{\Phi_B}^2}{2e^2\pi R^2} + T_W 2\pi R \ . \tag{2}$$

The stable configuration is the one that minimizes the tension: $R_{\rm V} = \sqrt[3]{2}\, n^{2/3} e^{-2/3} {T_{\rm W}}^{-1/3}$ and $T_{\rm V} = 3\sqrt[3]{2}\, n^{2/3} e^{-2/3} {T_{\rm W}}^{2/3}$.

And now comes a natural question. Is this wall vortex related to the ANO vortex? The answer is yes. When the number of flux quanta n is enough large, the radius of the ANO vortex increases while the thickness of the wall remain constant. So we expect that in the large n limit the ANO vortex is well approximated by the wall vortex of Figure 2.

3. The wall vortex

We now want to make a more ambitious passage. We consider a generic potential where the Coulomb phase $q = 0$ is *not* necessary a vacuum of the theory (see (B) of Figure 1) and we call the value of the potential at this point ε_0.

Suppose for the moment that the wall vortex is still a good approximation in the large n limit. At this stage this may seem a strange assumption. In fact the domain wall interpolating between the Coulomb and the Higgs phases exist and is stable only when the two phases are degenerate. But we want to take this as an assumption for the moment and work out the consequences. The energy of the tube must now be revised. If the radius grows with n (for the moment it does not matter how fast) the contribution coming from the Coulomb phase energy, that is $\varepsilon_0 \pi R^2$, dominates over the one coming from the tension. And so the tension becomes

$$T(R) = \frac{{\Phi_B}^2}{2\pi R^2} + \pi \varepsilon_0 R^2 \, . \tag{3}$$

Minimizing with respect to R we obtain tension and radius of the wall vortex[a]

$$T_{\rm V} = 2\sqrt{2}\pi \, \frac{\sqrt{\varepsilon_0}}{e} \, n \, , \qquad R_{\rm V} = \sqrt[4]{2} \, \frac{1}{e^{1/2} {\varepsilon_0}^{1/4}} \sqrt{n} \, . \tag{4}$$

Now we can write our conjecture: *Consider the Abelian Higgs model (1) with a general potential that has a true vacuum at $|q| = q_0 \neq 0$ and a Coulomb phase with energy density $V(0) = \varepsilon_0 \neq 0$. Call $T_{\rm V}(n)$ the tension of the vortex with n units of magnetic flux. The conjecture is that in the large n limit the ANO vortex becomes a wall vortex and*

$$\lim_{n \to \infty} T_{\rm V}(n) = 2\sqrt{2}\pi \, \frac{\sqrt{\varepsilon_0}}{e} \, n \, . \tag{5}$$

[a]Note that this is very similar to what happens in the MIT bag model.[7]

Now we make a non-trivial check of the result (4) using the famous example solved by Bogomol'nyi[2]. For the particular choice of the BPS potential the tension is exactly determined for all n:

$$V(|q|) = \frac{e^2}{2}(|\phi|^2 - \xi)^2 \quad \Longrightarrow \quad T_{\rm BPS} = 2\pi n \xi \ . \tag{6}$$

Solving the model with our trick the result in fact coincide with the BPS tension. This could hardly be just a coincidence, is the signal that we are on the right track!

We give now an intuitive argument to explain the formation of the domain wall in the large n limit. Consider the profile function $q(r)$ for the scalar field. We want to see that in the large n limit it becomes a step function. In order to have a properly defined limit we have to rescale the lengths so the $R_{\rm V}$ remain constant while sending n to infinity (remember that the radius $R_{\rm V}$ grows as $\sqrt{n}$). The essential ingredients are now the behaviour of $q(r)$ at the boundaries. Near zero $q(r)$ is proportional to r^n. At $r \to \infty$ the profile $q(r)$ deviates from q_0 by a quantity proportional to $-\exp(-\sqrt{n}r/m_{\rm H})$, where $m_{\rm H}$ is the mass of the Higgs boson and the $\sqrt{n}$ arise from the rescaling of the lengths. The profile function $q(r)$ must interpolate between the two asymptotic behaviors. The polynomial and the exponential, as n grows, *squeeze* the profile $q(r)$ into a step function (see Figure 3).

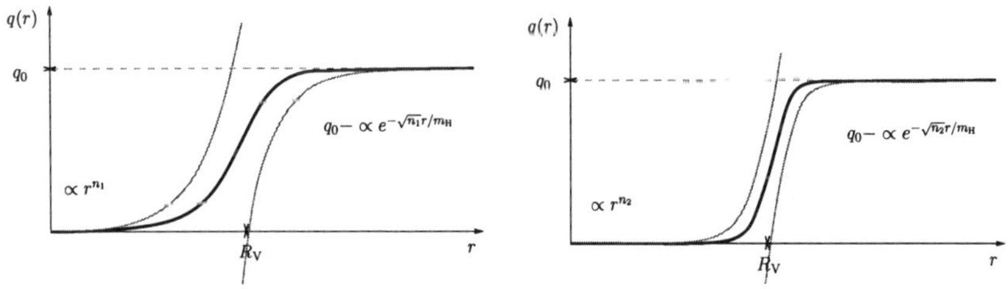

Fig. 3. The scalar field profile function $q(r)$ for different values of n ($n_1 \ll n_2$). Near zero is proportional to r^n while at $r \to \infty$ it reach q_0 with a negative deviation proportional to $\exp(-\sqrt{n}r/m_{\rm H})$. As n goes to infinity the two asymptotic formulas squeeze the profile function into a step function.

In Ref.[6] we have used a numerical code to solve the differential equation that determine the profile functions and vortex tension for large number of magnetic flux. In Figure 4 there are the three plots of the tension per unit of flux $\mathcal{T}(n) = T_{\rm V}(n)/n$, for $\beta = 1/16, 1, 16$ respectively ($\beta = m_{\rm H}/m_\gamma$

where $m_{\rm H}$ is the mass of the Higgs boson and m_γ the mass of the photon). Figure 4 is consistent with the ordinary expectation of type I and type II

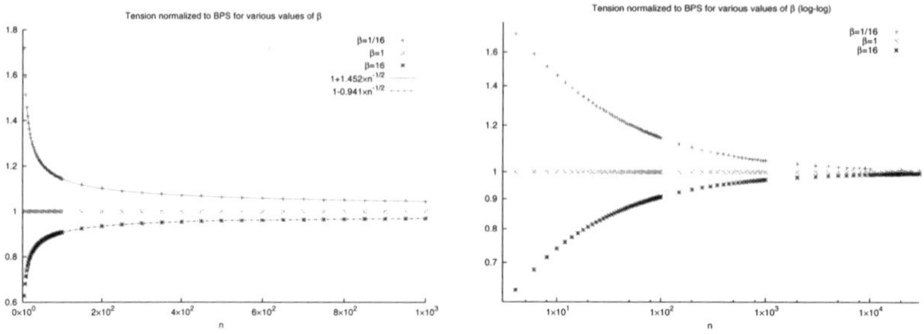

Fig. 4. The tension per unit of flux for three different quartic potentials: type I, BPS and type II. The left panel is a linear plot where a $1/n$ fit perfectly matches with the deviations. The right panel is a log-log plot where more data can be plotted.

superconducting vortices. If we take for example $\beta = 1/16$ (type I vortices) the derivative $\mathrm{d}\mathcal{T}/\mathrm{d}n$ is negative and this means that there is attraction. For $\beta = 16$ the function $\mathcal{T}$ grows up to 1 and this means that there is repulsion (type II vortices).

There is a nice interpretation of the moduli space of n BPS vortices in the large n limit. The tension formula (3) refers to a circle of radius R. If we substitute the circle with a generic surface of area $\mathcal{A}$, we obtain

$$T(R) = \frac{(2\pi n)^2}{2e^2\mathcal{A}} + \varepsilon_0 \mathcal{A} \, . \tag{7}$$

Thus we see that the energy minimization has a moduli space of solutions. Any surface that has the same area of the circular wall vortex has also the same tension. The moduli space of the wall vortex is thus the set of surfaces in the plane with fixed area. Note that, as expected from the large n limit, this is an infinite dimensional moduli space.

4. $\mathbf{Z}_N$ strings

We now try to see if the wall vortex can teach us something about large N Yang-Mills theory.[5] A $SU(N)$ gauge theory with only fields in the adjoint representation in the confining phase admit in general stable k-strings. k-strings are flux tubes that confines a quark and an antiquark in a representation with charge k with respect to the center of the gauge group.

First we recall a well established results regarding the k-string tension at large N. In the large N limit the interaction between two fundamental strings vanishes like $\frac{1}{N^2}$.[8] Thus the tension of the k-string, when k is fixed and N becomes large, is equal to k times the tension of the fundamental string plus subleading corrections

$$T_V(k) = kT_V(1) + \mathcal{O}\left(1/N^2\right) \ . \tag{8}$$

The tension of the fundamental string is of order 1 in the large N limit.

Now consider the $U(1)$ generated by the element of the Cartan subalgebra diag$(1,\ldots,1,-(N-1))$. This generator, up to gauge invariance, is the only one that can reach every element of the center of the gauge group when exponentiated. We now make the following working assumption: *k-strings are ANO vortices in a magnetic theory of the $U(1)$ generated by* diag$(1,\ldots,1,-(N-1))$. The important point is that we can work out a conclusion even if we don't know anything about this theory, in particular we don't know the potential and neither if is weakly or strongly coupled. Nevertheless the results of the previous section imply the following: at large k the ANO vortices become wall vortices and their tension is linear in k. This is always the case, independently on the potential and the coupling of this hypothetical dual $U(1)$ theory.

Given this assumption there is only one possibility in order to match the free string limit (8) with the wall vortex limit: the k-string must become BPS in the large N limit and their tension exactly a straight line!

4.1. *Lattice data and saturation limit*

The saturation limit is that in which N is sent to infinity while keeping $k/N = x$ fixed. The best way to compare the string tensions for different N is to rescale the ratio $R(k,N) = T_V(k)/T_V(1)$ with a $1/N$ factor. The quantity we are interested in is thus $\mathcal{R}(x,N) = R(xN,N)/N$. The good thing about $\mathcal{R}(x,N)$ is that it can be plotted in a graph where $0 < x < 1$ and $0 < \mathcal{R}(x,N) < 0.5$ for every N. In the large N limit $\mathcal{R}(x,N)$ should saturate and converge to a continuous line. For example the sine formula and the Casimir formula in the saturation limit become respectively: $\frac{1}{N}\frac{\sin\left(\frac{\pi k}{N}\right)}{\sin\left(\frac{\pi}{N}\right)} \longrightarrow \frac{1}{\pi}\sin(\pi x) + \mathcal{O}\left(1/N^2\right)$ and $\frac{1}{N}\frac{k(N-k)}{N-1} \longrightarrow x(1-x) + \mathcal{O}(1/N)$. An important thing to note is that both the sine and the Casimir formula saturate from above, that is the $\mathcal{O}(1/N)$ corrections are positive . Our formula is instead

$$\min(x, 1-x) + \mathcal{O}(1/N) \ , \tag{9}$$

and the saturation is from below.

Finally we confront with the lattice experiments that are plotted Figure 5. Up to now the largest N for which computations have been done is $N = 8$.[9] These data are still unprecise and far from the saturation limit and is of course premature to give a definitive interpretation. The sine formula seems to be consistent with the data. But the data also seem to indicate a saturation from below (negative $1/N$ corrections) and the appearance of an angle in the shape of the $N = 8$ data. This is consistent with (9).

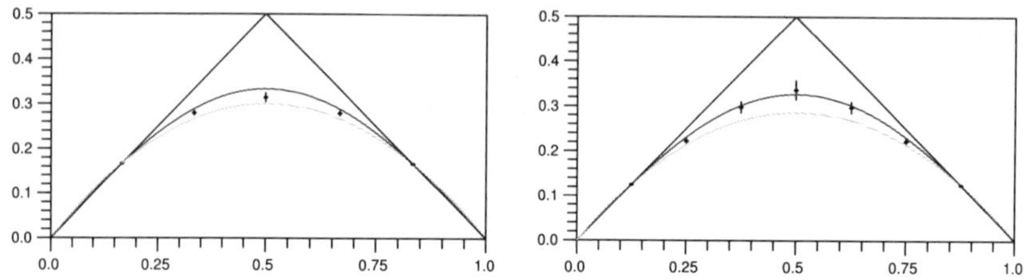

Fig. 5. The most recent lattice result on k-string tension. The left and right panels refers respectively to $SU(6)$ and $SU(8)$.

Acknowledgments

I thank K. Konishi for many discussions on the subject and S. B. Gudnason for the collaboration to some of the works[6]. This work is supported by the Marie Curie Excellence Grant under contract MEXT-CT-2004-013510.

References

1. A. A. Abrikosov, Sov. Phys. JETP **5** (1957) 1174 [Zh. Eksp. Teor. Fiz. **32** (1957) 1442]. H. B. Nielsen and P. Olesen, Nucl. Phys. B **61** (1973) 45.
2. E. B. Bogomolny, Sov. J. Nucl. Phys. **24** (1976) 449 [Yad. Fiz. **24** (1976) 861].
3. G. 't Hooft, Nucl. Phys. B **72** (1974) 461.
4. S. Bolognesi, Nucl. Phys. B **730** (2005) 127 [arXiv:hep-th/0507273].
5. S. Bolognesi, Nucl. Phys. B **730** (2005) 150 [arXiv:hep-th/0507286].
6. S. Bolognesi and S. B. Gudnason, Nucl. Phys. B **741** (2006) 1 [arXiv:hep-th/0512132]; arXiv:hep-th/0606065.
7. A. Chodos, R. L. Jaffe, K. Johnson, C. B. Thorn and V. F. Weisskopf, Phys. Rev. D **9** (1974) 3471.
8. A. Armoni and M. Shifman, Nucl. Phys. B **671** (2003) 67 [arXiv:hep-th/0307020].
9. B. Lucini, M. Teper and U. Wenger, JHEP **0406** (2004) 012 [arXiv:hep-lat/0404008].

AN OPEN-CLOSED STRING DUALITY IN FIELD THEORY?

DAVID TONG

Department of Applied Mathematics and Theoretical Physics
University of Cambridge, UK
Email: d.tong@damtp.cam.ac.uk
http://www.damtp.cam.ac.uk/user/tong/

I present an open-string description of solitonic domain walls in a semi-classical field theory. I speculate on the possibility for an open-closed string duality in this setting.

1. Introduction

Everybody loves a good D-brane. These objects have underpinned much of the progress in high-energy theoretical physics for the past ten years, culminating in the miraculous AdS/CFT correspondence which, at its heart, relies on the equivalence of open and closed string descriptions of D-brane dynamics. The purpose of this talk is to describe semi-classical D-brane objects in simple field theories, by which I mean gauge fields interacting with scalars and fermions, decoupled from the complications of gravity. I will then explain how, in certain regimes, an open-string description of the dynamics becomes viable[1].

So what is a D-brane? If we simply define it to be a dynamical surface on which strings can end, then Nature offers several examples, most notably in the arena of fluid dynamics. For example, the AB-interface in superfluid ^{3}He is a D-brane on which a vortex string may end. More prosaically, one may view a cumulonimbus cloud base as a D-brane on which a tornado or a water spout may terminate.

However, there is one crucial feature that these D-branes do not share with those found in string theory: there is no regime in which an open-string description holds. One would have a very hard time convincing a meteorologist that, as two clouds approach, their dynamics is governed by quantum effects of virtual tornadoes stretched between them. Yet this is precisely what happens for D-branes in string theory. Indeed, it is the key feature that governs many of the fascinating properties of D-branes. And, as we shall see, it is also what happens for the solitons in this talk.

2. The field theory and its spectrum

Since we are working with common-or-garden field theories, both the strings and the D-branes must appear as solitonic objects. We will consider a theory in $d = 3+1$ dimensions that contains both vortex strings and domain walls in its solitonic spectrum. The simplest example is a $U(1)$ gauge theory, coupled to N complex scalar fields q_i, $i = 1, \ldots, N$, each of charge $+1$. There is also a single, real, neutral scalar field ϕ. The scalar potential is given by

$$V = \frac{e^2}{2}(\sum_{i=1}^{N} |q_i|^2 - v^2)^2 + \sum_{1=1}^{N}(\phi - m_i)^2|q_i|^2 \tag{1}$$

This is the truncation of a theory with $\mathcal{N} = 2$ supersymmetry. While the bosonic action above will suffice to describe the classical solitons, when we come to the open string description we will work with the full $\mathcal{N} = 2$ theory, complete with fermions.

When the masses m_i are distinct, the theory (1) has N isolated vacua given by $\phi = m_i$ and $|q_j|^2 = v^2\delta_{ij}$ for $i = 1, \ldots, N$. Excitations around each vacuum are gapped. The photon has a mass $M_\gamma = ev$, where e^2 is the gauge coupling. The complex scalars have a mass $M_q = |m_i - m_j|$ for $i \neq j$. The quantum theory also has another, more nefarious, scale: it is the Landau pole $\Lambda \sim ev\, \exp(+1/e^2N)$, above which the theory ceases to make sense. In the following we will take the limit $e^2 \to \infty$[a]. In this limit, the classical theory reduces at energies $E \ll M_\gamma$ to a bunch of interacting massive scalars, described by a sigma-model with potential. Sigma-models in four-dimensions are, of course, non-renormalizable; this is the price we've paid for eliminating the Landau pole. At energies $E \ll v$ the theory is weakly coupled. We will implicitly assume an ultra-violet completion at energies $E \gg v$.

[a]It is not clear if this limit is necessary, but it evades certain subtleties regarding boojums[2] — energy associated to the end point of the open string — which scale as $1/e^2$ and must otherwise be treated with care when quantizing the open string. The strong coupling limit $e^2 \to \infty$ is usually considered sick in theories not displaying asymptotic freedom because $\Lambda \to M_\gamma$. To avoid these problems we simply take the UV cut-off of the theory to lie in the region $v, m_i \ll \mu_{UV} \ll M_\gamma$. This means that we ignore all dynamics of the massive photon, and the theory should really be thought of as a proxy for the massive sigma-model of interest.

3. The solitons

Our theory admits both string and domain wall solitons. Let's start with the strings which are vortices of the type introduced into high-energy physics over 30 years ago by Nielsen and Olesen[3]. The vortices are supported by the phase of the condensed scalar q_i, which winds around the $z = x^1 + ix^2$ plane transverse to the string. They are 1/2-BPS, with tension $T_{\text{vort}} = 2\pi v^2$ and width $L_{\text{vort}} = 1/ev$. Notice that in the sigma-model limit $e^2 \to \infty$, the vortex becomes singular as befits a fundamental string.

The second soliton is a domain wall whose existence is guaranteed by the multiple, isolated vacuum states. We impose the vacuum $\phi = m_i$ at $x^3 = -\infty$ and $\phi = m_j$ at $x^3 = +\infty$. The tension of the wall is $T_{\text{wall}} = v^2 \Delta m$ where $\Delta m = |m_i - m_j|$. In the $e^2 \to \infty$ limit of interest, the width of the wall is $L_{\text{wall}} = 1/\Delta m$. From now on we order the masses $m_1 > m_2 > \ldots > m_N$.

A single domain wall, interpolating between neighboring vacua $\phi = m_i$ and $\phi = m_{i+1}$, has a collective coordinate X^3, describing the center of mass. However, the domain wall also has a second, internal, collective coordinate[4]. To see this note that the original theory is invariant under a $U(1)^{N-1}$ flavor symmetry, in which the phases of the q_i fields are rotated independently. In each vacuum, only a single $q_i \neq 0$, and the rotation $q_i \to e^{i\sigma} q_i$ coincides with the gauge action and does not lead to any new physical state. However, in the center of the domain wall both q_i and q_{i+1} are non zero, and the global action $q_i \to e^{i\sigma} q_i$ with $q_{i+1} \to e^{-i\sigma} q_{i+1}$ does now sweep out a new physical configuration. The center of mass X^3 and the phase collective coordinate σ parameterize the domain wall moduli space $\mathbf{R} \times \mathbf{S}^1$. The low-energy dynamics of the domain wall is determined by promoting the collective coordinates to dynamical degrees of freedom in order to describe long wavelength fluctuations of the position and internal phase orientation.

4. Wall dynamics: Bulk description

We now consider the scattering of two parallel domain walls using the classical equations of motion. We call this the "bulk description".

In order to have two walls (as opposed to a wall and anti-wall) we need at least three vacua. We therefore choose $N \geq 3$ and set $\phi = m_{i-1}$ at $x^3 = -\infty$ and $\phi = m_{i+1}$ at $x^3 = +\infty$. With these boundary conditions there are domain wall solutions which have the profile of two domain walls, separated by an arbitrary distance[7] R. In between the two walls, the fields lie exponentially close to the middle vacuum configuration $\phi = m_i$.

The system with two walls has 4 collective coordinates, arising from the position and phase degree of freedom for each wall. The moduli space naturally factors into the center of mass and relative degrees of freedom: $\mathcal{M}_{2-\mathrm{wall}} \cong \mathbf{R} \times (\mathbf{S}^1 \times \mathcal{M}_{\mathrm{cigar}})/\mathbf{Z}_2$, where $\mathcal{M}_{\mathrm{cigar}}$ is a cigar like manifold describing the relative position and phase of the wall. The asymptotic, cylindrical regime of the cigar corresponds to far separated domain walls; the tip of the cigar corresponds to the configuration in which the domain walls sit on top of each other.

The low-energy scattering of the domain walls is described by a sigma-model on $\mathcal{M}_{cigar}$. The metric may be computed explicitly[8], but is not required to understand what happens when two walls collide. Geodesics on the moduli space simply round the tip of the cigar, corresponding to two walls approaching and rebounding in finite time, with their relative phase shifted by π. In particular, the key feature of the dynamics is so obvious that it is almost not worth stating: our domain walls cannot pass.

We need to analyze the regime of validity of this bulk calculation. Since the computation was purely classical, we must ensure that quantum effects can be consistently ignored. Higher order quantum corrections to the classical action will appear as an expansion in derivatives (∂/v). For domain walls of width $L_{\mathrm{wall}} = 1/\Delta m$, these may be consistently ignored when $\Delta m/v \ll 1$.

5. Wall dynamics: Open string description

We will now present a very different, dual, perspective on domain wall scattering, which we call the "open string description". Recall that the $d = 2+1$ dimensional worldvolume of the domain wall houses a periodic scalar σ. We may exchange this in favor of a photon using the duality map is $\partial_\alpha \sigma \sim \epsilon_{\alpha\beta\gamma} F^{\beta\gamma}$. In these variables, the low-energy effective action for the walls is a $d = 2+1$ $U(1)$ gauge theory. Including the fermion zero modes, this theory has $\mathcal{N} = 2$ supersymmetry (4 supercharges).

The existence of a $U(1)$ gauge field living on the domain wall is strikingly reminiscent of D-branes in string theory. But, so far, the derivation of the gauge field is rather cheap because nothing is charged under it. In fact, one can show that the vortex string can terminate on the domain wall, where its end is electrically charged[5,6]. There are a number of ways to see this. One is to write down the explicit solution for the string ending on the domain wall which can be found analytically in $e^2 \to \infty$ limit[5]. Another is to study the "BIon' spike solution emanating from the wall[5,6].

In the "open string description" of domain wall scattering, we will not consider the classical profile of the domain walls in four-dimensions, nor the bulk field theory interactions between them as they overlap. Instead we will treat the two domain walls as free objects, each described by a $U(1)$ gauge theory, and each free to roam along the x^3 line. In particular, they may move past each other unimpeded. The only interaction between the two walls comes from the quantum effects of open vortex strings stretched between the walls. Note that the quantum open string effects must be very powerful for they must, ultimately, stop the domain walls from passing each other. Let us see how this occurs. (More details can be found in [1]).

To study the effect of open strings, we must first ask what state the stretched open string gives rise to. The lowest open string mode has electric charge $(+1,-1)$ under the two gauge fields on the domain walls, and bare mass[b] $T_{\rm vort}R$. We also know that the stretched open string is 1/2 BPS in the domain wall worldvolume, ensuring that the lowest mode lies in a short representation of the supersymmetry algebra. There are two possibilities: a vector multiplet or a chiral multiplet. We are used to seeing light vector multiplets, and the associated non-abelian gauge symmetry enhancement, appearing in string theory when two identical branes coincide. However, this cannot be the case for us since our domain walls are not identical. For example, they have different tensions. So we conclude that the lowest mode of the open string gives rise to a charged chiral multiplet in $d=2+1$ dimensions.

I therefore claim that the relative dynamics of two domain walls is governed by a $d=2+1$, $\mathcal{N}=2$ supersymmetric $U(1)$ gauge theory coupled to a single chiral multiplet. However, there is a subtlety. Integrating out a charged chiral multiplet in $d=2+1$ dimensions induces a Chern-Simons coupling. Integrating *in* a charged chiral multiplet must also therefore induce a Chern-Simons coupling, $\kappa A\wedge F$, where $\kappa=-1/2$.

The Chern-Simons term gives a mass to the gauge field on the wall. By supersymmetry, it also gives a mass to the separation of the walls R. This suggests that, in this open string description, the walls cannot be separated. But, of course, we must look at the quantum theory and, in

[b] As stressed in [9], the physical mass of the state is infinite due to a logarithmic divergence associated to both the gauge field and the wall separation R. This is the familiar infra-red divergence in $d=2+1$ dimensions that occurs also for the D2-brane dynamics in string theory. It means that we cannot excite the modes of the open string on-shell, but this does not concern us here. Rather we are interested in the virtual effect of these states on the dynamics of the massless brane modes. And, as we shall see, these are perfectly finite.

particular, the effect of integrating out the open string mode. This gives a finite renormalization: $\kappa \rightarrow \kappa_{\text{eff}} = -\frac{1}{2} + \frac{1}{2}\text{sign}\,(R)$, where $\text{sign}\,(R)$ is the sign of the mass of the fermion in the chiral multiplet. We learn that we can separate the walls in the positive $R > 0$ direction at no cost of energy. Of course, this is not surprising: we have simply integrated in the chiral multiplet, and immediately integrated it out again, leaving us with no effective Chern-Simons coupling. But if we try to pass the domain walls through each other and separate them in the opposite direction $R < 0$, the Chern-Simons coupling bites, and the flat direction is lifted. The final effect of the quantum open strings on the domain wall dynamics is best summarized in the famous words of Gandalf: you cannot pass.

One may study the low-energy dynamics of the Chern-Simons theory in more detail. After transforming the dual photon σ, one finds that the long wavelength interactions of the massless modes R and σ are again described by a sigma-model with target space given by the cigar.

Finally, let us decide when the open-string approach is valid. We have integrated in the lightest mode of an open stretched string, and ignored the tower of higher excitations. This requires $T_{\text{vort}}R \ll \Delta m$, where Δm is the excitation gap for internal modes of the string. If we also require that $R \gg L_{\text{wall}} = 1/\Delta m$, then we are left with the condition for the validity of the open string description: $\Delta m/v \gg 1$. This is the opposite regime to where we performed the classical bulk computation[c].

6. Discussion

What are the implications of a field theory soliton whose dynamics admit both a bulk and open string description? Let me start by addressing what is, perhaps, the most familiar aspect of D-branes. Ask the average man on the street what characterizes a D-brane and he will answer that the tension goes as the inverse string coupling, $T \sim 1/g_s$. Is this true for our D-branes? In fact, it's hard to tell since we don't have a good handle on the string coupling g_s in these theories. We may evocatively write the tension of the domain wall as $T_{\text{wall}} = (\Delta m/v)\,(1/\alpha'^{3/2})$ with $\alpha' = 2\pi/T_{\text{vortex}}$, suggesting that the string coupling is $g_s = v/\Delta m$. Is this plausible? For $g_s \ll 1$, the closed loop of string appears to be the lightest object in the field theory.

[c]Indeed, the two calculations differ in the details. For example, in the classical computation, interactions are exponentially suppressed in the separation $\exp(-R\Delta m)$, while the open string approach has power-law interactions $1/R$. In fact, the latter is not to be trusted: the infinite tower of higher open string modes may possibly sum to produce an exponential suppression after all.

However, it is precisely in this regime that the sigma-model breaks down and one must study the UV completion of the theory, including potential renormalization to closed loops of string. It certainly seems possible that there exists a 4d little string theory completion of the theory. In this case, the relationship $g_s = v/\Delta m$ is very reasonable and is entirely analogous to similar expressions that appear in 6d little string theories in the double scaled limit.

In fact, the question of a UV completion has bearing on a more important issue. I referred to the classical scattering of domain walls as the "bulk calculation" rather than the more familiar "closed string" regime. Is this latter phrase appropriate? Can the bulk sigma-model fields be thought of as quantized loops of closed vortex string? I don't know the answer, but only if a UV completion exists can we think of the two descriptions of the domain wall dynamics as reflecting an underlying open-closed string duality, with the two methods above corresponding to suitable lowest-mode truncations of a modular invariant annulus partition function.

If such an open-closed string duality is indeed at play for our solitonic vortex strings, one may try to be more ambitious and study an AdS/CFT-like correspondence in theories without gravity. On a practical level, Maldacena's big breakthrough can be thought of as introducing a factor of N, the number of D-branes, to allow two, seemingly mutually exclusive, regimes of validity to overlap. In the present case it seems difficult to implement this. On a trivial level, our domain walls are not identical objects and bringing many of them together does not improve the validity of the bulk calculation. On a more fundamental level, questions of entropy matching between the two sides appear much more of a hurdle in theories without gravity and the associated thinning of degrees of freedom.

References

1. D. Tong, JHEP **0602**, 030 (2006) [arXiv:hep-th/0512192].
2. N. Sakai and D. Tong, JHEP **0503**, 019 (2005) [arXiv:hep-th/0501207].
3. H. B. Nielsen and P. Olesen, Nucl. Phys. B **61** (1973) 45.
4. E. R. C. Abraham and P. K. Townsend, Phys. Lett. B **291**, 85 (1992).
5. J. P. Gauntlett, R. Portugues, D. Tong and P. K. Townsend, Phys. Rev. D **63**, 085002 (2001) [arXiv:hep-th/0008221].
6. M. Shifman and A. Yung, Phys. Rev. D **67**, 125007 (2003) [arXiv:hep-th/0212293].
7. J. P. Gauntlett, D. Tong and P. K. Townsend, Phys. Rev. D **64**, 025010 (2001) [arXiv:hep-th/0012178].
8. D. Tong, Phys. Rev. D **66**, 025013 (2002) [arXiv:hep-th/0202012].
9. M. Shifman and A. Yung, arXiv:hep-th/0603236.

ANALOG OF BULK–BRANE DUALITY IN FIELD THEORY

M. SHIFMAN

William I. Fine Theoretical Physics Institute, University of Minnesota, Minneapolis, MN 55455, USA

A. YUNG

Petersburg Nuclear Physics Institute, Gatchina, St. Petersburg 188300, Russia

We consider (3+1)-dimensional $\mathcal{N} = 2$ supersymmetric QED with two flavors of fundamental hypermultiplets. Dynamics of the 4D bulk are mapped at low energies onto a 3D effective theory on the domain wall, namely 3D $\mathcal{N} = 2$ SQED with two matter superfields and a weak–strong coupling constant relation in 4D and 3D, respectively. Integrating out massive matter (strings in the bulk theory) one generates a Chern–Simons term on the wall world volume and an interaction between the wall and antiwall that scales as a *power* of distance. This talk is based on Ref. 1.

Keywords: Supersymmetry, flux tubes, domain walls, holographic duality

1. Introduction

Duality means that one and the same dynamical system can be described using two different theories. Usually, when one description is at weak coupling, the other is at strong coupling and *vice versa*. This makes duality extremely hard to prove. Exceptions are theories where one can find exact solutions and continue them from weak to strong coupling. The most well-known example is the Seiberg–Witten electromagnetic duality in $\mathcal{N} = 2$ gauge theories. In other cases duality is conjectured, e.g. the AdS/CFT correspondence,[2] and then indirectly verified. On the other hand, once duality is confirmed, it becomes a powerful tool to study theories at strong coupling using their weakly coupled duals.

In this talk we discuss a purely field-theoretic example of a low-energy "holographic" duality. We consider (3+1)-dimensional $\mathcal{N} = 2$ gauge theory with domain walls and flux tubes (strings) as our bulk model. One of the spatial dimensions is compactified, so that our four-dimensional space is in fact $R^3 \times S^1$. We show that a system of two parallel walls $W\bar{W}$ and strings stretched between these walls has a dual description in terms of (2+1)-dimensional U(1) gauge theory on the world volume of domain walls. In three dimensions the dual model has $\mathcal{N} = 2$ (four supercharges). Moreover, the weak coupling regime in the bulk theory corresponds to strong coupling in the world-volume theory and *vice versa.*

2. Domain walls, strings and their junctions

The bulk theory which we will work with is $\mathcal{N} = 2$ SQED with 2 flavors and the Fayet–Iliopoulos term. The bosonic part of the action is

$$S = \int d^4x \left\{ \frac{1}{4g^2} F_{\mu\nu}^2 + \frac{1}{g^2} |\partial_\mu a|^2 + \bar{\nabla}_\mu \bar{q}_A \nabla_\mu q^A + \bar{\nabla}_\mu \tilde{q}_A \nabla_\mu \bar{\tilde{q}}^A \right.$$

$$\left. + \frac{g^2}{8} \left(|q^A|^2 - |\tilde{q}_A|^2 - \xi \right) + \frac{g^2}{2} \left| \tilde{q}_A q^A \right|^2 + \frac{1}{2} (|q^A|^2 + |\tilde{q}^A|^2) \left| a + \sqrt{2} m_A \right|^2 \right\}, \tag{1}$$

where ξ is the coefficient in front of the Fayet-Iliopoulos term, the index $A = 1, 2$ is the flavor index; and the mass parameters m_1, m_2 are assumed to be real. In addition we will assume

$$\Delta m \equiv m_1 - m_2 \gg g\sqrt{\xi}\,, \qquad \Delta m \ll (m_1 + m_2)/2\,. \tag{2}$$

There are two vacua in this theory:

$$a = -\sqrt{2} m_1, \qquad q_1 = \sqrt{\xi}, \qquad q_2 = 0\,; \tag{3}$$

$$a = -\sqrt{2} m_2, \qquad q_1 = 0, \qquad q_2 = \sqrt{\xi}\,. \tag{4}$$

The vacuum expectation value (VEV) of the field $\tilde{q}$ vanishes in both vacua.

A BPS domain wall interpolating between the two vacua of our bulk theory was explicitly constructed in Ref. 3. Its tension can be identified as the $(1,0)$ central charge of the supersymmetry algebra,

$$T_{\rm w} = \xi\, \Delta m\,. \tag{5}$$

The wall solution has a three-layer structure: in the two outer layers (which have width $O((g\sqrt{\xi})^{-1})$) the squark fields drop to zero exponentially; in the inner layer the field a interpolates between its two vacuum values. The thickness of this inner layer is

$$R = \frac{4\Delta m}{g^2 \xi}\,. \tag{6}$$

The moduli space is described by two bosonic coordinates: one is responsible for translation, the other one is a U(1) phase parameter σ. Its origin is as follows.[3] The bulk theory at $\Delta m \neq 0$ has U(1)×U(1) flavor symmetry corresponding to two independent rotations of two quark flavors. In both vacua only one quark develops a VEV, only one of these two U(1)'s is broken. The corresponding phase is eaten by the Higgs mechanism. However, on the wall both quarks have nonvanishing values, breaking both U(1) groups. One of the two phases is eaten by the Higgs mechanism. The other one becomes a Goldstone mode living on the wall.

The U(1) phase σ discussed above can be dualized à la Polyakov into a U(1) gauge field in (2+1) dimensions. Thus, the world-volume theory on the wall is a U(1) gauge theory. The bosonic part of the (2+1)-dimensional action is

$$S_{2+1} = \int d^3x \left\{ \frac{1}{2e^2} \left(\partial_n a_{2+1}\right)^2 - \frac{1}{4e^2} \left(F^{(2+1)}_{mn}\right)^2 \right\}, \tag{7}$$

where a real scalar field a_{2+1} is related to the wall position z_0 as follows

$$a_{2+1} = 2\pi\xi\, z_0 , \tag{8}$$

and e^2 is the coupling constant of the effective U(1) theory on the wall,

$$e^2 = 4\pi^2 \frac{\xi}{\Delta m} . \tag{9}$$

A characteristic scale of massive excitations in the world volume theory is of the order of the inverse thickness of the wall $1/R$. The dimensionless parameter that characterizes the coupling strength in the world-volume theory is $e^2 R$,

$$e^2 R = \frac{16\pi^2}{g^2}. \tag{10}$$

The weak coupling regime in the bulk theory corresponds to strong coupling on the wall.[3]

Our bulk theory has magnetic Abrikosov–Nielsen–Olesen (ANO) flux tubes. These objects are half-critical, much in the same way as the domain walls above, and satisfy first order equations. The magnetic flux of the minimal-winding flux tube is 4π, while its tension is given by the $(1/2, 1/2)$ central charge,

$$T_{\rm s} = 2\pi\xi\, . \tag{11}$$

Let us consider a flux tube ending on a domain wall. The 1/2 BPS flux tube ending on the domain wall is semi-infinite and aligned perpendicular to the wall. This configuration was studied in Refs. 3, 4, 5, 6. The string-wall junction is a 1/4-BPS object satisfying first order equations. The solution to these equations in the bulk theory (1) was found in Ref. 3. At large distance r from the string end-point it has the form of a wall solution with the collective coordinates z_0 and σ depending on the world-volume coordinates x_1 and x_2 as follows:

$$z_0 = -\frac{1}{\Delta m} \ln r + \text{const}\,, \tag{12}$$

where $r = \sqrt{x_1^2 + x_2^2}$. The Polyakov-dualized σ becomes an electric field which at large distances from the string-wall junction is given by

$$F_{0i}^{2+1} = \frac{e^2}{2\pi} \frac{x_i}{r^2}\,, \qquad i = 1, 2\,. \tag{13}$$

The magnetic flux from the string penetrates into the wall. The string end-point is seen as an electric charge in the dual 3D QED.

In the world-volume theory *per se*, the fields (12) and (13) can be considered as produced by classical point-like charges which interact in a standard way with the electromagnetic field A_n and the scalar field a_{2+1}. However, due to long range logarithmic forces this theory has IR divergences. To make the world volume theory well defined we have to consider both strings and antistrings attached to the wall both from the left and from the right.[7,1] We have the following set of the electric and scalar charges associated with the string end-points:

$$\begin{aligned} n_e &= +1, \quad \text{incoming flux,} \\ n_e &= -1, \quad \text{outgoing flux,} \end{aligned} \tag{14}$$

while their scalar charges are

$$\begin{aligned} n_s &= +1, \quad \text{string from the right,} \\ n_s &= -1, \quad \text{string from the left.} \end{aligned} \tag{15}$$

Clearly, the bending of the wall produced by two string attached from different sides of the wall tends to zero at large r from the string end-points and produces no IR divergence.

3. Quantizing strings on the wall

A quantum version of the world-volume theory with charged matter fields representing strings of the bulk theory can be obtained proceeding along the lines of Ref. 8. A novel element is introduction of two types of charged matter, to represent both, strings attached from the left and from the right of the wall.

We compactify the $x_3 = z$ direction in our bulk theory on a circle of length L, see Fig. 1. Then we consider a pair "wall plus antiwall" oriented in the $\{x_1, x_2\}$ plane, separated by a distance l in the perpendicular direction. The wall and antiwall experience attractive forces. However, their interaction (due to an overlap of their profile functions) is exponentially suppressed at large separations, and in what follows we neglect exponentially suppressed effects. If so, our theory continues to have four conserved supercharges ($\mathcal{N} = 2$ supersymmetry in (2+1) dimensions) as was the case for the isolated single wall. The quantum version of the world volume

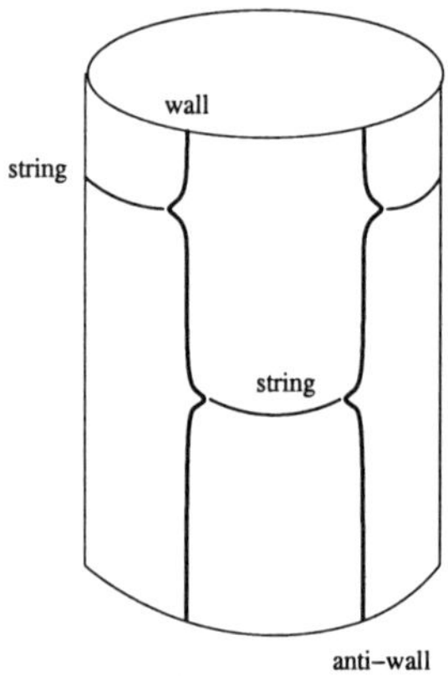

Fig. 1. A wall and antiwall connected by strings on the cylinder. The circumference of the circle (the transverse slice of the cylinder) is L.

theory is completely determined by the charge assignment (14), (15) and $\mathcal{N} = 2$ supersymmetry. The charged matter fields have the opposite electric charges. The bosonic part of the action has the form

$$S_{\rm bos} = \int d^3x \left\{ -\frac{1}{4e^2} F^-_{mn} F^{-\,mn} + \frac{1}{2e^2} \left(\partial_n a_-\right)^2 + |\mathcal{D}_n s|^2 + \left|\tilde{\mathcal{D}}_n \tilde{s}\right|^2 \right.$$
$$\left. - 2a_-^2 \, \bar{s}\, s - 2(m - a_-)^2 \, \bar{\tilde{s}}\, \tilde{s} - e^2 \left(|s|^2 - |\tilde{s}|^2\right)^2 \right\}. \qquad (16)$$

The electric charges of strings with respect to the field A^-_n are $\pm\sqrt{2}$.

4. Physics of the world-volume theory

Let us integrate out the string multiplets S and $\tilde{S}$ and study the effective theory for the U(1) gauge supermultiplet at scales below the string masses m_s. Integration over the charged matter fields in (16) leads to the generation of the Chern–Simons term and a potential for a_- via the corresponding D-term,

$$S_{2+1} = \int d^3x \left\{ \frac{1}{2e^2(a_-)}(\partial_n a_-)^2 - \frac{1}{4e^2(a_-)}(F^-_{mn})^2 \right. \\ \left. + \frac{1}{2\pi}\varepsilon_{nmk} A^-_n \partial_m A^-_k + \frac{e^2(a_-)}{8\pi^2}(2a_- - m)^2 \right\} . \tag{17}$$

The most dramatic effect in (17) is the generation of a potential for the field a_- which corresponds to the separation l between the walls. The vacuum of (17) is located at

$$\langle a_- \rangle = \frac{m}{2}, \qquad l = \frac{L}{2}. \tag{18}$$

There are two extra solutions at $a_- = 0$ and $a_- = m$, but they lie outside the limits of applicability of our approach.

We see that the wall and antiwall are pulled apart to be located at the opposite sides of the cylinder. Moreover, the potential is quadratically rising with the deviation from the equilibrium point (18). As we mentioned in Sect. 3, the wall and anti-wall interact with exponentially small potential due the overlap of their profiles. However, these interactions are negligibly small at $l \gg R$ as compared to the interaction in Eq. (17). The interaction potential in (17) arises due to virtual pairs of strings. Note, that if the wall-antiwall interactions were mediated by particles they would have exponential fall-off at large separations l (there are no massless particles in the bulk). Quadratically rising potential would never be generated. In our case the interactions are due to virtual pairs of extended objects – strings. Strings are produced as *rigid* objects stretched between the walls.

The presence of the potential for the scalar field a_- in Eq. (17) makes this field massive, with mass

$$m_a = \frac{e^2}{\pi}. \tag{19}$$

By supersymmetry, the photon is no longer massless too, it should acquire the same mass. This is associated with the Chern–Simons term in (17).

5. Conclusions

This work is devoted to a field-theoretic implementation of a holographic duality. If the bulk 4D world was a cylinder with walls $W\bar{W}$ parallel to each other along the cylinder axis, and were we the wall dwellers, we would establish that our 3D world is governed by $\mathcal{N} = 2$ 3D SQED with the Chern–Simons term. Dynamics of the 4D bulk theory and the bulk set-up on the one hand, and dynamics of the 3D theory (16) on the other hand, are in one-to-one correspondence in the low-energy limit. The charged particles the wall dwellers would discover in their 3D world would reflect two types of strings (and antistrings) stretched between the walls.

Quantum effects — virtual string loops — generate a mass gap proportional to e^2 in the world volume theory. Technically, the mass gap is due to an induced Chern–Simons term. This term gives mass to the 3D (dual) photon. $\mathcal{N} = 2$ supersymmetry imposes the same mass on other members of the vector supermultiplet. Thus, at the quantum level, the wall and anti-wall on the cylinder start interacting. Their interaction is non-exponential; rather it depends on the interwall distance as the square of the distance. The walls are stabilized on the opposite sides of the cylinder. This is the vacuum state in the world-volume theory. This feature is absolutely remarkable. Any interaction between the walls generated by exchange of localized states (particles) must die off exponentially in the interwall distance.

An obvious goal for future work is generalizing the bulk–brane duality we found, to non-Abelian models.

Acknowledgments: This work was supported in part by DOE grant DE-FG02-94ER408 and RFBR grant No.06-02-16364a.

References

1. Mikhail Shifman and Alexei Yung, *Bulk–Brane Duality in Field Theory,* hep-th/0603236.
2. J. M. Maldacena, Adv. Theor. Math. Phys. **2**, 231 (1998) [hep-th/9711200]; S. S. Gubser, I. R. Klebanov and A. M. Polyakov, Phys. Lett. B **428**, 105 (1998) [hep-th/9802109]; E. Witten, Adv. Theor. Math. Phys. **2**, 253 (1998) [hep-th/9802150].
3. M. Shifman and A. Yung, Phys. Rev. D **67**, 125007 (2003) [hep-th/0212293].
4. M. Shifman and A. Yung, Phys. Rev. D **70**, 025013 (2004) [hep-th/0312257].
5. Y. Isozumi, M. Nitta, K. Ohashi and N. Sakai, Phys. Rev. D **71**, 065018 (2005) [hep-th/0405129].
6. N. Sakai and D. Tong, JHEP **0503**, 019 (2005) [hep-th/0501207].
7. R. Auzzi, M. Shifman and A. Yung, Phys. Rev. D **72**, 025002 (2005) [hep-th/0504148].
8. D. Tong, JHEP **0602**, 030 (2006) [hep-th/0512192].

SECTION 6

LATTICE METHODS

Convener J. Giedt

STAGGERED FERMIONS AND POWER-COUNTING

JOEL GIEDT

Fine Theoretical Physics Institute, University of Minnesota
116 Church St. S.E., Minneapolis, MN 55455 USA
E-mail: giedt@physics.umn.edu

I discuss the difficulties of lattice power-counting for staggered fermions. The recently formulated staggered fermion power-counting theorem is summarized.

Keywords: Lattice field theory; staggered fermions; power-counting theorem; renormalization; lattice quantum chromodynamics.

1. Motivations

To obtain an inductive proof of perturbative renormalizability, the principal ingredients are: (i) a power-counting theorem; (ii) a compatible subtraction scheme; (iii) a theorem that subtraction is equivalent to field, mass and coupling renormalization. The role of the power-counting theorem is to establish that diagrams are finite, once subtractions implemented. For the power-counting theorem to be useful, it should possess an easily computed UV index (or degree, of divergence) that only depends on asymptotic properties of propagators and vertices. Also, the UV index computation should be inductive, relying on the indices of subdiagrams of lower order.

As is well known,[1] in lattice perturbation theory one encounters the following subtlety: propagators and vertices are periodic on the reciprocal lattice $(2\pi/a)\mathbf{Z}^4$ ("momentum space"; a is the lattice spacing). Therefore, a UV index based on large momentum behavior makes no sense. Reisz's solution is just this: to consider scaling of propagators and vertices with $a \to a/\lambda$. He is able to contruct a lattice power-counting theorem that applies to many theories.[1] However, staggered fermions[2–4] violate the conditions of Reisz's power-counting theorm.[5] Until recently,[6] this was an unsolved problem.

Staggered fermions are efficient to simulate; their mass is multiplicatively renormalized, so they are in some sense "chiral." That is, the chiral limit does not require fine-tuning. For this reason, they have been popular with the large collaborations that propose to obtain "first principles" results for QCD, such as in refs. 7, 8. However, the absence of a power-counting theorem has been the source of some consternation.[10–12] 1-loop is easy to power-count in the momentum space "taste" (or "flavor") basis. For higher loops this is much harder. Without a power-counting theorem that extends to all orders, there is no rigorous proof of renormalizability by pre-Wilsonian methods. One should note, however, that a Wilsonian analysis, based on non-irrelevant operators allowed by the staggered fermion symmetries, indicates that the continuum limit can be obtained by a suitable adjustment of the parameters and normalization of the bare theory.[13] Still, it is reassuring to have an old-fashioned proof of perturbative renormalizability.

2. The conflict with Reisz's theorem

2.1. *A Reisz review*

One partitions the domain of integration, depending on the values of the line momenta (the momenta flowing through propagators), which we write as:

$$\ell_i = C_{ij} k_j + D_{ij} q_j, \tag{1}$$

where $k_1, \ldots, k_L$ are the loop momentum, $q_1, \ldots, q_E$ are the external momentum and $\ell_1, \ldots, \ell_I$ are the loop momentum. Consider a subset J of the line momenta. Let $\epsilon \ll 1$ and consider the domain of k (collectively denoting the loop momenta) where:

• For $\ell_i \in J$, ℓ_i is ϵ-*close* to a pole in some Brillouin zone (i.e., not necessarily the first zone). I.e., $\exists z_i \in \mathbf{Z}^4$ s.t.

$$||\ell_i - (2\pi/a) z_i|| \leq (\pi/a)\epsilon. \tag{2}$$

• For $\ell_i \notin J$, ℓ_i is ϵ-*far* from all poles. I.e., $\nexists z_i \in \mathbf{Z}^4$ s.t. (2) holds. Rather, $\forall z_i \in \mathbf{Z}^4$

$$||\ell_i - (2\pi/a) z_i|| > (\pi/a)\epsilon. \tag{3}$$

Having this decomposition in mind, Reisz introduces the following resolution of identity:

$$1 = 1_B(\ell) \equiv \Theta^B_\epsilon(\ell) + \sum_{z \in \mathbf{Z}^4} \Theta\left(\frac{\pi}{a}\epsilon - \left|\left|\ell - \frac{2\pi}{a} z\right|\right|\right). \tag{4}$$

Here, Θ is Heaviside's unit step function. Θ_ϵ^B is another step function:

$$\Theta_\epsilon^B(\ell) = \begin{cases} 0 & \text{if } ||\ell - \frac{2\pi}{a}z|| < \frac{\pi}{a}\epsilon \text{ for some } z \in \mathbf{Z}^4, \\ 1 & \text{otherwise,} \end{cases} \tag{5}$$

introduced by Reisz. Then one inserts one such identity for each line momentum. It is easy to see that this leads to the desired partition of the Feynman integral, into ϵ-near and ϵ-far contributions. This is represented as:

$$\hat{I} = \sum_{J,z} \hat{I}_{Jz}. \tag{6}$$

This is a sum over the various sets $J = \{\ell_{i_1}, \ldots, \ell_{i_h}\}$ that are ϵ-near poles located at $z_1, \ldots, z_h$.

Consider a lattice propagator expressed in terms of a numerator N and denominator C:

$$N(\ell; m, a)/C(\ell; m, a). \tag{7}$$

For many lattice theories, the following bounds hold:

- Suppose $\ell_i \in J$ and $\ell_i \in \mathcal{B}_a$, where $\mathcal{B}_a$ is the first Brillouin zone:

$$\mathcal{B}_a \equiv (-\pi/a, \pi/a]^4. \tag{8}$$

Then there exists a constant $\alpha > 0$ s.t.:

$$C(\ell_i; m_i, a)^{-1} \leq \alpha(\ell_i^2 + m_i^2)^{-1}. \tag{9}$$

- If $\ell_i \notin J$, then there exists a constant $\gamma > 0$ s.t.:

$$C(\ell_i; m_i, a)^{-1} \leq \gamma a^2. \tag{10}$$

Note that in the first case ℓ_i is constrained to be in the first Brillouin zone (8). Introduce the set of 16 4-vectors with entries 0 or 1:

$$\mathcal{K} \equiv \{(0^4), (\underline{1, 0^3}), (\underline{1^2, 0^2}), (\underline{1^3, 0}), (1^4)\} \tag{11}$$

In the definition of the set of 4-vectors $\mathcal{K}$, powers indicate how many times a 0 or 1 appears. Underlining indicates that all permutations of entries are to be included. If the lattice theory enjoys translation symmetry in each of the 4 directions by a lattice spacing shift a, then the Feynman rules will be periodic under a shift in the loop momenta k by a reciprocal lattice vector: $k \to k + (2\pi/a)\eta$, $\eta \in \mathcal{K}$. Assuming that this translation invariance holds, Reisz proves that a choice of loop momenta can always be made such that all $\ell_i \in J$ lie in the first Brillouin zone $\mathcal{B}_a$.

Thus Reisz is able to bound the denominator of the Feynman integrand by the following continuum expression:

$$\alpha^h(\gamma a^2)^{I-h} \prod_{\ell_i \in J} (\ell_i^2 + m_i^2)^{-1}, \tag{12}$$

where h is the number of line momenta in the set J and I is the total number of line momenta (equivalently, the number of internal lines).

The problem is, staggered fermions do not satisfy (10). This is because they have 15 additional lattice poles[a] in $\mathcal{B}_a$. This fact was pointed out long ago by Lüscher.[5] Explicitly,

$$C = a^{-2} \sum_{\mu} \sin^2(\ell_\mu a) + m^2, \tag{13}$$

and the poles are at the minima $C = m^2$, $\ell \in (\pi/a)\mathcal{K}$. These give a contribution that does not vanish like a^2 in the continuum limit, but behave analogously to (9):

$$\begin{aligned} &C(\ell \approx (\pi/a)\eta; m, a)^{-1} \le \alpha(\boldsymbol{\pi}(\ell)^2 + m^2)^{-1}, \\ &\boldsymbol{\pi}(\ell) = \ell - (\pi/a)\eta, \quad \eta \in \mathcal{K}. \end{aligned} \tag{14}$$

That is, $\boldsymbol{\pi}$ is the projection into the *reduced Brillouin zone:*

$$\mathcal{B}_{2a} = (-\pi/2a, \pi/2a]^4. \tag{15}$$

The main task in power-counting for staggered fermions is to address the 15 additional poles.

3. The staggered fermion power-counting theorem

Recently it was shown how to overcome the difficulty posed by the additional poles.[6] The essential trick is to resolve identity on the *reduced reciprocal lattice* $(\pi/a)\mathbf{Z}^4$, and then to use transformed Feynman rules associated with shifts in loop momenta

$$k_i \to k_i + (\pi/a)A_i, \quad A_i \in \mathcal{K} \tag{16}$$

to eliminate explicit factors of π/a.

The resolution on $(\pi/a)\mathbf{Z}^4$ is the following. Define a step-function analogous to Reisz's:

$$\Theta_\epsilon^F(\ell) = \begin{cases} 0 \text{ if } ||\ell - \frac{\pi}{a}z|| < \frac{\pi}{a}\epsilon \text{ for some } z \in \mathbf{Z}^4, \\ 1 \text{ otherwise.} \end{cases} \tag{17}$$

[a]These are, by definition, the minima of the inverse propagator.

For any ℓ one can resolve identity as:

$$1 = 1_F(\ell) \equiv \Theta^F_\epsilon(\ell) + \sum_{z \in \mathbf{Z}^4} \Theta\left(\frac{\pi}{a}\epsilon - \left|\left|\ell - \frac{\pi}{a}z\right|\right|\right). \tag{18}$$

As above, Θ is Heaviside's unit step function. This resolution is useful for line momenta of staggered fermions, since it isolates the regions that are near staggered fermion poles.

After some work, one finds that the the power-counting theorem for staggered fermions takes the following form:

Definition. Suppose a lattice Feynman integral

$$\hat{I} = \int d^{4L}k\ F, \quad F = V/C. \tag{19}$$

Here V is the overall numerator and C is the overall denominator. Let $F_A = V_A/C_A$, $A \in \mathcal{K}^L = \times^L \mathcal{K}$ denote the transformed Feynman integrand under a π/a-valued shift in the loop momenta. That is:

$$\begin{aligned} V_A(k,q;m,a) &= V(k + (\pi/a)A, q; m, a), \\ C_A(k,q;m,a) &= C(k + (\pi/a)A, q; m, a). \end{aligned} \tag{20}$$

Generalize Reisz's UV degree as follows:

$$\begin{aligned} &\overline{\overline{\mathrm{degr}}}_{\hat{u}} F = \max_{A \in \mathcal{K}^L} \overline{\mathrm{degr}}_{\hat{u}} F_A, \quad \overline{\mathrm{degr}}_{\hat{u}} F_A = \overline{\mathrm{degr}}_{\hat{u}} V_A - \overline{\mathrm{degr}}_{\hat{u}} C_A; \\ &\overline{\overline{\mathrm{degr}}}_{\hat{H}} \hat{I} = 4d + \overline{\overline{\mathrm{degr}}}_{\hat{u}} F. \end{aligned} \tag{21}$$

Note that $u_1, \ldots, u_d$ parameterizes a Zimmermann subspace H. The UV degree $\overline{\mathrm{degr}}_{\hat{u}} X$ for any X is just the one defined by Reisz. In essence, it gives the scaling exponent under $u \to \lambda u$, $a \to a/\lambda$ with momenta orthogonal to the Zimmermann subspace held fixed.

Proposition. Suppose that

$$\overline{\overline{\mathrm{degr}}}_{\hat{H}} \hat{I} < 0 \quad \forall \quad H \in \mathcal{H}, \tag{22}$$

where $\mathcal{H}$ is the set of all Zimmermann subspaces. Then $\hat{I}$ converges, and

$$\lim_{a \to 0} \hat{I} = \sum_{A \in \mathcal{K}^L} \int_{-\infty}^{\infty} d^{4L}k\ \frac{P_A(k,q;m)}{E_A(k,q;m)}, \tag{23}$$

where

$$P_A(k,q;m) = \lim_{a \to 0} V_A(k,q;m,a), \quad E_A(k,q;m) = \lim_{a \to 0} C_A(k,q;m,a) \tag{24}$$

are just the continuum limits of the numerator and denominator resp. This indicates that various regions of loop momenta may contribute to the continuum limit, due to the presence of doublers in the fermion spectrum.

The generalization of Reisz's technique, to overcome the additional poles, allows one to prove the proposition. Some clarifying 1-loop example applications of the theorem are given in ref. 6.

4. Discussion

I have summarized the extension of Reisz's power-counting theorem to the case of staggered fermions. One thing that remains to be done is to use the staggered fermion power-counting theorem to prove perturbative renormalizability, following refs. 9, 14. Also, applications of the theorem to higher orders in perturbation theory should be explored in more detail. It is worth emphasizing that all of the manipulations that were performed here apply equally well to improved staggered fermion QCD.

Acknowledgements

This work was supported in part by the U.S. Department of Energy under grant No. DE-FG02-94ER-40823.

References

1. T. Reisz, "A Power Counting Theorem For Feynman Integrals On The Lattice," Commun. Math. Phys. **116** (1988) 81.
2. J. B. Kogut and L. Susskind, "Hamiltonian Formulation Of Wilson's Lattice Gauge Theories," Phys. Rev. D **11**, 395 (1975).
3. L. Susskind, "Lattice Fermions," Phys. Rev. D **16**, 3031 (1977).
4. H. S. Sharatchandra, H. J. Thun and P. Weisz, "Susskind Fermions On A Euclidean Lattice," Nucl. Phys. B **192** (1981) 205.
5. M. Lüscher, "Selected Topics In Lattice Field Theory," in "Fields, Strings and Critical Phenomena," Les Houches XLIX, 1988, eds. E. Brézin and J. Zinn-Justin, Elsevier, New York, 1989.
6. J. Giedt, "Power-counting theorem for staggered fermions," arXiv:hep-lat/0606003.
7. C. T. H. Davies *et al.* [HPQCD Collaboration], "High-precision lattice QCD confronts experiment," Phys. Rev. Lett. **92**, 022001 (2004) [arXiv:hep-lat/0304004].
8. C. DeTar, S. Gottleib, "Lattice Quantum Chromodynamics Comes of Age," Physics Today, February (2004) 45.

9. T. Reisz, "Lattice Gauge Theory: Renormalization To All Orders In The Loop Expansion," Nucl. Phys. B **318** (1989) 417.
10. K. Jansen, "Actions for dynamical fermion simulations: Are we ready to go?," Nucl. Phys. Proc. Suppl. **129**, 3 (2004) [arXiv:hep-lat/0311039].
11. D. H. Adams, "Testing universality and the fractional power prescription for the staggered fermion determinant," Nucl. Phys. Proc. Suppl. **140** (2005) 148 [arXiv:hep-lat/0409013].
12. L. Lellouch, "Flavour physics from lattice QCD," proceedings *Continuous advances in QCD,* ed. M. Peloso, FTPI, U. Minnesota, Minneapolis, May 11-14, 2006.
13. M. F. L. Golterman and J. Smit, "Self-energy and flavor interpretation of staggered fermions," Nucl. Phys. B **245** (1984) 61.
14. T. Reisz, "Renormalization Of Feynman Integrals On The Lattice," Commun. Math. Phys. **117** (1988) 79.

AT WHICH ORDER SHOULD WE TRUNCATE PERTURBATIVE SERIES?

Y. MEURICE

Department of Physics and Astronomy, University of Iowa,
Iowa City, IA 52242, USA
E-mail: yannick-meurice@uiowa.edu

Perturbative coefficients grow factorially with the order and one needs a prescription to truncate the series in order to obtain a finite result. A common prescription consists in dropping the smallest contribution at a given coupling and all the higher orders terms. We discuss the error associated with this procedure. We advocate a more systematic approach which consists in controlling the large field configurations in the functional integral. We summarize our best understanding of these issues for lattice QCD in the quenched approximation and their connection with convergence problems found in the continuum.

Keywords: QCD; Perturbation Theory; Lattice Gauge Theory

1. Introduction

Perturbation theory has played an essential role in developing and establishing the standard model of electroweak and strong interactions. The renormalizability of the theory guarantees that we can calculate the radiative corrections at any order in perturbation theory. On the other hand, a generalization of Dyson's argument [1] suggests that the perturbative series are divergent and one needs to truncate the series. In absence of a definite prescription to deal with this problem, one usually relies on the "rule of thumb" which consists in dropping the smallest contribution at a given coupling and all the higher order terms. Clearly, this procedure has a limited accuracy and it is not always obvious how to estimate the error or to decide if one needs to calculate one extra order.

The problem is particularly acute for QCD corrections because they are large even at low order. As emphasized by Z. Bern's talk [2], NLO corrections are important for multijet processes to be studied by the LHC. Another example [3] is the hadronic width of the Z^0 where the term of order α_s^3 is more than 60 percent of the term of order α_s^2 and contributes to one part in 1,000 of the total width (a typical experimental error for individual LEP experiments). It is not clear that the next term would improve the accuracy of the calculation. As these calculations can be extremely time consuming, it is necessary to address the lack of convergence of perturbative series in a more systematic way.

2. The rule of thumb

Consider a generic asymptotic series in a coupling λ with coefficients growing like $C_1 C_2^k \Gamma(k + C_3)$. If we assume that for sufficiently small λ the error made by truncating the series at order k is given by the order $k+1$ contribution, it is possible to show that the error is minimized by truncating the series at order $(\lambda|C_2|)^{-1} - C_3 - (1/2)$. The rule of thumb leads then to an error which is approximately

$$\sqrt{2\pi} C_1 (\lambda C_2)^{1/2 - C_3} e^{-\frac{1}{C_2 \lambda}} \tag{1}$$

This function is an approximate envelope for the accuracy curves at successive orders as illustrated in Fig. 1.

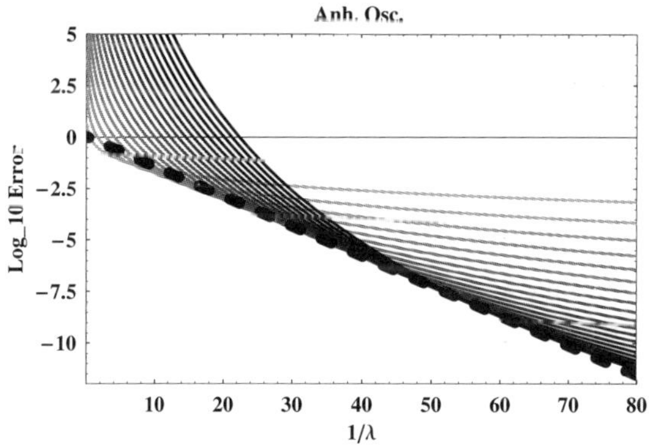

Fig. 1. Absolute value of the difference between the series and the numerical value for order 1 to 15 (in a Log_{10} scale) for the anharmonic oscillator as a function of $1/\lambda$. As the order increases, the curves get darker. The dash curve is $\ln\left((\sqrt{12}/\pi)e^{-\frac{1}{3\lambda}}\right)$

This type estimate is not always correct. For instance, for the ground state of the double-well potential, the instanton effect is much larger than this estimate. Often, neither the asymptotic behavior nor the non-perturbative effects are known. Consequently, it is worth trying to figure out a general method to handle the problem.

3. Quantum field theory with a large field cutoff

As can be seen in several examples where a path integral formulation is available, the factorial growth of perturbative series is related to configurations with arbitrary large fields. Large order coefficients are dominated by large field configurations for which the expansion of the exponential of the perturbation at that order is not a good approximation. In two non-trivial examples [4] where the modified coefficients can be calculated numerically, it was shown that a large field cutoff drastically affects the asymptotic behavior of the perturbative series. At this point we are only able to do numerical calculations for specific types of field cutoffs. Namely, we remove the tails of integration in configuration space. If ϕ_x denotes a scalar field at points x of a lattice, we impose the condition $|\phi_x| < \phi_{max}$ at every site. This procedure is quite convenient for a Monte Carlo calculation, but not to write modified Feynman rules because the additional condition is non-local in momentum space. At this point, a large field cutoff procedure that leads to simple Feynman rules remains to be found.

Quantum field theory with a large field cutoff is a subject in infancy. Numerical studies [5,6] show that, after appropriate rescalings, the transition between the large field cutoff regime and the small field cutoff regime can be described in good approximation by a universal function. This hints at a renormalization group explanation. It might be possible to construct an effective action, with couplings running with the field cutoff.

4. Optimization and interpolation

At a given order K and coupling λ, we can adjust $\phi_{max}(\lambda, K)$ in order to minimize or eliminate the discrepancy with the (usually unknown) correct value. As ϕ_{max} is varied, the curve (or the derivative) of the approximate expression crosses the numerical curve (or its derivative). The strong coupling can be used to calculate approximately this optimal $\phi_{max}(\lambda, K)$ and so this is a natural approach for the interpolation between the weak and strong coupling regimes. This procedure has been illustrated with two examples[7,8]. The calculation of the modified coefficients $a_k(\phi_{max})$ fall in three categories. Low k (the usual ones with exponentially small corrections calculable semi-classically); intermediate k (crossover; complicated but with universal features) and large k (power suppressed; no $k!$ behavior). The intuitive flow picture is that the beginning of the series corresponds to the behavior of the scaling variables near the gaussian fixed point. The large order, corresponds to the approach of the high-temperature/strong-coupling fixed point. The coefficients in the crossover (ϕ_{max} dependent) correspond to the crossover in the flows.

5. Lattice gauge theory with one plaquette

A simple example where the general ideas discussed above can be easily implemented is a $SU(N)$ lattice gauge theory with one plaquette [8]. The partition function reads.

$$Z(\beta, N) = \int \prod_{l \in p} dU_l e^{-\beta(1-\frac{1}{N} ReTrU_p)} , \tag{2}$$

with $\beta = 2N/g^2$. Specializing to $N = 2$, fixing the gauge to the identity for three of the links and integrating over the angular variables, we obtain

$$Z(\beta, 2) = (2/\beta)^{3/2} \frac{1}{\pi} \int_0^{2\beta} dt t^{1/2} e^{-t} \sqrt{1 - (t/2\beta)} \, . \tag{3}$$

Note that that due to the compactness of the group, the partition function has a large field cutoff, and also a large action cutoff which is gauge invariant. To construct a weak coupling expansion, we expand the square root in power of β^{-1}. However, the "coefficients" depend on β because the range of integration does. To get a regular series, we add the tails of integration because they are $e^{-2\beta} = e^{-8/g^2}$ effects, but this affects the asymptotic behavior of the series and the coefficients now grow factorially. The optimal order to truncate this series $2\beta - 1/2$. Incidently, this is also the order where the peak of the integrand moves out of the range of integration if it is kept below 2β as in the exact expression Eq. (3).

It is possible [8] to keep the finite bound of integration or to modify it in order to minimize the difference between the series at a given order and the original integral as explained in the previous section. However, if the order is large enough, the optimal cutoff becomes close to 2β. Note also that if we simply consider the regular perturbative series truncated using the rule of thumb, it is easy in this simple example to define and calculate the non-perturbative part of the integral. It consists in the higher order terms (to be calculated with the finite range of integration) minus the tails of integration that we have added.

6. The non-perturbative part of the plaquette

In the previous example, we can obtain a converging expansion by calculating the coefficients of the β^{-1} expansion keeping a finite range of integration. When there is more than one plaquette, this is much more complicated and we are presently developing practical methods to perform these calculations. In the rest of this talk, we discuss the regular perturbative series of the average plaquette of lattice QCD in the quenched approximation. We define the average plaquette

$$P(\beta) \equiv (1/\mathcal{N}_p)\left\langle \sum_p (1-(1/N)ReTr(U_p)) \right\rangle . \quad (4)$$

with $\mathcal{N}_p$ the total number of plaquettes. The β^{-1} series of P has been calculated [9,10] up to order 16. Despite the long series available, the factorial growth is not apparent. Instead, the series appears more like it has finite radius of convergence [11,12]. This is impossible because P takes different limits [13] when $g^2 \to 0^{\pm}$. The only plausible explanation is that the partition function has a zero in the complex β plane near $\beta \simeq 5.7$. Zeroes have been found [14] for imaginary values of β within predicted bounds [12] (and not found below the lower bound), however their numerical significance remains to be established .

A simplified model [15–17] that is capable of producing a series with coefficients growing factorially is

$$P \approx K \int_{t_1}^{t_2} dt e^{-\bar{\beta}t}\ (1-t\ 33/16\pi^2)^{-1-204/121} . \quad (5)$$

The new parameter $\bar{\beta}$ is related to the lattice β by a relation of the form $\bar{\beta} = \beta(1 + d_1/\beta + \dots)$. When deriving this expression as a sum of bubble diagrams, one realizes that $t_1 = 0$ corresponds to momenta at the UV cutoff and $t_2 = 16\pi^2/33$ corresponds to the Landau pole.

Note that the expression is quite similar to the one plaquette integral discussed in the previous section. This could in principle be compared with what would be obtained from the probability distribution for one plaquette after integrating over all the other links. Note also that in his ITEP lectures, M. Shifman emphasizes that it is necessary to introduce the gluon condensate in order to keep t_2 low enough and regularize the perturbative series, exactly as we advocated in lattice gauge theory.

Expanding $(1-t\ 33/16\pi^2)^{-1-204/121}$ in powers of t and extending the integration range to ∞, one finds that at leading order that coefficients of the $\bar{\beta}^{-1}$ expansion grow like $(33/16\pi^2)^k\ \Gamma[k + 204/121]$. According to the rule of thumb, we should truncate the series at an order $4.79\bar{\beta} - 2.19$. For $\beta = 6$ and d_1 small this means an order of about 25. For $d_1 = -3$, this order is lowered to 12. Using Eq. (1) and shifting to the lattice β, we conclude that the error is proportional to

$$(\beta)^{204/121-1/2} e^{-(16\pi^2/33)\beta} \ . \tag{6}$$

Except for the power -1/2, this expression depends on β as the fourth power of the two-loop renormalization group invariant scale. In practice, the dependence on the power of β is quite weak in the region of β where an empirical envelope for the accuracy curves can be seen. In Fig. 2, the thick line has been drawn using the one-loop formula formula with an adjustable normalization constant $1.3 \times 10^{10} \times e^{-(16\pi^2/33)\beta}$ and looks like a decent envelope. When the order increases, will the accuracy curves reach the straight

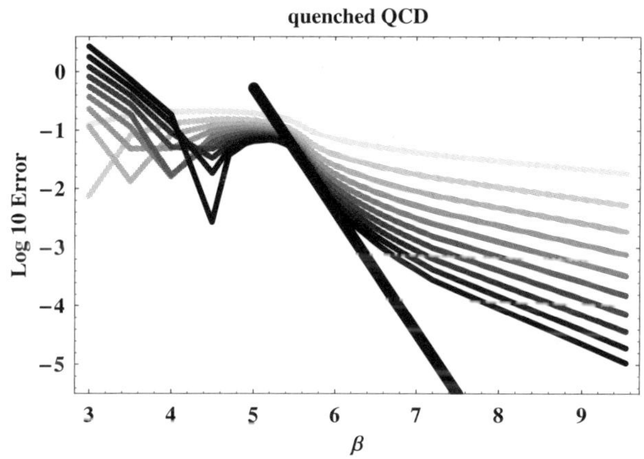

Fig. 2. Difference between the perturbative series at order 1 to 10 and the numerical value of the plaquette (in Log 10 scale). As the order increases, the curves get darker. The thick curve is discussed in the text.

line or will they maintain some curvature? Higher order extrapolations [11,10] favor the second possibility with a power 4 of the force scale. Lower power of this scale provide good fits of the accuracy curves at various orders. This question will be discussed into more detail in a forthcoming publication[18]

7. Conclusions

A better control of perturbative series of the standard model is necessary. A field cutoff drastically improves the asymptotic behavior of series. In simple examples, the field cutoff can be chosen to minimize the discrepancy with the uncut theory. There seems to be a connection between the crossover behavior of the perturbative coefficients and crossover behavior of the RG flows. Numerical methods need to be developed to implement large field or large action cutoffs. Analytic methods remain to be developed to estimate the various parameters determined empirically while studying the difference between the perturbative series and numerical values in lattice QCD.

References

1. F. Dyson, *Phys. Rev.* **85**, 631 (1952).
2. Z. Bern, A tour of the s-matrix: QCD, SYM and Supergravity, these Proceedings.
3. S. A. Larin, T. van Ritbergen and J. A. M. Vermaseren, *Phys. Lett.* **B320**, 159 (1994).
4. Y. Meurice, *Phys. Rev. Lett.* **88**, 141601 (2002).
5. L. Li and Y. Meurice, *J. Phys.* **A39**, 8681 (2006).
6. L. Li and Y. Meurice, *J. Phys. A* **38**, 8139 (2005).
7. B. Kessler, L. Li and Y. Meurice, *Phys. Rev.* **D69**, 045014 (2004).
8. L. Li and Y. Meurice, *Phys. Rev.* **D71**, 054509 (2005).
9. F. Di Renzo and L. Scorzato, *JHEP* **10**, 038 (2001).
10. P. E. L. Rakow, *PoS* **LAT2005**, 284 (2006).
11. R. Horsley, P. E. L. Rakow and G. Schierholz, *Nucl. Phys. Proc. Suppl.* **106**, 870 (2002).
12. L. Li and Y. Meurice, *Phys. Rev.* **D73**, 036006 (2006).
13. L. Li and Y. Meurice, *Phys. Rev. D* **71**, 016008 (2005).
14. A. Denbleyker and Y. Meurice, work in progress.
15. A. H. Mueller Talk given at Workshop on QCD: 20 Years Later, Aachen, Germany, 9-13 Jun 1992.
16. M. Shifman, *ITEP Lectures on Particle Physics and Field Theory*, 2001).
17. G. Burgio, F. Di Renzo, G. Marchesini and E. Onofri, *Phys. Lett.* **B422**, 219 (1998).
18. Y.Meurice, preprint in progress.

SECTION 7
GENERAL ASPECTS OF QCD AND GAUGE THEORIES

Convener M. Shifman

CONNECTING THE CHIRAL AND HEAVY QUARK LIMITS : FULL MASS DEPENDENCE OF FERMION DETERMINANT IN AN INSTANTON BACKGROUND

GERALD V. DUNNE

Department of Physics, University of Connecticut, Storrs, CT 06269, USA

This talk summarizes some recent progress in computing functional determinants in quantum field theory. The main example is the recent computation of the full quark mass dependence of the fermion determinant in an instanton background. Previously, only the small mass (chiral limit) and large mass (heavy quark limit) were known. Our results are found for any mass and interpolate smoothly between the two extremes. This is an example of a more general formalism whereby a functional determinant can be computed exactly when the relevant fluctuation operator can be separated into partial waves.

In this talk I summarize some recent progress [1; 2; 3] in computing one loop quantum vacuum polarization effects. Mathematically, this requires computing the determinant of a fluctuation operator, which typically describes one-loop physics or the quadratic fluctuations about a semiclassical solution. This is a difficult problem, but it is worth studying as it has important physical applications to computations of the effective action, the partition function or the free energy. It is also an interesting mathematical problem to learn about the spectral properties of partial differential operators. In this talk I focus mainly on the computation of fermion determinants in nontrivial background fields, which is an important challenge for both continuum and lattice quantum field theory. Explicit analytic results are known only for very simple backgrounds, and are essentially all variations on the original work of Heisenberg and Euler for constant background fields. For applications in quantum chromodynamics (QCD), an important class of background gauge fields are instanton fields, as these minimize the Euclidean gauge action within a given topological sector of the gauge field. Furthermore, instanton physics has many important phenomenological consequences. Thus, we are led to consider the fermion determinant, and the associated one-loop effective action, for quarks of mass m in an instanton background. Here, no exact results are known for the full mass dependence, although several terms have been computed analytically in the small mass [4; 5; 7] and large mass [6; 7] limits. Recently [1], with J. Hur, C. Lee and H. Min, we presented a new computation which is numerical, but essentially exact, that evaluates the one-loop effective action in a single instanton background, for any value of the quark mass (and for arbitrary instanton size parameter). The result is fully consistent with the known small and large mass limits, and interpolates smoothly between these limits. This could be of interest for the extrapolation of lattice results, obtained at unphysically large quark masses, to lower physical masses, and for various instanton-based phenomenology.

Our computational method can be adapted to many other determinant computations in which the background is sufficiently symmetric that the problem can be reduced to a product of one-dimensional radial determinants. While this is still a restricted set of background field configurations, it contains many examples of interest, the single instanton being one of the most obvious. The method is based on the Gelfand-Yaglom method for computing determinants of *ordinary* differential operators [8; 9; 10; 11]. But in higher-dimensional problems with *partial* differential operators, the naive generalization is divergent, even for simple separable problems where there is an infinite number of 1-D determinants to deal with. Physically, this divergence reflects the fact that in dimensions greater than one, one must confront renormalization. Our results can be viewed, in fact, as giving an extension of the Gelfand-Yaglom result to higher dimensional separable differential operators. More recently [3], these results have been obtained using the zeta function formalism for functional determinants.

First, recall that since the instanton background field is self-dual, we can deduce the fermion determinant from a computation of the determinant of the associated scalar Klein-Gordon operator. This is because self-dual gauge fields have the remarkable property that the Dirac and Klein-Gordon operators in such a background are isospectral [4; 12]. Thus [4; 7], the renormalized one-loop effective action of a Dirac spinor field of mass m (and isospin $\frac{1}{2}$), $\Gamma^F_{\rm ren}(A;m)$, is related to the corresponding scalar effective action, $\Gamma^S_{\rm ren}(A;m)$, for a complex scalar of mass m (and isospin $\frac{1}{2}$) by

$$\Gamma^F_{\rm ren}(A;m) = -2\,\Gamma^S_{\rm ren}(A;m) - \frac{1}{2}\ln\left(\frac{m^2}{\mu^2}\right) \quad , \tag{1}$$

where μ is the renormalization scale. The log term in (1) corresponds to the existence of a zero eigenvalue in the spectrum of the Dirac operator for a single instanton background. We consider an SU(2) single instanton background, in regular gauge [4].

The one-loop effective action must be regularized. We use a heavy regulator mass Λ, adapted to the Schwinger proper-time formalism, and later we relate this to dimensional regularization, as in the work of 't Hooft [4]. The renormalized effective action, in the minimal subtraction scheme, is defined as [4; 7]

$$\begin{aligned}\Gamma^S_{\rm ren}(A;m) &\equiv \lim_{\Lambda\to\infty}\left[\Gamma^S_\Lambda(A;m) - \frac{1}{12}\frac{1}{(4\pi)^2}\ln\left(\frac{\Lambda^2}{\mu^2}\right)\int d^4x\, {\rm tr}(F_{\mu\nu}F_{\mu\nu})\right]\\ &= \lim_{\Lambda\to\infty}\left[\Gamma^S_\Lambda(A;m) - \frac{1}{6}\ln\left(\frac{\Lambda}{\mu}\right)\right] \quad , \qquad (2)\end{aligned}$$

where we have subtracted the charge renormalization counter-term, and μ is the renormalization scale. By dimensional considerations, we introduce the modified scalar effective action $\tilde{\Gamma}^S_{\rm ren}(m\rho)$, which is a function of $m\rho$ only, defined by $\Gamma^S_{\rm ren}(A;m) = \tilde{\Gamma}^S_{\rm ren}(m\rho) + \frac{1}{6}\ln(\mu\rho)$, and concentrate on studying the $m\rho$ dependence of $\tilde{\Gamma}^S_{\rm ren}(m\rho)$. Without loss of generality we set the instanton scale $\rho = 1$ henceforth.

It is known from previous work that in the small mass [4; 5; 7] limit

$$\tilde{\Gamma}^S_{\rm ren}(m) \sim \alpha(1/2) + \frac{1}{2}\left(\ln m + \gamma - \ln 2\right)m^2 + \dots \quad , \qquad (3)$$

where $\alpha(1/2) = -\frac{5}{72} - 2\zeta'(-1) - \frac{1}{6}\ln 2 \simeq 0.145873$, and γ is Euler's constant. On the other hand, in the large mass [6; 7] limit,

$$\tilde{\Gamma}^S_{\rm ren}(m) \sim -\frac{\ln m}{6} - \frac{1}{75m^2} - \frac{17}{735m^4} + \frac{232}{2835m^6} - \frac{7916}{148225m^8} + \cdots \quad . \qquad (4)$$

The small mass expansion (3) follows from the known massless propagators in an instanton background. The large mass expansion (4) can be computed in several ways [6; 7], using heavy quark and/or heat kernel methods. These small and large mass limits are plotted as dashed curves in Figure 1. These dashed curves do not connect in the intermediate region $0.5 \leq m \leq 1$.

A beautiful result of Gelfand and Yaglom [8; 9; 10; 11] provides a spectacularly simple way to compute the determinant of a one-dimensional differential operator. For example, suppose $\mathcal{M}_1$ and $\mathcal{M}_2$ are two second order ordinary differential operators on the interval $r \in [0,\infty)$, with Dirichlet boundary conditions assumed. In practice we will choose $\mathcal{M}_1$ to be the operator of interest, and $\mathcal{M}_2$ to be the corresponding *free* operator. Then the ratio of the determinants is given by

$$\frac{\det \mathcal{M}_1}{\det \mathcal{M}_2} = \lim_{R\to\infty}\left(\frac{\psi_1(R)}{\psi_2(R)}\right) \quad , \qquad (5)$$

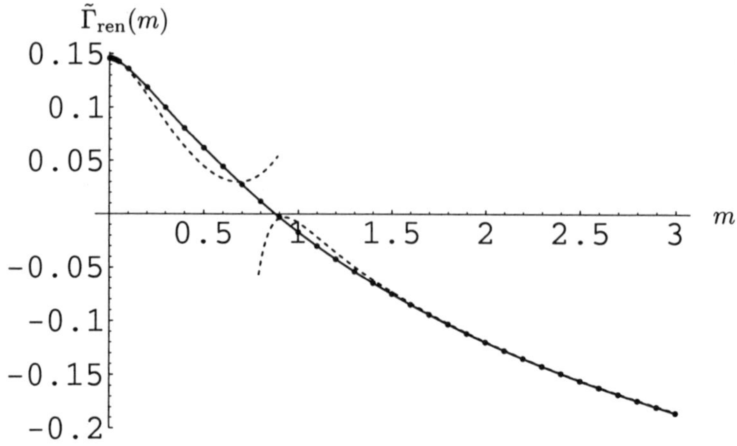

Fig. 1. Plot of the analytic small (3) and large (4) mass expansions (dashed lines) for $\tilde{\Gamma}^S_{\rm ren}(m)$. Note the gap in the region $0.5 \leq m \leq 1$, in which the two expansions do not match up. The solid line shows our exact result (10), valid for any mass.

where $\psi_i(r)$ (for $i = 1, 2$ labelling the two different differential operators, $\mathcal{M}_1$ and $\mathcal{M}_2$) satisfies the **initial value problem**

$$\mathcal{M}_i\, \psi_i(r) = 0, \qquad \text{with} \quad \psi_i(0) = 0 \quad \text{and} \quad \psi_i'(0) = 1 \quad . \tag{6}$$

Note that no direct information about the spectrum (either bound or continuum states, or phase shifts) is required in order to compute the determinant. All that is required is the integration of (6), which is straightforward to implement numerically.

Returning now to the instanton determinant problem, we can use the fact that the single instanton background has radial symmetry [4]. Thus, the regularized one-loop effective action can be reduced to a sum over partial waves of logarithms of determinants of radial ordinary differential operators. Each such radial determinant can be computed using the Gelfand-Yaglom result (5). However, this naïve sum over all partial waves is divergent. The physical challenge is to renormalize this divergent sum. This can be achieved as follows [1]. Split the l sum into two parts as :

$$\Gamma^S_\Lambda(A;m) = \sum_{l=0,\frac{1}{2},\ldots}^{L} \Gamma^S_{\Lambda,(l)}(A;m) + \sum_{l=L+\frac{1}{2}}^{\infty} \Gamma^S_{\Lambda,(l)}(A;m) \tag{7}$$

where L is a large but finite integer. The first sum involves low partial wave modes, and the second sum involves the high partial wave modes. We consider each sum separately, before recombining them to obtain our final expression (10).

The first sum in (7) is finite, so the cutoff Λ may be safely removed, and for each l the determinant can be computed using the Gelfand-Yaglom result (5). Including the degeneracy factor, the contribution from low l is:

$$\sum_{l=0,\frac{1}{2},\ldots}^{L} \Gamma_l^S(A;m) = \sum_{l=0,\frac{1}{2},\ldots}^{L} (2l+1)(2l+2)P(l) \quad , \tag{8}$$

where each $P(l)$ is computed numerically using (5). We find that $P(l) \sim O(1/l)$, which implies that the sum (8) has potentially divergent terms going like L^2, L and $\ln L$, as well as terms finite and vanishing for large L. Remarkably, we find below that these divergent terms are exactly canceled by the divergent large L terms found in the second sum in (7).

In the second sum in (7) we cannot take the large L and large Λ limits blindly, as each leads to a divergence. To treat these high l modes, we use radial WKB [13], which is a good approximation in precisely this limit. This means we can compute *analytically* the large Λ and large L divergences of the second sum in (7), using the WKB approximation for the corresponding determinants. This is a straightforward computation, the details of which can be found in [1; 13]. We find an analytic expression for large L:

$$\sum_{l=L+\frac{1}{2}}^{\infty} \Gamma_{\Lambda,(l)}^S(A;m) \sim \frac{1}{6}\ln\Lambda + 2L^2 + 4L - \left(\frac{1}{6} + \frac{m^2}{2}\right)\ln L \tag{9}$$
$$+ \left[\frac{127}{72} - \frac{1}{3}\ln 2 + \frac{m^2}{2} - m^2\ln 2 + \frac{m^2}{2}\ln m\right] + O\left(\frac{1}{L}\right)$$

It is important to identify the physical role of the various terms in (9). The first term is the expected logarithmic counterterm, $\frac{1}{6}\ln\Lambda$, which is subtracted as in (2) to define the renormalized action. Since both Λ and μ are measured in units of $1/\rho$, this also explains the origin of the $\frac{1}{6}\ln\mu$ term in the relation between $\Gamma_{\rm ren}^S$ and $\tilde{\Gamma}_{\rm ren}^S$. The next three terms give quadratic, linear and logarithmic divergences in L. These divergences exactly cancel against corresponding divergences in the first sum in (7), which were found in our numerical data. This holds for any mass m. It is a highly nontrivial check on this WKB computation that these divergent terms have the correct coefficients to cancel these divergences. Note that the $\ln L$ coefficient, and the finite term, are mass dependent, and these cancellations indeed occur for all m.

We now combine the numerical results for the low partial wave modes with the radial WKB results for the high partial wave modes to obtain the minimally subtracted renormalized effective action $\tilde{\Gamma}^{S}_{\text{ren}}(m)$ as

$$\tilde{\Gamma}^{S}_{\text{ren}}(m) = \lim_{L\to\infty}\left\{ \sum_{l=0,\frac{1}{2},\dots}^{L} (2l+1)(2l+2)P(l) + 2L^2 + 4L \right.$$
$$\left. -\left(\frac{1}{6}+\frac{m^2}{2}\right)\ln L + \left[\frac{127}{72} - \frac{1}{3}\ln 2 + \frac{m^2}{2} - m^2\ln 2 + \frac{m^2}{2}\ln m\right]\right\}. \quad (10)$$

In Figure 1 we plot these results for $\tilde{\Gamma}^{S}_{\text{ren}}(m)$, and compare them with the analytic small and large mass expansions in (3)-(4). The agreement is perfect. Thus, our *angular momentum cutoff regularization and renormalization* procedure of using the Gelfand-Yaglom result numerically for the low partial wave modes, and radial WKB for the exact large L behavior for the high partial wave modes, produces a finite renormalized effective action which interpolates precisely between the previously known small and large mass limits. This is interesting, as all three computations (small mass [4], large mass [6; 7], and all mass [1]) are independent. Our result also provides a simple interpolation formula [1] which can be translated into the instanton scale dependence at fixed quark mass, and this may be of phenomenological use.

This computational method is versatile and can be adapted to a large class of previously insoluble computations of one-loop functional determinants in nontrivial backgrounds in various dimensions of space-time, as long as the spectral problem of the given system can be reduced to that of partial waves. For example, it leads to a new computation [2], beyond the analytic thin-wall regime, of the false vacuum decay rate in a four dimensional self-interacting scalar field theory. Here the classical "bounce" solution is radial in 4d Euclidean space, so our technique is ideally suited to the computation of the fluctuation determinant about this bounce. An interesting feature here is the presence of a negative mode and zero modes. Interestingly, a different regularization method, also based on Gelfand-Yaglom, has previously been applied to false vacuum decay problems [14; 15]. The correspondence between these two approaches is explained in [3]. In other applications, a 2d instanton problem is solved in [16], and the fluctuation problem for gravitational false vacuum decay is considered in [17].

Finally, we note that our method provides an extension of the Gelfand-Yaglom result for ODE's to the case of separable PDE's. As noted by Forman [10] for the 2d radial disc problem, the naive extension does not work because the product over angular momenta diverges, even though the result for each angular momentum is finite. Physically, this is because in higher dimensions renormalization is required. Our method incorporates renormalization and yields a finite physical answer for the determinant of the separable partial differential operator.

Acknowledgments:

I thank my collaborators on this work, J. Hur, C. Lee and H. Min, and the US DOE for support through the grant DE-FG02-92ER40716.

References

1. G. V. Dunne, J. Hur, C. Lee and H. Min, Phys. Rev. Lett. **94**, 072001 (2005) [arXiv:hep-th/0410190]; Phys. Rev. D **71**, 085019 (2005) [arXiv:hep-th/0502087].
2. G. V. Dunne and H. Min, Phys. Rev. D **72**, 125004 (2005) [arXiv:hep-th/0511156].
3. G. V. Dunne and K. Kirsten, "Functional Determinants For Radial Operators," arXiv:hep-th/0607066.
4. G. 't Hooft, Phys. Rev. D **14**, 3432 (1976) [Err. D **18**, 2199 (1978)].
5. R. D. Carlitz and D. B. Creamer, Ann. Phys. **118**, 429 (1979).
6. V. A. Novikov, M. A. Shifman, A. I. Vainshtein and V. I. Zakharov, Fortsch. Phys. **32**, 585 (1985).
7. O. K. Kwon, C. Lee and H. Min, Phys. Rev. D **62**, 114022 (2000) [arXiv:hep-ph/0008028].
8. I. M. Gelfand and A. M. Yaglom, J. Math. Phys. **1**, 48 (1960).
9. S. Levit and U. Smilansky, Proc. Am. Math. Soc. **65**, 299 (1977).
10. R. Forman, Invent. Math. **88**, 447 (1987).
11. H. Kleinert, *Path Integrals in Quantum Mechanics, Statistics, Polymer Physics, and Financial Markets*, (World Scientific, Singapore, 2004).
12. A. S. Schwarz, Phys. Lett. B **67**, 172 (1977); L. S. Brown, R. D. Carlitz and C. Lee, Phys. Rev. D **16**, 417 (1977); R. Jackiw and C. Rebbi, Phys. Rev. D **16**, 1052 (1977).
13. G. V. Dunne, J. Hur, C. Lee and H. Min, Phys. Lett. B **600**, 302 (2004) [arXiv:hep-th/0407222].
14. G. Isidori, G. Ridolfi and A. Strumia, Nucl. Phys. B **609**, 387 (2001) [arXiv:hep-ph/0104016].
15. J. Baacke and G. Lavrelashvili, Phys. Rev. D **69**, 025009 (2004) [arXiv:hep-th/0307202].
16. Y. Burnier and M. Shaposhnikov, Phys. Rev. D **72**, 065011 (2005) [arXiv:hep-ph/0507130].
17. G. V. Dunne and Q. h. Wang, Phys. Rev. D **74**, 024018 (2006) [arXiv:hep-th/0605176].

DYNAMICS OF WILSON-LOOPS IN QCD

F. LENZ

Institute for Theoretical Physics, University of Erlangen Nürnberg
Erlangen 91058, Germany
E-mail: flenz@theorie3.physik.uni-erlangen.de

A phenomenological analysis of the distribution of Wilson-loops in SU(2) Yang-Mills theory in terms of diffusion processes on the group manifold is presented. The relation between diffusion and Casimir scaling is discussed. The results are summarized in an effective Wilson-loop action and consequences for correlation functions of large Wilson-loops are indicated.

1. Introduction

In the attempts to understand confinement, loop variables - either Polyakov or Wilson-loops - play an important role. Their expectation value distinguishes in SU(N) Yang-Mills theories the low temperature confined from the high temperature deconfined plasma phase. In this talk I will report on an investigation of the dynamics of the Wilson-loops and will introduce as the central quantity to these investigations the quantum mechanical distribution of a single Wilson-loop [1]. I will describe a phenomenological analysis of kinematics and dynamics of this quantity and will show that despite the complexity of Yang-Mills theories the distribution function of Wilson-loops actually exhibits a surprisingly simple dynamics. I will conclude with a discussion of the Wilson-loop effective action and an extension of these investigations to the dynamics of two Wilson-loops.

2. Wilson-loop distributions and diffusion

The object of our investigations is the distribution of $SU(2)$ Wilson-loops of fixed geometry

$$W = \frac{1}{2}\mathrm{tr}\, P\, \exp\{ig \oint_C dx^\mu A_\mu(x)\}\,.$$

For rectangular loops and for sufficiently large times, $T \gg R$, vacuum expectation values of Wilson-loops determine the interaction energy $V(R)$

of static quarks

$$\langle 0|W|0\rangle \sim e^{-T\,V(R)}\,, \qquad V(R) \sim \begin{cases} g^2 \dfrac{1}{R}\,, & R \to 0 \\ \sigma R\,, & R \to \infty\,. \end{cases}$$

Vacuum expectation values of Wilson-loops of sufficiently large area $\mathcal{A}$ satisfy an area law with string tension σ.
The distribution of Wilson-loops is defined by

$$p(\omega) = \langle 0|\delta\left(\omega - W\right)|0\rangle\,. \tag{1}$$

In Fig.1 are shown Wilson-loop distributions for rectangular loops of various sizes obtained in SU(2) lattice gauge theory. The strong concentration of the distributions for small size loops at $\omega \approx 1$ and the broad distributions for large size loops suggest a description in terms of diffusion with the diffusion time t being related to the size of the loops.

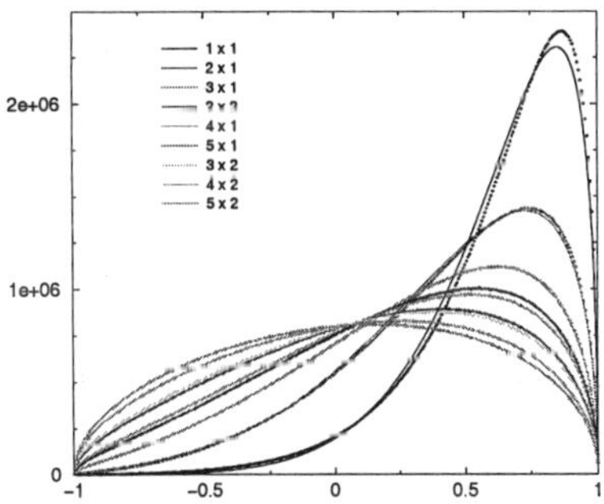

Fig. 1. Wilson-loop distribution for rectangular loops of different areas in SU(2) lattice gauge theory (32^4 lattice with lattice constant $a = 0.13$ fm). The solid lines are obtained from Eq. (3) with time t determined by the expectation values.

As a function of the size of the loop the value of the (untraced) Wilson-loop carries out a Brownian motion on the group manifold, S^3. The motion of an ensemble of these degrees of freedom is in turn described by a diffusion equation. Due to gauge invariance the diffusion is independent of the 2-nd polar angle and of the azimuthal angle on S^3. With

$$\omega = \cos\vartheta \tag{2}$$

the Laplace-Beltrami operator is given by

$$\Delta = \frac{1}{\sin^2\vartheta}\frac{d}{d\vartheta}\sin^2\vartheta\frac{d}{d\vartheta}$$

The diffusion equation on S^3 reads

$$(\frac{d}{dt} - \Delta)\, G(\vartheta, t) = \frac{1}{\sin^2 \vartheta}\, \delta(\vartheta)\delta(t)\,.$$

In terms of the eigenfunctions of the Laplace-Beltrami operator

$$\psi_n = \sqrt{\frac{2}{\pi}}\, \frac{\sin n\vartheta}{\sin \theta}, \quad -\Delta\, \psi_n = (n^2 - 1)\psi_n\,,$$

the solution of the diffusion equation is given by

$$G(\vartheta, t) = \frac{2}{\pi \sin \vartheta}\, \theta(t) \sum_{n=1}^{\infty} n \sin n\vartheta\, e^{-(n^2-1)t}\,. \tag{3}$$

In the limits of small and large loops the following closed expressions

$$G(\vartheta, t) \sim \begin{cases} t^{-3/2} \exp -\dfrac{\vartheta^2}{4t}\,, & t \to 0 \\[2ex] \dfrac{2}{\pi}\,, & t \to \infty \end{cases}$$

are obtained. For small loops the diffusion approaches the QED-distribution ($U(1)^3$) while for large loops the values of the Wilson-loops are distributed uniformly on S^3. For comparison with the results of the lattice gauge calculation the relation between "time" t and Wilson-loop size R has to be specified. The value of t is obtained by fitting for each size the Wilson-loop expectation value $\langle 0|W(R)|0\rangle = \exp(-3t)$ (see below Eq.(5)) to the numerical results. Fig.1 demonstrates the remarkable success of the diffusion model.

3. Casimir scaling and screening in confining phase

Given the Wilson-loop distribution (3), expectation values of other observables $\mathcal{O}(\vartheta)$

$$\langle \mathcal{O}(\vartheta) \rangle = \frac{2}{\pi} \int_0^{\pi} \sin \vartheta\; d\,\vartheta\; \mathcal{O}\,(\vartheta) \sum_{n=1}^{\infty} n \sin n\,\vartheta\; e^{-(n^2-1)t}$$

can be calculated. Here we consider Wilson-loops in higher representations of SU(2). The expectation value of the Wilson-loop operator in the 2j+1 representation

$$W_j(\vartheta) = \frac{1}{2j+1} \mathrm{tr}\;\exp 2i\vartheta \begin{pmatrix} -j & & \\ & \ddots & \\ & & j \end{pmatrix} \sim \psi_{2j+1}(\vartheta) \tag{4}$$

is given by

$$\langle W_j \rangle = e^{-4j(j+1)\cdot t} . \tag{5}$$

In the diffusion model, the logarithm of the Wilson-loop scales with the eigenvalues of the Laplace-Beltrami operator. In Fig. 2, the Wilson-loop operators (4) and the test of the scaling property (5) are shown.

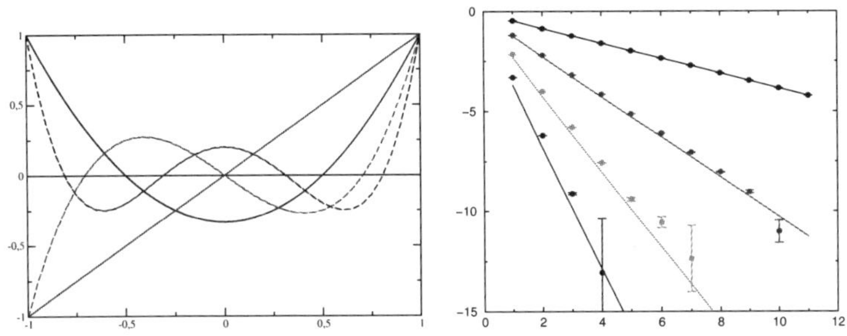

Fig. 2. Left: Wilson-loop operators $W_j(\vartheta)$ (Eq. (4)) in representations $j = 1/2,\ 1,\ 3/2,\ 2$ as a function of $\cos\vartheta$. Right: Logarithm of the corresponding Wilson-loop expectation values as function of the area (in lattice units) The straight lines are scaled by the ratios of the Casimir operators.

The scaling behavior (5) of the Wilson-loop vacuum expectation values is known as Casimir scaling and has been observed in lattice calculations for both SU(2) [2] and SU(3) [3] Yang-Mills theories. Casimir scaling is intimately related to diffusion in the group space. The Laplace-Beltrami operator on SU(N) coincides with the quadratic Casimir operator. Therefore exact scaling implies diffusion in group space and vice versa.

For asymptotically large loops, Casimir scaling must break down. Instead, complete screening of adjoint charges and screening of half-integer to fundamental charges must take place. Indications of a transition to the screening regime have been obtained in [4,5] and a an accurate determination of the breaking of the adjoint string has been achieved in [6].

The diffusion model provides a natural setting for incorporating complete screening of Wilson-loops in integer and partial screening in half-integer representations. Vanishing of asymptotically large half-integer loops is guaranteed if the diffusion process leads to a distribution which is symmetric around $\theta = \pi/2$. The half integer Wilson-loop operators are odd under reflections at this point (cf. Fig. 2). Vanishing of the large loops in integer representations on the other hand follows from the special property that the Wilson-loop operators coincide with the eigenfunctions of the Laplace-

Beltrami operator (cf. Eq. (4) and Fig. 2). This difference implies a distinction of the screening-properties of integer and half-integer loops in terms of a symmetry. A deviation from the simple diffusion model will in general destroy Casimir scaling and give rise to string breaking of Wilson-loops in integer representations. This will not be the case for half-integer loops provided the deviations from free diffusion are generated by a "drift term" $V(\vartheta)$ which respect the reflection symmetry i.e. if

$$V(\pi/2-\vartheta)=V(\vartheta-\pi/2)\,. \tag{6}$$

4. Wilson-loop effective action and approximate center symmetry

The success of the diffusion model suggests the following Wilson-loop effective action in the regime of Casimir scaling

$$S\sim\int dt\dot{\vartheta}^2\,,\quad 0\le\vartheta\le\pi\,.$$

S is also the action of a point particle in an infinite square well of size π. For large Wilson-loops, the area law requires $t\sim\mathcal{A}$. For circular loops of radius ρ, which we will consider in the following, the effective action therefore reads

$$S_{\text{eff}}[\vartheta]=\frac{3}{8\pi\sigma}\int\frac{\mathrm{d}\rho}{\rho}\left(\frac{\mathrm{d}\vartheta}{\mathrm{d}\rho}\right)^2\,. \tag{7}$$

To relate this effective action to Yang-Mills theory we choose the gauge where the azimuthal gauge field component is constant along the circle and diagonal in color space

$$A_\varphi^a=\delta_{a,3}\,\frac{1}{g\pi\rho}\,a(t,\rho,z)\,. \tag{8}$$

In this gauge, (circular) Wilson-loops $W(t,\rho,z)=\cos(a(t,\rho,z))$ are elementary degrees of freedom. The relevant part of the Yang-Mills action is given by [7]

$$S_{\text{YM}}=\int\rho\,\mathrm{d}\rho\,\mathrm{d}\varphi\,\mathrm{d}z\,\mathrm{d}t \tag{9}$$

$$\left\{\frac{1}{2\pi^2g^2\rho^2}\left[(\partial_\rho a)^2+(\partial_z a)^2+(\partial_t a)^2\right]+\mathcal{L}'(a,A_\perp)\right\}\,, \tag{10}$$

with $\mathcal{L}'$ denoting the dynamics of the other components $A_\perp$ of the gauge field and their interaction with the Wilson-loop variables a. The Lagrangian in (9) is symmetric under reflections of the Wilson-loop variables

$$\tilde{Z}: \frac{\pi}{2} - a(t,\rho,z) \to a(t,\rho,z) - \frac{\pi}{2} \tag{11}$$

if accompanied by appropriate changes in the other fields. This symmetry cannot be realized exactly. Unless

$$\rho \to 0 : a(t,\rho,z) \to 0\,, \tag{12}$$

a singular magnetic field $\sim \delta(\rho^2)$ of infinite action is generated. The boundary condition (12) prevents the symmetry $\tilde{Z}$ to be realized. Forced by this boundary condition the distribution of Wilson-loops for small ρ is concentrated around $a = 0$ (cf. Fig. 1). With increasing loop size, the influence of the boundary condition decreases and the symmetry is approximatively restored as in diffusion. Thus the Wilson-loop expectation value is a measure of the violation of the center symmetry (11)

$$\langle 0|W(t,\rho,z)|0\rangle \sim \exp(-\sigma\pi\rho^2)\,. \tag{13}$$

5. Correlation function of large Wilson-loops

The approximate realization of the center symmetry (11) for large Wilson-loops makes their dynamics similar to those of Polyakov loops. The center symmetry protects Polyakov loops from decaying and suppresses the decay of Wilson-loops (Eq. (13)). In turn this allows us to relate the corresponding correlation functions. With the correlator of Polyakov loops of circumference β given by

$$\langle 0|P_\beta(d)P_\beta(0)|0\rangle \sim \exp(-\sigma\beta d)\,,$$

we expect the following structure of the correlator of large Wilson-loops

$$\frac{\langle 0|\,W(d,\rho)\,W(0,\rho)|0\rangle}{\langle 0|\,W^2(0,\rho)|0\rangle} - 1 \approx d_1 e^{-\sigma 2\pi(\rho d - \rho^2)} + (c_1 e^{-m_{gb} d} + ..). \tag{14}$$

The subtraction on the left hand side eliminates the ground state contribution while the second term on the right hand side accounts for the $\exp(-\sigma\pi\rho^2)$ suppressed contributions (cf. Eq. (13)) from the decay of Wilson-loops into glueball states (m_{gb} - lowest glueball mass). The first term dominates for

$$2\pi\rho d - 2\pi\rho^2 < \frac{m_{gb} d}{\sigma}\,.$$

In this regime the correlation function of loops of different sizes is predicted to depend only on the combination $\rho d - \rho^2$. We have calculated the correlation function of quadratic Wilson-loops on the lattice [8]. In this case, we expect the correlation function of Wilson-loops of different sizes to depend only on the variable $\xi = 2Rd - R^2$ as long as $\xi < 0$. Fig.3 strongly supports the description of the correlation function (14) based on the approximate center symmetry (11). It displays a rather abrupt transition around ($\xi = 0$) induced by the decay of the Wilson-loops into glueball states. In essence this is the Gross-Ooguri transition [9] between two (gauge-) string configurations.

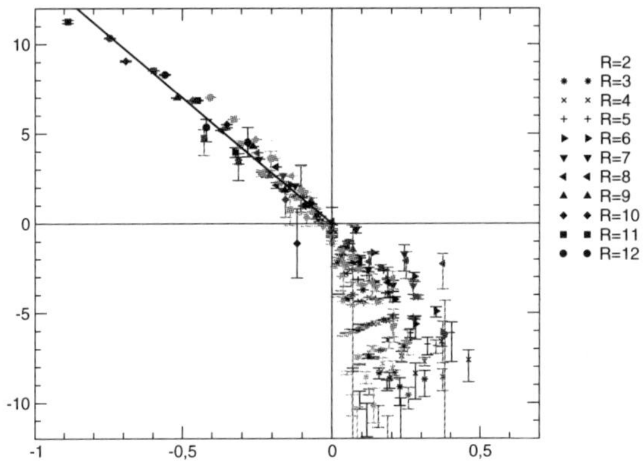

Fig. 3. Logarithm of correlation function of quadratic Wilson-loops ($R \times R$) as a function of ξ

6. Summary

In this talk I have presented a phenomenological analysis of the distribution of Wilson-loops in SU(2) Yang-Mills theory in terms of a diffusion process on the group manifold. I have discussed the relation to the phenomena of Casimir scaling and string breaking. The results have been summarized into an effective action which exhibits an approximative center symmetry and consequences for the correlation functions of large Wilson-loops have been studied.

References

1. A. M. Brzoska, F. Lenz, J. W. Negele, M. Thies, Phys.Rev. D71 (2005) 034008

2. J. Ambjørn, P. Olesen and C. Peterson, *Nucl. Phys.* B **240**, 189 (1984).
3. G.S. Bali, *Phys. Rev.* D **62**, 114503 (2000).
4. P. de Forcrand and O. Philipsen, *Phys. Lett.* B **475**, 280 (2000).
5. K. Kallio and H. Trottier *Phys. Rev.* D **66**, 034503 (2002).
6. S. Kratochvila, P. de Forcrand, *Nucl. Phys.* B. Nucl. B **671**, 103, (2003)
7. A. M. Brzoska, F. Lenz, to be published
8. A. M. Brzoska, F. Lenz, D. Steinbacher, to be published
9. D. J. Gross, H. Ooguri, *Phys.Rev.* **D58**,106002, (1998)

PERTURBATIVE CALCULATION OF THE VEV OF THE MONOPOLE CREATION OPERATOR

A. KOVNER

Physics Department, University of Connecticut,
2152 Hillside Road, Storrs, CT 06269-3046

A. KHVEDELIDZE and D. McMULLAN

School of Mathematics and Statistics, University of Plymouth,
Drake Circus, Plymouth PL4 8AA, UK

We set up the calculation of the expectation value of the monopole creation operator in the confining phase of the 4d Georgi-Glashow model. We find that in the leading order of the perturbation theory the VEV vanishes as a power of the volume of the system. This is in accordance with our naive expectation. We expect that nonperturbative effects will introduce an effective infrared cutoff on the calculation making the VEV finite.

Keywords: Confinement, monopoles

1. Introduction

Common lore has it that the condensation of magnetic monopoles is responsible for confinement in nonabelian gauge theories[1]. Numerous lattice investigations of the monopoles in pure Yang-Mills theories are available in the literature [2]. Nevertheless, there are many open questions regarding the status of magnetic monopoles in confining theories. The main problem with the monopole condensation scenario is that no gauge invariant definition of the monopole operator can be given in most theories of interest. Moreover, no gauge invariant observable corresponding to magnetic charge exists at least in the SU(2) pure Yang-Mills theory [3]. Thus serious doubts remain regarding the significance of the monopoles for confinement in pure gluodynamics.

On the other hand there undoubtedly exist theories where magnetic monopoles can be given a proper gauge invariant meaning. A prime example of such a theory is the Georgi-Glashow model in 3+1 dimensions. The model comprises of the SU(2) gauge field coupled to the adjoint Higgs

$$L = -\frac{1}{4}F_{\mu\nu}^2 + (D_\mu H)^2 - V(H^2)\,. \tag{1}$$

The theory is believed to have two phases as a function of the Higgs potential $V(H^2)$. In the Higgs phase, where the VEV of the Higgs field is nonvanishing, the magnetic monopoles exist in the spectrum as heavy particles. Closer to the boundary of the confining phase they become light, and it is natural to expect that they condense in the confining phase of the theory. The magnetic charge (which, as opposed to the pure Yang Mills theory, does exist here as a physical observable) is then expected to be spontaneously broken in the confining phase.

The main object of this work is to set up the perturbative calculation of the monopole creation operator in this theory in the leading perturbative order[4].

2. The monopole operator

We construct a continuum gauge invariant operator which creates the Coulombic magnetic field (without the Dirac string) of the monopole and has the correct commutation relation with the magnetic charge density:

$$[M(x), B_k(y)] = \frac{i}{g}\frac{(x-y)_k}{(x-y)^3}\,M(x)\,,$$
$$[M(x), j_0^M(y)] = \frac{4\pi}{g}\,\delta^3(x-y)\,M(x)\,. \tag{2}$$

The solution of this condition is given by (up to an unimportant for our prsent purposes prefactor)

$$M(x) = \exp\left(i\int \mathrm{d}^3y\,\lambda_i(x-y)\hat{H}^a(y)E_i^a(y)\right), \tag{3}$$

with λ_i the classical vector potential of a point-like Dirac monopole:

$$\lambda_i(x) = \frac{1}{g}\,\epsilon_{ij}\,\frac{r_\perp^j}{r_\perp^2}\,(\cos\theta - 1)\,. \tag{4}$$

An important property of the monopole operator is that it is not rotationally invariant even though the magnetic field created by it is spherically symmetric. This is directly related to the fact that the Dirac string is present in the definition of the monopole operator. The direction of the string has to be specified and it affects the commutators of M with local gauge invariant quantities, like the spacial components of electric current. We stress that these commutators are not hopelessly divergent at the position of the string, but merely depend on its orientation.

3. The VEV

Consider the path integral calculation of $\langle M\rangle$:

$$\langle M\rangle = \int D\hat{H} DA \exp\left(-S[A]\right) M[A]$$

$$\sim \int DA \exp\left(-\frac{1}{4}\int \mathrm{d}^4x \left(F^a_{\mu\nu} - f^a_{\mu\nu}\right)^2\right), \qquad (5)$$

with

$$f^a_{0i} = \lambda_i(x)\hat{H}^a\,\delta(x_4), \qquad f_{ij} = 0. \qquad (6)$$

We are interested in calculating the expectation value deep in the confining phase, where the Higgs field is very heavy. In the first approximation we therefore neglect the Higgs contribution to the action.

At weak coupling the path integral can be calculated in the steepest descent approximation. This becomes obvious once one realises that a simple rescaling of fields $A \to \frac{1}{g}A$, leads to the appearance of the factor $1/g^2$ in front of the action in eq. (5).

Our aim is therefore to find a classical configuration which minimises the action in eq. (5). First we cast the path integral expression into a more intuitively appealing form by changing variables

$$A^a_\mu(x) \to A^a_\mu(x) - \lambda_\mu(\vec{x})\theta(-x_4)\hat{H}^a, \qquad (7)$$

with $\lambda_0 = 0$ and the spatial components of λ_i given by eq. (4). The path integral for the calculation of VEV of the monopole operator becomes

$$\langle M\rangle = \int DA \exp\left\{-\frac{1}{4}\left(F^a_{\mu\nu} - \bar{f}_{\mu\nu}\,\hat{H}^a - \left(\lambda_\nu D_\mu \hat{H} - \lambda_\mu D_\nu \hat{H}\right)^a \theta(-x_4)\right)^2\right\}, \qquad (8)$$

where $\bar{f}_{\mu\nu}$ is the magnetic field of a point-like 'tHooft–Polyakov monopole at $x_4 < 0$. We find that the action is minimized by the axially symmetric "dandelion" configuration which comprises of a magnetic charge density coming in from minus infinite time and spreading into four dimensions at the spacetime point of the annihilation of the monopole.

$$\hat{H}^a = \frac{x^a}{r}, \qquad A^a_i = \frac{f(z,x)}{g}\,\epsilon_{aib}\frac{x^b}{r^2}, \qquad (9)$$

where

$$r^2 = x_1^2 + x_2^2 + x_3^2, \quad z = \frac{x_4}{r}, \qquad x = \cos\theta. \qquad (10)$$

By numerically minimizing the action on a set of functions f of the form

$$1 - f(z, x, \alpha, \beta) = \rho(z, \alpha)(1 + x)^{1-\rho(z,\beta)} . \tag{11}$$

we find the minimum at $\alpha \approx 1.35$ and $\beta \approx 2.65$. The classical action of the dandelion is $S_{\text{dandelion}} = 2.17\frac{\pi}{g^2} \ln(\Lambda L)$ with Λ and L - the ultraviolet and infrared cutoffs.

Thus we conclude that in the leading order in perturbation theory the expectation value of the monopole operator is

$$\langle M \rangle = \exp\left(- \frac{\pi c}{g^2} \ln(\Lambda L)\right) \tag{12}$$

with constant $c \approx 2.17$.

Perturbatively we find that the VEV vanishes in the infinite volume. However our result eq. (12) is an intrinsically perturbative one, and as such we certainly do not expect it to stand beyond perturbation theory. Nonperturbative contributions will provide an infrared cutoff. It is thus very likely that the $\langle M \rangle \neq 0$ in the confining phase.

References

1. G. 'tHooft, Nucl. Phys. B190 (1981) 455.
2. M.N. Chernodub, Katsuya Ishiguro and Tsuneo Suzuki, e-Print Archive: hep-lat/0503027, Phys. Rev. D69 (2004) 094508;
   V.G. Bornyakov, P.Yu. Boyko, M.I. Polikarpov and V.I. Zakharov, Nucl. Phys. B672 (2003) 222.
3. A. Kovner, M. Lavelle and D. McMullan, JHEP12 (2002) 045.
4. A. Khvedelidze, A. Kovner and D. McMullan, e-Print Archive: hep-th/0512142;

TOWARDS THE REGGEON FIELD THEORY IN QCD

M. LUBLINSKY

Physics Department, University of Connecticut,
Storrs, CT 06269, USA
E-mail: lublinsky@phys.uconn.edu

We review some recent attempts to understand the high energy limit of QCD which treats projectile-target scattering in a symmetric manner. From theoretical point of view the problem is tightly related to inclusion of Pomeron loops in the evolution kernel. The ultimate goal is to derive a complete effective theory of QCD at high energies, known as Reggeon Field Theory.

Keywords: QCD, High energy, Reggeon Field Theory, JIMWLK

1. High energy scattering

Thanks to the diluteness of perturbative projectile, the high energy limit of DIS is understood relatively well. Recently the main theoretical effort has been shifted towards high energy scattering of dense objects, which could be hadrons and/or nuclei. The theory has phenomenological implications for the LHC, RHIC and TeVatron. The physics of DIS is linear in projectile's density but non-linear in the density of a dense target. At high energies we resum Pomeron fan diagrams using GLR-type non-linear evolution equations [1]. The modern version of the GLR equation is known as the JIMWLK evolution [2,3]. In contrast to DIS, in hadron-hadron collisions physics is non-linear both in projectile's and target's densities. We have to resum both up and down type Pomeron fan diagrams as well as Pomeron loops.

The S-matrix for a collision of a projectile $|P\rangle$ described by a parton density ($\rho^p = \rho^-; k^- > \Lambda$) with a target $\langle T|$ ($\rho^t = \rho^+; k^+ > \Lambda$) is given by

$$S(Y) = \langle T \langle P| \; \hat{s}(\rho^t, \rho^p) \; |P\rangle T\rangle . \tag{1}$$

Here $\hat{s}$ is an S-matrix operator acting on a direct product of two Hilbert spaces. It is convenient to define a projectile averaged s-matrix Σ^p:

$$\Sigma^p_{Y-Y_0}(\rho^t) = \langle P|\, \hat{s}(\rho^t, \rho^p)\, |P\rangle = \int D\rho^p\ \hat{s}(\rho^t, \rho^p)\ W^p_{Y-Y_0}[\rho^p]. \quad (2)$$

In the second equality we use a path integral representation for the matrix element. The weight functional W^p can be though of as a density matrix. Using the same representation for the target average we write

$$S(Y) = \int D\rho^t\ \Sigma^p_{Y-Y_0}[\rho^t]\ W^t_{Y_0}[\rho^t]. \quad (3)$$

In Eq. (3) we restored the rapidity index. The target is assumed to be evolved to rapidity Y_0 while the rest of the total rapidity Y is carried by the projectile. Obviously, the S-matrix must be independent of the choice of Y_0 (Lorentz invariance).

We are interested in the energy evolution of S. To this goal we can boost the projectile. Being accelerated the valence partons in the projectile emit extra gluons and the wavefunction changes $|P\rangle_Y \rightarrow |P\rangle_{Y+\delta Y}$. As a result we obtain a Wilson-type evolution equation

$$\frac{d\Sigma^p}{dY} = H^{HE}_{\Sigma}\ \Sigma^p. \quad (4)$$

Here H^{HE}_{Σ} is an effective Hamiltonian which should be found from the underlying QCD dynamics.

Alternatively, we could boost the target, which would result in emission of extra gluons in the target wavefunction: $|T\rangle_Y \rightarrow |T\rangle_{Y+\delta Y}$. This leads to an evolution equation for the target weightfunctional

$$\frac{dW^t}{dY} = H^{HE}_{W}\ W^t. \quad (5)$$

The Lorentz invariance implies $H^{HE}_{\Sigma} = H^{HE}_{W}$.

Finally the S-matrix evolution is given by universal Hamiltonian H^{RFT} which can act either to the left or to the right:

$$\frac{dS}{dY} = \int D\rho^t\ \Sigma^p_{Y-Y_0}[\rho^t]\ H^{RFT}[\rho^t, \delta/\delta\rho^t]\ W^t_{Y_0}[\rho^t]. \quad (6)$$

It is our prime goal to find the Hamiltonian of the Reggeon Field Theory H^{RFT}. We know H^{RFT} in two limits: dilute and dense

$$H^{KLWMIJ} = H^{RFT}(\rho \rightarrow 0); \qquad H^{JIMWLK} = H^{RFT}(\rho \rightarrow \infty). \quad (7)$$

The JIMWLK Hamiltonian was derived in Refs. [2,3] and the KLWMIJ Hamiltonian was derived in Ref. [4]. By interpolating between these two limits we built a model of evolution with Pomeron loops.

1.1. *KLWMIJ Hamiltonian (dilute limit)*

The KLWMIJ Hamiltonian can be derived by considering an expectation value of an operator $\mathcal{O}$ in the projectile's wave function.

$$\langle \mathcal{O} \rangle_Y \;\equiv\; \langle P|\, \hat{O}[\rho^p]\, |P\rangle_Y \;=\; \int D\rho^p\;\, \mathcal{O}[\rho^p]\;\, W^p_Y[\rho^p]\,. \tag{8}$$

The KLWMIJ Hamiltonian is extracted from the evolution of this matrix element:

$$\frac{d\,\langle \mathcal{O} \rangle_Y}{d\,Y} \;=\; \int D\rho^p\;\, \mathcal{O}[\rho^p]\;\, H^{KLWMIJ}\;\, W^p_Y[\rho^p]\,. \tag{9}$$

In order to determine the evolution law we have to boost the projectile's wave function (Fig. 1,a). In the dilute limit only one gluon is emitted as

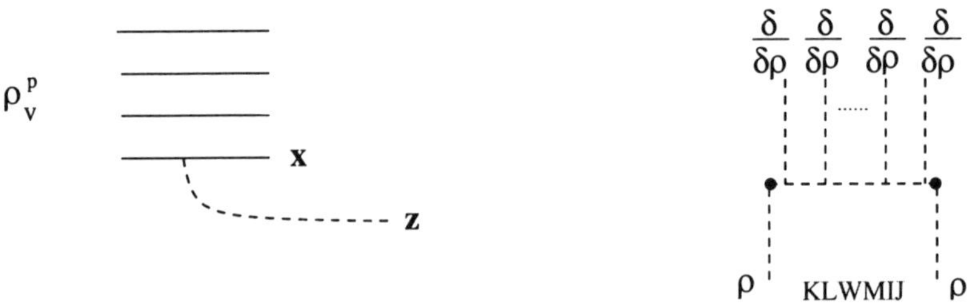

Fig. 1.

a result of the boost and the emission amplitude is linear in the number of emitters. This amplitude is given by the Weizsaker-Williams field of the emitted gluon

$$b^a_i(z) \;=\; \frac{g}{2\,\pi}\int d^2x\,\frac{(x\;-\;z)_i}{(x\;-\;z)^2}\;\rho^{p\,a}(x)\,. \tag{10}$$

In addition, due to appearance of extra gluon the charge density is shifted by the charge of this gluon

$$\rho^p \;\longrightarrow\; \rho^p \;+\; T^a\,; \qquad\qquad \hat{O}[\rho^p] \;\longrightarrow\; \hat{O}[\rho^p \;+\; T^a] \tag{11}$$

This shift can be accounted for by introducing a Dual Wilson line, which acts as a density shift operator

$$R(z) \;=\; P\,exp\left\{\int dx^-\;T^a\,\frac{\delta}{\delta\rho^a(z,\,x^-)}\right\}; \qquad R(z)\,\rho^a(z) \;=\; \rho^a(z)\;+\;T^a\,. \tag{12}$$

Taking into account the emission amplitude and absorption in the complex conjugate amplitude we arrive at the KLWMIJ Hamiltonian (Fig. 1,b)

$$H^{KLWMIJ}[\rho, \delta/\delta\rho] = \int_z b_i^a(z)\,[1 - R(z)]^{ab}\; b_i^b(z)\,. \qquad (13)$$

1.2. *JIMWLK Hamiltonian and beyond (dense limit)*

The JIMWLK Hamiltonian has a very similar structure (Fig. 2)

$$H^{JIMWLK}\left[\alpha^t, \frac{\delta}{\delta\alpha^t}\right] = \int_z b_i\left[\frac{\delta}{\delta\alpha^t}\right]\,[1 - S(z)]\, b_i\left[\frac{\delta}{\delta\alpha^t}\right]. \qquad (14)$$

Here S stands for the eikonal scattering matrix for projectile's gluon in the

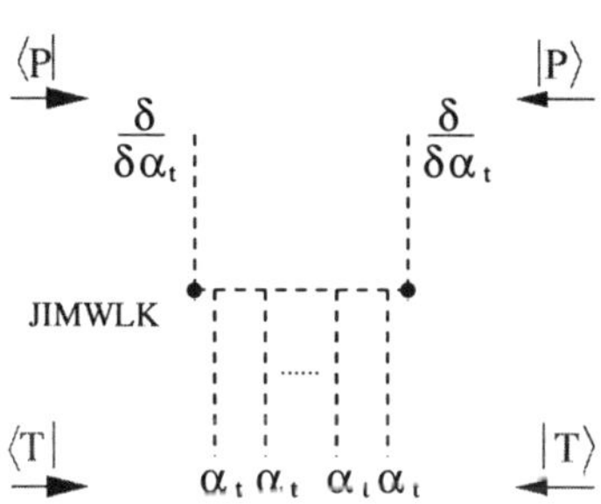

Fig. 2.

(strong) classical external field $\alpha^t \equiv A^+$ created by the target

$$S^{ab}(z) = P\,exp\left\{i\int dx^-\,\alpha^t(z, x^-)\right\}^{ab}. \qquad (15)$$

The field α^t is a solution of the light cone gauge Yang-Mills equations with source. In a simplified manner we write it as a Laplace equation

$$\Delta\,\alpha^t(z)\; '' = ''\;\rho^t(z)\,. \qquad (16)$$

Some corrections to the JIMWLK/KLWMIJ Hamiltonians can be accounted for by considering effects associated with a coherent emission of a single gluon. Instead of the Weizsaker-Williams field obtained for a single emitter we can solve a complete set of Yang-Mills equations for the field b created by the charges ρ_p

$$D_i[b]\,b_i^a = \rho_p^a\,.$$

This gives b as a non-linear functional of $\rho^p = \rho_v \geq 1$. While preserving

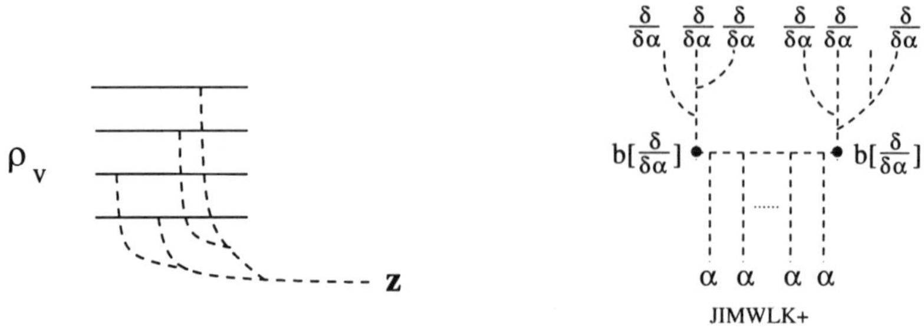

Fig. 3.

the very same structure of the Hamiltonian [5]

$$H^{JIMWLK+} \;=\; \int_z b_i \;[1 \;-\; S(z)]\; b_i \tag{17}$$

we account for non-linear effects in the projectile (Pomeron loops) as shown in Fig. 3. Similar results were lated obtained in Refs. [6,7].

1.3. *DDD - Dense Dilute Duality*

If we write again the JIMWLK and KLWMIJ Hamiltonians

$$\begin{aligned} H^{JIMWLK} &= \alpha_s \int_{x,y,z} \frac{(z-x)_i(z-y)_i}{(z-x)^2(z-y)^2} \frac{\delta}{\delta\alpha^a(x)} \left[1 \;-\; S(z)\right]^{ab} \frac{\delta}{\delta\alpha^b(y)} \\ H^{KLWMIJ} &= \alpha_s \int_{x,y,z} \frac{(z-x)_i(z-y)_i}{(z-x)^2(z-y)^2} \rho^a(x) \left[1 \;-\; R(z)\right]^{ab} \rho^b(y) \end{aligned} \tag{18}$$

we notice that they are mapped one into another by the symmetry transformation, which we named Dense Dilute Duality (DDD) [8]

$$i\,\alpha \rightarrow \frac{\delta}{\delta\,\rho}; \qquad\qquad \frac{\delta}{\delta\alpha} \rightarrow i\,\rho \qquad\qquad S \rightarrow R\,. \tag{19}$$

Furthermore, we were able to prove [8] that under the following three conditions the full and yet unknown RFT Hamiltonian must be self-dual under DDD

$$H^{RFT}(i\,\alpha\;,\delta/\delta\,\alpha) \;=\; H^{RFT}(\delta/\delta\,\rho,\; i\,\rho) \tag{20}$$

The required conditions are: (a) Lorentz Invariance; (b) Eikonal Approximation, (c) Projectile - Target Democracy. The last conditions means that the RFT Hamiltonian is universal. We speculate that the self-duality condition for H^{RFT} is equivalent to the t-channel unitarity of the scattering theory.

2. Reggeon field theory in QCD

In this section we summarize our knowledge and expectations regarding the full structure of RFT in QCD. The discussion is based on Ref. [9].

• RFT can be formulated as a Hamiltonian (2+1) dimensional interacting non-local field theory.

• The basic "quantum Reggeon field" is the unitary matrix R (S).

• Symmetry: DDD.

• The spectrum of the theory has two zero energy degenerate vacua ("Yang" and "Yin"), in which the DDD is spontaneously broken.

• Two degenerate towers of excitations are present. Excitations above "white" Yang vacuum are gluons while the excitation above "black" Yin vacuum have natural interpretation in terms of "holes".

The DDD symmetry and excitations are summarized in Fig. 4.

Fig. 4.

• The RFT is expected to have more symmetries. In particular the JIMWLK/KLWMIJ Hamiltonians each have $SU(N) \otimes SU(N)$ symmetries and at least one vector subgroup $SU_V(N)$ is expected to be present in the full RFT. Several Z_2 corresponding to C, P parities and signature will be also present. Though the RFT Hamiltonian is not expected to be fully conformal invariant, it might act as 2-d conformal invariant on a part of the Hilbert space.

3. Summary and outlook

- Some progress has been achieved in understanding high energy limit of QCD. We have derived new (KLWMIJ+) and extended previously known (JIMWLK+) evolution equations. A new concept of DDD and the Self-Duality theorem appear to be powerful tools as well as an important constraint on further extensions.
- We are in a quest for a complete QCD based RFT. The RFT must be Self-Dual under DDD transformation, which should pro-

vide both s- and t- channel unitarity. We are searching for additional symmetries like conformal symmetry which at the end should help us to establish integrability of the theory.

- Plenty of phenomenological applications relevant for present and near future colliders (LHC, HERA, RHIC, TeVatron) are waiting to be addressed in our framework.

Acknowledgments

All of the results presented were obtained in collaboration with Alex Kovner. I am most grateful to Alex for being a wonderful collaborator.

References

1. L.V. Gribov, E. Levin and M. Ryskin, *Phys. Rep.* **100** 1 (1983).
2. J. Jalilian Marian, A. Kovner, A.Leonidov and H. Weigert, *Nucl. Phys.***B504** 415 (1997); *Phys. Rev.* **D59** 014014 (1999); J. Jalilian Marian, A. Kovner and H. Weigert, *Phys. Rev.***D59** 014015 (1999); A. Kovner and J.G. Milhano, *Phys. Rev.* **D61** 014012 (2000) . A. Kovner, J.G. Milhano and H. Weigert, *Phys.Rev.* **D62** 114005 (2000); H. Weigert, *Nucl.Phys.* **A 703** (2002) 823.
3. E.Iancu, A. Leonidov and L. McLerran, *Nucl. Phys.* **A 692** (2001) 583; Phys. Lett. **B 510** (2001) 133; E. Ferreiro, E. Iancu, A. Leonidov, L. McLerran; *Nucl. Phys.***A703** (2002) 489.
4. A. Kovner and M.L., *Phys. Rev. D* **71** 085004 (2005).
5. A. Kovner and M.L., *JHEP* **0503** 001 (2005).
6. Y. Hatta, E. Iancu, L. McLerran, A. Stasto and D.N. Triantafyllopoulos, *Nucl. Phys. A* **764** 423 (2006).
7. I. Balitsky, *Phys. Rev. D* **72** 074027 (2005).
8. A. Kovner and M.L. *Phys. Rev. Lett.* **94** 181603 (2005).
9. A. Kovner and M.L., hep-ph/0512316, hep-ph/0604085.

LARGE N_c ORBIFOLD AND ORIENTIFOLD EQUIVALENCES: OLD AND NEW

MITHAT ÜNSAL

Department of Physics, Boston University,
590 Commonwealth Ave, Boston, MA 02215
E-mail:unsal@buphy.bu.edu

We provide an overview of the progress in large N_c equivalences among theories related to each other via an orbifold or orientifold projection. We emphasize the importance of symmetry realizations. The examples of (in)equivalences based on both orbifold and orientifold projections are given. Applications are chosen from matrix models, lattice gauge theory, $\mathcal{N} = 1$ supersymmetric gauge theories and their orbifold and orientifold projections.

Keywords: Large N_c equivalences, Spontaneous Symmetry Breaking

1. Orbifold and orientifold equivalences

In this talk, I will try to clarify issues on large N_c orbifold/orientifold equivalences. The equivalence is among pairs of theories which are related to each other via projections by discrete symmetries, called orbifold and orientifold projections.

First, I want to explain what I mean by large N_c equivalence. The large N_c equivalence of two theories (quantum field theory, or statistical mechanical system), in their neutral sectors (defined below), is a statement about the equivalence of *the theories*, not about a subclass of their perturbative Feynman diagrams, such as *planar diagrams*. If the theories in their neutral sectors coincide, so will their associated planar diagrams. The reverse statement is not necessarily true. The equivalence of theories, as discussed below, only relies on symmetry realizations. In particular, in the strongly coupled gauge theories, the perturbation theory, by construction, is blind to the spontaneously broken symmetries and can not be a good guide.

One of the main goals of this talk is to place the orbifold and orientifold equivalence on the same logical footing. In literature[1,2] it is often argued that there is a *conceptual* difference between the two, and it is emphasized that there is no twisted sector associated with orientifold equivalence unlike the orbifold case. Therefore, it is concluded that the orientifold equivalence does not fail due to the absence of the twisted sector, whereas the orbifold equivalence does fail in general because of the presence of the twisted sector. This is usually presented as a superiority of orientifolds over orbifolds.

Here, I will argue that there is a twisted sector associated with orientifolds as well, and furthermore, I will show that the orbifold and orientifold equivalences are conceptually on the same footing. Neither is superior. Both have twisted sectors, and the equivalence in *both* cases may fail or hold depending on symmetry realizations. The twisted sectors are not bad or harmful by any means, they are just nice order parameters for probing whether a symmetry is spontaneously broken or not. I will provide examples of equivalences and inequivalences arising from both orbifold and orientifold projections.

The usefulness of the large N_c equivalence is all about taking advantage of the better understood partners among pairs of theories. An old paradigm of such correspondence is Eguchi-Kawai (EK) equivalence, where a zero dimensional matrix model (in strongly coupled phase) is equivalent to a d-dimensional pure lattice gauge theory [3]. Obviously, the easier one to handle is the matrix model. A more recent manifestation of this idea is pairs of theories one of which is supersymmetric (and better understood) and the other is not supersymmetric and seems less under control.

Statement of equivalence, symmetries, order parameters

We consider a parent quantum field theory and a daughter theory obtained by an orbifold projection. The operators in the parent theory are charged under a discrete symmetry H_p used in projection. Similarly, operators in daughter theory are charged under symmetries, such as discrete translation symmetry H_d of quiver daughter theory. The dynamical statement of the Non-Perturbative Orbifold/Orientifold Equivalence (NPOE) is [4–6] that as long as

- the symmetry H_p used in the orbifold/orientifold projection is not spontaneously broken in parent theory,
- the symmetry H_d defining the neutral sector in daughter theory is not spontaneously broken,

the dynamics of the neutral sectors in parent and daughter theories coincide in the large N_c limit. Here, "coincide" means a well-defined, simple kinematic mapping of parameters and correlators which are even visible in perturbation theory. The neutral sectors are the ones that are invariant under the actions of H_p and H_d, in the parent and in the daughter theory, respectively. Clearly, this is not an equivalence of the theories in full. However, there is still much to learn in the neutral sectors of these theories. The equivalence does not say anything about charged (twisted) operators in these pairs of theories. Yet, the twisted operators are essential, since they are the order parameters of the symmetry realizations. If either H_p or H_d is spontaneously broken, the dynamics in the neutral sectors do not coincide, and there is no equivalence [4,5].

The equivalence owes its existence to the classical nature [7] of the large N_c limit. The quantum fluctuations are suppressed as $N_c \to \infty$, and consequently, the expectation values of products of operators factorize. The finite N_c corrections are typically $1/N_c^2$ for orbifolds and $1/N_c$ for orientifolds.

The criteria for the symmetry realizations can only be obtained by using non-perturbative techniques. This requires, first and foremost, a nonperturbative definition of the theory (i.e, lattice). We have learned the importance of the symmetry realizations from the lattice regularized loop equations, and the large N coherent state methods. The proof is done for a very large class of of Euclidean gauge theories and their orbifold projections.

Similarly, the orientifold projection is a Z_2 projection involving charge conjugation symmetry C. Either H_p *or* H_d involves or is C. An example of orientifold equivalence for the pure lattice gauge theory among $U(N)$, $SO(N)$ and $Sp(N)$ was proven[8] long time ago by using the loop equations.

EK and inverse EK reduction as an orbifold projection

Let me start with inverse EK, which is a new interpretation of the old story. [a]. Consider a $U(k^d N_c)$ matrix model of d unitary matrices U_μ, where $\mu = 1, \ldots d$ and action $S_p = \beta \sum_{\mu \neq \nu} \mathrm{Tr} U_\mu U_\nu U_\mu^\dagger U_\nu^\dagger + h.c.$. The model possesses a global $U(1)^d$ symmetry under which the matrices transform as $U_\mu \to e^{i\alpha_\nu \delta_{\mu\nu}} U_\mu$. We can perform a suitable $(Z_k)^d$ orbifold projection to generate a lattice gauge theory on a lattice with k^d sites. [b] The neutral operators in the parent are single trace operators (Wilson "loops"), neutral under $U(1)^d$, and the ones in daughter are single trace $(Z_k)^d$ (discrete translations) invariant observables on lattice. There is a one to one map among the neutral sectors.

What are the order parameters? In the parent, these are Polyakov "loops" charged under $U(1)^d$ and transforming as $\mathrm{Tr} U_\mu \to e^{i\alpha_\nu \delta_{\mu\nu}} \mathrm{Tr} U_\mu$. In the daughter, the order parameters are charged under lattice translation,

[a]The old EK reduction can be formulated as an orbifold projection as well. See Ref.6

[b]The orbifold projection of a theory can be pictured as a "quiver" or "moose" diagram or as a "theory space". There are various occasions, as above, in which the theory space, or quiver matrix theory is literally a Euclidean spacetime lattice.

i.e, they carry momentum. One example is $\mathrm{Tr}U_{\mu\nu,\mathbf{p}} = \sum_{\mathbf{n}\in Z_k^d} e^{\frac{i2\pi}{k}\mathbf{pn}}\mathrm{Tr}U_{\mu\nu,\mathbf{n}}$ where $U_{\mu\nu,\mathbf{n}} \equiv U_{\mu,\mathbf{n}}U_{\nu,\mathbf{n}+\mathbf{e}_\mu}U^\dagger_{\mu,\mathbf{n}+\mathbf{e}_\nu}U^\dagger_{\nu,\mathbf{n}}$. As usual, the question of equivalence boils down to the symmetry realizations. In the daughter, this amounts to the translation symmetry of lattice gauge theory, which does not break spontaneously as there are no competing effects. [c]

How about the parent matrix theory? There are indeed competing effects and spontaneous symmetry breaking takes place. The Haar measure, in the eigenvalue basis, gives a repulsive interaction among eigenvalues and the action provides an attractive interaction. In the strongly coupled phase $\beta < \beta_c$, the measure (or entropy) dominates, therefore the center symmetry is unbroken and there is equivalence. In the weak coupling phase $\beta \geq \beta_c$, the action dominates. Because of the sufficiently attractive interaction, eigenvalues clump, and Polyakov loops acquire a vacuum expectation value. In this phase, the center symmetry breaks down spontaneously[9], and there is no equivalence.

Generalization of these ideas tells us that the physics of large N_c gauge theories is independent of the volume, for sufficiently large volume. This is useful because it says, at the sufficiently large N_c, the finite volume effects are not present and one can use smaller lattices to extract information about the gauge theory. The recent results of Ref.10 are nice manifestations of these ideas. For a fuller discussion, see Ref.6

$\mathcal{N} = 1$ SYM and its orbifold projections

The $\mathcal{N} = 1$ SYM and its Z_k orbifold projections had been studied in the literature [11,1,12–14]. This is an example in which both parent and daughter theories are asymptotically free and confining gauge theories, and some daughters are non-supersymmetric. If such an equivalence holds, it would be a powerful tool for gaining non-perturbative data for non-susy theories[15–17].

The parent $U(kN_c)$ $\mathcal{N} = 1$ SYM theory has a non-anomalous discrete chiral symmetry Z_{2kN_c}. The Z_k projections for $k > 2$ can yield $U(N_c)^k$ chiral gauge theories with bifundamental fermions. The large N_c equivalence between parent and daughter theories does fail because of the spontaneous breaking of the discrete chiral symmetry Z_{2kN_c} down to Z_2 by the order parameter $\langle\lambda\lambda\rangle$ (the chiral condensate) in the parent theory[13].

Therefore, we are left with the $k = 2$ case only, where the parent is

[c] In literature, there are broad statements such as "the twisted sectors makes the non-perturbative equivalence invalid". A twisted operator acquiring a vacuum expectation value, in this example, would mean spontaneous breaking of translation symmetry on lattice, and would be detrimental for lattice gauge theory. Thankfully, this is not the case.

$U(2N)$ $\mathcal{N}=1$ SYM and the daughter is a $U(N)\times U(N)$ theory with bifundamental fermions. Ref.1, 11 argued that large N_c equivalence fails in this case, based on the mismatch of topological susceptibility, and on instanton zero modes when the theory is compactified on T^4 or $R^3\times S^1$. Based on this and related examples, the sudden death of nonperturbative orbifold equivalence was declared. However, in Ref.13, we showed that previous assertions of large N_c inequivalence, formulated on R^4, were based on incorrect mappings of vacuum energies, theta angles, or connected correlators between the two theories. With the correct identifications, there is no sign of any inconsistency.

More recently, "the mismatch" of the gauge and gravitational contribution to the chiral anomaly, as well as trace anomalies in the two theories were considered as evidence for failure of large N_c equivalence. If one assumes perturbative equivalence between planar diagrams[18], how can this be possible at all? The chiral and trace anomalies depend only on short distance perturbative physics (and not on the long distance, Z_2 symmetry realizations of the daughter theory). In fact, as shown in Ref.13, with the correct mapping of correlators, anomaly relations in the parent are mapped onto the correct anomaly equations in the daughter. I cannot do justice to the full discussion of $\mathcal{N}=1$ SYM and its Z_2 orbifold here. I refer the reader to Ref.13, 11.

Is Orientifold Equivalence Proven? Is it True?

The answer to the first question is no. To the second one, as always, it depends on symmetry realizations. Let me explain both.

In Ref.2, a particular application of orientifold equivalence is given. A mapping between bosonic sectors of $\mathcal{N}=1$ $U(N_c)$ SYM and one flavor $U(N_c)$ QCD with an antisymmetric representation Dirac fermion (as well as its multi-flavor generalizations) was suggested.

Even though this is presented as a parent-daughter orientifold equivalence, there is really no single projection whatsoever which takes one theory to the other. At first, this seems to be an unimportant fact. However, this eventually propagates to the claim of absence of twisted sectors, and thereof, the absence of symmetry breaking. This leads to the conclusion that the orientifold equivalence does not fail, and is usually presented as conceptual superiority of this construction. The claim is false.

Here, I will first present how this equivalence arises by using projections. [d] This eases the identification of twisted sectors. The parent is $SO(2N_c)$

[d] I am using orientifold projections as in Polchinski vol 1, pages 190-192 [19]. The orientifold

$\mathcal{N} = 1$ SYM. It has two nontrivial projections by $Z_2 = (-1)^F$, which amounts to solving $A_\mu = +JA_\mu J^T$ for the gauge field, and $\lambda = \pm J\lambda J^T$ for the gaugino, where $J = i\sigma_2 \otimes 1_N$. This yields two daughter theories. The first one is $U(N_c)$ $\mathcal{N} = 1$ SYM, the second is one flavor QCD with an antisymmetric Dirac fermion. The merit of this application is that the daughter-daughter equivalence, if true, is the closest one can get from a supersymmetric gauge theory to real QCD [2]. Let us now identify twisted and neutral sectors. Since the unitary group is complex, the neutral sector of the daughters are charge conjugation ($\mathcal{C}$) even sectors. Clearly, the twisted sector is the $\mathcal{C}$ odd sector. The parent-daughter equivalences, if true, imply daughter-daughter equivalence, and the above is an example of such. The equivalence relies on the $Z_2 = \mathcal{C}$ symmetry realization.

Is it possible that this Z_2 symmetry may break down? Yes. Many Z_2 symmetries appearing in nature do so. It is also possible that it does not. It is all about the dynamics. Clearly, there are many local order parameters which are odd under $\mathcal{C}$, and no theorem preferring one way or other.

In general, in strongly coupled gauge theories, it is hard to know whether a symmetry is spontaneously broken or not. Instead of dealing with the gauge theory example (which is not straightforward), I want to demonstrate in a simpler non-gauge theory that the orientifold equivalence may fail as well. Consider $U(N_c)$-algebra valued ϕ^4 model, with the potential $V = m^2\mathrm{Tr}(\phi\phi^*) + \lambda[\mathrm{Tr}(\phi\phi^*)]^2$, where $m^2 > 0$ and $\lambda > 0$. The traces are over $N \times N$ matrices, with $\phi = \phi^a t^a$ an adjoint representation hermitian scalar, $\phi^\dagger = \phi$. The Lagrangian has a Z_2 discrete charge conjugation symmetry, acting as $\mathcal{C} : \phi \to c\phi^*$ where $c = \pm 1$ is the charge conjugation assignment.

Let us write the Lagrangian more explicitly. A hermitian matrix can be written as the sum of a pure imaginary antisymmetric matrix A and a pure real symmetric matrix S. The matrix A takes values in the maximal $SO(N_c)$ subalgebra of the $U(N_c)$, S is in the coset $U(N_c)/SO(N_c)$. Therefore,

$$\mathrm{tr}(\phi\phi^*) = -\sum_{a\in SO(N)} \phi^a\phi^a + \sum_{a\in U(N)/SO(N)} \phi^a\phi^a \equiv (-\mathrm{tr}A^2 + \mathrm{tr}S^2)$$

projection of a complex group $U(N)$ by charge conjugation (and symplectic tensor) will yield a real (pseudo-real) group $SO(N)$ ($Sp(N)$). In the language of Feynman graphs, this corresponds to removing the arrows. Analogously, the massless open oriented strings with $U(N)$ group turns into unoriented open strings with $SO(N)$ ($Sp(N)$) group. The neutral (twisted) sector in parent $U(N)$ is $\mathcal{C}$-even ($\mathcal{C}$-odd) operators. There exists reverse projections, which takes real (pseudoreal) unoriented groups $SO(2N)$ ($Sp(2N)$) to the complex, oriented groups $U(N)$ [6]. In this case, the neutral sector in the daughter is defined by $\mathcal{C}$-even operators.

The potential, in terms of S and A variables may be written as $m^2(-\text{tr}A^2 + \text{tr}S^2) + \lambda(-\text{tr}A^2 + \text{tr}S^2)^2$. The statement of $\mathcal{C}$ invariance of the action, in terms of S and A translates into $\mathcal{C} : (S, A) \to (cS, -cA)$ where $c = \pm 1$. There are two orientifold projections of this model, obtained by removing degrees of freedom which are not invariant under $\mathcal{C}$ (or by solving the orientifold constraints $\phi = \pm\phi^* = \pm\phi^T$). A and S have opposite transformation properties under $\mathcal{C}$. Therefore, the two daughter theories are $SO(N)$ theories, with symmetric and antisymmetric representation scalars. The potentials are, respectively, $V = m^2\text{tr}S^2 + \lambda(\text{tr}S^2)^2$ and $V = -m^2(\text{tr}A^2) + \lambda(\text{tr}A^2)^2$. Clearly, there is no daughter-daughter equivalence because of the failure of symmetry realization conditions with their parent.

References

1. A. Gorsky and M. Shifman, *Phys. Rev.* **D67**, p. 022003 (2003).
2. A. Armoni, M. Shifman and G. Veneziano (2004).
3. T. Eguchi and H. Kawai, *Phys. Rev. Lett.* **48**, p. 1063 (1982).
4. P. Kovtun, M. Unsal and L. G. Yaffe, *JHEP* **07**, p. 008 (2005).
5. P. Kovtun, M. Unsal and L. G. Yaffe, *JHEP* **12**, p. 034 (2003).
6. P. Kovtun, M. Unsal and L. G. Yaffe, *to appear* (2006).
7. L. G. Yaffe, *Rev. Mod. Phys.* **54**, p. 407 (1982)
8. C. Lovelace, *Nucl. Phys.* **B201**, p. 333 (1982).
9. G. Bhanot, U. M. Heller and H. Neuberger, *Phys. Lett.* **B113**, p. 47 (1982).
10. R. Narayanan and H. Neuberger, *PoS* **LAT2005**, p. 005 (2006).
11. A. Armoni, A. Gorsky and M. Shifman, *Phys. Rev.* **D72**, p. 105001 (2005).
12. D. Tong, *JHEP* **03**, p. 022 (2003).
13. P. Kovtun, M. Unsal and L. G. Yaffe, *Phys. Rev.* **D72**, p. 105006 (2005).
14. J. L. F. Barbon and C. Hoyos, *JHEP* **01**, p. 114 (2006).
15. M. Schmaltz, *Phys. Rev.* **D59**, p. 105018 (1999).
16. M. J. Strassler (2001).
17. J. Erlich and A. Naqvi, *JHEP* **12**, p. 047 (2002).
18. M. Bershadsky and A. Johansen, *Nucl. Phys.* **B536**, 141 (1998).
19. J. Polchinski Cambridge, UK: Univ. Pr. (1998) 402 p.

BARYONS AND SKYRMIONS IN QCD WITH QUARKS IN HIGHER REPRESENTATIONS

S. BOLOGNESI

The Niels Bohr Institute,
Blegdamsvej 17, DK-2100 Copenhagen Ø, Denmark
E-mail: bolognesi@nbi.dk

We study the baryonic sector of QCD with quarks in the two index symmetric or antisymmetric representation. The minimal gauge invariant state that carries baryon number cannot be identified with the Skyrmion of the low energy chiral effective Lagrangian. Mass, statistics and baryon number do not match. We carefully investigate the properties of the minimal baryon in the large N limit and we find that it is unstable under formation of bound states with higher baryonic number. These states match exactly with the properties of the Skyrmion of the effective Lagrangian.

1. Introduction

The large N expansion is a major tool in the study of $SU(N)$ gauge theories.[1,2] In the usual limit quarks are taken in the fundamental representation and in particular the baryon is a gauge invariant composite of N quarks

$$\epsilon_{\alpha_1 \dots \alpha_N} Q^{\alpha_1} \dots Q^{\alpha_N} \ . \tag{1}$$

In the large N limit the baryon has a mass that scales like N and it can be identified with the Skyrmion of the chiral effective Lagrangian.[3,4]

Recently another kind of large N limit has received considerable attention. This is the case of quarks in the two index symmetric or antisymmetric representation.[5]

It has been noted[6] that, at least at a first glance, the identification between baryons and Skyrmions does not work in this limit. A natural choice for the baryon is the following

$$\epsilon_{\alpha_1 \alpha_2 \dots \alpha_N} \epsilon_{\beta_1 \beta_2 \dots \beta_N} Q^{\{\alpha_1 \beta_1]} Q^{\{\alpha_2 \beta_2]} \dots Q^{\{\alpha_N \beta_N]} \ . \tag{2}$$

This baryon is formed of N quarks and so the first guess is that its mass scales likes N in the large N limit. The mass of the Skyrmion scales like

F_π^2 where F_π is the pion decay constant. In the case of the quarks in higher representations F_π scales like N, so the mass of the Skyrmion scales like N^2, in contrast with the naive expectation for the baryon (2). This is the puzzle we are going to solve.[7]

The conventions we use in the paper are the following: $Q^{\{\alpha\beta\}}$ indicates a quark in the two index symmetric representation while $Q^{[\alpha\beta]}$ indicates a quark the two index antisymmetric representation. We use $Q^{\{\alpha\beta]}$ when the formula is applicable to both of them.

2. Effective lagrangians, anomalies and skyrmions

Consider asymptotically free theories with N colors and N_f Dirac fermions transforming according in the two index (anti)symmetric representation of the gauge group

$$\mathcal{L} = -\frac{1}{2}\mathrm{Tr}\ F_{\mu\nu}F^{\mu\nu} + \sum_{k=1}^{N_f} \overline{Q}^k (iD_\mu\gamma^\mu - m_k)Q_k \tag{3}$$

$F_{\mu\nu} = \partial_\mu A_\nu - \partial_\nu A_\mu + ig[A_\mu, A_\nu]$ is the field strength, g is the coupling constant, and the covariant derivative is $D_\mu = \partial_\mu - igA_\mu$. In order to have a well defined large N limit we take the product $g^2 N$ to be finite.

The dependence of the number of colors for the meson coupling can be evaluated using the planar diagram presented in Figure 1 and paying attention to the hadron wave function normalization. We will denote the

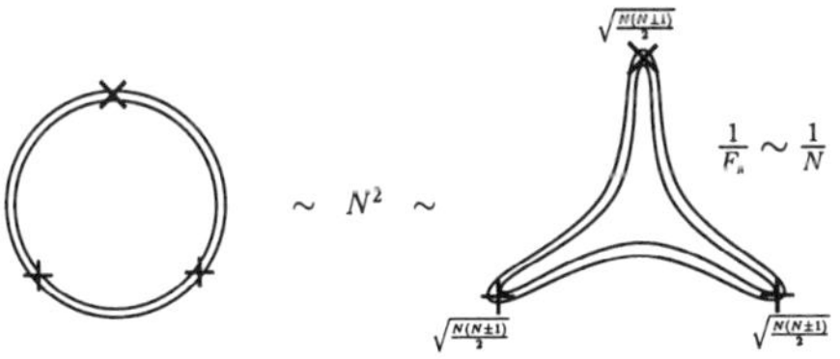

Fig. 1. The N dependence of the meson coupling F_π.

decay constant of the typical meson by F_π.

Here we will consider N_f massless flavors and hence the Lagrangian has a global symmetry $SU(N_f)_L \times SU(N_f)_R$. The chiral symmetry is expected to dynamically break to its maximal diagonal subgroup. The low energy effective Lagrangian describes the dynamics of the massless mesons, that are the Nambu-Goldstone bosons of the spontaneous chiral symmetry breaking.

Written in terms of the matrix $U(x) = \exp\left(\frac{i\,\pi(x)}{F_\pi}\right)$, where $\pi(x)$ is the Goldstone boson matrix, the effective Lagrangian is

$$S_{\text{eff}} = \frac{1}{16} F_\pi^2 \int d^4x \left\{ \text{Tr} \partial_\mu U \partial_\mu U^{-1} + \text{higher derivatives} \right\} + \Gamma_{\text{WZW}} \,. \qquad (4)$$

The Wess-Zumino-Witten term can be written as $\Gamma_{\text{WZW}} = n \int_{\mathcal{M}^5} \omega_{\mu\nu\rho\sigma\tau}$ where $\mathcal{M}^5$ is a five dimensional manifold whose boundary is the Minkowsky space and $\omega_{\mu\nu\rho\sigma\tau}$ is a properly normalized volume form. Quantum consistency of the theory requires n to be an integer. Matching with the underlying anomaly computations requires n to be equal to N in the case of the fundamental representation and $n = \frac{N(N\pm1)}{2}$ for the two index representation case.

The Skyrmion is a texture-like solution of the effective Lagrangian arising from the non-trivial third homotopy group of the possible configurations of the matrix $U(x)$ (namely $\pi_3(SU(N_f)) = \mathbf{Z}$). In the large N limit we can treat the effective Lagrangian as classical and, since the N dependence appears only as a multiplicative factor, the size and the mass of the Skyrmion scale respectively as N^0 and $\frac{N(N\pm1)}{2}$.[4] The baryon number of the Skyrmion is $\frac{N(N\pm1)}{2}$ times the baryon number of the quarks and the statistics is fermionic or bosonic accordingly if $\frac{N(N\pm1)}{2}$ is odd or even.

3. Baryons at large N

3.1. *The baryon in ordinary QCD*

Now we briefly review the large N behavior of the baryon in ordinary QCD. The gauge wave function (1) is antisymmetric under exchange of two quarks. Since the quarks are fermions, the total gauge function $\psi_{\text{gauge}}\psi_{\text{spin}}\psi_{\text{space}}$ must be antisymmetric under exchange of two quarks. The simplest choice is to take a completely symmetric spin wave function and a completely symmetric spatial wave function.[a]

$$\begin{array}{ccc} \psi_{\text{gauge}} & \psi_{\text{spin}} & \psi_{\text{space}} \\ - & + & + \end{array} \qquad (5)$$

In the large N limit the problem can be approximated by a system of free bosons in a mean field potential $V_{\text{mean}}\left(\frac{r}{R^*}\right)$ where R^* is the size of the baryon. The ground state is a Bose-Einstein condensate where the quarks

[a] In this section we consider for simplicity only one flavor. The results are independent on this detail.

are all in the ground state of the mean field potential. The large N behavior of the radius and the mass are respectively $R^* \sim \mathcal{O}(1)$, and $M \sim \mathcal{O}(N)$.

3.2. *The simplest baryon in higher representations*

In higher representations QCD the simplest baryon is (2). If we exchange two quarks, say for example $Q^{\{\alpha_1\beta_1]}$ and $Q^{\{\alpha_2\beta_2]}$, this is equivalent to the exchange of $\alpha_1\alpha_2$ in $\epsilon_{\alpha_1\alpha_2\ldots\alpha_N}$ and $\beta_1\beta_2$ in $\epsilon_{\beta_1\beta_2\ldots\beta_N}$. The result is that the gauge wave function is symmetric under exchange of two quarks. In order to have a total wave function that is antisymmetric under exchange, the spatial wave function ψ_{space} must be antisymmetric.

$$\begin{array}{ccc} \psi_{\text{gauge}} & \psi_{\text{spin}} & \psi_{\text{space}} \\ + & + & - \end{array} \tag{6}$$

In the large N limit the problem can be approximated by a system of free fermions in a mean field potential $V_{\text{mean}}(\frac{r}{R^*})$ where R^* is the size of the baryon. The ground state is a degenerate Fermi gas and is obtained by filling all the lowest energy states of the mean field potential up the Fermi surface. Now there are two kind of forces that enter in the game: 1) gauge forces that scales like N and are both repulsive and attractive; 2) Fermi zero temperature pressure that scales like $N^{4/3}$ and is only repulsive. This is sufficient to conclude that the mass of this baryon grows not faster than $N^{4/3}$ and not slower than N. We can thus immediately infer that the simplest baryon cannot be matched with the Skyrmion.

To know more precisely the properties of this baryon in the large N limit we have to make a mild assumption about the gauge forces. The energy as function of R is

$$E(R) \sim Nf(R) + \frac{N^{4/3}}{R}\,, \tag{7}$$

where $Nf(R)$ is the contribution of the gauge forces. The only assumption we make now is that $f(R)$ goes to infinity as $R \to \infty$. It is now easy to see that the mass of the baryon, that is the minimum of (7), goes to infinity faster than N. This imply that the mass per unit of baryonic number diverges as N goes to infinity.

The key point for the solution of the puzzle is the following. If we are able to find a baryon whose gauge wave function is completely antisymmetric under exchange of two quarks, it will be stable under the decay into this simplest baryon. This is because the fully antisymmetric baryon minimizes the mass per unit of baryonic number.

4. Construction of the stable baryons

In the following we will construct the only possible gauge wave function that gauge singlet and completely antisymmetric under exchange of two quarks. We will find that the required number of quarks, as expected from the Skyrmion analysis, is $\frac{N(N\pm1)}{2}$.

We start from $N = 2$ and symmetric representation. We want to construct a gauge invariant wave function that contains three quarks $Q^{\{\alpha_1\beta_1\}}$, $Q^{\{\alpha_2\beta_2\}}$, and $Q^{\{\alpha_3\beta_3\}}$ and is antisymmetric under exchange of two quarks. Then we "formally" split the high-representational quarks into two fundamental quarks. For example the quark $Q^{\{\alpha_1\beta_1\}}$ is splitted into two fundamental quarks q^{α_1} and q^{β_1}. Then we put the fundamental quarks in the diagram of Figure 2. Finally we saturate the indices with three antisym-

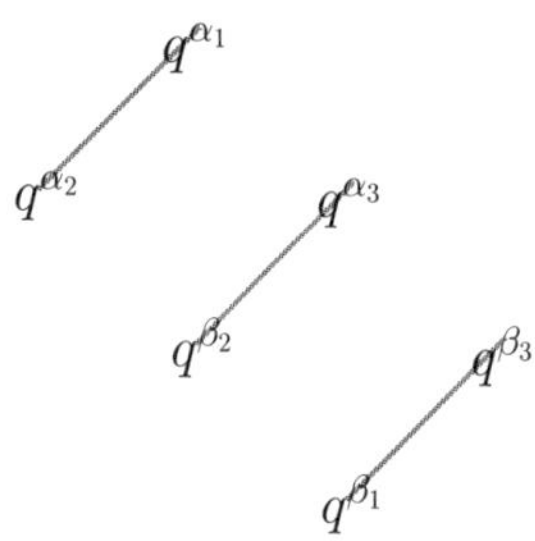

Fig. 2. Diagrammatic representation of the baryon for $N = 2$.

metric tensors ϵ that are the gray lines in Figure 2. The gauge wave function that correspond to the diagram is thus

$$\epsilon_{\alpha_2\alpha_1}\epsilon_{\beta_2\alpha_3}\epsilon_{\beta_1\beta_3}\, Q^{\{\alpha_1\beta_1\}}Q^{\{\alpha_2\beta_2\}}Q^{\{\alpha_3\beta_3\}}\ . \tag{8}$$

It can be easily verified that it is completely antisymmetric.

Now let us consider the first cases for the antisymmetric representation. For $N = 2$, we have $\frac{N(N-1)}{2} = 1$, and it is easy to find such a wave function $\epsilon_{\gamma\delta}\, Q^{[\gamma\delta]}$. For $N = 3$ we need a wave function that contains three quarks. To guess it using directly $Q^{[\alpha\beta]}$ is not easy, but we can use a trick. The antisymmetric representation for $N = 3$ is equivalent to the anti-fundamental $\widetilde{Q}_\gamma = \frac{1}{2}\epsilon_{\gamma\alpha\beta}Q^{[\alpha\beta]}$ and we know how to write a baryon for the anti-fundamental representation $\epsilon^{\gamma\rho\tau}\widetilde{Q}_\gamma\widetilde{Q}_\rho\widetilde{Q}_\tau$. Substituting the relation between $\widetilde{Q}_\gamma$ and $Q^{[\alpha\beta]}$ we obtain

$$\frac{1}{2}(\epsilon_{\gamma_1\delta_1\alpha}\epsilon_{\gamma_2\delta_2\beta} - \epsilon_{\gamma_2\delta_2\alpha}\epsilon_{\gamma_1\delta_1\beta})\, Q^{[\alpha\beta]}Q^{[\gamma_1\delta_1]}Q^{[\gamma_2\delta_2]}\ . \tag{9}$$

The correspondent diagram is given in Figure 3.

Fig. 3. Diagrammatic representation of the baryon for $N = 3$.

For a general number of colors we have the following proposition

Proposition 4.1. *There is one and only one gauge wave function that is gauge singlet and completely antisymmetric under exchange of two quarks. This wave function is composed by* $\frac{N(N\pm1)}{2}$ *quarks* $Q^{\{\alpha\beta]}$ *and is the completely antisymmetric subspace of the tensor product of* $\frac{N(N\pm1)}{2}$ *quarks.*

Proof. Let us consider the symmetric representation first. We need two basic facts to prove the proposition: 1) two indices of the same quark cannot belong to the same saturation line; 2) if q^{α_i} and q^{α_j} belong to the same saturation line, the two partners q^{β_i} and q^{β_j} cannot belong to the same saturation line . At this point we refer to the Figure 4 (A) where it can be easily seen that the two rules imply the existence of at least $\frac{N(N+1)}{2}$ quarks. I the case of the antisymmetric representation the only difference

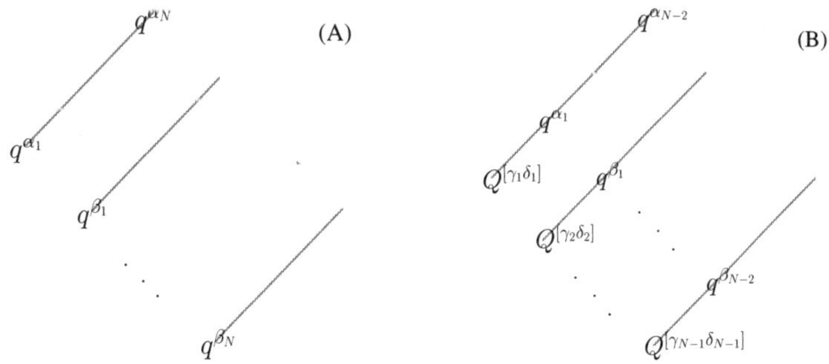

Fig. 4. Diagrammatic proof that the minimum number of quarks to form an antisymmetric baryon is $\frac{N(N+1)}{2}$.

is that now it is instead possible for a quark to have both indices in the same saturation line but one saturation line can contain at most one quark of this type. The Figure 4 (B) shows that now we need at least $\frac{N(N-1)}{2}$ quarks.

Now we need to prove the existence of this wave function. Consider the tensor product of certain number of quarks $Q^{\{\alpha\beta]}$, and take the subspace of the tensor product that is completely antisymmetric under exchange. This subspace is obviously closed under the action of the gauge group. If the number of quarks is greater than $\frac{N(N\pm1)}{2}$ this subspace has dimension zero. If the number of quarks is exactly $\frac{N(N\pm1)}{2}$ the antisymmetric subspace has dimension one and so it is gauge singlet. □

At a first glance it may seem that the one gluon exchange ruins the N^2 dependence of the mass of this fully antisymmetric baryon. The one gluon exchange is of order $1/N$ and, since we have $\frac{N(N\pm1)}{2}$ quarks, the statistical factor seems to be of order N^4. The solution of this puzzle is that the correct statistical factor is of order N^3 since only quarks that have at least one index in common can exchange a gluon.[8]

Acknowledgements

I thank especially F. Sannino who suggested me to work on this problem and also M. Shifman and T. Cohen for discussions. This work is supported by the Marie Curie Excellence Grant under contract MEXT-CT-2004-013510.

References

1. G. 't Hooft, Nucl. Phys. B **72** (1974) 461.
2. E. Witten, Nucl. Phys. B **160** (1979) 57.
3. T. H. R. Skyrme, Proc. Roy. Soc. Lond. A **260** (1961) 127.
4. E. Witten, Nucl. Phys. B **223** (1983) 422; Nucl. Phys. B **223** (1983) 433.
5. A. Armoni, M. Shifman and G. Veneziano, Nucl. Phys. B **667**, 170 (2003) [arXiv:hep-th/0302163]; Phys. Rev. D **71**, 045015 (2005) [arXiv:hep-th/0412203].
6. A. Armoni and M. Shifman, Nucl. Phys. B **670** (2003) 148 [arXiv:hep-th/0303109].
7. S. Bolognesi, arXiv:hep-th/0605065.
8. A. Cherman and T. D. Cohen, arXiv:hep-th/0607028.

SECTION 8

LIGHT QUARKS AND GLUONS

Convener A. Khodjamirian

MODEL INDEPENDENT DETERMINATION OF THE LOWEST RESONANCE OF QCD

H. LEUTWYLER

Institute for Theoretical Physics, University of Bern, Sidlerstr. 5, CH-3012 Bern, Switzerland
E-mail: leutwyler@itp.unibe.ch

I report on recent work done in collaboration with Irinel Caprini and Gilberto Colangelo.[1] We observe that the Roy equations lead to a representation of the $\pi\pi$ scattering amplitude that exclusively involves observable quantities, but is valid for complex values of s. At low energies, this representation is dominated by the contributions from the two subtraction constants, which are known to remarkable precision from the low energy theorems of chiral perturbation theory. Evaluating the remaining contributions on the basis of the available data, we demonstrate that the lowest resonance carries the quantum numbers of the vacuum and occurs in the vicinity of the threshold. Although the uncertainties in the data are substantial, the pole position can be calculated quite accurately, because it occurs in the region where the amplitude is dominated by the subtractions. The calculation neatly illustrates the fact that the dynamics of the Goldstone bosons is governed by the symmetries of QCD.

Pions play a crucial role whenever the strong interaction is involved at low energies – the Standard Model prediction for the muon magnetic moment provides a good illustration. The present talk concerns the remarkable theoretical progress made in low energy pion physics in recent years.

From the point of view of dispersion theory, $\pi\pi$ scattering is particularly simple: the s-, t- and u-channels represent the same physical process. As a consequence, the real part of the scattering amplitude can be represented as a dispersion integral over the imaginary part and the integral exclusively extends over the physical region.[2] The representation involves two subtraction constants which may be identified with the S-wave scattering lengths a_0^0, a_0^2. The projection of the amplitude on the partial waves leads to a dispersive representation for these, the Roy equations.

The pioneering work on the physics of the Roy equations was carried out more than 30 years ago.[3] The main problem encountered at that time was that the two subtraction constants occurring in these equations were not known. These constants dominate the dispersive representation at low energies, but since the data available at the time were consistent with a very broad range of S-wave scattering lengths, the Roy equation analysis was not conclusive.

The insights gained by means of chiral perturbation theory thoroughly changed the situation. Very accurate predictions for the scattering lengths are obtained by matching the chiral and dispersive representations in the interior of the Mandelstam triangle.[4] The Roy equations then determine the scattering amplitude throughout the low energy region – $\pi\pi$ scattering represents one of the very rare hadronic processes where theory is ahead of experiment. The existing low energy precision measurements[5–7] are consistent with the theoretical predictions and even offer a stringent test for one of these. Lattice results obtained with quark masses that are small enough for a controlled extrapolation to the values of physical interest to be within reach also confirm the predictions.[8,9]

In the following, I focus on the results obtained on this basis for the low energy properties of the isoscalar S-wave. The corresponding S-matrix element, $S_0^0 = \eta_0^0 \exp 2\,i\,\delta_0^0$, is related to the partial wave amplitude t_0^0 by

$$S_0^0(s) = 1 + 2\,i\,\rho(s)\,t_0^0(s)\,, \qquad \rho(s) = \sqrt{1 - 4M_\pi^2/s}\,. \tag{1}$$

The available phenomenological analyses are not in good agreement.[10] In fact, until ten years ago, the information about the $\sigma = f_0(600)$ was so shaky that this resonance was banned from the data tables. The work of Törnqvist and Roos[11] resurrected it, but the estimate of the Particle Data Group,[12] $M_\sigma - \frac{i}{2}\,\Gamma_\sigma = (400 - 1200) - i\,(300 - 500)$ MeV, indicates that it is not even known for sure whether the lowest resonance of QCD carries the quantum numbers of the σ or those of the ρ.

The positions of the poles represent universal properties of the strong interactions which are unambiguous even if the width of the resonance turns out to be large, but they concern the non-perturbative domain, where an analysis in terms of the local degrees of freedom of QCD – quarks and gluons – is not in sight. One of the reasons why the values for the pole position of the σ quoted by the Particle Data Group cover a very broad range is that all of these rely on the extrapolation of hand made parametrizations: the data are represented in terms of suitable functions on the real axis and the position of the pole is determined by continuing this representation into the complex plane. If the width of the resonance is small, the ambiguities inherent in the choice of the parametrization do not significantly affect the result, but the width of the σ is not small.

We have found a method that does not invoke parametrizations of the data. It relies on the following two observations:

1. The S-matrix has a pole on the second sheet if and only if it has a zero on the first sheet. In order to determine the pole position it thus suffices to have a reliable representation of the scattering amplitude on the first sheet.

2. The Roy equations hold not only on the real axis, but in a limited region of the first sheet. Since the pole from the σ occurs in that region, we do not need to invent a parametrization, but can rely on the explicit representation of the amplitude provided by these equations.

Roy established the validity of his equations from first principles for real values of s in the interval $-4M_\pi^2 < s < 60M_\pi^2$. Using known results of general quantum field theory,[13] we have demonstrated that these equations also hold for complex values of s, in the intersection of the relevant Lehmann-Martin ellipses.[1] The dash-dotted curve in Fig. 1 shows the domain of validity that follows from axiomatic field theory, while the full line depicts the slightly larger domain obtained under the assumption that the scattering amplitude obeys the Mandelstam representation. I emphasize that the boundary does not represent a singularity of the amplitude, but merely limits the region where the Roy equations hold in the form given. Modified representations with a much larger domain of validity can be found in Ref. 14 and in the references quoted therein.

For our analysis, it is essential that the dispersion integrals are dominated by the contributions from the low energy region: because the Roy equations involve two subtractions, the kernels fall off in proportion to $1/s'^3$ when s' becomes large. The left hand cut plays an important role here: taken by itself, the contribution from the right hand cut is sensitive to the poorly known high energy behaviour of $\mathrm{Im}\, t_0^0(s')$, but taken together with the one from the left hand cut, the high energy tails cancel.

The Roy equations thus provide us with an explicit representation of the function $S_0^0(s)$ for complex values of s, in terms of rapidly convergent dispersion integrals over the imaginary parts of the partial waves. In connection with the determination of the pole from the σ, the most important contribution is the one from the subtraction term. The dispersion integrals over the S- P- and D-waves generate a correction which can be evaluated with available phase shift analyses – in particular with the one obtained by solving the Roy equations.[4] The contributions from high energies and high

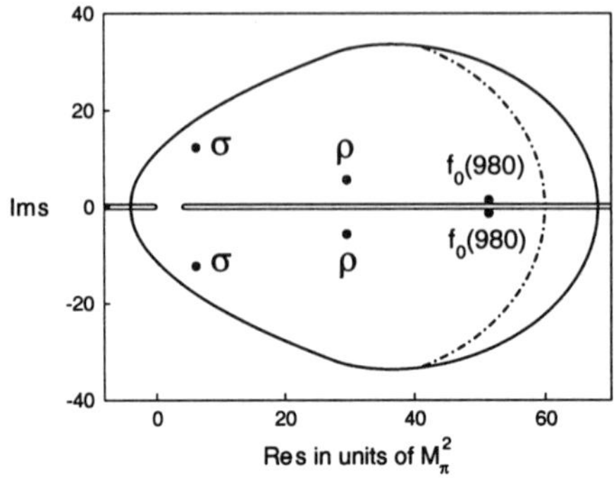

Fig. 1. Domain of validity of the Roy equations.

angular momenta can be estimated by means of the Regge representation of the scattering amplitude[15] – these contributions barely affect the pole position. For a brief account of the numerics, in particular also for a discussion of the error analysis, I refer to Ref. 1 – a more detailed report is in preparation.

For the central solution of the Roy equations, the function $S_0^0(s)$ contains two pairs of zeros in the domain of interest:

$$s_\sigma = (6.2 \pm i\,12.3)\,M_\pi^2\,, \quad s_{f_0} = (51.4 \pm i\,1.4)\,M_\pi^2\,.$$

These are indicated in Fig. 1, which may also be viewed as a picture of the second sheet – the dots then represent poles rather than zeros. For comparison, the figure also indicates the position of the zeros in $S_1^1(s)$, which characterize the ρ.

The higher one of the two pairs of zeros represents the well-established resonance $f_0(980)$, which sits close to the threshold of the transition $\pi\pi \rightarrow K\bar{K}$. The corresponding pole generates a spectacular interference phenomenon with the branch point singularity, which gives rise to a sharp drop in the elasticity. Our analysis adds little to the detailed knowledge of that structure.

The lower pair of zeros corresponds to a pole in the lower half of the second sheet at[1]

$$m_\sigma = \sqrt{s_\sigma} = 441\,^{+16}_{-\,8} - i\,272\,^{+9}_{-12.5}\ \mathrm{MeV}\,. \tag{2}$$

The error bars account for all sources of uncertainty. They are calculated by (a) estimating the uncertainties in the input used when solving the Roy equations and (b) following error propagation to determine the uncertainty in the result for the pole position. For details, I refer to Ref. 1.

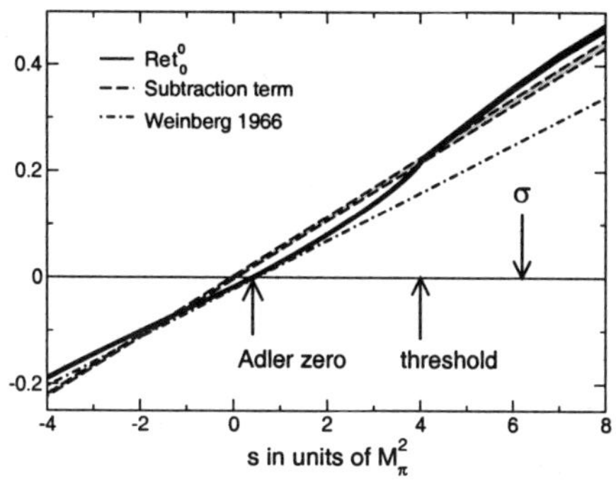

Fig. 2. Dominance of the subtraction term.

The reason why the pole position of the σ can be calculated rather accurately is that (a) the pole occurs at low energies and (b) there, the isoscalar S-wave is dominated by the subtraction term. This is illustrated in Fig. 2, where the real part of $t_0^0(s)$ is plotted versus s. The graph demonstrates that, in the region shown, the full amplitude closely follows the subtraction term. The final state interaction does generate curvature – in particular, the cusp at $s = 4M_\pi^2$ is visible – but since Goldstone bosons of low momentum interact only weakly, the contributions from the dispersion integrals amount to a small correction. In the language of chiral perturbation theory, the dispersion integrals only show up at NLO. Moreover, at leading order, the subtraction constants are determined by the pion decay constant. Dropping the dispersion integrals and inserting the lowest order predictions for the scattering lengths, the Roy equation for the isoscalar S-wave reduces to the well-known formula, which Weinberg derived 40 years ago,[16]

$$t_0^0(s) = \frac{2\,s - M_\pi^2}{32\,\pi\,F_\pi^2}\ , \tag{3}$$

and which is shown as a dash-dotted line in Fig.2.

The main feature at low energies is the occurrence of an Adler zero, $t_0^0(s_A) = 0$. At LO, the zero occurs at $s_A = \frac{1}{2}M_\pi^2$. The higher order corrections generate a small shift, which can be evaluated from our solution of the Roy equations. The uncertainties in the result, $s_A = (0.41 \pm 0.06)\,M_\pi^2$, are dominated by those in our predictions for the scattering lengths. While the Adler zero sits below the threshold, at a real value of s, the lowest zeros of the S-matrix, $S_0^0(s_\sigma) = 0$, occur in the region $\mathrm{Re}s_\sigma > 4M_\pi^2$, with an imaginary part that is about twice as large as the real part.

The Weinberg formula (3) explains why the S-matrix has a zero in the vicinity of the threshold. In this approximation, the zero of $S_0^0(s)$ occurs at $\sqrt{s} = 365 - i\,291$ MeV. The number differs from the "exact" result in Eq. (2) by about 20 percent. The dispersion integrals are essential for the partial wave to obey unitarity, but they only represent a correction. In view of this, the precision claimed in Eq. (2) is rather modest.

I conclude that the same theoretical framework that leads to incredibly sharp predictions for the threshold parameters of $\pi\pi$ scattering[4] also shows that the lowest resonance of QCD carries the quantum numbers of the vacuum. The pole sits in the vicinity of the threshold, but quite far from the real axis: the width of the σ is 3.7 times larger than the width of the ρ.

Since the Adler zero suppresses the scattering amplitude at low energies, the resonance is displaced: the phase of the isoscalar S-wave reaches 90° only when the energy becomes about twice as large as the mass of the σ. There is no such suppression in the amplitude of the decay $J/\psi \to \omega\pi\pi$. Indeed, the BES data[17] indicate that the peak in the S-wave projection of this amplitude does occur in the vicinity of m_σ. The observation might suggest that those data give a better handle on the σ than $\pi\pi$ scattering. In my opinion, this is not the case, however, because the loss in visibility is more than compensated by the gain in theoretical understanding: unitarity and crossing symmetry impose very strong constraints on the scattering amplitude, but only very weak ones on the decay amplitude. In particular, the distortion due to rescattering on the ω, the contributions from the inelastic channels $\pi\pi \to K\bar{K}$ and $\pi\pi \to \eta\eta$, as well as the curvature generated by the left hand cut need to be understood before a meaningful extrapolation of the data to the pole can be made. Incidentally, most pole determinations[a] assume that the left hand cut can be neglected. In view of the proximity of the pole, the validity of this assumption is doubtful.

The physics of the σ is governed by the dynamics of the Goldstone bosons: the properties of the interaction among two pions are relevant.[19] In quark model language, the wave function contains an important tetra-quark component.[20] The properties of the resonance $f_0(980)$ are also governed by Goldstone boson dynamics – two kaons in that case. Very recently, the method described here was applied to the case of πK scattering.[21] In this case, the analogue of the back-of-the-envelope calculation sketched above relies on the tree level approximation for the $I = \frac{1}{2}$ S-wave obtained from

[a] A notable exception is Ref. 18, where the two loop representation of chiral perturbation theory is used to obtain an estimate for the left hand discontinuity. The result of this calculation is in good agreement with our analysis.

the effective SU(3)×SU(3) lagrangian and yields $m_\kappa = 671 - i\,262$ MeV, remarkably close to the "exact" value, obtained from the solution of the Roy-Steiner equations, $m_\kappa = (658 \pm 13) - i\,(278.5 \pm 12)$ MeV. Evidently, the physics of the κ is very similar to the one of the σ.

It is a pleasure to thank M. Shifman and A. Vainshtein for a very enjoyable stay at Minneapolis.

References

1. I. Caprini, G. Colangelo and H. Leutwyler, Phys. Rev. Lett. **96** (2006) 132001.
2. S. M. Roy, Phys. Lett. B **36** (1971) 353.
3. J. L. Basdevant, C. D. Froggatt and J. L. Petersen, Nucl. Phys. B **72** (1974) 413.
4. G. Colangelo, J. Gasser and H. Leutwyler, Nucl. Phys. B **603** (2001) 125.
5. S. Pislak *et al.* [BNL-E865 Collaboration], Phys. Rev. Lett. **87** (2001) 221801; Phys. Rev. D **67** (2003) 072004.
6. B. Adeva *et al.* [DIRAC Collaboration], Phys. Lett. B **619** (2005) 50.
7. J. R. Batley *et al.* [NA48/2 Collaboration], Phys. Lett. B **633**, 173 (2006).
8. C. Aubin *et al.*, [MILC Collaboration], Phys. Rev. D **70** (2004) 114501.
9. S. R. Beane, P. F. Bedaque, K. Orginos and M. J. Savage [NPLQCD Collaboration], Phys. Rev. D **73** (2006) 054503.
10. For recent reviews, see for instance
    P. Minkowski and W. Ochs, AIP Conf. Proc. **814** (2006) 52.
    D. V. Bugg, AIP Conf. Proc. **814** (2006) 78.
    M. R. Pennington, arXiv:hep-ph/0604212.
11. N. A. Tornqvist and M. Roos, Phys. Rev. Lett. **76** (1996) 1575.
12. S. Eidelman *et al.* [Particle Data Group], Phys. Lett. B **592** (2004) 1.
13. A. Martin, *Scattering Theory: Unitarity, Analyticity and Crossing*, Lecture Notes in Physics, Vol. 3, (Springer-Verlag, Berlin, 1969).
14. S. M. Roy and G. Wanders, Nucl. Phys. B **141** (1978) 220.
15. B. Ananthanarayan, G. Colangelo, J. Gasser and H. Leutwyler, Phys. Rept. **353** (2001) 207.
16. S. Weinberg, Phys. Rev. Lett. **17** (1966) 616.
17. M. Ablikim *et al.* [BES Collaboration], Phys. Lett. B **598** (2004) 149.
    D. V. Bugg, arXiv:hep-ph/0603089.
18. Z. Y. Zhou, G. Y. Qin, P. Zhang, Z. G. Xiao, H. Q. Zheng and N. Wu, JHEP **0502** (2005) 043.
19. V. E. Markushin and M. P. Locher, Frascati Physics Series, **15** (1999) 229.
20. R. L. Jaffe, Phys. Rev. D **15** (1977) 267, 281.
21. S. Descotes-Genon and B. Moussallam, arXiv:hep-ph/0607133.

RESTORATION OF CHIRAL AND $U(1)_A$ SYMMETRIES IN EXCITED HADRONS IN THE SEMICLASSICAL REGIME

L. YA. GLOZMAN

Institute for physics/Theoretical physics, University of Graz,
Universitätsplatz 5, A-8010 Graz, Austria
E-mail: leonid.glozman@uni-graz.at

Restoration of chiral and $U(1)_A$ symmetries in excited hadrons is reviewed. Implications of the OPE as well as of the semiclassical expansion for this phenomenon are discussed. A solvable model of the 't Hooft type in 3+1 dimensions is presented, which demonstrates a fast restoration of both chiral and $U(1)_A$ symmetries at larger spins and radial excitations.

Keywords: Chiral and $U(1)_A$ symmetry restoration; Excited hadrons

1. Introduction

There are some phenomenological evidences that the highly excited hadrons, both baryons [1–3] and mesons [4,5] fall into approximate multiplets of $SU(2)_L \times SU(2)_R$ and $U(1)_A$ groups, for a short overview see ref. [6]. This is illustrated in Fig. 1, where the excitation spectrum of the nucleon as well as the excitation spectrum of π and f_0 (with the $\bar{n}n = \frac{\bar{u}u+\bar{d}d}{\sqrt{2}}$ content) mesons are shown. Starting from the 1.7 GeV region the nucleon (and delta) spectra show obvious signs of parity doubling. There are a couple of examples where chiral partners of highly excited states have not yet been seen. Their experimental discovery would be an important task. Similarly, in the chirally restored regime π and $\bar{n}n$ f_0 states must be systematically degenerate. If confirmed by discovery of still missing states, this phenomenon is referred to as effective chiral symmetry restoration or chiral symmetry restoration of the second kind.

By definition this effective chiral symmetry restoration means the following. All hadrons that are created by the given interpolator, J_α, appear as intermediate states in the two-point correlator,

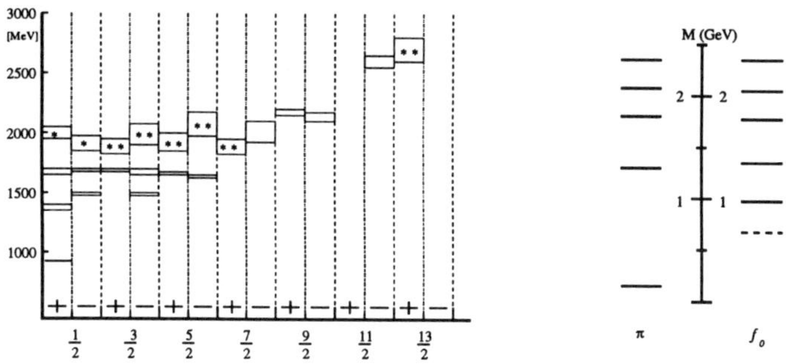

Fig. 1. Left panel: excitation spectrum of the nucleon (those resonances which are not yet established are marked by two or one stars according to the PDG classification). Right panel: pion and $n\bar{n}$ f_0 spectra.

$$\Pi = \imath \int d^4x \; e^{\imath qx} \langle 0|T\{J_\alpha(x) J_\alpha^\dagger(0)\}|0\rangle. \tag{1}$$

Consider two interpolating fields $J_1(x)$ and $J_2(x)$ which are connected by a chiral transformation (or by a $U(1)_A$ transformation), $J_1(x) = UJ_2(x)U^\dagger$. Then, in the Wigner-Weyl mode, $U|0\rangle = |0\rangle$, it follows from (1) that the spectra created by the operators $J_1(x)$ and $J_2(x)$ would be identical. We know that in QCD one finds $U|0\rangle \neq |0\rangle$. As a consequence the spectra of the two operators must be in general different. However, it happens that the noninvariance of the vacuum becomes unimportant (irrelevant) high in the spectrum. Then the spectra of both operators become close at large masses and asymptotically identical. This effective chiral symmetry restoration means that the role of the quark condensates that break chiral symmetry in the vacuum becomes progressively less important high in the spectrum. One could say, that the valence quarks in high-lying hadrons *decouple* from the QCD vacuum.

2. Chiral symmetry restoration and the quark-hadron duality

There is a heuristic argument that supports this idea [2]. The argument is based on the well controlled behaviour of the two-point function (1) at the large space-like momenta $Q^2 = -q^2$, where the operator product expansion (OPE) is valid and where all nonperturbative effects can be absorbed into condensates of different dimensions [7]. The key point is that all nonperturbative effects of the spontaneous breaking of chiral symmetry at large Q^2

are absorbed into the quark condensate $\langle \bar{q}q \rangle$ and other quark condensates of higher dimension. However, the contributions of these condensates to the correlation function are proportional to $(1/Q^2)^n$, where the index n is determined by the quantum numbers of the current J and by the dimension of the given quark condensate. Hence, at large enough Q^2 the two-point correlator becomes approximately chirally symmetric. At these high Q^2 a matching with the perturbative QCD (where no SBCS occurs) can be done.

Then we can invoke into analysis a dispersion relation. Since the large Q^2 asymptotics of the correlator is given by the leading term of the perturbation theory, then the asymptotics of the spectral density, $\rho(s)$,at $s \to \infty$ must also be given by the same term of the perturbation theory if the spectral density approaches a constant value (if it oscillates, then it must oscillate around the perturbation theory value). Hence both spectral densities $\rho_{J_1}(s)$ and $\rho_{J_2}(s)$ at $s \to \infty$ must approach the same value and the spectral function becomes chirally symmetric. This is definitely true in the asymptotic (jet) regime where the spectrum is strictly continuous. The conjecture of ref. [2] was that may be this is also true in the regime where the spectrum is still quasidiscrete and saturated mainly by resonances.

The question arises then what is the functional behaviour that determines approaching the chiral-invariant regime at large s? One would expect that OPE could help us. This is not so, however, for two reasons. First of all, we know phenomenologically only the lowest dimension quark condensate. But even if we knew all quark condensates up to a rather high dimension, it would not help us. This is because the OPE is only an asymptotic expansion. While such kind of expansion is very useful in the space-like region, it does not define any analytical solution which could be continued to the time-like region at finite s. This means that while the real (correct) spectrum of QCD must be consistent with OPE, there is an infinite amount of incorrect spectra that can also be consistent with OPE. Then, if one wants to get some information about the spectrum, one needs to assume something else on the top of OPE. Clearly a success then is crucially dependent on these additional assumptions, for the recent activity in this direction see refs. [8–10]. This implies that in order to really understand chiral symmetry restoration one needs a microscopic insight and theory that would incorporate *at the same time* chiral symmetry breaking and confinement.

3. Restoration of the classical symmetry in the semiclassical regime

A fundamental insight into phenomenon can be obtained from the semiclassical argument [6]. We know that the axial anomaly as well as the spontaneous breaking of chiral symmetry in QCD is an effect of quantum fluctuations of the quark field. The latter can generally be seen from the definition of the quark condensate, which is a closed quark loop. This closed quark loop explicitly contains a factor $\hbar$. The chiral symmetry breaking, which is necessarily a nonperturbative effect, is actually a (nonlocal) coupling of a quark line with the closed quark loop, which is a graphical representation of the Schwinger-Dyson (gap) equation. Hence chiral symmetry breaking in QCD manifestly vanishes in the classical limit $\hbar \to 0$.

At large n (radial quantum number) or at large angular momentum J we know that in quantum systems the *semiclassical* approximation *must* work. Physically this approximation applies in these cases because the de Broglie wavelength of particles in the system is small in comparison with the scale that characterizes the given problem. In such a system as a hadron the scale is given by the hadron size while the wavelength of valence quarks is given by their momenta. Once we go high in the spectrum the size of hadrons increases as well as the typical momentum of valence quarks. This is why a highly excited hadron can be described semiclassically in terms of the underlying quark degrees of freedom.

The physical content of the semiclassical approximation is most transparently given by the path integral. The contribution of the given path to the path integral is regulated by the action $S(\phi(x))$ along the path $\phi(x)$ (the fields $\bar{\psi}, \psi, A$ are collectively denoted as ϕ) through the factor $\sim e^{i\frac{S(\phi(x))}{\hbar}}$. The semiclassical approximation applies when the action in the system $S \gg \hbar$. In this case the whole amplitude (path integral) is dominated by the classical path $\phi_{cl}(x)$ (stationary point) and those paths that are infinitesimally close to the classical path. In other words, in the semiclassical case the quantum fluctuations effects are strongly suppressed and vanish asymptotically. Then the generating functional can be expanded in powers of $\hbar$ as

$$W(J) = W_0(J) + \hbar W_1(J) + ..., \tag{2}$$

where $W_0(J) = S(\phi_{cl}) + J\phi_{cl}$ and $W_1(J)$ represents contributions of the lowest order quantum fluctuations around the classical solution (determinant of the classical solution). The classical path, which is generated by W_0, is a tree-level contribution to the path integral and keeps chiral symmetries of the classical Lagrangian. Its contribution is of the order $(\hbar/S)^0$. The quantum fluctuations contribute at the orders $(\hbar/S)^1$ (the one loop order, generated by W_1), $(\hbar/S)^2$ (the two loops order), etc. The $U(1)_A$ as well as the spontaneous $SU(2)_L \times SU(2)_R$ breakings start from the one-loop order. However, in a hadron with large enough n or J, where action is large, the loop contributions must be relatively suppressed and vanish asymptotically. Then it follows that in such systems both the chiral and $U(1)_A$ symmetries should be approximately restored. This is precisely what we see phenomenologically.

While the argument above is solid, theoretically it is not clear *a-priori* whether isolated hadrons still exist at excitation energies where a semiclassical regime is achieved. However, the large N_c limit of QCD, while keeping all basic properties of QCD like asymptotic freedom, confinement and chiral symmetry breaking, allows for a significant simplification. In this limit it is known that all mesons represent narrow states. At the same time the spectrum of mesons is infinite. Then one can always excite a meson of any arbitrary large energy, which is of any arbitrary large size. In such a meson the action $S \gg \hbar$. Hence a description of this meson necessarily must be semiclassical. Actually we do not need the exact $N_c = \infty$ limit for this statement. It can be formulated in the following way. For any large $S \gg \hbar$ there always exist such N_c that the isolated meson with such an action does exist and can be described semiclassically. From the empirical fact that we observe multiplets of chiral and $U(1)_A$ groups high in the hadron spectrum it follows that $N_c = 3$ is large enough for this purpose.

4. A solvable model of the 't Hooft type

While the argument presented above is general and solid enough, a detailed microscopical picture is missing. Then to see how all this works one needs a solvable field-theoretical model. Clearly the model must be chirally symmetric and contain the key elements, such as confinement and spontaneous breaking of chiral symmetry. Such a model is known, it is a generalized Nambu and Jona-Lasinio model (GNJL) with the instantaneous Lorentz-vector confining kernel [15–17]. This model can be considered as a generalization of the large N_c 't Hooft model (QCD in 1+1 dimensions) [18] to 3+1 dimensions. In both models the only interaction between quarks is the instantaneous infinitly raising Lorentz-vector linear potential. Then chiral symmetry breaking is described by the standard summation of the valence quarks self-interaction loops (the Schwinger-Dyson or gap equations), while mesons are obtained from the Bethe-Salpeter equation for the quark-antiquark bound states.

An obvious advantage of the GNJL model is that it can be applied in 3+1 dimensions to systems of arbitrary spin. In 1+1 dimensions there is no spin, the rotational motion of quarks is impossible. Then it is known that the spectrum represents an alternating sequence of positive and negative parity states and chiral multiplets never emerge. In 3+1 dimension, on the contrary, the quarks can rotate and hence can always be ultrarelativistic and chiral multiplets should emerge naturally [3]. Restoration of chiral symmetry in excited heavy-light mesons has been previously studied with the quadratic confining potential [12]. Here we report our results for excited light-light mesons with the linear potential [19].

An effective chiral symmetry restoration means that (i) the states fall into approximate multiplets of $SU(2)_L \times SU(2)_R$ and the splittings within the multiplets ($\Delta M = M_+ - M_-$) vanish at $n \to \infty$ and/or $J \to \infty$; (ii) the splitting within the multiplet is much smaller than between the two subsequent multiplets [4–6]. Note that within the present model the axial anomaly is absent so the mechanism of the $U(1)_A$ symmetry breaking and restoration is exactly the same as of $SU(2)_L \times SU(2)_R$.

The condition (i) is very restrictive, because the structure of the chiral multiplets for the $J = 0$ and $J > 0$ mesons is very different [4,5]. For the $J > 0$ mesons chiral symmetry requires a *doubling* of states with some quantum numbers. Given the set of quantum numbers I, J^{PC}, the multiplets of $SU(2)_L \times SU(2)_R$ for the $J = 0$ mesons are

$$
\begin{aligned}
(1/2,1/2)_a &: 1, 0^{-+} \longleftrightarrow 0, 0^{++} \\
(1/2,1/2)_b &: 1, 0^{++} \longleftrightarrow 0, 0^{-+},
\end{aligned}
\tag{3}
$$

while for the mesons of even spin, $J > 0$, they are

$$
\begin{aligned}
(0,0) &: 0, J^{--} \longleftrightarrow 0, J^{++} \\
(1/2,1/2)_a &: 1, J^{-+} \longleftrightarrow 0, J^{++} \\
(1/2,1/2)_b &: 1, J^{++} \longleftrightarrow 0, J^{-+} \\
(0,1) \oplus (1,0) &: 1, J^{++} \longleftrightarrow 1, J^{--}
\end{aligned}
\tag{4}
$$

and for odd J they are

$$
\begin{aligned}
(0,0) &: 0, J^{++} \longleftrightarrow 0, J^{--} \\
(1/2,1/2)_a &: 1, J^{+-} \longleftrightarrow 0, J^{--} \\
(1/2,1/2)_b &: 1, J^{--} \longleftrightarrow 0, J^{+-} \\
(0,1) \oplus (1,0) &: 1, J^{--} \longleftrightarrow 1, J^{++}.
\end{aligned}
\tag{5}
$$

Restoration of the $U(1)_A$ symmetry would mean a degeneracy of the opposite spatial parity states with the same isospin from the distinct $(1/2, 1/2)_a$ and $(1/2, 1/2)_b$ multiplets of $SU(2)_L \times SU(2)_R$. Note that within the present model there are no vacuum fermion loops. Then since the interaction between quarks is flavor-blind the states with the same J^{PC} but different isospins from the distinct multiplets $(1/2, 1/2)_a$ and $(1/2, 1/2)_b$ as well as the states with the same J^{PC} but different isospins from $(0, 0)$ and $(0, 1) \oplus (1, 0)$ representations are degenerate.

In the Table below we present masses (in units $\sqrt{\sigma}$) of $I = 1$ mesons with $J = 0, 1$ and $J = 6$. The excited states fall into approximate chiral multiplets and a very fast restoration of both $SU(2)_L \times SU(2)_R$ and $U(1)_A$ symmetries with increasing J and essentially more slow restoration with increasing of n is seen.

chiral multiplet	J^{PC}	radial excitation n 0	1	2	3	4	5	6
$(1/2, 1/2)_a$	0^{-+}	0.00	2.93	4.35	5.49	6.46	7.31	8.09
$(1/2, 1/2)_b$	0^{++}	1.49	3.38	4.72	5.80	6.74	7.57	8.33
$(1/2, 1/2)_a$	1^{+-}	2.68	4.03	5.15	6.14	7.01	7.80	8.53
$(1/2, 1/2)_b$	1^{--}	2.78	4.18	5.32	6.30	7.17	7.96	8.68
$(0, 1) \oplus (1, 0)$	1^{--}	1.55	3.28	4.56	5.64	6.57	7.40	8.16
$(0, 1) \oplus (1, 0)$	1^{++}	2.20	3.73	4.95	5.98	6.88	7.69	8.43
$(1/2, 1/2)_a$	6^{-+}	6.88	7.61	8.29	8.94	9.55	10.1	10.7
$(1/2, 1/2)_b$	6^{++}	6.88	7.61	8.29	8.94	9.56	10.1	10.7
$(0, 1) \oplus (1, 0)$	6^{++}	6.83	7.57	8.25	8.90	9.51	10.1	10.7
$(0, 1) \oplus (1, 0)$	6^{--}	6.83	7.57	8.26	8.90	9.52	10.1	10.7

In Fig. 2 the rates of the symmetry restoration against the radial quantum number n and spin J are shown. It is seen that with the fixed J the splitting within the multiplets ΔM decreases asymptotically as $1/\sqrt{n}$, dictated by the asymptotic linearity of the radial Regge trajectories. This property is consistent with the dominance of the free quark loop logarithm at short distances.

In Fig. 3 the angular Regge trajectories are shown. They exhibit deviations from the linear behavior. This fact is obviously related to the chiral symmetry breaking effects for lower mesons.

In the limit $n \to \infty$ and/or $J \to \infty$ one observes a complete degeneracy

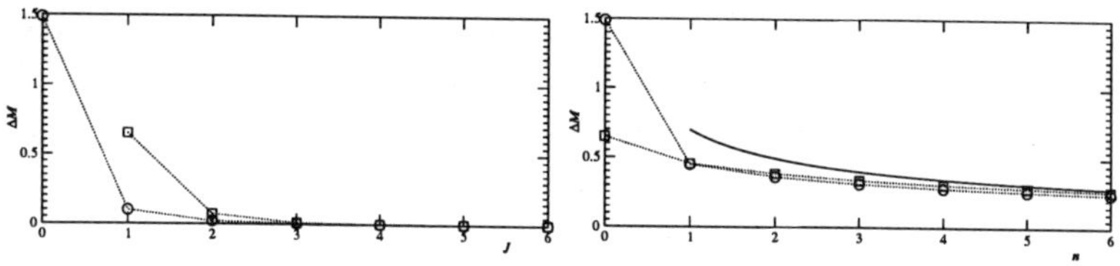

Fig. 2. Mass splittings in units of $\sqrt{\sigma}$ for mesons in the multiplets $(1/2, 1/2)_a$ and $(1/2, 1/2)_b$ (circles) and within the multiplet $(0,1) \oplus (1,0)$ (squares) against J for $n = 0$ and against n for $J = 0$ and $J = 1$, respectively. The full line in the latter plot is $0.7\sqrt{\sigma/n}$

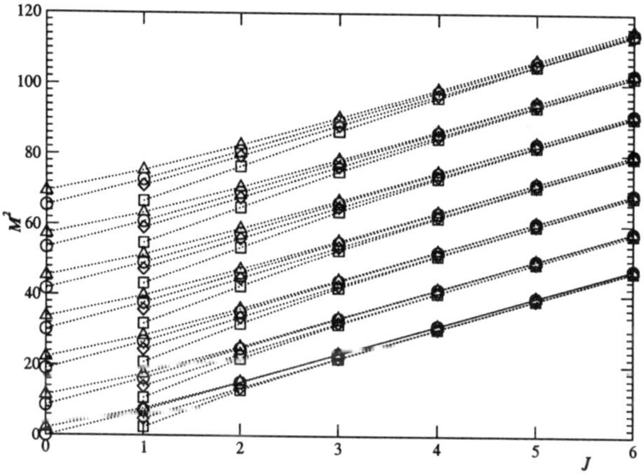

Fig. 3. Angular Regge trajectories for isovector mesons with M^2 in units of σ. Mesons of the chiral multiplet $(1/2, 1/2)_a$ are indicated by circles, of $(1/2, 1/2)_b$ by triangles, and of $(0,1) \oplus (1,0)$ by squares (J^{++} and J^{--} for even and odd J, respectively) and diamonds (J^{--} and J^{++} for even and odd J, respectively).

of all multiplets, which means that the states fall into

$$[(0,1/2) \oplus (1/2,0)] \times [(0,1/2) \oplus (1/2,0)]$$

representation that combines all possible chiral representations for the systems of two massless quarks [5]. This means that in this limit the loop effects disappear completely and the system becomes classical [6,11].

A few words about physics which is behind these results are in order. In highly excited hadrons a typical momentum of valence quarks is large. Consequently, the chiral symmetry violating Lorentz-scalar dynamical mass of quarks, which is a very fast decreasing function at larger momenta, becomes small and asymptotically vanishes [1,11,20]. Consequently, chiral and $U(1)_A$ symmetries get approximately restored. Exactly the same reason implies a decoupling of these hadrons from the Goldstone bosons [3,20]. Namely, the coupling of the valence quarks to Goldstone bosons is constraint by the axial current conservation, i.e. it must satisfy the Goldberger-Treiman relation. Then the coupling constant must be proportional to the Lorentz-scalar dynamical mass of valence quarks and vanishes at larger momenta. This represents a microscopical mechanism of decoupling which is required by the general considerations of chiral symmetry in the Nambu-Goldstone mode [21,22].

This work was supported by the Austrian Science Fund (projects P16823-N08 and P19168-N16).

References

1. L. Ya. Glozman, Phys. Lett. B **475**, 329 (2000).
2. T. D. Cohen and L. Ya. Glozman, Phys. Rev. D **65**, 016006 (2002); Int. J. Mod. Phys. A **17**, 1327 (2002).
3. L. Ya. Glozman, Phys. Lett. B **541**, 115 (2002).
4. L. Ya. Glozman, Phys. Lett. B **539**, 257 (2002).
5. L. Ya. Glozman, Phys. Lett. B **587**, 69 (2004).
6. L. Ya. Glozman, Int. J. Mod. Phys. A. **21**, 475 (2006).
7. M. A. Shifman, A. I. Vainstein and V. I. Zakharov, Nucl. Phys. B **147**, 385 (1979).
8. S.S. Afonin et al, J. High. Energy. Phys. **04**, 039 (2004).
9. M. Shifman, hep-ph/0507246.
10. O. Cata, M. Golterman, S. Peris, hep-ph/0602194.
11. L. Ya. Glozman, A. V. Nefediev, J.E.F.T. Ribeiro, Phys. Rev. D **72**, 094002 (2005).
12. Yu. S. Kalashnikova, A. V. Nefediev, J.E.F.T. Ribeiro, Phys. Rev. D **72**, 034020 (2005).
13. T. DeGrand, Phys. Rev. D **64**, 074024 (2004).
14. T. D. Cohen, hep-ph/0605206.
15. J. R. Finger and J. E. Mandula, Nucl. Phys. B **199**, 168 (1982).
16. A. Le Yaouanc, L. Oliver, O. Pene, and J. C.Raynal, Phys. Rev. D **29**, 1233 (1984); **31**, 137 (1985).
17. P. Bicudo and J. E. Ribeiro, Phys. Rev. D **42**, 1611 (1990); **42**, 1625 (1990).
18. G. t Hooft, Nucl. Phys. B **75**, 461 (1974).
19. R. F. Wagenbrunn, L. Ya. Glozman, hep-ph/0605247.
20. L. Ya. Glozman, A. V. Nefediev, Phys. Rev. D **73**, 074018 (2006).
21. T. D. Cohen and L. Ya. Glozman, hep-ph/0512185.
22. R. L. Jaffe, D. Pirjol, A. Scardicchio, hep-ph/0602010.

WHY MASSLESS PIONS DO NOT PRECLUDE EFFECTIVE CHIRAL RESTORATION IN THE HADRON SPECTRUM

THOMAS D. COHEN

Department of Physics, University of Maryland,
College Park, Maryland 20742, USA
E-mail: cohen@physics.umd.edu

A simple model is presented which illustrates the idea of effective chiral restoration in the hadronic spectrum. This model demonstrates that there is no underlying "no-go" theorem based on general principles of chiral symmetry and its breaking which prevents effective chiral restoration from occurring.

1. Effective Chiral restoration

As has been known for more than four decades the dynamics of strong interactions has an approximate chiral symmetry which is spontaneously broken by some mechanism. Whatever this mechanism is, it has a characteristic scale. One might expect that hadrons with an excitation energy significantly larger than this scale to become largely insensitive to the dynamics of spontaneous chiral symmetry breaking. If this occurs, highly excited hadronic states will fall into patterns which closely approximate the chiral multiplets one would see in the unbroken phase. In baryon spectrum either nearly degenerate parity partners or larger approximate chiral multiplets. The conjecture that this is what occurs in the hadronic spectrum has been dubbed, "effective restoration of chiral symmetry" [1–5]. There is some evidence for this pattern in the highly excited hadrons—there are numerous examples of parity doublets and some evidence for larger chiral multiplets[2–5]. One can argue about the degree to which this empirical evidence is compelling.

2. A theoretical challenge

It remains an open question as to whether this phenomenon occurs. Given this situation it is important to test the idea using whatever theoretical tools are at our disposal.

Recently a paper entitled, *Why massless pions preclude $SU(2)_L \times SU(2)_R$ restoration in the hadron spectrum*, presented analysis based on general properties of chiral symmetry and its breaking which at first glance appears to bear on the issue[6]. The basic argument consisted of:

(i) Reproducing the standard result of Callen, Coleman, Wess and Zumino on allowed representations in the non-linear realization of chiral symmetry[8]. There are no representations with states of opposite parity.

(i) Showing that for a linear realization of chiral symmetry there are no constraints on the masses of opposite parity members of the representation and no constraints on couplings if the system is in the broken phase. This is demonstrated explicitly via field redefinitions involving pion fields.

Based on this reasoning it was argued that chiral restoration in the spectrum was excluded[6]: *"We conclude that parity doubling cannot be a manifestation of chiral symmetry. Once the symmetry is manifested in the Nambu-Goldstone mode, through the appearance of pions, it cannot be "restored", i.e., implemented in the Wigner-Weyl mode, somewhere in the spectrum. Roughly speaking, when the axial charge acts on a hadron, it creates zero energy pions instead of transforming the hadron into a degenerate hadron of the opposite parity."* While a subsequent version[7] of the paper makes somewhat more nuanced statements—and has a more nuanced title—the general sense may have been created that effective chiral symmetry breaking has been excluded—*i.e.*, that a "no -go" theorem exists preventing effective chiral restoration from occurring in the hadron spectrum.

However, this is not the case—there is not a no-go theorem. The principle purpose of this talk is to clear up any misapprehensions on this subject which may have resulted from the argument of ref. [6].

3. A counterexample

The most direct way to disprove a putative theorem is via counterexample. Note that the preceding argument only used the most basic of properties of chiral symmetry and its breaking; it made no specific use of the QCD lagrangian. Thus, one can demonstrate that massless pions do *not* preclude effective chiral restoration—*i.e.*, there is not a no-go theorem—by constructing a model which has both massless pions and effective chiral restoration. This is easy to do[9].

Consider a model with the following characteristics:

(i) It has an infinite number of sigma and pion fields.
(ii) The couplings between the mesons are considered in the weak coupling limit.
(iii) The theory is manifestly chirally symmetric.
(iv) Depending on the sign of a coupling constant it breaks the symmetry through an expectation value of the lowest sigma field.
(v) The coupling of higher mass states to the lowest chiral fields (i.e., the chiral condensate) decrease with increasing mass.

Conditions (i) and (ii) become exact in the large N_c limit of QCD. Condition (ii) renders the theory calculable—the theory can be treated classically in this limit.

The Lagrangian for the theory is given by

$$\begin{aligned}\mathcal{L} = & \sum_j \frac{1}{2}\left(\partial^\mu \sigma_j\, \partial_\mu \sigma_j \, + \partial^\mu \vec{\pi}_j \cdot \partial_\mu \vec{\pi}_j\right) \\ & - \frac{m_o^2}{2}\left(\alpha(\sigma_1^2 + \vec{\pi}_1 \cdot \vec{\pi}_1) + \frac{g}{2m_o^2}(\sigma_1^2 + \vec{\pi}_1 \cdot \vec{\pi}_1)^2\right) \\ & - \frac{m_o^2}{2}\sum_{j=2}^{\infty}\left(j^2(\sigma_j^2 \, + \vec{\pi}_j \cdot \vec{\pi}_j) + \frac{g}{j\, m_o^2}(\sigma_1 \sigma_j + \vec{\pi}_1 \cdot \vec{\pi}_j)^2\right) \\ & - \frac{g}{2j}\sum_{j=2}^{\infty} \sigma_1^2 + \vec{\pi}_1 \cdot \vec{\pi}_1)\,(\sigma_j^2 + \vec{\pi}_j \cdot \vec{\pi}_j)\,. \end{aligned} \tag{1}$$

The constant m_o has dimensions of mass; α and g are dimensionless. The constant α controls chiral symmetry breaking: for $\alpha < 0$ chiral symmetry is spontaneously broken. The factors of $1/j$ impose a decreased coupling strength of the higher-lying fields from the dynamics of chiral symmetry breaking—as one would expect if effective chiral restoration is to occur.

Note that chiral dynamics remains important even in the weak coupling limit. The reason is simple. In the broken phase $\langle \sigma_1 \rangle$ is given by:

$$\langle \sigma_1 \rangle = m_o \sqrt{\frac{-\alpha}{g}}\,. \tag{2}$$

Thus as g becomes small, the expectation value becomes large which in turn compensates for the weak coupling. This behavior is completely consistent with what one has in large N_c QCD where $f_\pi \sim N_c^{1/2}$. The fact that chiral dynamics is in play at weak coupling is clear from Fig. 1. For $\alpha < 0$ there is a Goldstone pion and the pion and sigma fields are split in mass. This fact implies an important point. The theory has a single chiral symmetry, and not one for each pair of fields—as one might naively think happens at weak

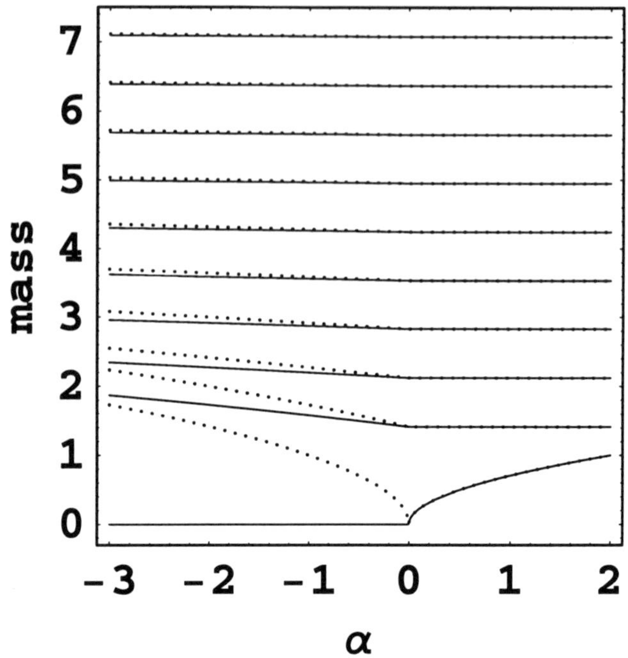

Fig. 1. The mass spectrum of the model in Eq. (1) in units of the mass parameter m_o. The solid lines correspond to pions while the dotted lines correspond to σ-mesons.

coupling. Note that although only one field acquires a vacuum expectation value, all of the pairs are split to one degree or another. Thus, the chiral symmetry is broken everywhere in the spectrum and the theory is in the Goldstone mode throughout.

However, despite being in the Goldstone mode, it is clear from Fig. 1 that as one goes up in the spectrum, the pion and sigma masses becomes closer and closer to being degenerate; as one goes infinitely high in the spectrum the degeneracy becomes exact. This is precisely what is meant by effective chiral restoration in the hadronic spectrum. Thus, the model in Eq. (1) demonstrates clearly that effective chiral restoration is not ruled out on general grounds.

4. Discussion

As noted above, ref. [7] reaches more nuanced conclusions than ref. [6], noting that "parity doubling cannot be a consequence of the $SU(2) \times SU(2)$

symmetry of QCD alone. If it occurs, it must be a manifestation of additional dynamics beyond chiral symmetry." However, the conclusions of ref. [7] are no real help in answering the key question of whether effective chiral restoration occurs. Indeed they are essentially obvious and provide the *starting point* of the analysis in refs. [1–5]. The fundamental issue for parity doublets is whether there is a natural mechanism by which the effects which split them are small, *i.e.*, are the coefficients which split them suppressed for dynamical reasons deeply connected to the underlying chiral symmetry. General arguments based on the structure of chiral symmetry of the sort of refs. [6,7] are irrelevant to this central question.

However, refs. [1–5] made very clear the dynamical reason why effective chiral restoration could occur: As the excitation energies of the hadrons become much larger than the energy scales associated with the dynamics of spontaneous chiral symmetry the hadrons decouple from these effects yielding effective restoration. There remains an open issue associated with this argument. While the general effect must occur in spectral functions at high enough mass, it is not obvious that observable discrete resonances still exist at these high masses.

Part of the reason for possible confusion is because there is an important distinction between qualitative and quantitative effects associated with spontaneous chiral symmetry breaking. There are certain qualitative effects which clearly differentiate between being in the Goldstone phase (for states at an energy scale where effective chiral restoration might occur in the spectrum) from being in the Wigner phase. These distinctions can be made very clear in the case of exact chiral symmetry—*i.e.*, where there is no explicit breaking—and, for the purpose of illustration this limit will be taken. For the sake of argument, suppose further that (narrow) high-lying resonant states exist high in the spectrum. Then if the system is in the Wigner phase, chiral multiplets will exist and the axial matrix element between even parity states and odd parity states in the multiplet are generally nonzero and are fixed by group theory. However, in the Goldstone phase they are identically zero.

To illustrate this let us consider the jth pion and sigma meson multiplets in the model given in Eq. (1). Focus on the matrix element:

$$\langle \sigma_j, \vec{p}_\sigma | A^a_\mu | \pi^a_j, \vec{p}_\pi \rangle = \left(p_{\sigma\,\mu} + p_{\pi\,\mu} \right) f(q^2) \, , \quad q_\mu = p_{\sigma\,\mu} - p_{\pi\,\mu} \, . \tag{3}$$

By symmetry $f(0) = 1$ in the Wigner phase. Next consider taking the 4-divergence of Eq. (3).

$$\partial^\mu \langle \sigma_j, \vec{p}_\sigma | A^a_\mu | \pi^a_j, \vec{p}_\pi \rangle = i \left(m_{\sigma\,j}^{\,2} - m_{\pi\,j}^{\,2} \right) f(q^2) \, . \tag{4}$$

Since chiral symmetry is taken to be exact here, the left-hand side of Eq. (4) is exactly zero. On the other hand, since effective chiral restoration is only approximate, $m_{\sigma j}$ is not exact equal to $m_{\pi j}$, and the equality can only be satisfied if $f(q^2) = 0$. This feature is generic and does not depend on the details of the model.

This shows that certain properties of states in the two phases remain qualitatively distinct no matter how high one goes in the spectrum and regardless of whether effective chiral restoration occurs in the spectrum. The existence of a splitting between opposite parity states in a "would be" multiplet are in this category. The splitting is zero in the Wigner phase and non-zero in the Goldstone phase. However, there the *size* of the splitting between "would be" partners in the Goldstone phase is a quantitative matter and is *not* fixed by any general qualitative properties; it depends entirely on the dynamical details of the theory. Nothing prevents this from being numerically small on the scale of hadronic physics—and there is good reason to suppose that it becomes small at high mass. It is for this reason there is not a no-go theorem for effective chiral restoration even though the system remains in the Wigner phase throughout.

Acknowledgments

This work was supported by the United States Department of Energy. The author thanks L. Ya. Glozman for numerous interesting discussions on this subject.

References

1. L. Ya. Glozman, Phys. Lett. B **475**, 329 (2000).
2. T. D. Cohen and L. Ya. Glozman, Phys. Rev. D **65**, 016006 (2002); Int. J. Mod. Phys. A **17**, 1327 (2002).
3. L. Ya. Glozman, Phys. Lett. B **539**, 257 (2002); *ibid*, **587**, 69 (2004).
4. L. Ya. Glozman, Phys. Lett. B **541**, 115 (2002).
5. L. Ya. Glozman, Int. J. Mod. Phys. A., in press (hep-ph/0411281); L. Ya. Glozman, A. V. Nefediev, J.E.F.T. Ribeiro, Phys. Rev. **D 72**, 094002 (2005).
6. R. L.Jaffe, D. Pirjol, A. Scardicchio, hep-ph/0511081 v1.
7. R. L.Jaffe, D.Pirjol, A. Scardicchio, Phy. Rev. Lett. hep-ph/0511081 v2.
8. S. R. Coleman, J. Wess and B. Zumino, Phys. Rev. **177**, 2239 (1969); C. C. Callan, S. R. Coleman, J. Wess and B. Zumino, Phys. Rev. **177**, 2247 (1969).
9. T. D. Cohen and L. Ya. Glozman, hep-ph/0512185.

QCD GLUEBALL SUM RULES AND VACUUM TOPOLOGY

HILMAR FORKEL

Institute for Theoretical Physics, University of Heidelberg
D-69102 Heidelberg, Germany
and
IFT-UNESP
01405-900 - São Paulo, SP, Brazil
E-mail: forkel@ift.unesp.br

Several key problems of QCD sum rules in the spin-0 glueball channels are resolved by implementing nonperturbative short-distance physics from direct instantons and topological charge screening. A lattice-based instanton size distribution and the IR renormalization of the nonperturbative Wilson coefficients are also introduced. Results of a comprehensive quantitative sum rule analysis are reviewed and their implications discussed.

1. Introduction

The gluonium states of QCD have remained intriguing for almost four decades[1]. Their "exotic" nature reflects itself not least in several long-standing problems which the QCD sum rule approach faces in the spin-0 glueball channels[2]. In the scalar (0^{++}) glueball correlator, in particular, the departure from asymptotic freedom sets in at unusually small distances[3] and the perturbative Wilson coefficients of the standard operator product expansion (OPE) proved inadequate to establish consistency both among the 0^{++} glueball sum rules and with an underlying low-energy theorem[4]. Although nonperturbative contributions due to direct (i.e. small) instantons[5] were early candidates for the missing short-distance physics[6], insufficient knowledge of the instanton size distribution[5,7] prevented their quantitative implementation at the time. Only recently, the derivation of the exact instanton contributions (to leading order in $\hbar$), their duality continua and the corresponding Borel sum-rule analysis[4] showed that direct instantons indeed solve the mentioned key problems in the scalar glueball channel. Below we will outline a more thorough and systematic treatment[9] which eliminates artefacts of earlier approximations, significantly modifies the sum-rule results in both spin-0 channels and improves their reliability. Moreover, we implement for the first time topological charge screening contributions to the OPE in the 0^{-+} channel[9] and show how those resolve earlier sum-rule inconsistencies[10]. Finally, we review several new predictions for spin-0 glueball properties.

2. Correlators and sum rules

The spin-0 glueball sum rules are based on the scalar (0^{++}) and pseudoscalar (0^{-+}) correlations functions

$$\Pi_G(-q^2) = i \int d^4x \, e^{iqx} \langle 0|T \, O_G(x) \, O_G(0) \, |0\rangle \tag{1}$$

where O_G with $G \in \{S, P\}$ are the gluonic interpolating fields (of lowest mass dimension) $O_S = \alpha_s G^a_{\mu\nu} G^{a\mu\nu}$ and $O_P = \alpha_s G^a_{\mu\nu} \tilde{G}^{a\mu\nu}$ with $\tilde{G}_{\mu\nu} \equiv (1/2) \varepsilon_{\mu\nu\rho\sigma} G_{\rho\sigma}$ (the definition in Ref. 9 contains a typo). The zero-momentum limit of the correlator (1) is governed by the low-energy theorems (LETs)

$$\Pi_S \left(q^2 = 0\right) = \frac{32\pi}{b_0} \left\langle \alpha G^2 \right\rangle \tag{2}$$

in the scalar[11] and (for three light flavors and $m_{u,d} \ll m_s$)

$$\Pi_P \left(q^2 = 0\right) = (8\pi)^2 \frac{m_u m_d}{m_u + m_d} \langle \bar{q} q \rangle \tag{3}$$

in the pseudoscalar channel[12]. (Note that Eq. (3) vanishes in the chiral limit.) Consistency with the low-energy theorems places stringent constraints on the sum rules which cannot be satisfied without nonperturbative short-distance physics[4,9].

Contact with the hadronic information in the glueball correlators is established by means of the dispersive representation

$$\Pi_G \left(Q^2\right) = \frac{1}{\pi} \int_0^\infty ds \frac{\mathrm{Im}\, \Pi_G(-s)}{s + Q^2} \tag{4}$$

where the necessary number of subtractions is implied but not written explicitly. The standard sum-rule description of the spectral functions

$$\mathrm{Im}\, \Pi_G^{(ph)}(s) = \mathrm{Im}\, \Pi_G^{(pole)}(s) + \mathrm{Im}\, \Pi_G^{(cont)}(s) \tag{5}$$

contains one or two resonance poles in zero-width approximation, $\mathrm{Im}\, \Pi^{(pole)}(s) = \pi \sum_{i=1}^2 f_{Gi}^2 m_{Gi}^4 \delta\left(s - m_{Gi}^2\right)$, and the local-duality continuum $\mathrm{Im}\, \Pi_G^{(cont)}(s) = \theta(s - s_0) \, \mathrm{Im}\, \Pi_G^{(IOPE)}(s)$ from the IOPE discontinuities in the "duality range" which starts at an effective threshold s_0.

In order to write down QCD sum rules, the Borel-transformed IOPE and dispersive representations of the correlators - weighted by k powers of $-Q^2$ - are matched in the fiducial τ-region (τ is the Borel variable) and rearranged in terms of the continuum-subtracted Borel moments $\mathcal{R}_{G,k}$ as

$$\begin{aligned}\mathcal{R}_{G,k}(\tau; s_0) &\equiv \frac{1}{\pi}\int_0^{s_0} ds s^k \operatorname{Im}\Pi_G^{(IOPE)}(s)\, e^{-s\tau} \\ &= \sum_{i=1}^{2} f_{Gi}^2 m_{Gi}^{4+2k} e^{-m_{Gi}^2\tau} - \delta_{k,-1}\Pi_G^{(ph)}(0).\end{aligned} \tag{6}$$

The pole contributions of interest (and for $k = -1$ the crucial subtraction terms) are then isolated on the RHS, and the hadronic parameters m_{Gi}, f_{Gi} and s_0 can be determined numerically.

3. IOPE

Our theoretical framework, the instanton-improved operator product expansion (IOPE), factorizes the correlators at large, spacelike momenta $Q^2 \equiv -q^2 \gg \Lambda_{QCD}$ into contributions from "hard" field modes (with momenta $|k| > \mu$) in the Wilson coefficients $\tilde{C}_D(Q^2)$ and "soft" field modes (with $|k| \leq \mu$) in the "condensates" $\left\langle \hat{O}_D \right\rangle_\mu$ of operators $\hat{O}_D$ with increasing dimension D. Previous glueball sum rules based on the OPE with purely perturbative Wilson coefficients were plagued by notorious inconsistencies between the predictions of different moment sum rules and by massive LET violations. Moreover, the soft nonperturbative condensate contributions were exceptionally small.

We have therefore analyzed *hard* nonperturbative contributions to the Wilson coefficients. They are strongly channel dependent and due to direct instantons and topological charge screening. The instanton contributions to the spin-0 coefficients[9],

$$\Pi_G^{(I+\bar{I})}(x^2) = \frac{2^8 3}{7}\int d\rho n(\rho)\frac{1}{\rho^4}\, {}_2F_1\left(4, 6, \frac{9}{2}, -\frac{x^2}{4\rho^2}\right), \tag{7}$$

are large (while those in the 2^{++} tensor channel vanish) and add to the unit-operator coefficients $\tilde{C}_0^{(G)}$. The imginary part of their Fourier transform at timelike momenta generates the Borel moments[4]

$$\begin{aligned}\mathcal{R}_k^{(I+\bar{I})}(\tau) = &-2^7\pi^2\delta_{k,-1}\int d\rho n(\rho) - 2^4\pi^3\int d\rho \\ &\times n(\rho)\rho^4\int_0^{s_0} ds s^{k+2} J_2\left(\sqrt{s}\rho\right) Y_2\left(\sqrt{s}\rho\right) e^{-s\tau}\end{aligned} \tag{8}$$

which are similar or larger in size than the perturbative ones. The evaluation of these moments requires as the sole input the (anti-) instanton distribution $n(\rho)$ which is implemented by means of a lattice-based Gaussian-tail parametrization with the correct small-ρ behavior,

$$n_g(\rho) = \frac{2^{18}}{3^6\pi^3}\frac{\bar{n}}{\bar{\rho}}\left(\frac{\rho}{\bar{\rho}}\right)^4 \exp\left(-\frac{2^6}{3^2\pi}\frac{\rho^2}{\bar{\rho}^2}\right) \tag{9}$$

(for $N_c = N_f = 3$), which was introduced in Ref. 9 and shown to prevent several artefacts of the oversimplified "spike" approximation $n(\rho) = \bar{n}\delta(\rho - \bar{\rho})$ on which all previous direct-instanton calculations had relied.

As an additional benefit, the realistic size distribution allows for a gauge-invariant IR renormalization which excludes large instantons with size $\rho > \mu^{-1}$ from the Wilson coefficients,

$$n(\rho) \to \tilde{n}(\mu;\rho) \equiv \theta_\beta\left(\rho - \mu^{-1}\right) n(\rho) \tag{10}$$

(θ_β is a soft step function). The instanton-induced μ dependence turns out to be relatively weak for $\mu < \bar{\rho}^{-1}$, as necessary to compensate its perturbative counterpart. Neglect of this renormalization, although common practice in perturbative Wilson coefficients, would significantly contaminate the results, e.g. by missing the reduction of the direct-instanton density

$$\bar{n} = \int_0^\infty d\rho n(\rho) \to \int_0^\infty d\rho \tilde{n}(\mu;\rho) \equiv \bar{n}(\mu). \tag{11}$$

Another important renormalization effect is the reduction of the instanton contributions to the pseudoscalar relative to the scalar sum rules.

The instanton's self-duality causes a strongly repulsive contribution to the 0^{-+} channel, with seemingly detrimental impact on the sum rules[10]: the glueball signal disappears and both unitarity and the LET (3) are badly violated. The origin of these problems can be traced to the neglect of topological charge screening[9]. Due to their high channel selectivity and a small screening length $\lambda_D \sim m_{\eta'}^{-1} \sim 0.2$ fm, the model-independent screening correlations[14]

$$\Pi_P^{(scr)}(x) \simeq -2^8\pi^2\left(\xi\gamma_{\eta_0}\right)^2 \langle \eta_0(x)\, \eta_0(0)\rangle \tag{12}$$

(ξ is the overall topological charge density ($= \bar{n}$ for instantons) and η_0 the flavor-singlet part of the η') affect almost exclusively the 0^{-+} Wilson coefficients. They arise from the axial-anomaly induced attractive (repulsive) interaction between topological charge lumps of opposite (equal) sign due to η_0 exchange[13] and, after correcting for $\eta_0 - \eta_8$ mixing, add the terms

$$\mathcal{R}_{P,k}^{(scr)}(\tau) = -\delta_{k,-1}\left(\frac{F_{\eta'}^2}{m_{\eta'}^2} + \frac{F_\eta^2}{m_\eta^2}\right) + F_{\eta'}^2 m_{\eta'}^{2k} e^{-m_{\eta'}^2\tau} + F_\eta^2 m_\eta^{2k} e^{-m_\eta^2\tau} \tag{13}$$

to the pseudoscalar IOPE moments.

They strongly reduce the direct-instanton induced repulsion and resolve the disastrous problems mentioned above[9]: positivity of the spectral function is restored, the four 0^{-+} Borel sum rules (6) are stable and contain consistent pseudoscalar glueball information. (All previous analyses had discarded the $k = -1$ sum rule and thereby missed valuable first-principle information and LET consistency checks.)

4. Results and discussion

We have implemented direct instanton and topological charge screening contributions into the OPE coefficients of the spin-0 glueball correlators, evaluated their duality continuua and demonstrated how these contributions resolve the problems of previous QCD sum rule analyses[9]. A lattice-based instanton size distribution and the gauge-invariant IR renormalization of the nonperturbative Wilson coefficients were also introduced. Quark admixtures, and thereby quarkonium mixing effects, enter through quark loops, the instanton size distribution and the condensates.

In the scalar channel, the sizeable direct instanton contributions are indispensable for mutually and LET consistent sum rules. Their improved treatment reduces our earlier (spike-distribution based) result for the 0^{++} glueball mass to $m_S = 1.25 \pm 0.2$ GeV. (The mass stays well beyond 1 GeV, however, in contrast to obsolete predictions based on purely perturbative coefficients.) This value is somewhat smaller than the quenched lattice results[15] (which will probably be reduced by light-quark effects) and consistent with the broad glueball state found in a comprehensive K-matrix analysis[16]. The systematics among our different Borel moments likewise indicates a rather large width of the scalar glueball, $\Gamma_S \gtrsim 0.3$ GeV.

Our prediction for the scalar glueball decay constant, $f_S = 1.05 \pm 0.1$ GeV, is several times larger than the value obtained when ignoring the nonperturbative Wilson coefficients. This implies an exceptionally small 0^{++} glueball size, in agreement with several lattice results[17]. Another stringent, OPE- and sum-rule-independent consistency check of the instanton contributions and their f_S enhancement provide numerical simulations in an instanton ensemble[18] (ILM). For $|x| \lesssim 0.5$ fm the scalar ILM and IOPE correlators are indeed very similar, and the ILM prediction $f_S^{(ILM)} = 0.8$ GeV is similarly large. This indicates a robust instanton effect and rules out that the large f_S "may signal some eventual internal inconsistencies in the treatment of the instanton contributions"[19]. A subsequent model-independent confirmation of the f_S enhancement was supplied by the first direct (quenched) lattice result[20] $f_S = 0.86 \pm 0.18$ GeV which is consistent both with our prediction and the ILM value (the latter is practically independent of quark quenching[18]).

Our prediction for f_S implies enhanced partial widths for radiative J/ψ and Υ decays into scalar glueballs and is therefore relevant for experimental glueball searches, e.g. in the CLEO and BES data on $\Upsilon \to \gamma f_0$ and other decay branches. Since the exceptionally small size and large decay constant are particular to the *scalar* glueball and since only part of the scalar decay constant contributes to the radiative production rates[21], however, the above results are not ruled out by experimental data on decays into 0^{-+} glueballs (cf. Ref. 19).

In the pseudoscalar (0^{-+}) glueball correlator we have identified and implemented a new type of nonperturbative contributions to the Wilson coefficients, due to topological charge screening. Roughly speaking, the screening effects "unquench" the direct instanton contributions, thereby restoring unitarity, the axial Ward identity and the resonance signals. Consistency among all moment sum rules and with the underlying LET is also achieved, and the resulting mass prediction $m_P = 2.2 \pm 0.2$ GeV lies inside the range of quenched and unquenched lattice data. The coupling $f_P = 0.6 \pm 0.25$ GeV is somewhat enhanced by the topological short-distance physics, affecting radiative production rates, the $\gamma\gamma \to G_P \pi^0$ cross section at high momentum transfers and other glueball signatures.

The crucial impact of the nonperturbative Wilson coefficients on both spin-0 glueball correlators is particularly evident in the interplay between their subtraction constants. Indeed, the notorious consistency problems which plagued previous 0^{++} glueball sum rules were primarily caused by the large LET-induced subtraction constant $\Pi_S(0) \simeq 0.6$ GeV^4 (cf. Eq. (2)): it cannot be matched by perturbative Wilson coefficients (since $\Pi_S^{(pert)}(0) = 0$) and requires the direct instanton contribution $\Pi_{S/P}^{(I,\bar{I})}(0) = \pm 2^7 \pi^2 \bar{n}_{dir} \simeq 0.63$ GeV^4. At first sight this seems to imply a conflict with the much smaller LET subtraction constant $\Pi_P(0) \simeq -0.02$ GeV^4 (cf. Eq. (3)) in the 0^{+-} channel, however, since the instanton contributions to both spin-0 correlators are equal (up to a sign). Here the topological screening contributions $\Pi_P^{(scr)}(0) \simeq 0.59$ GeV^4 from Eq. (13) prove indispensable: they restore consistency by canceling most (and in the chiral limit, where $\Pi_P(0) \to 0$, all) of the instanton contributions: $\Pi_P^{(scr)}(0) + \Pi_P^{(I,\bar{I})}(0) \simeq \Pi_P(0)$.

To summarize: contrary to naive expectation, the nonperturbative contributions to the OPE of the spin-0 glueball correlators reside primarily in the Wilson coefficients (i.e not in the condensates) and are closely related to the topological vacuum structure. The nonperturbative short-distance physics resolves long-standing consistency problems of the associated QCD sum rules and generates a rather diverse set of new glueball predictions. A large part of the 0^{++} glueball mass and binding originates from direct instantons, for example, while their net effects in the 0^{+-} channel are smaller and more subtle, due to cancellations between instanton and topological charge screening contributions.

This work was supported by FAPESP and CNPq of Brazil.

References

1. M. Gell-Mann, Acta Phys. Aust. Suppl. **9**, 733 (1972); H. Fritzsch and M. Gell-Mann, 16th Int. Conf. High-Energy Phys., Chicago, Vol. 2, 135 (1972).
2. S. Narison, Nucl. Phys. **B509**, 312 (1998) and references therein.
3. V.A. Novikov, M.A. Shifman, A.I. Vainsthein, and V.I. Zakharov, Nucl. Phys. **B191**, 301 (1981).
4. H. Forkel, Phys. Rev. D **64**, 034015 (2001).
5. T. Schäfer and E.V. Shuryak, Rev. Mod. Phys. **70**, 323 (1998). For an introduction see H. Forkel, *A Primer on Instantons in QCD*, hep-ph/0009136
6. V.A. Novikov, M.A. Shifman, A.I. Vainsthein, and V.I. Zakharov, Nucl. Phys. **B165**, 67 (1980).
7. A. Ringwald and F. Schremmp, Phys. Lett. **B 459**, 249 (1999).
8. D. Harnett and T.G. Steele, Nucl. Phys. **A 695**, 205 (2001).
9. H. Forkel, Phys. Rev. D **71**, 054008 (2005); Braz. J. Phys. **34**, 875 (2004); AIP Conf. Proc. **739**, 434 (2004).
10. A. Zhang and T.G. Steele, Nucl. Phys. **A 728**, 165 (2003).
11. V.A. Novikov, M.A. Shifman, A.I. Vainshtein and V.I. Zakharov, Nucl. Phys. **B191**, 301 (1981).
12. H. Leutwyler and A. Smilga, Phys. Rev. D **46**, 5607 (1992).
13. P. Di Vecchia and G. Veneziano, Nucl. Phys. **B 171**, 253 (1980).
14. N.J. Dowrick and N.A. McDougall, Phys. Lett. **B 285**, 269 (1992); H. Kikuchi and J. Wudka, Phys. Lett. **B 284**, 111 (1992).
15. W. Lee and D. Weingarten, Phys. Rev. D **61**, 014015 (2000) and references therein; C. Morningstar and M. Peardon, Phys. Rev. D **60**, 034509 (1999).
16. V.V. Anisovich, AIP Conf. Proc. **717**, 441 (2004), arXiv:hep-ph/0310165.
17. N. Ishii, H. Suganuma, and H. Matsufuru, Phys. Rev. D **66**, 94506 (2002); P. de Forcrand and K.-F. Liu, Phys. Rev. Lett. **69**, 245 (1992); R. Gupta et al., Phys. Rev. D **43**, 2301 (1991).
18. T. Schäfer and E.V. Shuryak, Phys. Rev. Lett. **75**, 1707 (1995).
19. S. Narison, arXiv:hep-ph/0512256.
20. Y. Chen et al., Phys. Rev. D **73**, 014516 (2006).
21. X.-G. He, H.-Y. Jin and J. P. Ma, Phys. Rev. D **66**, 74015 (2002).

COUNTING RULES, HOLOGRAPHIC WAVE FUNCTIONS, MEROMORPHIZATION AND QUARK-HADRON DUALITY

A. V. RADYUSHKIN[1,2,*]

[1] *Physics Department, Old Dominion University, Norfolk, VA 23529, USA*
[2] *Theory Center, Jefferson Lab, Newport News, VA 23606, USA*
E-mail: radyush@jlab.org

We start with study of the meson form factor $F_M(Q^2)$ constructed using recently proposed holographic light-front wave functions. We find that its asymptotic behavior is generated by soft Feynman mechanism rather than by short distance dynamics. This causes a very late onset of the $1/Q^2$ asymptotic behavior at unaccessible momenta $Q^2 \gtrsim 10\ \mathrm{GeV}^2$. We show that this model can be also obtained within the Migdal's "meromorphization" approach if one applies it to 3-point function for scalar currents made of scalar quarks. For spin-1/2 quarks, we demonstrated that resulting form factor $F_M^{\mathrm{spinor}}(Q^2)$ has $1/Q^4$ asymptotic behavior, but owing to late onset of this asymptotic pattern, $F_M^{\mathrm{spinor}}(Q^2)$ imitates the $1/Q^2$ behavior in the few GeV^2 region. In the meromorphization approach, just like in the local quark-hadron duality model for the pion form factor, adding the $\mathcal{O}(\alpha_s)$ correction to the spectral function brings in the hard pQCD contribution that has the dimensional counting $1/Q^2$ behavior at large Q^2. At accessible Q^2, the $\mathcal{O}(\alpha_s)$ term is a rather small fraction of the total result. In this scenario, both hard and soft contributions are present, and the "observed" quark counting rules for hadronic form factors is an approximate and transitional phenomenon resulting from long-distance dynamics, Feynman mechanism in its preasymptotic regime.

1. Introduction

Experimental evidence that form factors of hadrons consisting of n quarks have behavior close to $(1/Q^2)^{n-1}$ provokes expectations that there is a fundamental and/or easily visible reason for such a phenomenon, scale invariance being the most natural suspect[1]. Hard rescattering in a theory with dimensionless coupling constant provides a specific dynamical mechanism[2] that produces a scale invariant behavior. Perturbative QCD predicts the $(\alpha_s/Q^2)^{n-1}$ asymptotic behavior[3–6]. In the pion case, the pre-

[*] Also at Bogoliubov Laboratory of Theoretical Physics, JINR, Dubna, Russia

diction is $F_\pi(Q^2) = (2\alpha_s/\pi)s_0/Q^2$, where $s_0 = 4\pi^2 f_\pi^2 \approx 0.7\,\text{GeV}^2$ is a constant close to $m_\rho^2 \approx 0.6\,\text{GeV}^2$. This indicates that the pQCD asymptotics *is not* the large-Q^2 limit of the phenomenologically successfull VMD fit $F_\pi^{\text{VMD}}(Q^2) \sim 1/(1+Q^2/m_\rho^2)$, but rather looks like $\mathcal{O}(\alpha_s)$ correction to it. Also, the smallness of α_s/π undermines attempts to describe available data solely by pQCD hard mechanism. The growing consensus is that at available Q^2 form factors are dominated by soft contributions described by nonforward parton densities[7] $\mathcal{F}(x,Q^2)$ (or generalized parton distributions $H(x,\xi;Q^2)$ for zero skewness ξ), and successful fits were obtained[8–10] in models with $\mathcal{F}(x,Q^2) = f(x)\,e^{-Q^2 g(x)}$ having exponential behavior for large Q^2 at fixed x. A powerlike asymptotics in this case appears only after integration over x, i.e., it is governed by the Feynman mechanism[11], and is determined by the $x \to 1$ behavior of $f(x)$ in contrast to the hard mechanism for which the subprocess amplitude already has the $(1/Q^2)^{n-1}$ power behavior not affected by subsequent x-integrations.

Another new development is related to applications of AdS/CFT construction to QCD and claims[12,13] that this framework ("AdS/QCD") provides a nonperturbative explanation of quark counting rules. Given explicit expressions with the basic scale Λ fixed from fitting the hadron masses, it is straightforward to check the structure of AdS/QCD results for form factors and their potential to describe the features of existing data. This is one of the goals of the present investigation. Another is to study the recently proposed interpretation[14] of AdS/QCD results in terms of light-front wave functions, which opens a possibility to find out whether, in terms of the light-cone momenta $x, \mathbf{k}_\perp$, the AdS/QCD quark counting corresponds to large-$\mathbf{k}_\perp$ hard mechanism or, as we will show, to the $x \to 1$ soft Feynman/Drell-Yan[11,15] mechanism. In view of yet another recent observation[16] that some of the results of the AdS/QCD coincide with those of Migdal's program[17] we apply the extension[18] of this "meromorphization" idea to the 3-current correlators, and establish a connection between this approach and holographic light-front wave functions of Ref.[14].

The meromorphization procedure has features similar to the "local quark-hadron duality" model[19] that succesfully describes the pion form factor data and gives a unified description of its soft and hard parts. The soft term $F_\pi^{\text{LD}}(Q^2)$ in this approach dominates at accessible Q^2, but has the $1/Q^4$ asymptotic behavior. Due to its late onset, the curve $Q^2 F_\pi^{\text{LD}}(Q^2)$ has a wide plateau in a few GeV^2 region, i.e., $F_\pi^{\text{LD}}(Q^2)$ imitates there the $1/Q^2$ behavior. Thus, the desired $1/Q^2$ result for accessible Q^2 is obtained because the nonperturbative term has a faster, $1/Q^4$ asymptotic fall-off.

2. Holographic wave functions and Feynman mechanism

In the hard-wall approximation[12], the expression for the elastic form factor in the holographic duality model is given by

$$F(Q^2) = \int_0^{1/\Lambda} \frac{d\zeta}{\zeta^3} \, \Phi_{P'}(\zeta) J(Q,\zeta) \Phi_P(\zeta) \ , \tag{1}$$

where $J(Q,\zeta) = \zeta Q K_1(\zeta Q)$ stays for EM current, and $\Phi_P(\zeta)$, $\Phi_{P'}(\zeta)$ for initial and final states, with $\Phi(\zeta) = C\zeta^2 J_L(\beta_{L,k}\zeta\Lambda)$, $\beta_{L,k}$ being the kth root of the Bessel function $J_L(x)$. On the other hand, in the light-cone (LC) formalism, the meson form factor is given by a 3-dimensional integral. In particular, the 2-body part of a meson form factor is given by

$$F_{(2)}(Q^2) = \int_0^1 dx \int d^2\mathbf{b}_\perp e^{i\bar{x}\mathbf{b}_\perp\cdot\mathbf{q}_\perp} \left|\Psi_2(x,\mathbf{b}_\perp)\right|^2 \equiv \int_0^1 dx \, \mathcal{F}_{(2)}(x,Q^2) \ , \tag{2}$$

where $\bar{x} \equiv 1-x$ and $\mathcal{F}_{(2)}(x,Q^2)$ is the 2-body part of GPD. In Ref.[14] it was noticed that if $\Psi(x,\mathbf{b}_\perp) \sim \sqrt{x\bar{x}}\,\phi\,(b\sqrt{x\bar{x}})$, then LC formula converts into the 1-dimensional integral (1) with correct $J(Q,\zeta)$, provided that $\zeta = b\sqrt{x\bar{x}}$. This gives the "holographic" model[14] for light-front wave functions. For the lowest meson, $L=0, k=1$ and mass $M = \beta_{0,1}\Lambda$. Then

$$\Psi_M(x,b) = \frac{M\sqrt{x\bar{x}/\pi}}{\beta_{0,1}J_1(\beta_{0,1})} \, J_0(\sqrt{x\bar{x}}Mb)\,\theta(\sqrt{x\bar{x}}\,b \le \beta_{0,1}/M) \ . \tag{3}$$

The $\mathbf{k}_\perp$ counterpart of this wave function is given by

$$\widetilde{\Psi}_M(x,\mathbf{k}_\perp) = \frac{M}{\sqrt{\pi x\bar{x}}} \, \frac{J_0(\beta_{0,1}k_\perp/\sqrt{x\bar{x}}M)}{M^2 - k_\perp^2/x\bar{x}} \ . \tag{4}$$

For small $k_\perp$, this wave function is close to $\exp[-k_\perp^2/2M^2x\bar{x}]/\sqrt{x\bar{x}}$. For large $k_\perp$, it oscillates with the magnitude decreasing as $1/k_\perp^{5/2}$. This result contradicts the statement made in Ref.[13] that the AdS/QCD construction corresponds to a purely power-law large-$k_\perp$ behavior.

Explicit calculation of the lowest state form factor gives

$$F_M(Q^2) = \frac{4M^2}{Q^2[\beta_{0,1}J_1(\beta_{0,1})]^2}\left[1 - \frac{4M^2}{Q^2} + \frac{9}{8}\left(\frac{4M^2}{Q^2}\right)^2 + \mathcal{O}(M^6/Q^6)\right] . \tag{5}$$

Though this result has a monopole-like structure, the scale $4M^2$ is evidently too large if $M = m_\rho$: the curve for $Q^2F_M(Q^2)$ is far from being flat in the accessible region $Q^2 \lesssim 10$ GeV^2.

The easiest way to see which mechanism is responsible for the $1/Q^2$ asymptotics is to approximate $\widetilde{\Psi}_M(x,\mathbf{k}_\perp)$ by a Gaussian function $\widetilde{\Psi}_G(x,\mathbf{k}_\perp) \sim \exp[-k_\perp^2/2M^2x\bar{x}]/\sqrt{x\bar{x}}$. In the impact parameter space,

$\Psi_G(x,b) \sim \sqrt{x\bar{x}}\exp[-b^2M^2x\bar{x}/2]$. Integrating over *all* (including small) values of $\mathbf{b}_\perp$ in the form factor formula (2) gives $\mathcal{F}_G(x,Q^2) = e^{-\bar{x}Q^2/4xM^2}$, a function that, for any fixed x, vanishes exponentially at large Q^2. The power-law asymptotics is obtained only after integrating over x, which gives

$$F_G(Q^2) = \sum_{n=1}^{\infty}(-1)^{n-1}n!\left(\frac{4M^2}{Q^2}\right)^n , \tag{6}$$

a result that has the $\mathcal{O}(M^2/Q^2)$ large-Q^2 behavior and structure similar to that of Eq. (5). The crucial role of integration over the $x \sim 1$ region in getting the power-law behaviour is evident: if the integration is restricted to $x < x_0$, the outcome vanishes like $\exp[-\bar{x}_0Q^2/4x_0M^2]$ for large Q^2. Thus, the GPD $\mathcal{F}_G(x,Q^2)$ has the same $\mathcal{F}(x,Q^2) = f(x)\,e^{-Q^2g(x)}$ structure as those considered in Refs.[8–10]. In Ref.[21] it is shown that the large-Q^2 asymptotics of the meson form factor in the model of Ref.[14] with original, nonapproximated wave function is also governed by the *soft* Feynman mechanism, with the power-law asymptotics determined by the $x \to 1$ behavior of the prefactor $f(x)$ accompanying a decreasing function of $\bar{x}Q^2/x\Lambda^2$. In the present case, $f(x) = 1$, which gives $F_M(Q^2) \sim 1/Q^2$. Clearly, $f(x) = \mathcal{F}(x,Q^2=0)$ is the parton distribution function, i.e., the model of Ref.[14] gives a constant, x-independent parton distribution $f(x) = 1$.

3. Meromorphization and local quark-hadron duality

One may question interpretation of the holographic variable ζ as a particular product $b\sqrt{x\bar{x}}$ of light-cone variables. We are going to show that the picture similar to that of Ref.[14] emerges also within the approach related to Migdal's program[17] of Padé approximating the correlators $\Pi(p^2)$ of hadronic currents calculated in perturbation theory. Recently, it was demonstrated[16] that some of Migdal's results coincide with those of the holographic approach. Migdal's program involves "meromorphization"

$$\Pi(p^2) = \frac{1}{\pi}\int_0^\infty \frac{\rho(s)}{s-p^2}\,ds \Rightarrow \Pi_{\mathcal{M}}(p^2) = \Pi(p^2) - \frac{1}{\pi\mathcal{Q}(p^2)}\int_0^\infty \frac{\rho(s)\,\mathcal{Q}(s)}{s-p^2}\,ds \tag{7}$$

that substitutes the original correlator $\Pi(p^2)$ by a function $\Pi_{\mathcal{M}}(p^2)$ in which the cut of the original correlator for real positive p^2 is eliminated by the second term in Eq. (7), with zeros of $\mathcal{Q}(p^2)$ at timelike p^2 generating the poles interpreted as hadronic bound states. Explicit Padé construction gives $\mathcal{Q}(p^2) \Rightarrow J_0(\beta_{0,1}\sqrt{p^2}/M)$, with M being the mass of the lowest state. In the lowest order, $\rho(s) = \rho_0\,\theta(s)$, with $\rho_0 = 1/16\pi$ for $j = \varphi\varphi$. The coupling constant f_M^2 of the lowest state is given by

$$f_M^2 = \lim_{p^2\to M^2}(M^2 - p^2)\,\Pi_{\mathcal{M}}(p^2) = \frac{4\rho_0 M^2}{\pi[\beta_{0,1}J_1(\beta_{0,1})]^2}\,. \tag{8}$$

For form factors, it was suggested[18] to "meromorphize" the 3-point function $T(p_1^2, p_2^2, Q^2)$ by adding the term

$$\frac{1}{\pi^2\mathcal{Q}(p_1^2)\mathcal{Q}(p_2^2)}\int_0^\infty ds_1\int_0^\infty ds_2\,\frac{\rho(s_1,s_2,Q^2)\,\mathcal{Q}(s_1)\,\mathcal{Q}(s_2)}{(s_1-p_1^2)(s_2-p_2^2)}\,, \tag{9}$$

that removes its cuts and substitutes them by poles at the same locations as in $\Pi_{\mathcal{M}}(p^2)$. The elastic form factor of the lowest state is extracted using

$$f_M^2 F_M(Q^2) = \lim_{p_1^2\to M^2}\lim_{p_2^2\to M^2}(p_1^2-M^2)(p_2^2-M^2)\,T_{\mathcal{M}}(p_1^2,p_2^2,Q^2)\,. \tag{10}$$

The spectral densities $\rho(s_1, s_2, Q^2)$ can be calculated[20] using the Cutkosky rules and light-cone variables:

$$\rho(s_1,s_2,Q^2) = \rho_0\int_0^1 dx\,\frac{n(x)}{x\bar{x}}\int d^2\mathbf{k}_\perp\,\delta\left(s_1 - \frac{\mathbf{k}_\perp^2}{x\bar{x}}\right)\delta\left(s_2 - \frac{(k_\perp + \bar{x}\mathbf{q}_\perp)^2}{x\bar{x}}\right), \tag{11}$$

Taking scalar $\varphi\varphi$ currents for hadronized vertices gives $n(x) = 1$ and

$$f_M^2 F_M^{\text{scalar}}(Q^2) = \frac{\rho_0}{\pi^2[\mathcal{Q}'(M^2)]^2}\int_0^1\frac{dx}{x\bar{x}}\int d^2\mathbf{k}_\perp \times \frac{\mathcal{Q}(\mathbf{k}_\perp^2/x\bar{x})}{M^2 - \mathbf{k}_\perp^2/x\bar{x}}\,\frac{\mathcal{Q}((\mathbf{k}_\perp+\bar{x}\mathbf{q}_\perp)^2/x\bar{x})}{M^2 - (\mathbf{k}_\perp+\bar{x}\mathbf{q}_\perp)^2/x\bar{x}}\,, \tag{12}$$

i.e., a LC form factor expression. Using $\mathcal{Q}(s) = J_0(\beta_{0,1}\sqrt{s}/M)$ and Eq. (8) for f_M^2, we obtain that the relevant wave function $\widetilde{\Psi}_{\mathcal{M}}^{\text{scalar}}(x, \mathbf{k}_\perp)$ coincides with the holographic wave function (4), thus supporting interpretation in terms of light-cone variables x, b proposed in Ref.[14].

Using spin-1/2 quarks and vector currents j_α, j_β gives tensor amplitude $T^\mu_{\alpha\beta}(p_1, p_2)$. As a simple example, we took the projection $T^\mu_{\alpha\beta}n_\mu n^\alpha n^\beta$, where n is a lightlike vector satisfying $(nq) = 1$. The numerator factor is then $n(x) = 6x\bar{x}$. Explicit calculation gives

$$F_M^{\text{spinor}}(Q^2) = \frac{2M^2}{Q^2[\beta_{0,1}J_1(\beta_{0,1})]^2}\left[0 + 24\,\frac{M^2}{Q^2} - 288\,\frac{M^4}{Q^4} + \mathcal{O}(M^6/Q^6)\right] \tag{13}$$

The first term vanishes, and the leading term has $1/Q^4$ behavior, the result that could be traced to an extra $\bar{x}$ factor in GPD $\mathcal{F}(x, Q^2)$. Due to slow convergence, the $1/Q^4$ asymptotics establishes only above $20\,\text{GeV}^2$. As a

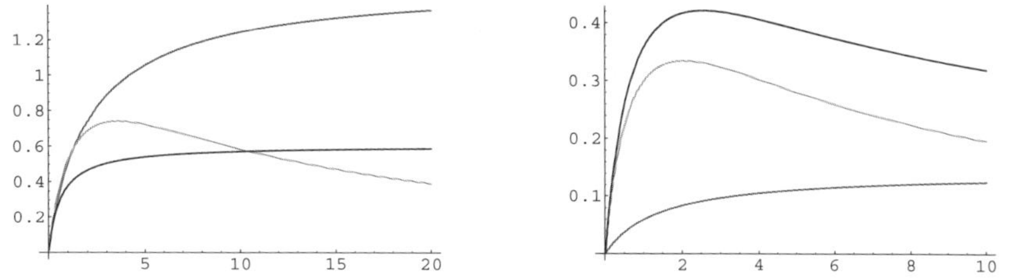

Fig. 1. Left: Meson form factors: $Q^2F_M(Q^2)$ (upper curve), $Q^2F_M^{\rm spinor}(Q^2)$ (middle curve); for comparison also shown $Q^2/(1+Q^2/m_\rho^2)$ (lower curve).
Right: Pion form factor in local quark-hadron duality model: $Q^2F_\pi^{\rm LD(\alpha_s)}$ (lower curve), $Q^2F_\pi^{\rm LD}$ (middle curve), total contribution (upper curve).

result, the $F_M^{\rm spinor}(Q^2)$ curve in the region of a few GeV^2 imitates the "power counting" $1/Q^2$ behavior much more successfully than Eq. (5) that displays its nominal $1/Q^2$ asymptotics only well outside the few GeV^2 region.

Referring to the "established" $1/Q^2$ behavior of meson form factors, one has in mind data on the pion EM form factor which resemble monopole form $F_\pi(Q^2) \sim 1/(1+Q^2/m_\rho^2)$. In fact, the data are well described by our local quark-hadron duality model[19], in which the pion form factor is obtained by integrating the perturbative spectral density $\rho(s_1, s_2, Q^2)$ with $\theta(s_i \le s_0)$ weights, where $s_0 = 4\pi^2 f_\pi^2 \approx 0.7\,\mathrm{GeV}^2$ is the duality interval. This prescription is analogous to that derived in the meromorphization approach, where density was integrated with weights proportional to $J_0(\beta_{0,1}\sqrt{s_i}/M)/(s_i - M^2)$. Explicit expression for the pion form factor in the local duality model is known from Ref.[19]:

$$F_\pi^{\rm LD}(Q^2) = 1 - \frac{1+6s_0/Q^2}{(1+4s_0/Q^2)^{3/2}} = \frac{6s_0^2}{Q^4} - \frac{40s_0^3}{Q^6} + \frac{210s_0^4}{Q^8} + \mathcal{O}(s_0^4/Q^8)\,. \quad (14)$$

Again, $F_\pi^{\rm LD}(Q^2)$ behaves like $1/Q^4$ for large Q^2, but convergence is slow, and for a few Q^2 the $F_\pi^{\rm LD}(Q^2)$ curve successfully imitates the $1/Q^2$ behavior and goes very close to existing and preliminary experimental data[22].

Both in the meromorphization and local duality approaches one should include higher order α_s corrections to spectral densities. Among $\mathcal{O}(\alpha_s)$ contributions, there are gluon-exchange diagrams whose asymptotic large-Q^2 behavior is determined by the hard pQCD mechanism. As a result, the leading $1/Q^2$ term of the spectral density can be written in pQCD-like form. Substituting it into the local duality relation gives the result coinciding with the pQCD hard gluon exchange contribution $F_\pi^{\rm pQCD}(Q^2) = 8\pi\alpha_s f_\pi^2/Q^2 = 2(s_0/Q^2)(\alpha_s/\pi)$ calculated for the asymptotic shape of the pion DA. A very good approximation for the local duality model prediction for $F_\pi^{\mathrm{LD}(\alpha_s)}$ at all Q^2 is given by a simple interpolation formula[23] $F_\pi^{\mathrm{LD}(\alpha_s)}(Q^2) = (\alpha_s/\pi)/(1+Q^2/2s_0)$ between the $Q^2 = 0$ value $F_\pi^{\mathrm{LD}(\alpha_s)}(0) = \alpha_s/\pi$ (fixed by the Ward identity) and the large-Q^2 asymptotic behavior. With $\alpha_s/\pi \approx 0.1$, the $\mathcal{O}(\alpha_s)$ term is a $\lesssim 30\%$ correction to the $\mathcal{O}(\alpha_s^0)$ term in the $Q^2 \leq 4$ GeV^2 region, and their sum is in very good agreement with data[22].

In this scenario, both soft and hard contibutions are present, and the "observed" quark counting appear because the numerically larger soft term (having faster decrease at asymptotically large Q^2) approximately imitates the $1/Q^2$ behavior in the experimentally accessible region.

References

1. V. A. Matveev, R. M. Muradian and A. N. Tavkhelidze, Lett. Nuovo Cimento **7**, 719 (1973).
2. S. J. Brodsky and G. R. Farrar, Phys. Rev. Lett. **31**, 1153 (1973).
3. G. R. Farrar and D. R. Jackson, Phys. Rev. Lett. **43**, 246 (1979)
4. V. L. Chernyak, A. R. Zhitnitsky and V. G. Serbo, JETP Lett. **26**, 594 (1977); Sov. J. Nucl. Phys. **31**, 552 (1980)
5. A. V. Radyushkin, JINR report R2-10717 (1977), arXiv:hep-ph/0410276 (English translation); A. V. Efremov and A. V. Radyushkin, Theor. Math. Phys. **42**, 97 (1980); Phys. Lett. B **94**, 245 (1980)
6. G. P. Lepage and S. J. Brodsky, Phys. Lett. B **87**, 359 (1979); Phys. Rev. Lett. **43**, 545 (1979) Phys. Rev. D **22**, 2157 (1980)
7. A. V. Radyushkin, Phys. Rev. D **58**, 114008 (1998)
8. A. V. Belitsky, X. d. Ji and F. Yuan, Phys. Rev. D **69**, 074014 (2004)
9. M. Diehl, T. Feldmann, R. Jakob and P. Kroll, Eur. Phys. J. C **39**, 1 (2005)
10. M. Guidal, M. V. Polyakov, A. V. Radyushkin and M. Vanderhaeghen, Phys. Rev. D **72**, 054013 (2005)
11. R. P. Feynman, *Photon-Hadron Interactions*, Benjamin, New York (1972).
12. J. Polchinski and M. J. Strassler, JHEP **0305**, 012 (2003)
13. S. J. Brodsky and G. F. de Teramond, Phys. Lett. B **582**, 211 (2004)
14. S. J. Brodsky and G. F. de Teramond, Phys. Rev. Lett. **96**, 201601 (2006)
15. S. D. Drell and T. M. Yan, Phys. Rev. Lett. **24**, 181 (1970).
16. J. Erlich, G. D. Kribs and I. Low, arXiv:hep-th/0602110.
17. A. A. Migdal, Annals Phys. **109**, 365 (1977)
18. H. G. Dosch, J. Kripfganz and M. G. Schmidt, Phys. Lett. B **70**, 337 (1977); Nuovo Cim. A **49**, 151 (1979)
19. V. A. Nesterenko and A. V. Radyushkin, Phys. Lett. B **115**, 410 (1982);
20. A. V. Radyushkin, Acta Phys. Polon. B **26**, 2067 (1995)
21. A. V. Radyushkin, arXiv:hep-ph/0605116.
22. J. Volmer *et al.* [The Jefferson Lab F(pi) Collaboration], Phys. Rev. Lett. **86**, 1713 (2001); T. Horn, Jefferson Lab seminar, April 2006.
23. A. V. Radyushkin, Nucl. Phys. A **532**, 141 (1991)

HIGH-ENERGY EFFECTIVE ACTION FROM SCATTERING OF QCD SHOCK WAVES

I. BALITSKY

Physics Dept, Old Dominion Univ., Norfolk, VA 23529, and
Theory Group, JLab, 12000 Jefferson Ave, Newport News, VA 23606
E-mail: balitsky@jlab.org

At high energies, the relevant degrees of freedom are Wilson lines - infinite gauge links ordered along straight lines collinear to the velocities of colliding particles. The effective action for these Wilson lines is determined by the scattering of QCD shock waves. I develop the symmetric expansion of the effective action in powers of strength of one of the shock waves and calculate the leading term of the series. The corresponding first-order effective action, symmetric with respect to projectile and target, includes both up and down fan diagrams and pomeron loops.

Keywords: Effective action, small-x evolution, Wilson lines.

1. Introduction

It is widely believed that the relevant degrees of freedom for the description of high-energy scattering in QCD are Wilson lines - infinite straight-line gauge factors (for a review see Ref. 1). The particles with different rapidities perceive each other as Wilson lines so these lines may serve as the relevant degrees of freedom for high-energy scattering. The goal of this approach is to rewrite the original functional integral over gluons (and quarks) as a 2+1 theory with the effective action written in terms of the dynamical Wilson lines. For a given interval of rapidity, the effective action is an amplitude of scattering of two QCD shock waves, see Fig. 1. Indeed, let us integrate over the gluons in this interval of rapidity $\eta_1 > \eta > \eta_2$ leaving the gluons with $\eta > \eta_1$ (the "right-movers') and with $\eta < \eta_2$ (the "left-movers') intact (to be integrated over later). Due to the Lorentz contraction, the left-moving and the right-moving gluons shrink to the two gluon "pancakes" or shock waves. The result of the integration over the rapidities $\eta_1 > \eta > \eta_2$ is the effective action which depends on the Wilson lines made from the left-and right-movers.

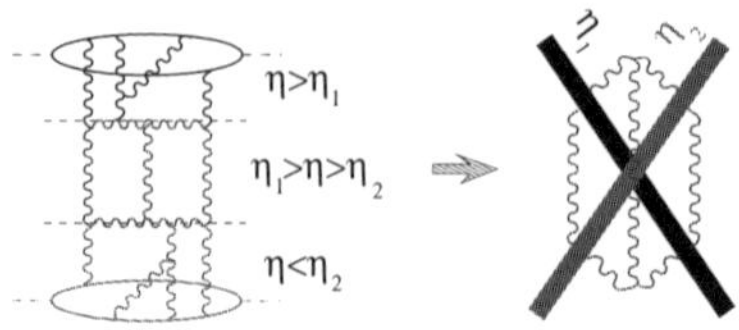

Fig. 1. High-energy effective action as an amplitude of the collision of two shock waves.

2. Rapidity factorization

The main technical tool of the shock-wave approach to the high-energy scattering is the rapidity factorization developed in Ref. 2. Consider a functional integral for the typical scattering amplitude

$$\int DAJ(p_A)J(p_B)J(-p'_A)J(-p'_B)\ e^{iS(A)} \tag{1}$$

where the currents $J(p_A)$ and $J(p_B)$ describe the two colliding particles (say, photons). We use Sudakov variables $k = \alpha p_1 + \beta p_2 + k_\perp$ and the notations $x_\bullet = p_1^\mu x_\mu = \sqrt{\frac{s}{2}}x^-$, $x_* = p_2^\mu x_\mu = \sqrt{\frac{s}{2}}x^+$ Here p_1 and p_2 are the light-like vectors close to p_A and p_B: $p_A = p_1 + \frac{p_A^2}{s}p_2$, $p_B = p_2 + \frac{p_B^2}{s}p_1$.

Let us take some "rapidity divide" η_1 such that $\eta_A > \eta_1 > \eta_B$ and integrate first over the gluons with the rapidity $\eta > \eta_1$, see Fig. 2a. From

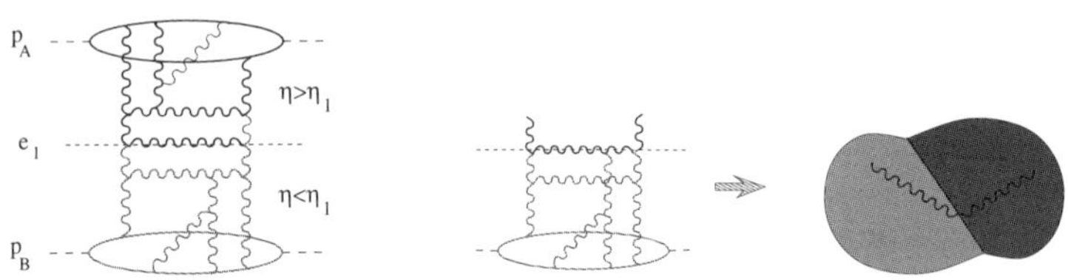

Fig. 2. Rapidity factorization (a) and shock wave in the temporal gauge (b).

the viewpont of such particles, the fields with $\eta < \eta_1$ shrink to a shock wave so the result of the integration is presented by Feynman diagrams in the shock-wave background. In the covariant gauge, this shock wave has the only non-vanishing component $A_\bullet$ which is concentrated near $x_* = 0$. In order to write down factorization we need to rewrite the shock wave in the temporal gauge $A_0 = 0$. In such gauge the most general form of a shock-wave background has the form (see Fig. 2b)

$$A^i = \mathcal{U}_1^i\theta(x_*) + \mathcal{U}_2^i\theta(-x_*), \quad A_\bullet = A_* = 0 \tag{2}$$

where $\mathcal{U}_1^i = U_1^\dagger \frac{i}{g}\partial_i U_1$ and $\mathcal{U}_2^i = U_2^\dagger \frac{i}{g}\partial_i U_2$ are the pure gauge fields (filling the half-spaces $x_* > 0$ and $x_* < 0$). There is a redundant gauge symmetry $U_1(x_\perp) \to U_1(x_\perp)\Omega(x_\perp)$, $\quad U_2(x_\perp) \to U_2(x_\perp)\Omega(x_\perp)$ related to the fact that gauge invariant objects like the color dipole depend only on the product $U_{1z}U_{2z}^\dagger$.

The generating functional for the Green functions in the Eq. (2) background is given by[2]

$$\int DAJ(p_A)J(-p'_A)\ e^{iS(A)+i\int d^2z_\perp (0,F_{*i},0)_z^a(\mathcal{U}_1^{ai}-\mathcal{U}_2^{ai})_z}, \tag{3}$$

$$(0,F_{ei},0)_z \equiv \int_{-\infty}^{\infty} du\ [0,ue]_z F_{ei}(ue+z_\perp)[ue,0]_z \ = \ [0,\infty e]_z(i\frac{\partial}{\partial z^i}$$

$$+\ gA_i(\infty e+z_\perp))[\infty e,0]_z -\ [0,-\infty e]_z(i\frac{\partial}{\partial z^i}+gA_i(-\infty e+z_\perp))[-\infty e,0]_z$$

where $F_{ei} \equiv e^\mu F_{\mu i}$ and $(0,F_{\mu i},0)^a \equiv 2\mathrm{tr}\ t^a(0,F_{\mu i},0)$. [a]

To complete the factorization formula one needs to integrate over the remaining B fields with rapidities $\eta < \eta_1$:

$$\int DAJ(p_A)J(-p'_A)e^{-iS(A)}\ e^{iS(A)}J(p_B)J(-p'_B) \ = \ \int DAJ(p_A)J(-p'_A)$$

$$\times \int DBJ(p_B)J(-p'_B)e^{iS(A)+iS(B)+i\int d^2z_\perp (0,F_{e_1 i},0)_z^a(0,G_{e_1 i},0)_z^a} \tag{4}$$

where the Wilson-line operators $(0,F_{e_1 i},0)_z^a$ and $(0,G_{e_1 i},0)_z^a$ are the operators (3) made from A and B fields, respectively.

3. The effective action

3.1. *Scattering of QCD shock waves*

Applying the factorization formula (4) two times, one gets (see Fig. 3):

$$\int DA\ J(p_A)J(p_B)J(-p'_A)J(-p'_B)\ e^{iS(A)} \ = \tag{5}$$

$$\int DAJ(p_A)J(-p'_A)e^{iS(A)}\int DB\ J(p_B)J(-p'_B)\ e^{iS(B)}$$

$$\times\ \int DC\ e^{iS(C)+i\int d^2z_\perp\{[0,A_{e_1 i},0]_z^a[0,C_{e_1 i},0]_z^a+(0,C_{e_2 i},0)_z^a(0,B_{e_2 i},0)_z^a]\}}$$

where the slope is $e_1 = p_1 + e^{-\eta_1}p_2$ for the [...] Wilson lines and $e_2 = p_1 + e^{-\eta_2}p_2$ for the (...) ones.

---

[a]The sum over the Latin indices $i,j...$ runs over the two transverse components while the sum over Greek indices runs over the four components as usual

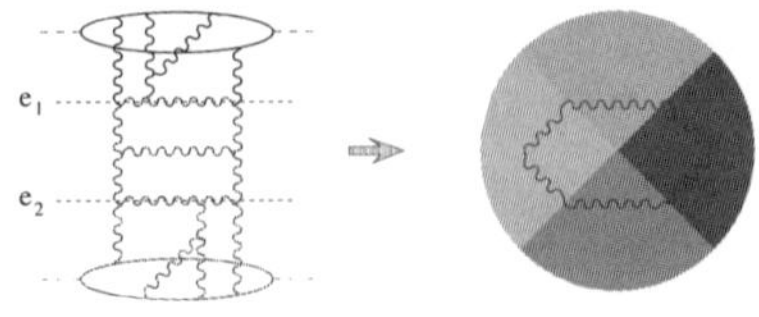

Fig. 3. Effective action as a scattering of two shock waves.

The functional integral over the central range of rapidity $\eta_1 > \eta > \eta_2$ is determined by the integral over C field with the sources made from "external" A and B fields: $(0, A_{e_1 i}, 0)_z = (\mathcal{V}_1^{ai} - \mathcal{V}_2^{ai})_z$ and $(0, B_{e_2 i}, 0)_z = (\mathcal{U}_1^{ai} - \mathcal{U}_2^{ai})_z$ where

$$\begin{aligned} V_{1,2}(z_\perp) &= [0, \pm\infty e_1]_z[\pm\infty e_1 + z_\perp, \pm\infty e_1 + \infty e_\perp] \\ U_{1,2}(z_\perp) &= [0, \pm\infty e_2]_z[\pm\infty e_2 + z_\perp, \pm\infty e_2 + \infty e_\perp] \end{aligned} \tag{6}$$

Since there is no field strength $F_{\mu\nu}$ at infinite time the direction of $e_\perp$ does not matter.

The result of the integration over the C field (the last line in the Eq. (5) is an effective action for the $\eta_1 > \eta > \eta_2$ interval of rapidity. The amplitude (1) is then the integral over A and B fields with this effective action.

3.2. *Expansion in commutators*

The effective action is defined by the last line in the functional integral (5) (hereafter we switch back to the usual notation A_μ for the integration variable and $F_{\mu\nu}$ for the field strength)

$$\begin{aligned} e^{iS_{\text{eff}}(V_1,V_2,U_1,U_2;\eta_1-\eta_2)} &= \int DA \exp\Big(iS(A) \\ &+ i\int d^2z_\perp\Big\{(\mathcal{V}_1^{ai} - \mathcal{V}_2^{ai})_z[0, F_{\bullet i}, 0]_z^a + (\mathcal{U}_1^{ai} - \mathcal{U}_2^{ai})_z(0, F_{*i}, 0)_z^a\Big\}\Big) \end{aligned} \tag{7}$$

Taken separately, the sources $\sim \mathcal{U}_i$ create a shock wave $\mathcal{U}_{1i}\theta(x_*) + \mathcal{U}_{2i}\theta(-x_*)$ and those $\sim \mathcal{V}_i$ create $\mathcal{V}_{1i}\theta(x_\bullet) + \mathcal{V}_{2i}\theta(-x_\bullet)$ In QED, the two sources $\mathcal{U}_i$ and $\mathcal{V}_i$ do not interact (in the leading order in α) so the sum of the two shock waves

$$\bar{A}_i^{(0)} = \mathcal{U}_{1i}\theta(x_*) + \mathcal{U}_{2i}\theta(-x_*) + \mathcal{V}_{1i}\theta(x_\bullet) + \mathcal{V}_{2i}\theta(-x_\bullet), \quad \bar{A}_\bullet^{(0)} = \bar{A}_*^{(0)} = 0 \tag{8}$$

is a classical solution. In QCD, the interaction between these two sources is described by the commutator $g[\mathcal{U}_i, \mathcal{V}_k]$ (the coupling constant g corresponds to the three-gluon vertex). We take the trial configuration in the form of a (improved) sum of the two shock waves (see Eq. (10) below) and expand the "deviation" of the full QCD solution from the QED-type ansatz (8) in powers of commutators $[U, V]$.

As demonstrated in Ref. 3, the gluon field at space (or time) infinity is a pure-gauge field which has the form of a sum of the shock waves plus a correction proportional to their commutator. Technically, for a pair of pure gauge fields $\mathcal{U}_i(x_\perp)$ and $\mathcal{V}_i(x_\perp)$ we define $\mathcal{W}_i(x_\perp) = \mathcal{U}_i(x_\perp) + \mathcal{V}_i(x_\perp) + gE_i(x_\perp; U, V)$ as a pure gauge field satisfying the equation $(i\partial_i + g[\mathcal{U}_i + \mathcal{V}_i,)E^i = 0$. In the first order in $[U, V]$ this field has the form

$$E_i^a(U,V) \;=\; -(x_\perp|U\frac{p^k}{p_\perp^2}U^\dagger + V\frac{p^k}{p_\perp^2}V^\dagger - \frac{p^k}{p_\perp^2}|^{ab}[\mathcal{U}_i, \mathcal{V}_k]^b - i \leftrightarrow k) \tag{9}$$

where $[\mathcal{U}_i, \mathcal{V}_k]^a \equiv 2\mathrm{Tr}t^a[\mathcal{U}_i, \mathcal{V}_k]$.

The zero-order approximation for the solution of the classical equations for the functional integral (7) can be taken as a superposition of pure gauge fields in the forward, backward, left, and right quadrants of the space.

$$\bar{A}_\bullet^{(0)} = \bar{A}_*^{(0)} = 0, \qquad \bar{A}^{(0)i} = \mathcal{W}_F^i(x_\perp)\theta(x_*)\theta(x_\bullet) \tag{10}$$
$$+\ \mathcal{W}_L^i(x_\perp)\theta(-x_*)\theta(x_\bullet) + \mathcal{W}_R^i(x_\perp)\theta(x_*)\theta(-x_\bullet) + \mathcal{W}_B^i(x_\perp)\theta(-x_*)\theta(-x_\bullet)$$

where

$$\begin{aligned} \mathcal{W}_F^i &= \mathcal{U}_1^i + \mathcal{V}_1^i + E_F^i, \quad & \mathcal{W}_L^i &= \mathcal{U}_2^i + \mathcal{V}_1^i + E_L^i \\ \mathcal{W}_R^i &= \mathcal{U}_1^i + \mathcal{V}_2^i + E_R^i, \quad & \mathcal{W}_B^i &= \mathcal{U}_2^i + \mathcal{V}_2^i + E_B^i \end{aligned} \tag{11}$$

and $E_F^i(U_1, V_1)$, $E_L^i(U_2, V_1)$, $E_R^i(U_1, V_2)$, and $E_B^i(U_2, V_2)$ are given by Eq. (9). It is easy to demonstrate that the Lipatov vertex of gluon emission by the colliding shock waves is

$$L_i \;=\; 2(E_F^i - E_L^i - E_R^i + E_B^i) \;=\; 2(\mathcal{W}_F^i - \mathcal{W}_L^i - \mathcal{W}_R^i + \mathcal{W}_B^i) \tag{12}$$

(in the first order in $[U, V]$ expansion). As usually, the effective action in the leading LLA order is a product of two Lipatov vertices times the infinitesimal rapidity interval

$$S_{\mathrm{eff}} = \frac{\alpha_s \Delta\eta}{4}\int d^2z_\perp L_i^a(z_\perp)L^{ia}(z_\perp) \tag{13}$$

The corresponding functional integral for the finite rapidity interval has the form[3]

$$e^{iS_{\mathrm{eff}}(U_1(x),U_2(x),V_1(x),V_2(x);\eta_1-\eta_2)} = \int DV_j(x,\eta)DU_j(x,\eta)\Big|_{U_j(x,\eta_2)=U_j(x)}^{V_j(x,\eta_1)=V_j(x)}$$
$$\times\ e^{i\int d^2x[\mathcal{V}_{1i}^a(x)-\mathcal{V}_{2i}^a(x)][\mathcal{U}_1^{ai}(x,\eta_1)-\mathcal{U}_2^{ai}(x,\eta_1)]}$$
$$\times\ e^{\int_{\eta_2}^{\eta_1}d\eta\int d^2x[-i(\mathcal{V}_{1i}^a-\mathcal{V}_{2i}^a)\frac{\partial}{\partial\eta}(\mathcal{U}_1^{ai}-\mathcal{U}_2^{ai})\ +\ \frac{\alpha_s}{4}L_i^a(V,U)L^{ai}(V,U)]} \tag{14}$$

It can be demonstrated that the functional integral (14) reproduces the non-linear BK evolution[4] in the case of one small source (when either $\mathcal{U}_i$ or $\mathcal{V}_i$ is small).

3.3. *Gauge-invariant form of the effective action*

The product of two Lipatov vertices (13) can be rewritten it in the gauge-invariant "diamond" form of trace of four Wilson lines at $x_{\bullet,*} = \pm\infty$ as suggested in a recent paper [5](see also [6,7]):

$$\begin{aligned}
\frac{1}{8}L_i^a(U,V)L^{ai}(U,V) &= \mathrm{tr}\{[-\infty p_1, F_{\bullet i}, \infty p_1]_{\infty p_2}[\infty p_2, F_{*i}, -\infty p_2]_{\infty p_1} \\
&\times [\infty p_1, -\infty p_1]_{-\infty p_2}[-\infty p_2, \infty p_2]_{-\infty p_1} + \mathrm{tr}[-\infty p_1, \infty p_1]_{\infty p_2} \\
&\times [\infty p_2, F_{*i}, -\infty p_2]_{\infty p_1}[\infty p_1, F_{\bullet i}, -\infty p_1]_{-\infty p_2}[-\infty p_2, \infty p_2]_{-\infty p_1} \\
&+ \mathrm{tr}[-\infty p_1, \infty p_1]_{\infty p_2}[\infty p_2, -\infty p_2]_{\infty p_1}[\infty p_1, F_{\bullet i}, \infty p_1]_{-\infty p_2} \\
&\times [-\infty p_2, F_{*i}, \infty p_2]_{-\infty p_1} + \mathrm{tr}[-\infty p_1, F_{\bullet i}, \infty p_1]_{\infty p_2}[\infty p_2, -\infty p_2]_{\infty p_1} \\
&\times [\infty p_1, -\infty p_1]_{-\infty p_2}[-\infty p_2, F_{*i}, \infty p_2]_{-\infty p_1}\}
\end{aligned} \tag{15}$$

where the transverse arguments in all Wilson lines are $x_\perp$. The structure of the effective action in the functional integral (14) is presented in Fig. 4. Note that the two terms in the exponent in the effective action, shown in Fig. 4, are both local in $x_\perp$ but differ with respect to the longitudinal coordinates: the first (kinetic) term is made from the Wilson lines located at $x_+ = 0$ or $x_- = 0$ while the second term is made from the Wilson lines at $x_\pm = \pm\infty$. Unfortunately, the transition between these Wilson lines is nonlocal in $x_\perp$ (see Eq. (9)) and so the resulting effective action is a non-local function of the dynamical variables U and V.

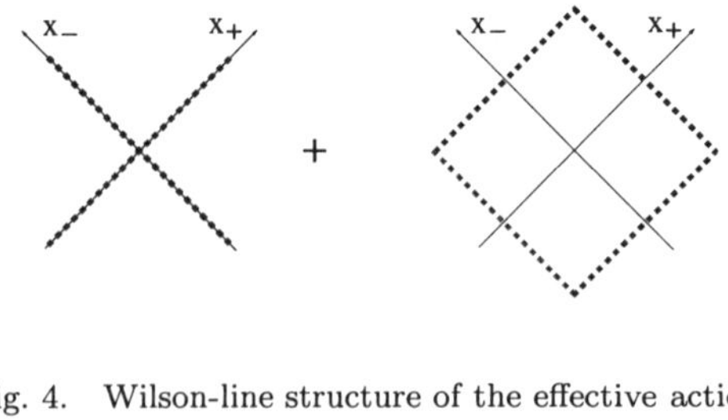

Fig. 4. Wilson-line structure of the effective action.

4. Conclusion

The functional integral (14) gives an example of the effective 2+1 theory for high-energy scattering in QCD. It is only a model - the genuine effective action for the $2+1$ high-energy theory of Wilson lines must include all the contributions $\sim [U,V]^n$. However, this model is correct in the case of weak projectile fields and strong target fields, and vice versa. In terms of Feynman diagrams, the effective action (14) includes both "up" and "down" fan ladders and the pomeron loops. In the dipole language, it describes both multiplication and recombination of dipoles (see the discussion in Refs.8, 9).

In conclusion I would like to emphasize that the effective action (14) summarizes all present knowledge about the high-energy evolution of Wilson lines in a way symmetric with respect to projectile and target and hence it may serve as a starting point for future analysis of high-energy scattering in QCD.

Acknowledgments

This work was supported by contract DE-AC05-84ER40150 under which the Southeastern Universities Research Association (SURA) operates the Thomas Jefferson National Accelerator Facility.

References

1. I. Balitsky, *"High-Energy QCD and Wilson Lines"*, In *Shifman, M. (ed.): At the frontier of particle physics, vol. 2*, p. 1237-1342 (World Scientific, Singapore,2001) [hep-ph/0101042]
2. I. Balitsky, *Phys. Rev. Lett.* **81**, 2024 (1998), *Phys. Rev.* **D60**, 014020 (1999).
3. I. Balitsky, *Phys. Rev.* **D70**, 114030 (2004), *Phys. Rev.* **D72**, 074027 (2005).
4. I. Balitsky, *Nucl. Phys.* **B463**, 99 (1996); Yu.V. Kovchegov, *Phys. Rev.* **D60**, 034008 (1999), *Phys. Rev.* **D61**, 074018 (2000).
5. Y. Hatta, E. Iancu, L. McLerran, A. Stasto, and D.N. Triantafyllopoulos *Nucl.Phys.***A764**, 423 (2006).
6. A. Kovner and M. Lublinsky, *Phys.Rev.***D71**, 085004(2005); *Phys.Rev.Lett.***94**, 181603(2005); *JHEP*, 0503:001(2005)
7. A.H. Mueller, A.I. Shoshi, and S.M.H. Wong, *Nucl.Phys.***B715**, 440 (2005).
8. Y. Hatta, E. Iancu, L. McLerran, and A. Stasto, *Nucl.Phys.***A762**, 272 (2005).
9. C. Marquet, A.H. Mueller, A.I. Shoshi, and S.M.H. Wong, *Nucl.Phys.***A762**, 252 (2005).

WEAKLY BOUND DIQUARKS AND EFIMOV HYPERONS IN QCD

J. A. O. MARINHO, E. GAMBIN and T. FREDERICO*

Departamento de Física, Instituto tecnológico de Aeronáutica,
12.228-900, São José dos Campos, SP, Brazil
**E-mail: tobias@ita.br*
www.ita.br

We study an extended color gauge invariant Quantum Chromodynamics (QCD) effective model with quarks, gluons and scalar color anti-triplets. We show that in a condensed phase of diquarks, the gluon and the photon acquire dynamical masses. In the BEC phase and near the zero-binding diquark transition, it is possible the onset of the Thomas-Efimov effect for some three-quark systems, like the hyperonic quartet-spin channel. We also evaluate the running coupling constant in lowest perturbative order in the extended color gauge model. Our calculation suggests that the freezing of the coupling constant can occur for a small number of quark and diquark flavors.

Keywords: Colored scalar fields; non-abelian gauge group; gluon mass; asymptotic freedom; Thomas-Efimov effects.

1. Introduction

It is a wide spread idea that diquark degrees of freedom in anti-triplet color states ($[qq]^{\bar{3}_c}$) can play a role in low-energy QCD and hadron structure [1–3]. The energy involved in the stability of the color anti-triplet scalar diquark configuration is about the nucleon and delta mass difference which gives $\sim$300 MeV. Near the QCD phase transition $T_c \sim 200$MeV and the diquark dissolves. Indeed, it was estimated that the zero binding line of the $[qq]^{\bar{3}_c}$ in the phase diagram - (T, μ) plane - is somewhat above the phase boundary of QCD [4]. Therefore, below the QCD phase transition it is possible that correlated quarks keep their identity and appear as building blocks of hadrons as for example in the interpretation of light scalar mesons as a tetraquark nonet [5].

Although it has been recognized that correlated quarks in color triplet/anti-triplet and spin 1 states may also have some importance[1], we are going to investigate a color gauge invariant extended gauge invariant Lagrangian containing quarks, gluons and scalar fields. Within a chiral effective Lagrangian, Hong and collaborators [6] have already introduced a color anti-triplet diquark coupled to the gauge field. Moreover, to get some interesting physics of our model at finite temperature and density, we will address the situation where vector diquarks are near zero-binding [4].

Our aim in this work is to study some consequences of the extended renormalizable QCD model with scalar diquarks, looking to two distinct effects: i) the gluon mass generation in a condensed phase of diquarks and its implications to the formation of Efimov hyperons and ii) the contribution of strongly correlated color anti-triplet scalar diquarks to the running coupling constant.

2. Extended QCD model with diquarks

The Lagrangian of the color anti-triplet/triplet scalars and quarks coupled to the gluon field is given by:

$$L = -\frac{1}{4}G^{a,\mu\nu}G^a_{\mu\nu} \mp \bar{\Psi}\left(i\not{D}_{(q)} - m_q\right)\Psi + \left(D^\mu_{(s)}\phi\right)^\dagger \left(D_{(s)\mu}\phi\right) - \phi^\dagger \hat{m}_s^2 \phi, \quad (1)$$

where $\phi(x)$ are the scalar fields with mass matrix $\hat{m}_s$, $G^a_{\mu\nu}$ is the gauge field tensor and Ψ the quark field. The covariant derivative for scalar and quark fields are written as:

$$\left(D_{(s)\mu}\phi\right)_i = \partial_\mu \phi_i - igA^a_\mu (T^a_{ij})^* \phi_j \quad and \quad \left(D_{(q)\mu}\psi\right)_i = \partial_\mu \psi_i + igA^a_\mu T^a_{ij}\psi_j \quad (2)$$

Note that the diquark field belongs to the color conjugate representation. The matrices T^a (a=1 to 8) satisfy $[T^a, T^b] = if^{abc}T^c$ with f^{abc} the structure constants of the group. The gauge field tensor is written as

$$G^a_{\mu\nu} = \partial_\mu A^a_\nu - \partial_\nu A^a_\mu - gf^{abc}A^b_\mu A^c_\nu, \quad (3)$$

in terms of the gluon field A^a_μ.

The model of Eq. (1) with the covariant derivatives from Eq. (2) is gauge invariant and renormalizable, once the gauge fixing term is added. Our calculations for the running coupling constant will proceed in the Feynman gauge.

3. Gluon and photon masses in the scalar condensate

The first direct consequence of the effective QCD Lagrangian, Eq. (1), is the generation of a mass for the gluon in a medium with a finite density of colored triplet and anti-triplet scalars. The gluon mass term is originated in the following part of the Lagrangian density:

$$L_{A^2} = g^2 A^{\mu a} A^b_\mu \phi^\dagger_j (T^a)_{ij} {T^b}^*_{ik} \phi_k \ . \tag{4}$$

In a colorless medium with a finite density of colored scalars, the particular Lagrangian term, (4), of the extended QCD model implies in a gluon mass of:

$$m_g^2 = 2g^2 \langle \phi^\dagger_j (T^a)_{ij} {T^a}^*_{ik} \phi_k \rangle = 2g^2 \langle \phi^\dagger \phi \rangle (T^a)_{ij} {T^a}^*_{ij} = 4g^2 \langle \phi^\dagger \phi \rangle \ , \tag{5}$$

where we assumed a colorless medium. (There is no summation over a in Eq. (5).) The average value of the square of the scalar field is $\langle \phi^\dagger \phi \rangle = \frac{1}{3}\sum_{flavor,i} \langle {\phi^f_i}^\dagger \phi^f_i \rangle$, where the superscript f indicates the diquark flavor. We estimate $\sum_{flavor,i} m^2_{s,f} \langle {\phi^f_i}^\dagger \phi^f_i \rangle \sim \varepsilon$ (energy density of the BEC phase). Using diquark mass, $m_s \sim 0.6$ GeV, then $\langle \phi^\dagger \phi \rangle \sim \varepsilon/(3m_s^2)$ and with that $m_g \sim g\sqrt{4\varepsilon/(3m_s^2)} \sim 0.17\, g\sqrt{\varepsilon/\varepsilon_0}$ GeV ($\varepsilon_0 = 1\text{GeV/fm}^3$ an energy density enough for the quark-gluon plasma formation according to lattice calculations [7]). The gluon mass is roughly inversely proportional to the mass of the scalars. Therefore in the route of chiral symmetry restoration, if the scalar condensate survives due to the formation of a relativistic Bose-Einstein condensate of diquarks [8], the gluon mass will increase, and matter will be opaque to the gluon-exchange interaction.

The photon can also acquires a mass in an uncharged medium of condensed diquarks:

$$m_\gamma^2 = 6\frac{e^2}{N_s} \langle \phi^\dagger \phi \rangle Tr\left[\hat{Q}^\dagger \hat{Q}\right] \ , \tag{6}$$

where N_s is the flavor number and $\hat{Q}$ is the charge matrix. The estimative of the photon mass is $m_\gamma \sim \sqrt{e^2 \varepsilon \sum_f q_f^2/(N_s m_s^2)}$. The photon mass is about $m_\gamma \sim 0.04\sqrt{\sum_f q_f^2\ \varepsilon/(\varepsilon_0 N_s)}$ GeV, which can be of the order of few tens of MeV near the chiral phase transition. The dilepton decay of the massive photons can be an indicator of the formation of the light scalar condensate (formed by diquarks with u, d and s quarks). We may speculate that for light scalar condensate energy density of $\epsilon = \epsilon_0$ the gluon mass is about 0.4 GeV (for $\alpha_{QCD} \sim 1$) while the photon mass in the medium is around 0.01 GeV.

4. Thomas-Efimov effect in the BEC phase

The quarks will interact through a short ranged one-gluon-exchange force in the presence of a Bose-Einstein condensate and towards the zero-binding diquark limit [4] (large scattering length a) it may be possible that the three-body Thomas [9] and Efimov [10] physics appears in this phase. The three-quark system formed by the quarks of a scalar or vector diquark and a third one can access a state of maximum symmetry in the coordinate space or s-wave, and besides that, two ingredients are essential for the onset of the Thomas-Efimov physics [11]: large scattering length and finite range interaction. The Thomas-Efimov limit comes for $|a|/r_0 \to \infty$ (r_0 effective range). In the Thomas-Efimov limit an infinite number of bound states with energy ratio ~ 515 between consecutive states. For a fixed interaction range the states condense at zero energy -Efimov effect- which will be of interest here.

It is possible that near the zero-binding diquark transition (large scattering length) in the BEC phase with finite range interaction (due to the gluon finite mass), the Thomas-Efimov physics is triggered with the appearance of long-range triquark correlations with the consequent enhancement of three-quark collisions. We suppose that besides the flavor antisymmetric scalar diquarks ($I = 0$ for u and d or $\overline{3}_f$ for u, d and s) also the flavor symmetric ($I = 1$ for u and d or 6_f for u, d and s) vector diquarks are weakly bound and the quark masses are degenerated. Indeed, as the hyperfine interaction from the one-gluon-exchange splits the vector and scalar diquarks, being attractive for the scalar channel, it is even natural to think that the vector diquark dissolves more easily than the scalar one, with the increase of temperature and density, making possible the coexistence of a phase of condensed scalars and zero binding vector diquarks.

Just to point out where this phenomena can be triggered with u, d and s quarks we pick one three-quark state channel where Thomas-Efimov physics may occur among other configurations: the state with zero color, flavor symmetric and spin 3/2. The wave function in this hyperon state has a totally symmetric configuration space component and therefore it is analogous to a three-boson case. Therefore, looking in this particular hyperonic channel the interacting three-quark system achieves ideal conditions for the Thomas-Efimov physics to arise. Near the zero-binding diquark transition long-range three-quark correlations appear and three-quark collisions in hyperonic channels will be enhanced.

5. Running coupling constant

In the following we are going to calculate the contribution of correlated quarks and antiquarks to the running coupling constant. We calculate the β-function in order g^2. For that purpose we obtain the renormalization constants of the fields , i.e., $\Psi_B = \sqrt{Z_2}\Psi$, $A^a_{B\mu} = \sqrt{Z_3}A^a_\mu$, $\phi_B = \sqrt{Z^s_2}\phi$

(the subindex B means bare operators), which obey several identities, and among them:

$$g_B = g\mu^{\frac{\epsilon}{2}} \frac{Z_1}{Z_2\sqrt{Z_3}} = g\mu^{\frac{\epsilon}{2}} \frac{Z_1^s}{Z_2^s\sqrt{Z_3}} = \ldots \tag{7}$$

They are indeed valid as a consequence of Ward identities [12]. In particular the last term in the above equality comes from the gluon-scalar-scalar vertex. The running coupling constant is derived from renormalization group invariance expressed by $\frac{dg_B}{d\mu} = 0$.

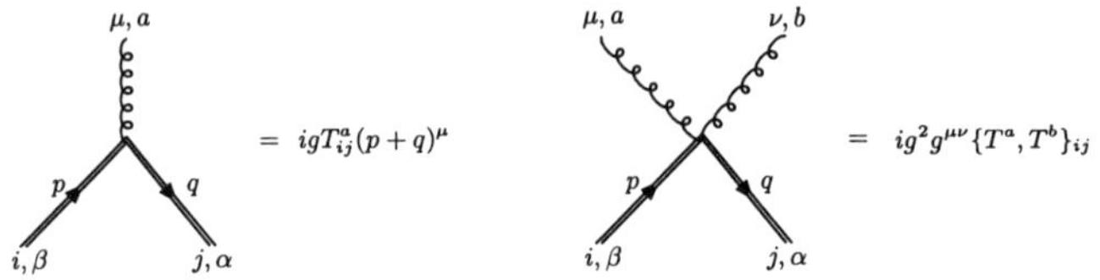

Fig. 1. Feynman rules for scalar-gluon coupling.

The scalars contribute to the standard calculation of the quark running coupling constant through the renormalization of the gluon field, which fluctuates in a pair of scalar-antiscalar or when a scalar bubble emerges in the gluon propagation. The Feynman rules for the interaction of the gluon field with scalars from the effective Lagrangian, Eq. (1), are shown in fig. 1.

Fig. 2. $O(g^2)$ scalar contributions to vacuum polarization effects in the gluon propagation. Left diagram ($d1$): vacuum fluctuation in scalar pairs. Right diagram ($d2$): scalar-bubble.

The corrections to the gluon propagator in order g^2 due to the vacuum polarization coming from scalar fluctuations are shown in fig. 2.

A straightforward calculation with dimensional regularization gives the scalar contribution to the gluon polarization tensor as:

$$\Pi^{\mu\nu}_{ab}(d1 + d2) = \frac{g^2}{48\pi^2\varepsilon} N_s \delta_{ab}[(p^\mu p^\nu - p^2 g^{\mu\nu})] \ , \tag{8}$$

and together with the standard results for QCD with quarks [12], we obtain a new value for the gluon renormalization constant:

$$Z_3 = 1 + \frac{g^2}{8\pi^2\varepsilon}(5 - \frac{2}{3}N_q - \frac{1}{6}N_s) \ . \tag{9}$$

Taking into account the values of Z_1 and Z_2 [12] we can evaluate α_{QCD} for a model with quarks and scalars, or only scalars as matter field. The result is

$$\alpha_{QCD}^{q+s}(Q^2) = \frac{4\pi}{(11 - \frac{2}{3}N_q - \frac{1}{6}N_s)\log\frac{Q^2}{\Lambda^2}} \ , \tag{10}$$

obtained from the beta-function:

$$\beta(g) = -\frac{g^3}{(4\pi)^2}(11 - \frac{2}{3}N_q - \frac{1}{6}N_s) \ . \tag{11}$$

Supposing only $[qq]^{\overline{3}_c}$ configurations forming the scalars $N_s = N_q(N_q - 1)/2$, the zero of the lowest order calculation of the β-function appears at $11 - \frac{2}{3}N_q^* - \frac{1}{12}N_q^*(N_q^* - 1) = 0$ giving $N_q^* \approx 8.5$, approaching the freezing of the coupling constant to the physical flavor number. It is suggestive that the consideration of explicit correlated quarks (as effective fields) approaches the Banks and Zaks [13] expansion near to the hexa-flavor world.

6. Conclusions

We study some consequences of the possibility that systems of correlated quarks in the color triplet/anti-triplet channel and $J^P = 0^+$ have an independent dynamical role in QCD. For that purpose we make use of an extended color local gauge invariant effective model with quarks, gluons and $\overline{3}_c$ scalars. The scalar fields can represent different configurations of strongly correlated few-quark systems.

We show that in a condensed phase of diquarks, the gluon and the photon acquire dynamical masses. The photon decay into low mass dileptons will be one of the signatures of the condensed phase, with estimate mass of few tens of MeV. Moreover, it is remarkable that the conditions for the onset of the Thomas-Efimov physics, which occurs near the zero-binding diquark transition (large scattering length) in the BEC phase where the one-gluon exchange interaction has a finite range, are fulfilled for particular color/flavor/spin configurations of three-quark systems. Under the above conditions it appears long-range triquark correlations with the consequent enhancement of three-quark collisions in some channels. It is the case of the hyperonic quartet-spin channel of the $\{1_c, 1_f\}$ representation. The possibility that other color/flavor/spin three-quark configurations exhibit this effect will be investigated in the future.

We also evaluate the running coupling constant in lowest perturbative order in the extended color gauge model with scalar fields representing the strongly correlated scalar quark pairs. Considering only scalars from strongly correlated $[qq]^{\bar{3}_c}$ pairs which forms fifteen diquarks flavors, $N_s =$ 15 from $N_q = 6$ (u, d, s, c, b, t), the running coupling constant is still dominated by gluon self-interactions and asymptotic freedom is verified at short distances. However, the diquark correlation implies in $N_q^* = 8.5$ for the zero of the beta-function in lowest perturbative order, approaching the freezing of the coupling to the physical world.

The effective Lagrangian of this work can be used as well to model interactions between quarks and scalars for application in exotic hadron phenomenology, however a word of caution should be given, once the scalars represent composite fields, one should be careful in applying the model to avoid subtle double-counting.

Acknowledgments

We thank the Brazilian agencies FAPESP, CAPES and CNPq for financial support.

References

1. R. L. Jaffe, Nucl. Phys. Proc. Suppl **142** (2005) 343; Phys.Rep. **409** (2005) 1.
2. F. Wilczek, arXiv:hep-ph/0409168
3. R. L. Jaffe, Phys. Rev. **D72** (2005) 074508.
4. E. V. Shuryak and I. Zahed, Phys. Rev. **C70** (2004) 021901(R).
5. L. Maiani, F. Piccinini, A. D. Polosa, V. Riquer, *Phys. Rev. Lett.* **93** (2004) 212002.
6. D. K. Hong, Y. J. Sohn, I. Zahed, *Phys. Lett.* **B596** (2004) 191.
7. F. Karsch, Nucl. Phys. **A698** (2002) 199.
8. Y. Nishida and H. Abuki, Phys. Rev. **D 72** (2005) 096004.
9. L.H. Thomas, Phys. Rev. **47** (1935) 903.
10. V. Efimov, Phys. Lett. **B 33** (1970) 563; Nucl. Phys. **A362** (1981) 45; V. Efimov, Comm. Nucl. Part. Phys. **19** (1990) 271.
11. S.K. Adhikari, A. Delfino, T. Frederico, I.D. Goldman, L. Tomio, Phys. Rev. **A37** (1988) 3666 .
12. Lewis H. Ryder, *Quantum Field Theory*, Cambridge University Press (1985 & 1996).
13. T. Banks and A. Zaks, Nucl. Phys. **B196** (1982) 189.

SECTION 9

LARGE N

Convener T. Cohen

BARYONS AND LARGE N_c IN HAPPY RESONANCE

R.F. LEBED

Department of Physics, Arizona State University,
Tempe, AZ 85287-1504, USA
E-mail: richard.lebed@asu.edu

I discuss recent developments in the large N_c treatment of unstable baryon resonances and the scattering amplitudes in which they appear. These include pion photoproduction, extension to three-flavor processes, decoupling of large N_c artifacts, and combination of this approach with results of chiral symmetry.

1. Introduction

At the time of this writing, the 2006 *Review of Particle Physics*[1] has just landed on my desk. Of this tome's 1232 pages, almost 10% list properties of baryon resonances. And yet, no consistent picture has yet been developed that predicts their rich spectroscopy with any degree of accuracy and consistency. Quark models accommodate a number of observed multiplets but predict numerous others unsupported by experimental evidence, while treating resonances as meson-baryon bound states explains some of the resonant branching fractions but fails to reproduce the multiplet structure. Both of these approaches derive from the underlying QCD theory only by employing heuristic arguments that are often obscure. And lattice gauge theory, while not suffering this ambiguity, has far to develop before it will be able to take on such an intricate tangle of unstable states.

Since the beginning of 2002, Tom Cohen and I have been developing a method[2–13] that treats unstable baryon resonances consistently in the $1/N_c$ expansion of QCD as broad, unstable states [masses lying $O(N_c^0)$ above the ground-state baryons of mass $O(N_c^1)$, and widths of $O(N_c^0)$]. The $1/N_c$ expansion is a well-defined field-theoretical limit of QCD-like theories (specifically, QCD with N_c rather than 3 colors) and therefore escapes the criticisms of the previous paragraph. On the other hand, in its basic form the $1/N_c$ expansion tells only how to count powers of $1/N_c$ and does not by itself provide a means of computing dynamical quantities. Even so, it provides an organizing principle of remarkable power,[14] especially for baryons.

This formalism has been outlined in a number of earlier conference proceedings,[15,16] the first of which was the preceding installment of this CAQCD series.[15] Since space here is limited, I provide only a sketchy description of the origin of the method in Sec. 2 and refer the interested reader to the prior write-ups for more detail, and in Sec. 3 summarize findings since the last CAQCD conference.[7–13]

2. The scattering method

Chiral soliton models, particularly the Skyrme model, achieved a degree of fashionability in the early to mid-1980's, owing to a series of papers[17] discussing their topological character and connection to large N_c. Prominent in this era were a number of works noting the model independence of a number of their predictions, particularly linear relations between meson-baryon scattering amplitudes (*e.g.*, $S_{11} = S_{31}$).[18,19] As gradually became clear, these results are consequences of an underlying symmetry imposed by the soliton's hedgehog configuration, which is characterized by the quantum number K, where $\mathbf{K} \equiv \mathbf{I} + \mathbf{J}$; the scattering amplitudes, labeled by good I and J quantum numbers, are then obtained by forming linear combinations of the K-labeled amplitudes in an exercise of "Clebschology."

Subsequently it was recognized that the underlying K conservation can be expressed, via crossing from the s channel to the t channel, in terms of the rule $I_t = J_t$.[20,21] Nevertheless, the connection of these results to large N_c depended upon identifying the hedgehog configuration and its excitations with this limit; it is certainly a reasonable approach, because the ground-state baryons are built from the hedgehog, which exhibits the maximal symmetry between I and J characterizing the ground-state baryons. Nevertheless, to see true compatibility with large N_c one needs a formalism that studies real meson-baryon scattering in the large N_c limit; this is provided by the consistency condition approach developed in the early 1990's.[22] One offshoot of this program[23] showed the $I_t = J_t$ rule to be an immediate consequence, and moreover that amplitudes with $|I_t - J_t| = n$ are suppressed by at least $1/N_c^n$. The pieces were then in place to show that 1) the old linear amplitude relations based on K are true large N_c results,[2] 2) the relations apply also to the baryon resonances embedded in these amplitudes,[2] and 3) $O(1/N_c^n)$ corrections can be incorporated[6] by including suppressed amplitudes with $|I_t - J_t| = n$. Moreover, while the ground-state band of baryons, $I = J = \frac{1}{2}, \frac{3}{2}, \ldots$, in the naive quark and chiral soliton pictures for large N_c is the same, the irreducible excited baryon multiplets of the quark picture turn out to be reducible collections of multiplets labeled by K ("compatibility").[3] The underlying K conservation leads to a predictive pattern[2] of allowed and forbidden decays [explaining, *e.g.*, the dominance of the ηN decay of $N(1535)$, since the $K{=}0$ pole in the $I{=}J{=}\frac{1}{2}$ negative-parity channel couples only to ηN], tells what types of resonant multiplets are allowed,[2,3,5] and respects the broadness of resonances and configuration mixing between resonances of the same quantum numbers.[4]

3. New developments

Pion Photoproduction. The same approach that applies to meson-baryon scattering may be employed as well for other cases of physical interest. As long as the quantum numbers and $1/N_c$ suppressions of the fields coupling to the baryon are known, the same methods apply. Here one has in mind such processes as photoproduction,[8] electroproduction, or real or virtual Compton scattering. Photons, for example, carry both isovector and isoscalar quantum numbers, the former dominating[24] in baryon couplings by a factor of N_c. Amplitudes that include the leading [relative $O(N_c^0)$] and first subleading [$O(1/N_c)$] isovector and the leading [$O(1/N_c)$] isoscalar amplitudes then produces linear relations among multipole amplitudes with relative $O(1/N_c^2)$ corrections.[8] While the relations obtained this way that reflect the dominance of isovector over isoscalar amplitudes agree quite impressively with data (*i.e.*, the amplitudes have the same shape as functions of photon energy), a number of other relations, particularly for magnetic multipoles, superficially appear to fare badly. However, in those cases the threshold behaviors still agree quite well, followed by seemingly divergent behavior in the respective resonant regions. These discrepancies appear to be due to the slightly different placement [at $O(\Lambda_{\rm QCD}/N_c) = O(100\,{\rm MeV})$] of resonances that are degenerate in the large N_c limit but with different I, J values. When a comparison of the amplitude relations is performed by taking *on-resonance* couplings[1], the linear relations good to $O(1/N_c^2)$ do indeed produce agreement to within about 1 part in 9.[8]

Three Flavors. The work summarized in the previous section all referred to systems either containing only u and d quarks, or those in which any heavier quarks are inert spectators. In nature, however, there appears to be some evidence that not only the baryon ground-state band but resonances as well exhibit a degree of SU(3) flavor symmetry. The first task in such analysis in the $1/N_c$ expansion is to have the group theory under control, which requires tables of SU(3) Clebsch-Gordan coefficients

(CGC) as functions of N_c; the standard ladder operator methods may be employed to derive them.[7] Since baryons at large N_c have the quantum numbers of N_c quarks, their SU(3) representations are much larger than for $N_c = 3$; for example, the analogue to the 3-color, 3-flavor octet is called "**8**": $(p, q) = [1, \frac{1}{2}(N_c - 1)]$, which has $O(N_c^2)$ states. One finds that the linear amplitude relations connecting resonances of different I and J for two flavors contain SU(3) CGC in the 3-flavor case, and therefore link together members of distinct SU(3) multiplets with various J values.[9] Such a phenomenon is known from chiral soliton models, and occurs for the same reason: The group theory is inherited from large N_c! Once a single resonance is found and its quantum numbers are measured, large N_c tells what other states of different J, I, and strangeness should be degenerate to $O(\Lambda_{\rm QCD}/N_c) = O(100\,{\rm MeV})$ [not counting the additive $\sim$150 MeV contribution for each s-quark] in both mass and width. This is a very useful diagnostic when one has a candidate exotic baryon, such as was the Θ^+.[5,9]

But it is also useful in identifying SU(3) partners of nonexotic resonances. For example, the $N(1535)$ should have strange partners[10] that are also η-philic; and in fact, there exists the S_{01} state $\Lambda(1670)$ that lies only 5 MeV above the $\eta\Lambda$ threshold (the phase space for $\pi\Sigma$ is 6 times larger), and yet its branching ratio to this channel is 10–25%. Other convincing examples[10] following the large N_c reasoning populate the sector of Λ and Σ resonances, but in many cases the uncertainty on branching ratios, or even the existence of the resonances themselves, is questionable. For example, large N_c makes definite statements about the spectroscopy and decays of Ξ and Ω resonances as well, but too little is known about them experimentally to make definitive comparisons.

As noted in the previous section, the familiar quark model multiplets at large N_c such as the SU(6)×O(3) ("**70**",1^-) actually form collections of distinct irreducible multiplets in large N_c. In the case just mentioned, one finds 5 such multiplets, labeled by $K = 0, \frac{1}{2}, 1, \frac{3}{2}, 2$, whose masses can differ at $O(N_c^0)$. Those with $K = 0, 1, 2$ define the multiplets with nonstrange members (*e.g.*, "**8**", "**10**"), while those with $K = \frac{1}{2}, \frac{3}{2}$ define multiplets whose states of maximal hypercharge have a strange quark (*e.g.*, "**1**").

An unexpected result arises in the SU(3) group theory, in the form a a theorem restricting which meson-baryon states have leading-order [$O(N_c^0)$, by unitarity] CGC.[9,10] One can prove that SU(3) CGC for $baryon_1$ + *meson* $\leftrightarrow$ $baryon_2$ can be $O(N_c^0)$ only if $Y_{\rm meson} = Y_{B_2,\,\rm max} - Y_{B_1,\,\rm max}$, which is to say that the meson must have a hypercharge equal to the difference of the tops of the two baryon representations. In particular, the dominant two-body decay mode of a Λ resonance in an "**8**" is predicted to be $\pi\,\Sigma(1192)$, while one in a "**1**" prefers $\overline{K}N$. Evidence for this remarkable result appears in, *e.g.*, $\Lambda(1520)$ D_{03}, where the coupling constant to $\overline{K}N$, once the threshold $p^{2L+1} \to p^5$ factor is taken into account, is 4–5 [$= O(N_c^1)$] times larger than that for $\pi\Sigma$. Indeed, the SU(3) content of $\Lambda(1520)$ is traditionally assigned to be dominantly singlet.

The SU(3) theorem is also the key ingredient in the proof that the $I_t = J_t$ rule holds for three-flavor as well as two-flavor processes, and moreover that processes with strangeness exchange are suppressed (the $Y_t = 0$ rule).[11] This result predicts, for example, that the process $K^- p \to \pi^+ \Sigma^-$ is suppressed in cross section by $1/N_c$ compared to $K^- p \to K^- p$.

Decoupling Spurious States. Since, as noted, large N_c baryons inhabit much larger SU(3) representations and also allow much higher spins than for $N_c = 3$, most of these states must be $N_c > 3$ artifacts and need to be decoupled from the theory as spurious if they appear in amplitudes for physical $N_c = 3$ processes. The effects of such states must disappear smoothly and exactly at the value of N_c where they become spurious.[12] It would be extremely coincidental for such decoupling to occur through cancellation among dynamical quantities, which depend sensitively upon nonperturbative effects, not least of which are the precise values of quark masses. We therefore argue[12] that the only sources of true decoupling are group-theoretical in nature, either because the states of interest have isospin or strangeness values too high to reach in conventional $N_c = 3$ meson-baryon scattering, or lie in states in SU(3) multiplets that decouple when $N_c = 3$. Such decoupling appears through factors of $1 - 3/N_c$, a very special type of $1/N_c$ correction. For example, for $N_c > 3$ the "**1**" contains states with Ξ quantum numbers, and the SU(3) coupling for $\overline{K}\Sigma$ to this state indeed contains a factor of $\sqrt{1 - 3/N_c}$.

Interplay of Chiral and Large N_c Limits. Since meson-baryon scattering at threshold has also been studied in the context of another well-known expansion of strong interaction physics, the chiral expansion, it is natural to ask how the chiral and large N_c limits cooperate and compete.[13] The large N_c counting constraints for scattering amplitudes formally hold at all energies, while the chiral results become increasingly better the closer one approaches threshold. Among these results are the famed Weinberg-Tomozawa relation (between $I = 0, 2$ πN scattering lengths). Other well-known results relevant to the chiral π in meson-baryon scattering are the Adler-Weisberger

(AW) and Goldberger-Oehme-Miyazawa (GMO) cross section sum rules.

From the point of view of the $1/N_c$ expansion, scattering lengths are simply the derivatives of scattering amplitudes at zero meson energy. One finds that the combination corresponding to the WT and other chiral-limit relations are $O(N_c^0)$ but very well satisfied experimentally (as expected for threshold results), while most combinations that are $O(1/N_c)$ are not as well satisfied, but still have magnitudes consistent with the naive counting [that is, a dimensionless $O(1/N_c)$ combination is not larger than, say, 2/3].

The AW and GMO sum rules are not chiral limit results (They sum over all allowed energies), but nevertheless refer to chiral π's. In these cases one encounters the effect of the noncommutativity of the two limits: In particular, both fail if the large N_c limit is taken prior to the chiral limit, a consequence of the fact that the Δ-N mass difference is $O(1/N_c)$ and therefore falls below m_π for sufficiently large N_c. In that case the Δ becomes stable and must be treated on the same footing as the N.

4. Conclusions

The $1/N_c$ expansion provides a formalism in which to analyze meson-baryon scattering processes, including their rich spectrum of embedded resonances. The resonances appear in multiplets that bear a similarity to, but are more fundamental than, the old SU(6)×O(3) quark model multiplets. These multiplets have distinct dominant decay channels, it e.g., by preferring $\overline{K}N$ over $\pi\Lambda$ or $\pi\Sigma$. They are constrained by preferring $I_t = J_t$, $Y_t = 0$ processes. The problem of removing spurious ($N_c > 3$) states from the theory has been solved, and the nature of the interplay between the chiral and large N_c limits in meson-baryon scattering has been explored.

A number of interesting formal and phenomenological issues remain to be resolved. For example, all results presented here hold for single-meson scattering; of course, constraints also occur for scattering with, *e.g.*, $\pi\pi N$ final states, but these are as yet unexplored. Moreover, the three-flavor scattering results presented thus far only scratch the surface of the detailed phenomenological analysis that is now possible. A reliable means now exists to shed light on one of the darker corners of particle physics.

Acknowledgments

I thank the organizers for their kind invitation to this interesting and productive meeting. This work was supported in part by the NSF under Grants No. PHY-0140362 and PHY-0456520.

References

1. Particle Data Group (W.-M. Yao *et al.*), it J. Phys. G **33**, 1 (2006).
2. T.D. Cohen and R.F. Lebed, *Phys. Rev. Lett.* **91**, 012001 (2003); *Phys. Rev.* D **67**, 012001 (2003).
3. T.D. Cohen and R.F. Lebed, *Phys. Rev.* D **68**, 056003 (2003).
4. T.D. Cohen, D.C. Dakin, A. Nellore, and R.F. Lebed, *Phys. Rev.* D **69**, 056001 (2004).
5. T.D. Cohen and R.F. Lebed, *Phys. Lett.* B **578**, 150 (2004).
6. T.D. Cohen, D.C. Dakin, A. Nellore, and R.F. Lebed, *Phys. Rev.* D **70**, 056004 (2004).
7. T.D. Cohen and R.F. Lebed, *Phys. Rev.* D **70**, 096015 (2004).
8. T.D. Cohen, D.C. Dakin, R.F. Lebed, and D.R. Martin, *Phys. Rev.* D **71**, 076010 (2005).
9. T.D. Cohen and R.F. Lebed, *Phys. Lett.* B **619**, 115 (2005).
10. T.D. Cohen and R.F. Lebed, *Phys. Rev.* D **72**, 056001 (2005).
11. R.F. Lebed, *Phys. Lett.* B **639**, 68 (2006).
12. T.D. Cohen and R.F. Lebed, *Phys. Rev.* D **74**, 036001 (2006).
13. T.D. Cohen and R.F. Lebed, `hep-ph/0608038` (accepted for publication in *Phys. Rev.* D).
14. R.F. Lebed, *Czech. J. Phys.* **49**, 1273 (1999) [nucl-th/9810080].
15. R.F. Lebed, `hep-ph/0406236`, published in *Continuous Advances in QCD 2004*, edited by T. Gherghetta, World Scientific, Singapore, 2004.
16. R.F. Lebed, `hep-ph/0601022`, published in *NStar 2005*, edited by S. Capstick *et al.*, World Scientific, Singapore (2006); `hep-ph/0509020`, published in *Proceedings of the International Conference on QCD and Hadronic Physics*, edited by K.-T. Chao *et al.*, *Int. J. Mod. Phys.* A **21**, 877; `hep-ph/0501021`, published in *Large N_c QCD 2004*, edited by J.L. Goity *et al.*, World Scientific, Hackensack, NJ, USA (2005); T.D. Cohen, `hep-ph/0501090`, published in *Proceedings of the 10th International Conference on Structure of Baryons*, edited by M. Guidal *et al.*, *Nucl. Phys.* A **755**, 40 (2005).
17. E. Witten, *Nucl. Phys.* **B223**, 433 (1983); G.S. Adkins, C.R. Nappi, and E. Witten, *Nucl. Phys.* **B228**, 552 (1983); G.S. Adkins and C.R. Nappi, *Nucl. Phys.* **B249**, 507 (1985).
18. A. Hayashi, G. Eckart, G. Holzwarth, H. Walliser, *Phys. Lett.* B **147**, 5 (1984).
19. M.P. Mattis and M. Karliner, *Phys. Rev.* D **31**, 2833 (1985); M.P. Mattis and M.E. Peskin, *Phys. Rev.* D **32**, 58 (1985); M.P. Mattis, *Phys. Rev. Lett.* **56**, 1103 (1986); *Phys. Rev.* D **39**, 994 (1989); *Phys. Rev. Lett.* **63**, 1455 (1989); *Phys. Rev. Lett.* **56**, 1103 (1986).
20. J.T. Donohue, *Phys. Rev. Lett.* **58**, 3 (1987); *Phys. Rev.* D **37**, 631 (1988).
21. M.P. Mattis and M. Mukerjee, *Phys. Rev. Lett.* **61**, 1344 (1988).
22. R.F. Dashen and A.V. Manohar, *Phys. Lett.* B **315**, 425 (1993); R.F. Dashen, E. Jenkins, and A.V. Manohar, *Phys. Rev.* D **49**, 4713 (1994).
23. D.B. Kaplan and M.J. Savage, *Phys. Lett.* B **365**, 244 (1996); D.B. Kaplan and A.V. Manohar, *Phys. Rev.* C **56**, 76 (1997).
24. E. Jenkins, X. Ji, and A.V. Manohar, *Phys. Rev. Lett.* **89**, 242001 (2002).

LARGE N GAUGE THEORIES – NUMERICAL RESULTS

R. NARAYANAN

Department of Physics, Florida International University,
Miami, FL 33199, USA
E-mail: rajamani.narayanan@fiu.edu

H. NEUBERGER

Rutgers University, Department of Physics and Astronomy,
Piscataway, NJ 08855, USA
E-mail: neuberg@physics.rutgers.edu

Some physical results in four dimensional large N gauge theories on a periodic torus are summarized.

Keywords: Large N QCD, deconfinement, phase transitions.

1. Introduction

The large N limit of four dimensional non-abelian gauge theories is interesting from the view point of QCD phenomenology [1] and string theory [2]. Lattice QCD is a useful technique for extracting fundamental results in the large N limit of QCD. Fermions are naturally quenched in the 't Hooft limit of large N QCD and this significantly reduces the computational cost in a lattice calculation. In addition, there is a concept of continuum reduction [3], namely, physics does not depend on the size of box l^4 for $l > l_c$ and l_c is a physical critical size. These two observations have been used to extract physical results in the large N limit of QCD using numerical techniques on the lattice.

2. Phases of large N QCD

Large N QCD on a continuum torus l^4 has several phases [4] depending upon the size of the torus as shown in Fig. 1. The continuum action has $U^4(1)$ symmetries associated with the Polyakov loops in the four directions and the various phases correspond to the number of directions in which this

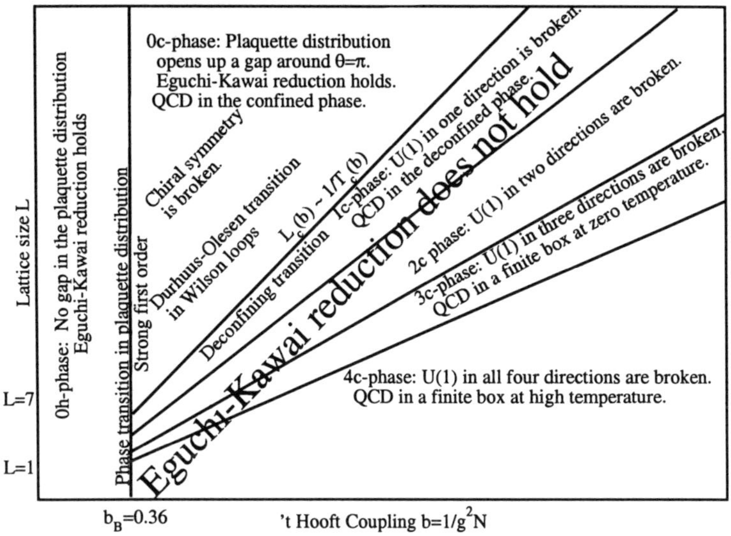

Fig. 1. The various phases of four dimensional large N QCD as viewed from the lattice.

symmetry is broken. The continuum limit is obtained by going to the top-right corner of Fig. 1 and different approaches to this corner will result in one of the five continuum phases.

The 0h-phase present for $b < 0.36$ for all L is an unphysical phase that does not survive the continuum limit. The 0h to 0c transition is associated with the single plaquette operator opening up a gap around π in its eigenvalue distribution. Gauge fields come in disconnected pieces in all the Xc-phases due to the presence of the gap in the single plaquette operator [5].

The 0c-phase is the confined phase of large N QCD and the 1c-phase is the deconfined phase and $l_c = 1/t_c$. An immediate consequence of continuum reduction is that large N QCD does not feel temperatures below t_c [6]. Numerical analysis has shown that lattice spacing effects are small in the 0c-phase and it is sufficient to work at $L \sim 9$ and $N \sim 30$ to extract continuum results. Therefore, numerical computations can be performed on a serial computer and a cluster of computers can be efficiently used to generate statistics in a Monte-Carlo calculation.

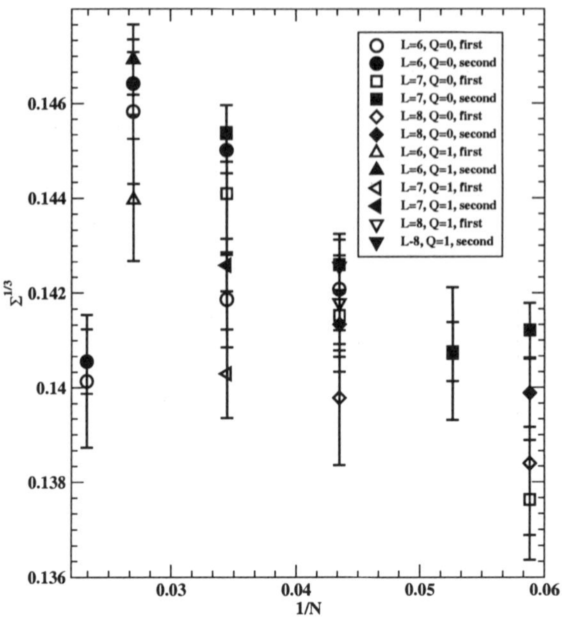

Fig. 2. Extraction of the chiral condensate in large N lattice QCD at a fixed lattice coupling.

3. Chiral symmetry breaking in finite volume

Since physics does not depend on the box size in 0c-phase, one should show that chiral symmetry is spontaneously broken in finite volume in order to properly reproduce physics in this phase [7]. The order of limits are important and one has to take the large N limit before taking the quark mass to zero at finite phsyical volume. The low lying spectrum of the massless Dirac operator shows evidence for spontaneous chiral breaking since the eigenvalues, $i\lambda$, scales like $z = \lambda \Sigma N l^4$ with z obeying a universal distribution. The chiral condesate, Σ, is independent of l. Results of a calculation of the chiral condensate on the lattice is shown in Fig. 2. Results from chiral random matrix theory [8] were used to extract the chiral condensate at a fixed N, L and lattice coupling. Two lowest non-zero eigenvalues in the $Q = 0$ and $Q = 1$ topological sectors were used to show consistency. The plot shows that there is a limit as $N \to \infty$ and this limit is independent of L. Results obtained at different lattice couplings yield $\frac{l_c^3}{N}\{\bar{\psi}\psi\} \approx (0.65)^3$.

4. Pions in large N QCD

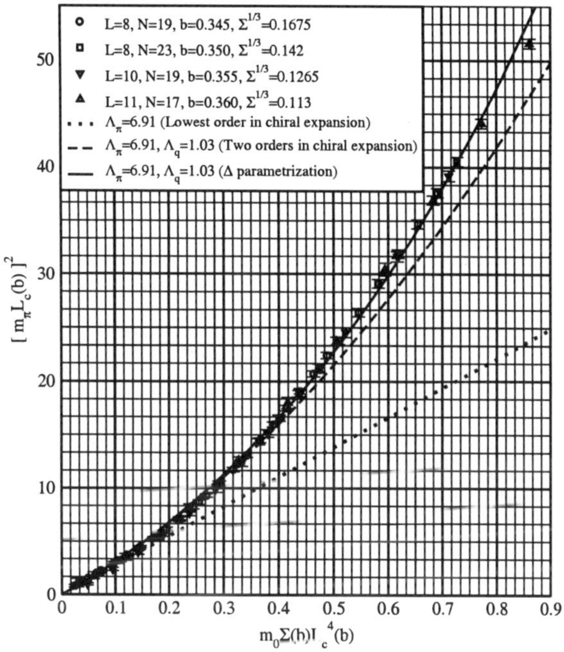

Fig. 3. The pion mass as a function of quark mass in large N QCD.

Since chiral symmetry is broken in large N QCD even in finite volume, one should be able to observe massless pions in finite volume. This result emerges in the following manner [9]. Properties of a single quark in a background gauge field $A_\mu(x)$ cannot depend on a shift of $A_\mu(x) \to A_\mu(x) + \frac{p_\mu}{l}$ for arbitrary p_μ since the $U^4(1)$ symmetries associated with the Polyakov loops are not broken in the 0c-phase. But the propagator of a non-singlet meson will depend on p_μ, if one quark sees $A_\mu(x) \to A_\mu(x) + \frac{p_\mu}{2l}$ and the other sees $A_\mu(x) \to A_\mu(x) - \frac{p_\mu}{2l}$ as their respective gauge fields. This is referred to as the quenched momentum prescription [10] for the computation of meson propagators in the large N limit of QCD. One can use the results for the chiral condensate, $\Sigma(b)$, and critical lattice size, $L_c(b)$, to plot the pion mass as a function of $m < \bar{\psi}\psi >$. The results fall on a single universal curve as shown in fig. 3 and $f_\pi l_c = \frac{1}{\sqrt{2\Lambda_\pi}} \approx 0.269$.

5. Chiral symmetry restoration

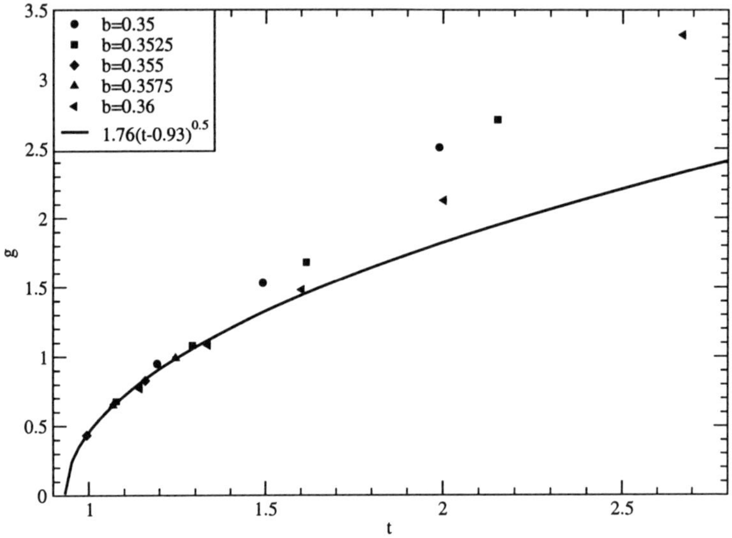

Fig. 4. The gap in the quark spectrum as a function of temperature in large N QCD.

The 0c to 1c phase transition is the confinement-deconfinement phase transition since the $U(1)$ symmetry associated with one of the Polyakov loops is broken. This transition is first order since there is a latent heat associated with the single plaquette [11]. The fermion determinant does matter in the 1c-phase and it picks the correct boundary conditions for fermions in the broken direction, namely, anti-periodic with respect to the Polyakov loop [12]. The lowest eigenvalue of the Dirac operator with the correct boundary conditions can be used to study the gap in the 1c-phase and one finds that chiral symmetry is restored in the 1c-phase [12]. Furthermore, the chiral transition is first order as shown in Fig. 4. If one were to super-cool the 1c-phase into the 0c-phase, a second order transition with a square root singularity would be observed at $0.93/l_c$.

6. Phase transition in the Wilson loop operator

Non-abelian gauge theories in the confined phase are strongly interacting at large distances and weakly interacting at short distances. If this is seen as a phase transition in some observable, one could use the universal behavior of this transition to connect the low energy physics of QCD to the high energy physics of QCD. The Wilson loop operator as a function of its size is the most likely candidate to study this transition within the 0c-phase of large N QCD [13]. Such a transition exists in two dimensional large N QCD [14] and it is claimed that the transition in four dimensional QCD is in the same universality class. This transition is referred to as the Durhuus-Olesen phase transition in Fig 1.

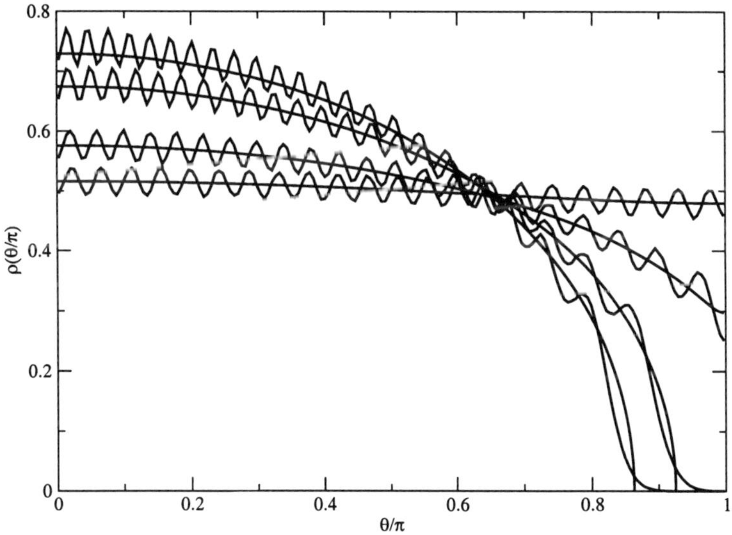

Fig. 5. The phase transition in the expectation value of the Wilson loop operator as a function of its size in large N QCD. The distribution of the eigenvalues of the smeared Wilson loop operator are compared to the Durhuus-Olesen distributions. Results are shown for loops of sizes $l/l_c = 0.740, 0.660, 0.560, 0.503$. They match with $k = 4.03, 2.30, 1.41, 1.15$ where k is the dimensionless area in two dimension large N QCD. $k = 2$ is the critical area.

Wilson loop operators suffer from a perimeter divergence in four dimensions and one has to eliminate it when defining this operator if one were to study its eigenvalue distribution. One possible way to suppress the perimeter divergence is to use smeared operators on the lattice. Numerical studies of the eigenvalue distribution of the smeared Wilson loop operator on the lattice shows clear evidence for a phase transition as a function of the size of the Wilson loop [13]. The universality class is the same as the one found in two dimensional large N QCD as shown in Fig. 5 and the critical loop size is roughly $0.6l_c$.

Acknowledgements

R. N. would like to thank the organizers of CAQCD-06 for a stimulating atmosphere. R. N. acknowledges partial support by the NSF under grant number PHY-0300065 and partial support from Jefferson Lab. The Thomas Jefferson National Accelerator Facility (Jefferson Lab) is operated by the Southeastern Universities Research Association (SURA) under DOE contract DE-AC05-84ER40150. H. N. acknowledges partial support by the DOE under grant number DE-FG02-01ER41165 at Rutgers University.

References

1. A. V. Manohar, arXiv:hep-ph/9802419.
2. O. Aharony, S. S. Gubser, J. M. Maldacena, H. Ooguri and Y. Oz, Phys. Rept. **323**, 183 (2000) [arXiv:hep-th/9905111].
3. J. Kiskis, R. Narayanan and H. Neuberger, "Does the crossover from perturbative to nonperturbative physics in QCD Phys. Lett. B **574**, 65 (2003) [arXiv:hep-lat/0308033].
4. R. Narayanan and H. Neuberger, PoS **LAT2005**, 005 (2006) [arXiv:hep-lat/0509014].
5. J. Kiskis, R. Narayanan and H. Neuberger, Phys. Rev. D **66**, 025019 (2002) [arXiv:hep-lat/0203005].
6. T. D. Cohen, Phys. Rev. Lett. **93**, 201601 (2004) [arXiv:hep-ph/0407306].
7. R. Narayanan and H. Neuberger, Nucl. Phys. B **696**, 107 (2004) [arXiv:hep-lat/0405025].
8. J. J. M. Verbaarschot and T. Wettig, Ann. Rev. Nucl. Part. Sci. **50**, 343 (2000) [arXiv:hep-ph/0003017].
9. R. Narayanan and H. Neuberger, Phys. Lett. B **616**, 76 (2005) [arXiv:hep-lat/0503033].
10. D. J. Gross and Y. Kitazawa, Nucl. Phys. B **206**, 440 (1982).
11. J. Kiskis, arXiv:hep-lat/0507003.
12. R. Narayanan and H. Neuberger, Phys. Lett. B **638**, 546 (2006) [arXiv:hep-th/0605173].
13. R. Narayanan and H. Neuberger, JHEP **0603**, 064 (2006) [arXiv:hep-th/0601210].
14. B. Durhuus and P. Olesen, Nucl. Phys. B **184**, 461 (1981).

SECTION 10
MULTIPARTON AMPLITUDES

Convener Z. Bern

MHV VERTICES AND ON-SHELL RECURSION RELATIONS

PETER SVRČEK

Physics Department, Stanford University/SLAC,
Palo Alto CA 94305/94309, USA
E-mail: svrcek@stanford.edu

In 2003, Witten proposed a topological string theory in twistor space that is dual to a weakly coupled gauge theory. This has lead to new developments in computing gauge theory scattering amplitudes. In these proceedings, we review two new methods, the MHV vertices construction and the on-shell recursion relations.

Keywords: Gauge Theories, Scattering Amplitudes, Twistors.

1. Introduction

The idea that a gauge theory should be dual to a string theory goes back to 't Hooft [33]. 't Hooft considered $U(N)$ gauge theory in the large N limit while keeping $\lambda = g_{YM}^2 N$ fixed. He observed that the perturbative expansion of Yang-Mills can be reorganized in terms of Riemann surfaces, which he interpreted as an evidence for a hypothetical dual string theory with string coupling $g_s \sim 1/N$.

In 1997, Maldacena proposed a concrete example of this duality [40]. He considered the maximally supersymmetric Yang-Mills theory and conjectured that it is dual to type IIB string theory on $AdS_5 \times S^5$. This duality led to many new insights from string theory about gauge theories and vice versa. At the moment, we have control over the duality only for strongly coupled gauge theory. This corresponds to the limit of large radius of $AdS_5 \times S^5$ in which the string theory is well described by supergravity. However, QCD is asymptotically free, so we would also like to have a string theory description of a weakly coupled gauge theory.

In weakly coupled field theories, the natural object to study is the perturbative S matrix. The perturbative expansion of the S matrix is conventionally computed using Feynman rules. Starting from early studies of de

Witt [28], it was observed that scattering amplitudes show simplicity that is not apparent from the Feynman rules. For example, the maximally helicity violating (MHV) amplitudes can be expressed as simple holomorphic functions.

Recently, Witten proposed a string theory that is dual to a weakly coupled $\mathcal{N} = 4$ gauge theory [48]. The perturbative expansion of the gauge theory is related to D-instanton expansion of the string theory. The string theory in question is the topological open string B-model on a Calabi-Yau supermanifold $\mathbb{CP}^{3|4}$, which is a supersymmetric generalization of Penrose's twistor space.

At tree level, evaluating the instanton contribution has led to new insights about scattering amplitudes. 'Disconnected' instantons give the MHV diagram construction of amplitudes in terms of Feynman diagrams with vertices that are suitable off-shell continuations of the MHV amplitudes [22]. The 'connected' instanton contributions express the amplitudes as integrals over the moduli space of holomorphic curves in twistor space [45]. Surprisingly, the MHV diagram construction and the connected instanton integral can be related via localization on the moduli space [27].

The study of twistor structure of scattering amplitudes has inspired new developments in perturbative Yang-Mills theory itself. At tree level, this has led to recursion relations for on-shell amplitudes [19]. At one loop, unitarity techniques [13,12] have been used to find new ways of computing $\mathcal{N} = 4$ [18], $\mathcal{N} = 1$ [21] and $\mathcal{N} = 0$ [17] amplitudes Yang-Mills amplitudes.

In these proceedings, we will describe the MHV vertices and the on-shell recursion relations for tree level gauge theory amplitudes. We will briefly comment on recent developments with one-loop scattering amplitudes. For a more detailed account of these techniques and further references we refer the reader to the lecture notes [26].

2. Helicity amplitudes

2.1. *Spinors*

Recall that the complexified Lorentz group is locally isomorphic to

$$SO(3,1,\mathbb{C}) \cong Sl(2,\mathbb{C}) \times Sl(2,\mathbb{C}), \tag{1}$$

hence the finite dimensional representations are classified as (p,q) where p and q are integer or half-integer. The negative and positive chirality spinors transform in the representations $(1/2,0)$ and $(0,1/2)$ respectively. We write generically $\lambda_a, a = 1,2$ for a spinor transforming as $(1/2,0)$ and $\tilde{\lambda}_{\dot{a}}, \dot{a} = 1,2$ for a spinor transforming as $(0,1/2)$.

The spinor indices of type $(1/2, 0)$ are raised and lowered using the antisymmetric tensors ϵ_{ab} and ϵ^{ab} obeying $\epsilon_{12} = 1$ and $\epsilon^{ac}\epsilon_{cb} = \delta^a{}_b$

$$\lambda^a = \epsilon^{ab}\lambda_b \qquad \lambda_a = \epsilon_{ab}\lambda^b. \tag{2}$$

Given two spinors λ and λ', both of negative chirality, we can form the Lorentz invariant product

$$\langle \lambda, \lambda' \rangle = \epsilon_{ab}\lambda^a \lambda'^b. \tag{3}$$

Similarly, we lower and raise the indices of positive chirality spinors with the antisymmetric tensor $\epsilon_{\dot{a}\dot{b}}$ and its inverse $\epsilon^{\dot{a}\dot{b}}$. For two spinors $\tilde{\lambda}$ and $\tilde{\lambda}'$, both of positive chirality we define the antisymmetric product

$$[\tilde{\lambda}, \tilde{\lambda}'] = -[\tilde{\lambda}', \tilde{\lambda}] = \epsilon_{\dot{a}\dot{b}}\tilde{\lambda}^{\dot{a}}\tilde{\lambda}'^{\dot{b}}. \tag{4}$$

The vector representation of $SO(3, 1, \mathbb{C})$ is the $(1/2, 1/2)$ representation. Thus a momentum vector $p_\mu, \mu = 0, \ldots, 3$ can be represented as a bi-spinor $p_{a\dot{a}}$ with one spinor index a and $\dot{a}$ of each chirality. For any vector, the relation between p_μ, and $p_{a\dot{a}}$ is

$$p_{a\dot{a}} = p_\mu \sigma^\mu_{a\dot{a}} = p_0 + \vec{\sigma} \cdot \vec{p}, \tag{5}$$

where $\sigma^\mu_{\dot{a}a}$ are Pauli matrices. It follows that,

$$p_\mu p^\mu = \det(p_{a\dot{a}}). \tag{6}$$

Hence, p_μ is lightlike if the corresponding determinant is zero. This is equivalent to the rank of the 2×2 matrix $p_{a\dot{a}}$ being less than or equal to one. So p^μ is lightlike precisely, when it can be written as a product

$$p_{a\dot{a}} = \lambda_a \tilde{\lambda}_{\dot{a}} \tag{7}$$

for some spinors λ_a and $\tilde{\lambda}_{\dot{a}}$.

Usually, we describe massless gluons with their momentum vector p^μ and polarization vector ϵ^μ. For a positive helicity gluon, we take

$$\epsilon^+_{a\dot{a}} = \frac{\mu_a \tilde{\lambda}_{\dot{a}}}{\langle \mu, \lambda \rangle}, \tag{8}$$

where μ is any negative chirality spinor that is not a multiple of λ. To get a negative helicity polarization vector, we take

$$\epsilon^-_{a\dot{a}} = \frac{\lambda_a \tilde{\mu}_{\dot{a}}}{[\tilde{\lambda}, \tilde{\mu}]}, \tag{9}$$

where $\tilde{\mu}$ is any positive chirality spinor that is not a multiple of $\tilde{\lambda}$.

2.2. *Scattering amplitudes*

Let us consider scattering of massless particles in four dimensions. Consider the situation with n particles of momenta $p_1, p_2, \ldots, p_n$. For particles with spin, the scattering amplitude is a function of the momenta p_i and the wavefunctions ψ_i of each particle

$$A(p_1, \psi_1; \ldots; p_n, \psi_n). \tag{10}$$

Here, A is linear in each of the wavefunctions ψ_i. The spinors determine the momenta $p_i = \lambda_i \tilde{\lambda}_i$ and the wavefunctions $\psi_i(\lambda_i, \tilde{\lambda}_i, h_i)$. In summary, a general scattering amplitude of massless particles can be written as

$$A = (2\pi)^4 \delta^4 \left(\sum_i \lambda_i^a \tilde{\lambda}_i^{\dot{a}} \right) A(\lambda_i, \tilde{\lambda}_i, h_i), \tag{11}$$

where we have written explicitly the delta function of momentum conservation.

2.3. *Maximally helicity violating amplitudes*

To make the discussion more concrete, we consider tree level scattering of n gluons in Yang-Mills theory. These amplitudes are of phenomenological importance. The multijet production at LHC will be dominated by tree level QCD scattering.

Consider Yang-Mills theory with gauge group $U(N)$. Recall that tree level scattering amplitudes are planar and lead to single trace interactions

$$A = g^{n-2}(2\pi)^4 \delta^4 \left(\sum_i^n p_i \right) \mathcal{A}(1, 2, \ldots, n) \mathrm{Tr}(\mathrm{T}_1 \mathrm{T}_2 \ldots \mathrm{T}_\mathrm{n}) + \text{permutations}. \tag{12}$$

Here, g is the coupling constant of the gauge theory. In the rest of the lecture notes we will consider the 'reduced color ordered amplitude' $\mathcal{A}(1, 2, \ldots, n)$.

The scattering amplitude with n incoming gluons of the same helicity vanishes. So does the amplitude, for $n \geq 3$, with $n - 1$ incoming gluons of one helicity and one of the opposite helicity. The first nonzero amplitudes, the maximally helicity violating (MHV) amplitudes, have $n - 2$ gluons of one helicity and two gluons of the other helicity. Suppose that gluons r, s have negative helicity and the rest of gluons have positive helicity

$$\mathcal{A}(r^-, s^-) = g^{n-2} \frac{\langle \lambda_r, \lambda_s \rangle^4}{\prod_{k=1}^n \langle \lambda_k, \lambda_{k+1} \rangle}. \tag{13}$$

The amplitude $\mathcal{A}(r^+, s^+)$ with gluons r, s of positive helicity and the rest of the gluons of negative helicity follows from (13) by exchange $\langle\rangle \leftrightarrow []$. The amplitude $\mathcal{A}$ is sometimes called 'holomorphic' because it depends on the 'holomorphic' spinors λ_i only.

3. Scattering amplitudes in twistor space

To get from momentum space into twistor space, we make the following transformation

$$\tilde{\lambda}_{\dot{a}} \rightarrow i\frac{\partial}{\partial \mu^{\dot{a}}}$$
$$\frac{\partial}{\partial \tilde{\lambda}^{\dot{a}}} \rightarrow i\mu_{\dot{a}}. \tag{14}$$

Making this substitution we have arbitrarily chosen to Fourier transform $\tilde{\lambda}$ rather than λ. This choice breaks the symmetry between positive and negative helicities. The amplitude with n_1 positive helicity and n_2 negative helicity gluons has different description in twistor space from an amplitude with n_2 positive helicity gluons and n_1 negative helicity gluons.

For signature $++--$, where $\tilde{\lambda}$ is real, the transformation from momentum space scattering amplitudes to twistor space scattering amplitudes is made by a simple Fourier transform that is familiar from quantum mechanics

$$\mathcal{A}(\lambda_i, \mu_i) = \int \prod_{j=1}^{n} \frac{d^2\tilde{\lambda}_j}{(2\pi)^2} \exp(i[\mu_j, \tilde{\lambda}_j])\mathcal{A}(\lambda_i, \tilde{\lambda}_i). \tag{15}$$

In an n gluon scattering process, after the Fourier transform into twistor space, the external gluons are associated with points P_i in the projective twistor space. The scattering amplitudes are functions of the twistors P_i, that is, they are functions defined on the product of n copies of twistor space, one for each particle.

Let us see what happens to the tree level MHV amplitude with $n-2$ gluons of positive helicity and 2 gluons of negative helicity, after Fourier transform into twistor space. We recall that the MHV amplitude with negative helicity gluons r, s is

$$A(\lambda_i, \tilde{\lambda}_i) = (2\pi)^4\delta^4(\sum_i \lambda_i\tilde{\lambda}_i)f(\lambda_i), \tag{16}$$

where

$$f(\lambda_i) = g^{n-2}\frac{\langle\lambda_r, \lambda_s\rangle^4}{\prod_k\langle\lambda_k, \lambda_{k+1}\rangle}. \tag{17}$$

Hence, we can rewrite the amplitude as

$$A(\lambda_i, \tilde{\lambda}_i) = \int d^4x \exp\left(ix_{b\dot{b}}\sum_i \lambda_i^b\tilde{\lambda}_i^{\dot{b}}\right) f(\lambda_i), \tag{18}$$

where we expressed the delta function of momentum conservation as an integral. The transform of the amplitude to twistor space is [42]

$$A(\lambda_i, \mu_i) = \int d^4x \prod_{j=1}^{n} \delta^2(\mu_{j\dot{a}} + x_{a\dot{a}}\lambda_j^a) f(\lambda_i). \tag{19}$$

This equation has a simple geometrical interpretation. Pick some $x^{a\dot{a}}$ and consider the equation

$$\mu_{\dot{a}} + x_{a\dot{a}}\lambda^a = 0. \tag{20}$$

The solution set for $x = 0$ is a $\mathbb{RP}^1$ or $\mathbb{CP}^1$ depending on whether the variables are real or complex. This is true for any x as the equation lets us solve for $\mu_{\dot{a}}$ in terms of λ^a. So (λ^1, λ^2) are the homogeneous coordinates on the curve.

Going back to the amplitude (19), the δ-functions mean that the amplitude vanishes unless $\mu_{j\dot{a}} + x_{a\dot{a}}\lambda_j^a = 0, j = 1, \ldots n$, that is, unless some curve of degree one determined by $x_{a\dot{a}}$ contains all n points (λ_j, μ_j). The result is that the MHV amplitudes are supported on genus zero curves of degree one. This is a consequence of the holomorphy of these amplitudes.

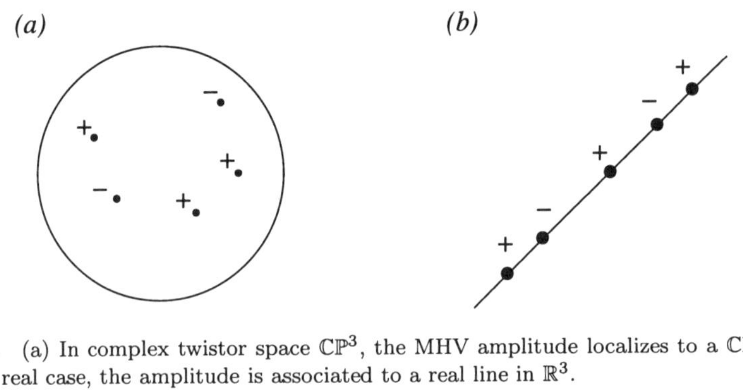

Fig. 1. (a) In complex twistor space $\mathbb{CP}^3$, the MHV amplitude localizes to a $\mathbb{CP}^1$. (b) In the real case, the amplitude is associated to a real line in $\mathbb{R}^3$.

The general conjecture is that an l-loop amplitude with p gluons of positive helicity and q gluons of negative helicity is supported on a holomorphic curve in twistor space. The degree of the curve is determined by

$$d = q - 1 + l. \tag{21}$$

The genus of the curve is bounded by the number of the loops

$$g \leq l. \tag{22}$$

The MHV amplitudes are a special case of this for $q = 2, l = 0$. Indeed the conjecture in this case gives that MHV amplitudes are supported in twistor space on a genus zero curve of degree one.

The natural interpretation of this is that the curve is the worldsheet of a string. In some way of describing the perturbative gauge theory, the amplitudes arise from coupling of the gluons to a string [48].

4. MHV diagrams

In this section, we start with a motivation of the MHV diagrams construction of amplitudes from basic properties of twistor correspondence. We then go on to discuss simple examples and extensions to loop amplitudes.

Recall that MHV scattering amplitudes are supported on $\mathbb{CP}^1$'s in twistor space. Each such $\mathbb{CP}^1$ can be associated to a point $x^{a\dot{a}}$ in Minkowski space

$$\mu_{\dot{a}} + x_{a\dot{a}}\lambda^a = 0. \tag{23}$$

So, in a sense, we can think of MHV amplitudes as local interaction vertices [22]. To take this analogy further, we can try to build more complicated amplitudes from Feynman diagrams with vertices that are suitable off-shell continuations of the MHV amplitudes, fig. 2. MHV amplitudes are functions of holomorphic spinors λ_i only. Hence, to use them as vertices in Feynman diagrams, we need to define λ for internal off-shell momenta $p^2 \neq 0$.

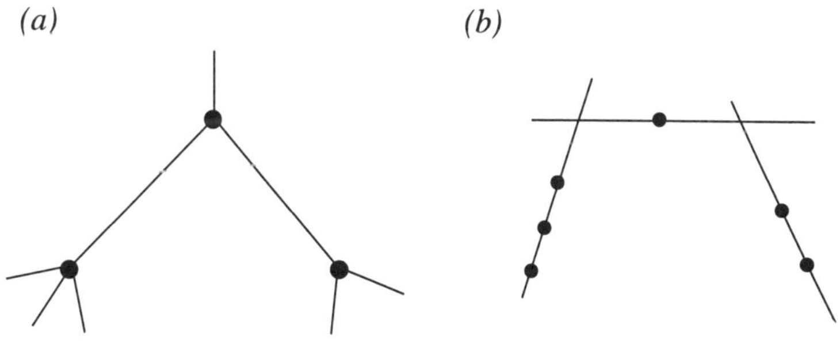

Fig. 2. Two representations of a degree three MHV diagram. (a) In Minkowski space, the MHV vertices are represented by points. (b) In twistor space, each MHV vertex corresponds to a line. The three lines pairwise intersect.

To motivate the off-shell continuation, notice that for on-shell momentum $p^{a\dot{a}} = \lambda^a\tilde{\lambda}^{\dot{a}}$, we can extract the holomorphic spinors λ from the momentum p by picking arbitrary anti-holomorphic spinor $\eta^{\dot{a}}$ and contracting it with $p^{a\dot{a}}$. This gives λ^a up to a scalar factor

$$\lambda^a = \frac{p^{a\dot{a}}\eta_{\dot{a}}}{[\tilde{\lambda}, \eta]}. \tag{24}$$

For off-shell momenta, this strategy almost works except for the factor $[\tilde{\lambda}, \eta]$ in the denominator which depends on the undefined spinor $\tilde{\lambda}$. Fortunately, $[\tilde{\lambda}, \eta]$ scales out of Feynman diagrams, so we take as our definition

$$\lambda^a = p^{a\dot{a}}\eta_{\dot{a}}. \tag{25}$$

This is clearly well-defined for off-shell momentum. We complete the definition of the MHV rules, by taking $1/k^2$ for the propagator connecting the MHV vertices.

For further work on MHV vertices construction of tree-level gluon amplitudes, see [39,49,51,14]. MHV vertices have many generalizations; in particular, to amplitudes with fermions and scalars [30,31,38,50,47], with Higgses [29,2] and with electroweak vector-boson currents [10]. For an attempt to generalize MHV vertices to gravity, see [32,43,1]. Recently there have been attempts to prove MHV vertices construction using light-cone gauge [41] and using a new twistor action [15].

Loop Amplitudes

Similarly, one can compute loop amplitudes using MHV diagrams. This has been carried out for the one loop MHV amplitude in $\mathcal{N} = 4$ [16] and $\mathcal{N} = 1$ [44,5] Yang-Mills theory, in agreement with the known answers.

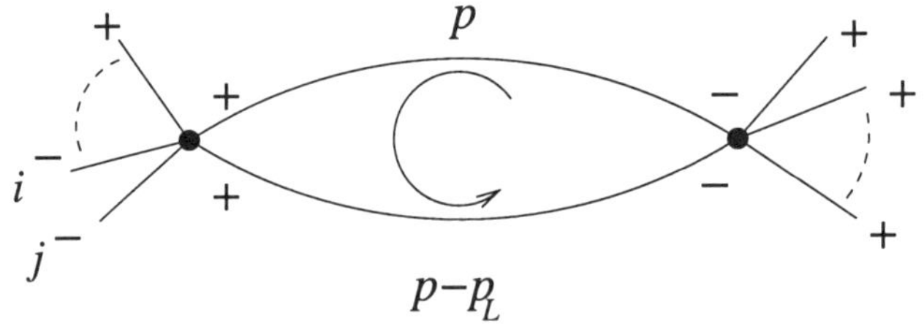

Fig. 3. Schematic representation of a hypothetical twistor string computation of one-loop MHV amplitude. The picture shows a diagram in which the negative helicity gluons i^-, j^- are on the same MHV vertex.

The expression for an MHV diagram contributing to the one-loop MHV amplitude is just what one would expect for a one-loop Feynman diagram with MHV vertices, fig. 3. There are two MHV vertices, each coming with two negative helicity gluons. The vertices are connected with two Feynman propagators that absorb two negative helicities, leaving two negative helicity external gluons

$$\mathcal{A}^{loop} = \sum_{\mathcal{D},h} \int \frac{d^{4-2\epsilon}p}{(2\pi)^4} \mathcal{A}_L(\lambda_k, \lambda_p, \lambda_{p-p_L}) \frac{1}{p^2(p-p_L)^2} \mathcal{A}_R(\lambda_k, \lambda_p, \lambda_{p-p_L}). \tag{26}$$

The off-shell spinors entering the MHV amplitudes $\mathcal{A}_L, \mathcal{A}_R$ are determined in terms of the momenta of the internal lines

$$\lambda_p^a = p^{a\dot{a}}\eta_{\dot{a}}, \qquad \lambda_{p-p_L}^a = (p-p_L)^{a\dot{a}}\eta_{\dot{a}}, \tag{27}$$

which is the same prescription as for tree level MHV diagrams. The sum in (26) is over partitions $\mathcal{D}$ of the gluons among the two MHV diagrams that preserve the cyclic order and over the helicities of the internal particles[a].

This calculation makes the twistor structure of one-loop MHV amplitudes manifest. The two MHV vertices are supported on lines in twistor space, so the amplitude is a sum of contributions, each of which is supported on a disjoint union two lines. In a hypothetical twistor string theory computation of the amplitude, these two lines are connected by open string propagators. This pictures agrees with studies of the twistor structure using differential equations [23], after taking into account the holomorphic anomaly of the differential equations [24,7].

Finally, we make a few remarks about the nonsupersymmetric one-loop MHV amplitudes. The $\mathcal{N} = 0$ MHV amplitudes are sums of cut-constructible terms and rational terms. The cut-constructible terms are correctly reproduced from MHV diagrams [4]. The rational terms are single valued functions of the spinors, hence they are free of cuts in four dimensions. Their twistor structure suggests that they receive contribution from diagrams in which, alongside with MHV vertices, there are new one-loop vertices coming from one-loop all-plus helicity amplitudes [23]. However, a suitable off-shell continuation of the one-loop all-plus amplitude has not been found yet. Recently, the rational part of one-loop QCD amplitudes has been computed using a generalization of the tree level recursion relations [11,8].

There has been recent progress in computing the rational part of some one-loop QCD amplitudes using a generalization of the tree level recursion relations.

---

[a]Similarly, the double-trace contribution to one-loop MHV amplitudes comes from Feynman diagrams with double-trace MHV vertices [34,35].

5. BCFW recursion relations

We have seen how tree-level amplitudes of gluons can be computed in a simple and systematic manner by using MHV diagrams. However, from the study of infrared divergencies of one-loop $\mathcal{N} = 4$ amplitudes of gluons, surprisingly simple and compact forms for many tree amplitudes were found in [9,46]. These miraculously simple formulas were given an explanation when a set of recursion relations for amplitudes of gluons was conjectured in [19]. The Britto-Cachazo-Feng-Witten (BCFW) recursion relations were later proven and extended in [20]. Here we review the BCFW proof of the general set of recursion relations.

Consider a tree-level amplitude $\mathcal{A}(1, 2, \ldots, n-1, n)$ of n cyclically ordered gluons, with any specified helicities.

In what follows, we single out two of the gluons for special treatment. Using the cyclic symmetry, without any loss of generality, we can take these to be the gluons k and n. We introduce a complex variable z, and let

$$\begin{aligned} p_k(z) &= \lambda_k(\tilde{\lambda}_k - z\tilde{\lambda}_n), \\ p_n(z) &= (\lambda_n + z\lambda_k)\tilde{\lambda}_n. \end{aligned} \tag{28}$$

We leave the momenta of the other gluons unchanged, so $p_s(z) = p_s$ for $s \neq k, n$. In effect, we have made the transformation

$$\tilde{\lambda}_k \to \tilde{\lambda}_k - z\tilde{\lambda}_n, \qquad \lambda_n \to \lambda_n + z\lambda_k, \tag{29}$$

with λ_k and $\tilde{\lambda}_n$ fixed. Note that $p_k(z)$ and $p_n(z)$ are on-shell for all z, and $p_k(z) + p_n(z)$ is independent of z. As a result, we can define the following function of a complex variable z,

$$\mathcal{A}(z) = \mathcal{A}(p_1, \ldots, p_{k-1}, p_k(z), p_{k+1}, \ldots, p_{n-1}, p_n(z)). \tag{30}$$

The right hand side is a physical, on-shell amplitude for all z. Momentum is conserved and all momenta are on-shell.

For any $z \neq 0$, the deformation (28) does not make sense for real momenta in Minkowski space, as it does not respect the Minkowski space reality condition $\tilde{\lambda} = \pm\bar{\lambda}$. However, (28) makes perfect sense for complex momenta or (if z is real) for real momenta in signature $++--$. In any case, we think of $\mathcal{A}(z)$ as an auxiliary function. In the end, all answers are given in terms of spinor inner products and are valid for any signature.

We claim three facts about $\mathcal{A}(z)$: (1) It is a rational function. (2) It only has simple poles. (3) It vanishes for $z \to \infty$.

These three properties of $\mathcal{A}(z)$ imply that it can be written as follows

$$\mathcal{A}(z) = \sum_{p \in \{\text{poles}\}} \frac{c_p}{z - z_p}, \tag{31}$$

where c_p is the residue at a given pole and the sum is over the whole set of poles. It turns out that, as we will see below, c_p is proportional to the product of two physical amplitudes with fewer gluons than $\mathcal{A}(z)$. Therefore, (31) provides a recursion relation for amplitudes of gluons.

Let us prove the three statements. (1) This is easy. Note that the original tree-level amplitude is a rational function of spinor products. Since the z dependence enters only via the shift $\tilde{\lambda}_k \to \tilde{\lambda}_k - z\tilde{\lambda}_n$ and $\lambda_n \to \lambda_n + z\lambda_k$, $\mathcal{A}(z)$ is clearly rational in z. (2) By definition, $\mathcal{A}(z)$ is constructed out of Feynman diagrams. The only singularities $\mathcal{A}(z)$ can have come from propagators. The propagators $1/P_{ij}(z)^2$ can have only a single, simple pole, which is located at $z_{ij} = P_{ij}^2/\langle\lambda_k|P_{ij}|\tilde{\lambda}_n]$. Finally, for a simple proof of (3) using Feynman diagram, see the original reference [20].

Now we can rewrite (31) more precisely as follows

$$\mathcal{A}(z) = \sum_{i,j} \frac{c_{ij}}{z - z_{ij}}, \tag{32}$$

where c_{ij} is the residue of $\mathcal{A}(z)$ at the pole $z = z_{ij}$. Finally, we have to compute the residues c_{ij}. To get a pole at $P_{ij}^2(z) = 0$, a tree diagram must contain a propagator that divides it into a "left" part containing all external gluons not in the range from i to j, and a "right" part containing all external gluons that are in that range. The contribution of such diagrams near $z = z_{ij}$ is $\sum_h \mathcal{A}_L^h(z)\mathcal{A}_R^{-h}(z)/P_{ij}(z)^2$, where $\mathcal{A}_L^h(z)$ and $\mathcal{A}_R^{-h}(z)$ are the amplitudes on the left and the right with indicated helicities. Since the denominator $P_{ij}(z)^2$ is linear in z, to obtain the function $c_{ij}/(z - z_{ij})$ that appears in (32), we must simply set z equal to z_{ij} in the numerator. When we do this, the internal line becomes on-shell, and the numerator becomes a product $\mathcal{A}_L^h(z_{ij})\mathcal{A}_R^{-h}(z_{ij})$ of physical, on-shell scattering amplitudes. More precisely we have,

$$\mathcal{A}_L^h(z_{ij}) = \mathcal{A}(p_{j+1}, \ldots, p_k(z_{ij}), \ldots, p_{i-1}, P_{ij}^h(z_{ij})), \tag{33}$$

and a similar expression for $\mathcal{A}_R^{-h}(z_{ij})$.

The formula (32) for the function $\mathcal{A}(z)$ therefore becomes

$$\mathcal{A}(z) = \sum_{i,j}\sum_{h} \frac{\mathcal{A}_L^h(z_{ij})\mathcal{A}_R^{-h}(z_{ij})}{P_{ij}(z)^2}. \tag{34}$$

To get the physical scattering amplitude $\mathcal{A}(1, 2, \ldots, n-1, n)$, we set z to zero in the denominator without touching the numerator. Hence,

$$\mathcal{A}(1, 2, \ldots, n-1, n) = \sum_{i,j}\sum_{h} \frac{\mathcal{A}_L^h(z_{ij})\mathcal{A}_R^{-h}(z_{ij})}{P_{ij}^2}. \tag{35}$$

This is the BCFW recursion relation [19,20].

Let us also mention that many generalizations of the BCFW recursion relations have been made, in particular, to include amplitudes with fermions and scalars [36,37] and to gravity amplitudes [6,25]. The recursion relations have also been generalized to amplitudes with massive particles [3], and most recently to one-loop amplitudes in pure YM [11,8].

Acknowledgements

We would like to thank the organizers of Continuous Advances in QCD 2006 for their hospitality and for their invitation to speak. The author was supported by NSF under grant PHY-0244728 and the DOE under contract DE-AC03-76SF00515.

References

1. Y. Abe, "An interpretation of multigraviton amplitudes," arXiv:hep-th/0504174.
2. S. D. Badger, E. W. N. Glover and V. V. Khoze, "MHV rules for Higgs plus multi-parton amplitudes," JHEP **0503**, 023 (2005) [arXiv:hep-th/0412275].
3. S. D. Badger, E. W. N. Glover, V. V. Khoze and P. Svrcek, "Recursion relations for gauge theory amplitudes with massive particles," arXiv:hep-th/0504159.
4. J. Bedford, A. Brandhuber, B. Spence and G. Travaglini, "Non-supersymmetric loop amplitudes and MHV vertices," arXiv:hep-th/0412108.
5. J. Bedford, A. Brandhuber, B. Spence and G. Travaglini, "A twistor approach to one-loop amplitudes in N = 1 supersymmetric Yang-Mills theory," Nucl. Phys. B **706**, 100 (2005) [arXiv:hep-th/0410280].
6. J. Bedford, A. Brandhuber, B. Spence and G. Travaglini, "A recursion relation for gravity amplitudes," arXiv:hep-th/0502146.
7. I. Bena, Z. Bern, D. A. Kosower and R. Roiban, "Loops in twistor space," [arXiv:hep-th/0410054].
8. C. F. Berger, Z. Bern, L. J. Dixon, D. Forde and D. A. Kosower, arXiv:hep-ph/0604195.
9. Z. Bern, V. Del Duca, L. J. Dixon and D. A. Kosower, "All non-maximally-helicity-violating one-loop seven-gluon amplitudes in N = 4 super-Yang-Mills theory," Phys. Rev. D **71**, 045006 (2005) [arXiv:hep-th/0410224].
10. Z. Bern, D. Forde, D. A. Kosower and P. Mastrolia, "Twistor-inspired construction of electroweak vector boson currents," arXiv:hep-ph/0412167.
11. Z. Bern, L. J. Dixon and D. A. Kosower, "On-shell recurrence relations for one-loop QCD amplitudes," arXiv:hep-th/0501240.
12. Z. Bern, L. J. Dixon, D. C. Dunbar and D. A. Kosower, "One loop n point gauge theory amplitudes, unitarity and collinear limits," Nucl. Phys. B **425**, 217 (1994) [arXiv:hep-ph/9403226].

13. Z. Bern, L. J. Dixon, D. C. Dunbar and D. A. Kosower, "Fusing gauge theory tree amplitudes into loop amplitudes," Nucl. Phys. B **435**, 59 (1995) [arXiv:hep-ph/9409265].
14. T. G. Birthwright, E. W. N. Glover, V. V. Khoze and P. Marquard, "Multi-gluon collinear limits from MHV diagrams," arXiv:hep-ph/0503063.
15. R. Boels, L. Mason and D. Skinner, arXiv:hep-th/0604040.
16. A. Brandhuber, B. Spence and G. Travaglini, "One-loop gauge theory amplitudes in N = 4 super Yang-Mills from MHV vertices," Nucl. Phys. B **706**, 150 (2005) [arXiv:hep-th/0407214].
17. A. Brandhuber, S. McNamara, B. J. Spence and G. Travaglini, JHEP **0510**, 011 (2005) [arXiv:hep-th/0506068].
18. R. Britto, F. Cachazo and B. Feng, "Generalized unitarity and one-loop amplitudes in N = 4 super-Yang-Mills," arXiv:hep-th/0412103.
19. R. Britto, F. Cachazo and B. Feng, "New recursion relations for tree amplitudes of gluons," arXiv:hep-th/0412308.
20. R. Britto, F. Cachazo, B. Feng and E. Witten, "Direct proof of tree-level recursion relation in Yang-Mills theory," arXiv:hep-th/0501052.
21. R. Britto, E. Buchbinder, F. Cachazo and B. Feng, "One-loop amplitudes of gluons in SQCD," arXiv:hep-ph/0503132.
22. F. Cachazo, P. Svrcek and E. Witten, "MHV vertices and tree amplitudes in gauge theory," JHEP **0409**, 006 (2004) [arXiv:hep-th/0403047].
23. F. Cachazo, P. Svrcek and E. Witten, "Twistor space structure of one-loop amplitudes in gauge theory," JHEP **0410**, 074 (2004) [arXiv:hep-th/0406177].
24. F. Cachazo, P. Svrcek and E. Witten, "Gauge theory amplitudes in twistor space and holomorphic anomaly," JHEP **0410**, 077 (2004) [arXiv:hep-th/0409245].
25. F. Cachazo and P. Svrcek, "Tree level recursion relations in general relativity," arXiv:hep-th/0502160.
26. F. Cachazo and P. Svrcek, PoS **RTN2005**, 004 (2005) [arXiv:hep-th/0504194].
27. S. Gukov, L. Motl and A. Neitzke, "Equivalence of twistor prescriptions for super Yang-Mills," arXiv:hep-th/0404085.
28. B. S. DeWitt, "Quantum Theory Of Gravity. Iii. Applications Of The Covariant Theory," Phys. Rev. **162** (1967) 1239.
29. L. J. Dixon, E. W. N. Glover and V. V. Khoze, "MHV rules for Higgs plus multi-gluon amplitudes," JHEP **0412**, 015 (2004) [arXiv:hep-th/0411092].
30. G. Georgiou and V. V. Khoze, "Tree amplitudes in gauge theory as scalar MHV diagrams," JHEP **0405**, 070 (2004) [arXiv:hep-th/0404072].
31. G. Georgiou, E. W. N. Glover and V. V. Khoze, "Non-MHV tree amplitudes in gauge theory," JHEP **0407**, 048 (2004) [arXiv:hep-th/0407027].
32. S. Giombi, R. Ricci, D. Robles-Llana and D. Trancanelli, "A note on twistor gravity amplitudes," JHEP **0407**, 059 (2004) [arXiv:hep-th/0405086].
33. G. 't Hooft, "A Planar Diagram Theory For Strong Interactions," Nucl. Phys. B **72**, 461 (1974).
34. M. x. Luo and C. k. Wen, "One-loop maximal helicity violating amplitudes

in N = 4 super Yang-Mills theories," JHEP **0411**, 004 (2004) [arXiv:hep-th/0410045].
35. M. x. Luo and C. k. Wen, "Systematics of one-loop scattering amplitudes in N = 4 super Yang-Mills theories," Phys. Lett. B **609**, 86 (2005) [arXiv:hep-th/0410118].
36. M. x. Luo and C. k. Wen, "Recursion relations for tree amplitudes in super gauge theories," arXiv:hep-th/0501121.
37. M. x. Luo and C. k. Wen, "Compact formulas for all tree amplitudes of six partons," arXiv:hep-th/0502009.
38. V. V. Khoze, "Gauge theory amplitudes, scalar graphs and twistor space," arXiv:hep-th/0408233.
39. D. A. Kosower, "Next-to-maximal helicity violating amplitudes in gauge theory," Phys. Rev. D **71**, 045007 (2005) [arXiv:hep-th/0406175].
40. J. M. Maldacena, "The large N limit of superconformal field theories and supergravity," Adv. Theor. Math. Phys. **2**, 231 (1998) [Int. J. Theor. Phys. **38**, 1113 (1999)] [arXiv:hep-th/9711200].
41. P. Mansfield, JHEP **0603**, 037 (2006) [arXiv:hep-th/0511264].
42. V. P. Nair, "A Current Algebra For Some Gauge Theory Amplitudes," Phys. Lett. B **214** (1988) 215.
43. V. P. Nair, "A note on MHV amplitudes for gravitons," arXiv:hep-th/0501143.
44. C. Quigley and M. Rozali, "One-loop MHV amplitudes in supersymmetric gauge theories," JHEP **0501**, 053 (2005) [arXiv:hep-th/0410278].
45. R. Roiban, M. Spradlin and A. Volovich, "On the tree-level S-matrix of Yang-Mills theory," Phys. Rev. D **70**, 026009 (2004) [arXiv:hep-th/0403190].
46. R. Roiban, M. Spradlin and A. Volovich, "Dissolving N = 4 loop amplitudes into QCD tree amplitudes," Phys. Rev. Lett. **94**, 102002 (2005) [arXiv:hep-th/0412265].
47. X. Su and J. B. Wu, "Six-quark amplitudes from fermionic MHV vertices," arXiv:hep-th/0409228.
48. E. Witten, "Perturbative gauge theory as a string theory in twistor space," Commun. Math. Phys. **252**, 189 (2004) [arXiv:hep-th/0312171].
49. J. B. Wu and C. J. Zhu, "MHV vertices and scattering amplitudes in gauge theory," JHEP **0407**, 032 (2004) [arXiv:hep-th/0406085].
50. J. B. Wu and C. J. Zhu, "MHV vertices and fermionic scattering amplitudes in gauge theory with quarks and gluinos," JHEP **0409**, 063 (2004) [arXiv:hep-th/0406146].
51. C. J. Zhu, "The googly amplitudes in gauge theory," JHEP **0404**, 032 (2004) [arXiv:hep-th/0403115].

QUANTUM MHV DIAGRAMS

A. BRANDHUBER and G. TRAVAGLINI
Department of Physics, Queen Mary, University of London
London E1 4NS, United Kingdom
E-mail: {a.brandhuber, g.travaglini}@qmul.ac.uk

Over the past two years, the use of on-shell techniques has deepened our understanding of the S-matrix of gauge theories and led to the calculation of many new scattering amplitudes. In these notes we review a particular on-shell technique developed recently, the quantum MHV diagrams, and discuss applications to one-loop amplitudes. Furthermore, we briefly discuss the application of D-dimensional generalised unitarity to the calculation of scattering amplitudes in non-supersymmetric Yang-Mills.

Keywords: Style file; LaTeX; Proceedings; World Scientific Publishing.

1. Introduction

Recently, remarkable progress has been made in understanding the structure of the S-matrix of four-dimensional gauge theories. This progress was prompted by Witten's proposal [1] of a new duality between $\mathcal{N}$=4 supersymmetric Yang-Mills (SYM) and the topological open string B-model with target space the Calabi-Yau supermanifold $\mathbf{CP}^{3|4}$, a supersymmetric version of Penrose's twistor space. In contrast to usual dualities, this novel duality relates two weakly-coupled theories, and as such it can in principle be tested by explicit computations on both sides.

One of the striking results of Witten's analysis is the understanding of the remarkable simplicity of scattering amplitudes in Yang-Mills and gravity – simplicity which is completely obscure in computations using textbook techniques – in terms of the geometry of twistor space. More precisely, Witten [1] observed that tree-level scattering amplitudes, when Fourier transformed to twistor space, localise on algebraic curves in twistor space. For the simplest case of the maximally helicity violating (MHV) amplitude, the curve is just a (complex) line.

The simple geometrical structure in twistor space of the amplitudes was also the root of further important developments of new efficient tools to calculate amplitudes. In [2], Cachazo, Svrček and Witten (CSW) proposed a novel perturbative expansion for amplitudes in YM, where the MHV amplitudes are lifted to vertices, and joined by scalar propagators in order to form amplitudes with an increasing number of negative helicity gluons. This method was shown to be applicable to one-loop amplitudes in supersymmetric [3–5] and non-supersymmetric theories [6], where a new infinite series of amplitudes in pure YM was calculated. In a different development, the new twistor ideas were merged with earlier applications of unitarity [7,8] and generalised unitarity [9], and led to highly efficient techniques to calculate one-loop amplitudes in $\mathcal{N}=4$ SYM [10,11] and in $\mathcal{N}=1$ SYM [12,13].

There are many important reasons for the interest in new, more powerful techniques to calculate scattering amplitudes. Besides improving our theoretical understanding of gauge theories at the perturbative level the most important reason is the need for higher precision in our theoretical predictions. The advent of the Large Hadron Collider requires the knowledge of perturbative QCD backgrounds at unprecedented precision to distinguish "old" physics from the sought for "new" physics, and there exist long wish-lists of processes that are yet to be computed. Traditional methods using Feynman rules are rather inefficient since they hide the simplicity of scattering amplitudes, intermediate expressions tend to be larger than the final formulas and one has to face the problem of the factorial growth of the number of diagrams which hampers the use of brute force methods. Therefore, techniques that directly lead to the simple final answers are desirable. In the following we want describe some of the novel "twistor string inspired" techniques, focusing on methods relevant for one-loop amplitudes.

2. Colour decomposition and spinor helicity formalism

Here we describe two ingredients that are essential in order to make manifest the simplicity of scattering amplitudes: the colour decomposition, and the spinor helicity formalism. We will later see how new twistor-inspired techniques merge fruitfully with these tools.

At tree level, Yang-Mills interactions are planar, hence an amplitude can be written as a sum over single-trace structures times partial or colour-stripped amplitudes,

$$\mathcal{A}_n^{\rm tree}\left(\{p_i, \epsilon_i, a_i\}\right) = \sum_{\sigma \epsilon S_n/\mathbf{Z}_n} \operatorname{Tr}\left(T^{a_{\sigma(1)}} \dots T^{a_{\sigma(n)}}\right) A_n\left(\sigma(p_1, \epsilon_1), \dots, \sigma(p_n, \epsilon_n)\right) . \tag{1}$$

Partial amplitudes do not carry any colour structure. They receive contributions only from Feynman diagrams with a fixed cyclic ordering of the external lines and have a simpler analytic structure than the full amplitude. At loop level, also multi trace structures appear, which are subleading in the $1/N$ expansion. However, for the one-loop gluon scattering amplitudes there exist a simple (linear) relation between the planar and non-planar terms and, therefore, we have to consider only planar ones.

The spinor helicity formalism, is largely responsible for the existence of very compact formulas of tree and loop amplitudes in massless theories. In four dimensions the Lorentz group is $\mathbf{SL}(2,\mathbf{C})$ and a Lorentz vector is equivalent to a bi-spinor, i.e. the four-momentum can be written as a 2×2 matrix $p_{a\dot{a}} = p_\mu \sigma^\mu_{a\dot{a}}$, $a, \dot{a} = 1, 2$, where a and $\dot{a}$ are left and right handed spinor indices, respectively. If p_μ describes a massless, on-shell particle then $p^2 = \det p_{a\dot{a}} = 0$ and the matrix $p_{a\dot{a}}$ can be written as a product of a left and a right-handed spinor

$$p_{a\dot{a}} = \lambda_a \widetilde{\lambda}_{\dot{a}} \,. \tag{2}$$

In real Minkowski space the left and right-handed spinors are related by complex conjugation $\widetilde{\lambda} = \overline{\lambda}$. It is useful to introduce the following Lorentz invariant brackets: $\langle ij\rangle = \epsilon^{ab}\lambda^i_a\lambda^j_b$ and $[ji] = \epsilon^{\dot{a}\dot{b}}\widetilde{\lambda}^i_{\dot{a}}\widetilde{\lambda}^j_{\dot{b}}$, in terms of which dot products of on-shell four-momenta can be rewritten as $2k_i \cdot k_j = \langle ij\rangle[ji]$.

The simplest non-vanishing tree-level gluon scattering amplitudes have exactly two negative helicity gluons, denoted by i and j, and otherwise only positive helicity gluons. Since amplitudes with zero or one negative helicity gluon vanish, these amplitudes are called maximally helicity violating (MHV). Up to a trivial momentum conservation factor they take the following form

$$A_{\mathrm{MHV}}(\lambda_i) = ig^{n-2}\frac{\langle ij\rangle}{\langle 12\rangle \ldots \langle (n-1)n\rangle\langle n1\rangle} \,. \tag{3}$$

The most remarkable fact about this formula, besides its simplicity, is that it is holomorphic (no $\widetilde{\lambda}$s appear), which has important consequences when the amplitude is transformed to twistor space.

3. Twistor space and MHV diagrams

In a nutshell, twistor space is obtained from the spinor variables $(\lambda_a, \widetilde{\lambda}_{\dot{a}})$ by a "half Fourier transform" (FT), *i.e.* a FT of one of the spinor variables, say $\widetilde{\lambda}_{\dot{a}} \to \mu_{\dot{a}}$. Twistor space (for complexified Minkowski space) is a four-dimensional complex space $(\lambda_a, \mu_{\dot{a}})$. However, amplitudes turn out to be homogeneous functions of the twistor variables and it is more natural to think about projective twistor space $(\lambda_a, \mu_{\dot{a}}) \sim (t\lambda_a, t\mu_{\dot{a}})$ with t a non-zero complex number. In twistor theory a central rôle is played by the incidence relation $\mu_{\dot{a}} + x_{a\dot{a}}\lambda^a = 0$ which describes the correspondence between Minkowski space and twistor space. In particular if we fix a point in Minkowski space $x_{a\dot{a}}$ this leads to two complex equations that describe a line in projective twistor space.

As mentioned earlier the MHV tree amplitudes are holomorphic, except for the overall momentum conservation factor $\delta^4(\sum \lambda_i\widetilde{\lambda}_i)$, and therefore their transformation to twistor space is particularly simple

$$\widetilde{A}_{\mathrm{MHV}}(\lambda, \mu) \sim A_{\mathrm{MHV}}(\lambda) \int \prod_i d\widetilde{\lambda}_i e^{i\mu_i\widetilde{\lambda}_i} e^{ix\lambda_i\widetilde{\lambda}_i}$$
$$\sim A_{\mathrm{MHV}}(\lambda) \prod_i \delta(\mu_i + x\lambda_i)\,, \tag{4}$$

which implies that MHV amplitudes are supported on a line in twistor space. For more complicated amplitudes the localisation properties are almost as simple, in particular an amplitude with Q negative helicity gluons localises on sets of $Q-1$ intersecting lines. These simple geometrical struc-

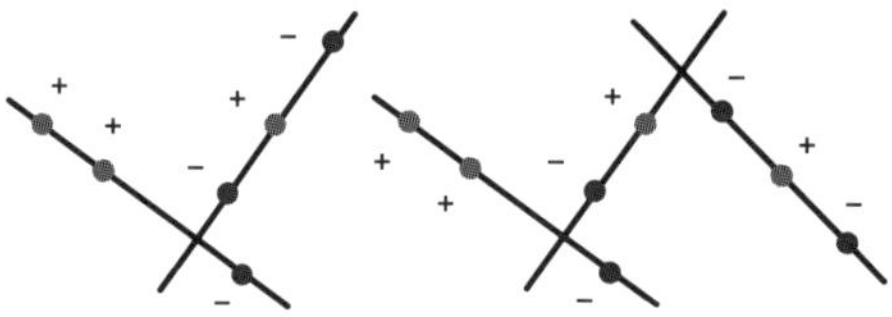

Fig. 1. Twistor space localisation of tree amplitudes with $Q = 3$ and $Q = 4$

tures in twistor space of the amplitudes led Cachazo, Svrček and Witten (CSW) to propose [2] a novel perturbative expansion for tree-level amplitudes in Yang-Mills using MHV amplitudes as effective vertices. It is indeed natural to think of an MHV amplitude as a local interaction, since the line in twistor space on which a MHV amplitude localises corresponds to a point in Minkowski space via the incidence relation. In the MHV diagrammatic method of [2], MHV vertices are connected by scalar propagators $1/P^2$ and all diagrams with a fixed cyclic ordering of external lines have to be summed. The crucial point is clearly to define an off-shell continuation of MHV amplitudes. CSW proposed to associate internal (off-shell) legs with momentum P the spinor $\lambda_{Pa} := P_{a\dot{a}}\eta^{\dot{a}}$, where η denotes an arbitrary reference spinor. we can assign a spinor to every internal momentum and insert this spinor in the Parke-Taylor formula (3) to define the off-shell MHV vertex. It can be shown [2] that after summing all diagrams the dependence on the reference spinor drops out. Proofs of the equivalence of MHV diagrams and usual Feynman diagrams at tree level have been presented in [14–17]. Interestingly, MHV rules for tree-level gravity have also been derived in [18].

4. From trees to loops

The success of the MHV method at tree level brings up the question if this can be extended to the quantum level, i.e. to loop amplitudes. The original prognosis from twistor string theory was negative because of the presence of unwanted conformal supergravity modes that spoil the duality with Yang-Mills at loop level. Remarkably, we will find perfect agreement between loop MHV diagrams (such as in Figure 2) and the standard results obtained with more standard methods.

For simplicity let us focus on the simplest one-loop amplitudes, the MHV one-loop amplitudes in $\mathcal{N} = 4$ super Yang-Mills. These amplitudes were

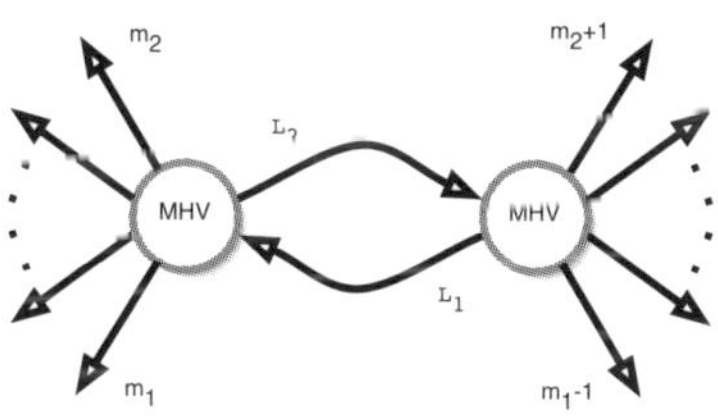

Fig. 2. MHV diagrams for the MHV one-loop amplitudes in $\mathcal{N} = 4$ SYM

computed in [7] using the four-dimensional cut-constructibility approach, which utilises the fact that one-loop amplitudes in supersymmetric theories can be reconstructed from their discontinuities. The result is surprisingly simple and can be expressed in terms of the so-called *2-mass easy box functions* $F^{2\text{me}}(s,t,P^2,Q^2)$

$$A_{\text{MHV}}^{1-\text{loop}} = A_{\text{MHV}}^{\text{tree}} \times \sum_{p,q} F^{2\text{me}}(s,t,P^2,Q^2) \quad (5)$$

It turns out that the MHV diagrammatic calculation [3] agree perfectly with the result of [7]. A few remarks are in order here:

- In the calculation it is crucial to decompose the loop momentum L in an on-shell part l plus a part proportional to the reference null-momentum η: $L = l + z\eta$.

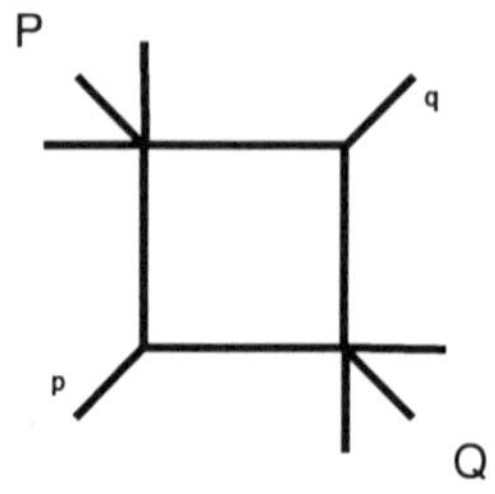

Fig. 3. A graphical representation of the 2-mass easy box function

- This decomposition naturally leads to *dispersion integrals* $\int dz/z(\cdots)$, which do not need subtractions.
- This calculation provides a check of the MHV method, e.g. the proof of covariance (η-independence) is highly non-trivial.
- This calculation incorporates a large number of conventional Feynman diagrams.

This approach readily applies to non-MHV amplitudes and theories with less supersymmetry. In [5,4] the method was applied to the case of MHV one-loop amplitudes in $\mathcal{N} = 1$ SYM and complete agreement with a calculation using the cut-constructibility approach [8] was found. Finally, in [6] the first new result from MHV diagrams at one-loop was obtained: the cut-containing part of the MHV one-loop amplitudes in pure Yang-Mills.

Having discussed these important checks of the MHV method, in the next Section we will argue that generic one-loop scattering amplitudes can be equivalently computed with MHV diagrams.

5. From loops to trees

In order to prove that MHV diagrams produce the correct result for scattering amplitudes at the quantum level, one has to

1. Prove the covariance of the result;

2. Prove that all physical singularities of the scattering amplitudes (soft, collinear and multiparticle) are correctly reproduced.

Indeed, if $A^{(n)}_{\text{MHV}}$ is the result of the calculation of an n-point scattering amplitude based on MHV diagrams, and A_{F} is the correct result (obtained using Feynman diagrams), the difference $A^{(n)}_{\text{MHV}} - A^{(n)}_{\text{F}}$ must be a polynomial which, by dimensional analysis, has dimension $4-n$. But such a polynomial cannot exist except for $n = 4$. This case can be studied separately, and we have already discussed the agreement with the known results for theories with $\mathcal{N} = 1, 2, 4$ supersymmetry.

In this section we will will report on the calculation of generic one-loop amplitudes with MHV diagrams [19]. Firstly, we will prove the covariance of the result. This proof relies on two basic ingredients: the locality of the MHV vertices, and a beautiful result by Feynman known as the Feynman Tree Theorem [20]. This theorem relates the contribution of a loop amplitude to those of amplitudes obtained by opening up the loop in all possible ways; strikingly, this theorem allows one to calculate loops from on-shell trees. Our next goal will be to show that soft and collinear singularities are correctly reproduced by MHV diagrams at one loop. Thus, for a complete proof of the MHV method at one loop, it would only remain to show the agreement of multiparticle singularities.

5.1. *The Feynman tree theorem and the proof of covariance*

We begin by briefly reviewing Feynman's Tree Theorem. This result is based on the decomposition of the Feynman propagator into a retarded (or advanced) propagator and a term which has support on shell. For instance,

$$\Delta_F(P) \;=\; \Delta_R(P) + 2\pi\delta(P^2 - m^2)\theta(-P_0)\;. \tag{6}$$

Suppose we wish to calculate a certain one-loop diagram $\mathcal{L}$, and let $\mathcal{L}_R$ be the quantity obtained from $\mathcal{L}$ by replacing all Feynman propagators by retarded propagators.[a] Clearly $\mathcal{L}_R = 0$, as there are no closed timelike curves in Minkowski space. This equation can fruitfully be used to work out the desired loop amplitude. Indeed, by writing the retarded propagator as a sum of Feynman propagator plus an on-shell supported term as dictated by (6), one finds that

$$\mathcal{L} \;=\; \mathcal{L}_{1-\text{cut}} + \mathcal{L}_{2-\text{cut}} + \mathcal{L}_{3-\text{cut}} + \mathcal{L}_{4-\text{cut}}\;. \tag{7}$$

Here $\mathcal{L}_{p-\text{cut}}$ is the sum of all the terms obtained by summing all possible diagrams obtained by replacing p propagators in the loop with delta functions. Each delta function cuts open an internal loop leg, and therefore a term with p delta functions computes a p-particle cut in a kinematical channel determined by the cut propagators (whose momentum is set on shell by the delta functions).

Feynman's Tree Theorem (7) states that a one-loop diagram can be expressed as a sum over all possible cuts of the loop diagram. The process of cutting puts internal lines on shell; the remaining phase space integrations

[a]The same argument works for advanced propagators.

have still to be performed, but these are generically easier than the original loop integration. Thus the Feynman Tree Theorem implies that one-loop scattering amplitudes can be determined from on-shell data alone. Interestingly, one can iterate this procedure and apply it to higher loop diagrams.

Now we apply the Feynman Tree Theorem to MHV diagrams. This possibility is guaranteed by the local character in Minkowski space of an MHV interaction vertex; thus a closed loop MHV diagram where propagators are replaced by retarded or advanced propagators vanishes, and one arrives at an equation identical to (7). Specifically, in [19] we used Feynman's Tree Theorem in order to prove that one-loop amplitudes calculated with MHV diagrams are covariant. More precisely, we were able to show that the sum of all possible p-cut MHV diagrams is separately covariant. The remaining Lorentz-invariant phase space integrations are also invariant, hence the full amplitude – given by summing over p-trees with $p = 1, \ldots, 4$ as indicated by (7) – is also covariant.

A sketch of the proof of covariance for the case of one-loop amplitudes with the MHV helicity configuration is as follows. The one-loop MHV diagrams contributing to an n-point MHV amplitude are presented in Figure 2. In Figure 4 we show the one-particle and two-particle cut diagrams

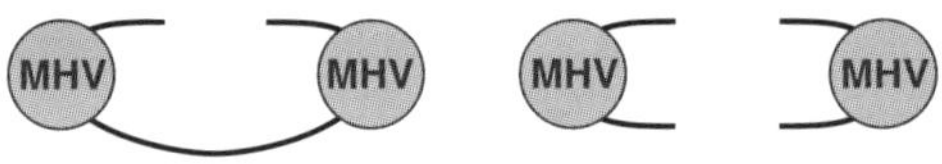

Fig. 4. One-particle and two-particle MHV diagrams contributing to the one-loop MHV scattering amplitude.

which are generated in the application of the Feynman Tree Theorem.[b] We start by focussing on one-particle cut diagrams. These one-particle cut-diagrams are nothing but tree-level diagrams, which are then integrated using a Lorentz invariant phase space measure. We now make the following important observation: these tree (one-cut) diagrams would precisely sum to a tree-level next-to-MHV (NMHV) amplitude with $n + 2$ external legs (which would then be covariant as shown in [2]), *if* we also include the set

[b]In our notation we only draw vertices and propagators (or cut-propagators) connecting them. It will be understood that we have to distribute the external gluons among the MHV vertices in all possible ways compatible with cyclic ordering, and the requirement that the two vertices must have the helicity configuration of an MHV amplitude. Moreover we will have to sum over all possible helicity assignments of the internal legs and, where required, over all possible particle species which can run in the loop.

of diagrams where the two legs into which the cut propagator is broken are allowed to be at the same MHV vertex. Such diagrams are obviously never generated by cutting a loop leg in MHV diagrams of the type depicted in Figure 2. These "missing" diagrams are drawn in Figure 5. MHV rules tell us, before any phase space integration is performed, that the sum of one-particle cut diagrams of Figures 4 and 5 generates a NMHV amplitude with $n+2$ external legs. Since the phase space measure is Lorentz invariant, it follows that the sum of one-particle cut diagrams, including the missing diagrams, is covariant.

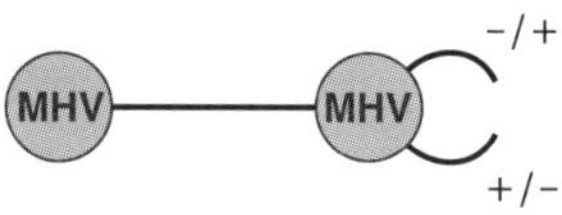

Fig. 5. In this Figure we represent "missing diagrams", mentioned in the text.

To complete the proof of covariance we have to justify the inclusion of these missing diagrams. Here we will present an explanation which relies on supersymmetry.[c] The diagrams where two adjacent and opposite helicity legs from the same MHV vertex are sewn together vanish when summed over particle species in a supersymmetric theory. Individual diagrams before summing over particle species diverge because of the collinearity of the momenta of the two legs, but the sum over particle species vanishes even before integration. So we discover that we could have actually included these diagrams from the start, since their contribution is zero. Hence one-particle cut diagrams of MHV one-loop amplitudes generate phase space integrals of tree-level NMHV amplitudes, and are, therefore, covariant.

Next we look at two-particle cut diagrams. These split the one-loop diagram of Figure 2 into two disconnected pieces (see the second diagram in Figure 4). These are two MHV amplitudes, because the two internal legs are put on shell by the Feynman cuts. Therefore, no η-dependence is produced by these two-particle cut diagrams.

Summarising, we have shown that Feynman one-particle and two-particle cut diagrams are separately covariant. By Feynman's Tree Theorem (7) we conclude that the physical one-loop MHV amplitude is covariant too.

The main lines of the proof of covariance we have discussed in the simple

[c]An alternative proof, not relying on supersymmetry, can be found in [19].

example of an MHV amplitude are easily generalised more complicated amplitudes. In particular, we notice that:

1. In a one-loop MHV diagram with v vertices and n external particles (contributing to an N^{v-2}MHV amplitude), the top-cut is necessarily the v-particle cut. This will always be η-independent by construction. Notice that this top cut will generically vanish if $v > 4$.

2. All p-particle cuts which are generated by the application of Feynman's Tree Theorem split each one-loop MHV diagram into p disconnected pieces when $p > 1$. In all such cases we see that amplitudes are produced on all sides of the cut propagators when the sum over all MHV diagrams is taken.

3. The case of a one-particle cut is special since it generates a connected tree diagram. Similarly to the example discussed earlier, one realises that by adding missing diagrams the one-cut diagrams group into N^{v-1}MHV amplitudes with $n+2$ external legs (which are of course covariant).

4. To see amplitudes appearing on all sides of the cuts, one has to sum over all one-loop MHV diagrams.

In this way one can show the covariance of the result of a MHV diagram calculation for amplitudes with arbitrary helicities.

5.2. *Collinear limits*

In this section we address the issue of reproducing the correct singularities of the scattering amplitudes from MHV diagrams. We begin with collinear limits. At tree level, collinear limits were studied in [2], and shown to be in agreement with expectations from field theory. Consider now a one-loop scattering amplitude $\mathcal{A}_n^{1-\mathrm{loop}}$. When the massless legs a and b become collinear, the amplitude factorises as [7,8,21,22]

$$\begin{aligned}
&\mathcal{A}_n^{1-\mathrm{loop}}(1,\ldots,a^{\lambda_a},b^{\lambda_b},\ldots,n) \xrightarrow{a\|b} \\
&\quad\sum_\sigma \Big[\mathrm{Split}_{-\sigma}^{\mathrm{tree}}(a^{\lambda_a},b^{\lambda_b})\,\mathcal{A}_{n-1}^{1-\mathrm{loop}}(1,\ldots,(a+b)^\sigma,\ldots,n) \\
&\qquad + \mathrm{Split}_{-\sigma}^{1-\mathrm{loop}}(a^{\lambda_a},b^{\lambda_b})\,\mathcal{A}_{n-1}^{\mathrm{tree}}(1,\ldots,(a+b)^\sigma,\ldots,n)\Big]\,.
\end{aligned} \tag{8}$$

$\mathrm{Split}^{\mathrm{tree}}$ are the gluon tree-level splitting functions, whose explicit forms can be found e.g. in [23]. $\mathrm{Split}^{1-\mathrm{loop}}$ is a supersymmetric one-loop splitting function. In [24] and [25] explicit formulae for this one-loop splitting function, valid to all orders in the dimensional regularisation parameter ϵ, were found.

The result of [25] is:

$$\mathrm{Split}^{1-\mathrm{loop}}_{-\sigma}(a^{\lambda_a}, b^{\lambda_b}) \;=\; \mathrm{Split}^{\mathrm{tree}}_{-\sigma}(a^{\lambda_a}, b^{\lambda_b})\; r(z)\;, \tag{9}$$

where, to all orders in ϵ, $r(z) =$

$$\frac{c_\Gamma}{\epsilon^2}\left(\frac{-s_{ab}}{\mu^2}\right)^{-\epsilon}\left[1 \,-\, {}_2F_1\left(1,-\epsilon,1-\epsilon,\frac{z-1}{z}\right) \,-\, {}_2F_1\left(1,-\epsilon,1-\epsilon,\frac{z}{z-1}\right)\right] \tag{10}$$

and

$$c_\Gamma \;=\; \frac{\Gamma(1+\epsilon)\Gamma^2(1-\epsilon)}{(4\pi)^{2-\epsilon}\Gamma(1-2\epsilon)}\;. \tag{11}$$

Here the parameter z is introduced via the relations $k_a := zk_P$, $k_b := (1-z)k_P$, where $k_P^2 \to 0$ in the collinear limit. Notice that in (8) we sum over the two possible helicities $\sigma = \pm$.

In [19] we were able to reproduce (8) from a calculation based on one-loop MHV diagrams; in particular we were able to re-derive the all-orders in ϵ expressions (9) and (10). Similarly to the tree-level, the different collinear limits at one-loop arise from different MHV diagrams. As an example, consider the $++ \to +$ collinear limit.

Consider first the diagrams where the two legs becoming collinear, a and b, are either a proper subset of the legs attached to a single MHV vertex, or belong to a four-point MHV vertex $\mathcal{A}_{4,\mathrm{MHV}}(a, b, l_2, -l_1)$ but the loop legs L_1 and L_2 are connected to different MHV vertices.[d] In this case we call s_{ab} a *non-singular* channel. Summing over all MHV diagrams where s_{ab} is a non-singular channel, one immediately sees that a contribution identical to the first term in (8) is generated.

Next we consider singular-channel diagrams (a prototypical one is shown in Figure 6), i.e. diagrams where the legs a and b belong to a four-point MHV vertex and the two remaining loop legs are attached to the same MHV vertex. The diagram shown in Figure 6 has been calculated in [3], and in [19] we showed that it precisely accounts for the second term in (8).

A similar analysis can be carried out for the other collinear limits, as well as for the soft limits. In all cases we found complete agreement with the known expressions for the all-orders in ϵ collinear and soft functions [19].

---

[d]Thus, even if L_1 and L_2 become null and collinear, nothing special happens to the sum of tree MHV diagrams on the right hand side of this four-point MHV vertex, precisely because L_1 and L_2 are not part of the same MHV vertex.

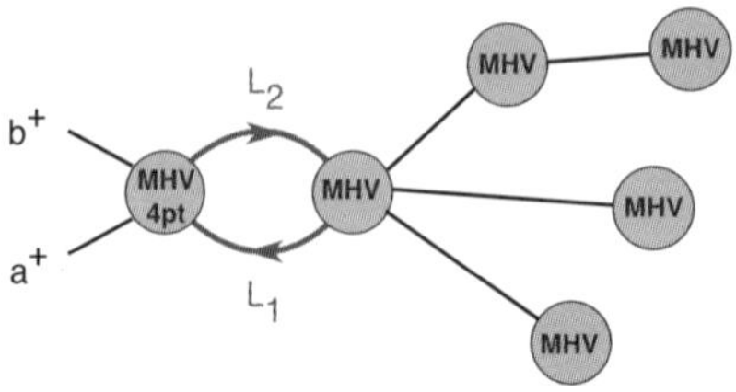

Fig. 6. A schematic example of a one-loop MHV diagram contributing to a generic non-MHV one-loop amplitude where s_{ab} is a singular channel. In the collinear limit $a \parallel b$, diagrams of this type generate the second term on the right hand side of (8).

6. Generalised unitarity

One-loop scattering amplitudes in supersymmetric gauge theories are linear combinations of scalar box, tensor triangle and tensor bubble integral functions with coefficients that are rational functions of the momenta and spinor variables. One wonders if these coefficients can be determined directly without performing any loop integration. The answer turns out to be yes and the main tool are unitarity (2-particle) cuts and generalised (3-particle and 4-particle) cuts [26,27]. In an $n-$particle cut of a loop diagram, n propagators are replaced by δ-functions which (partially) localise the loop integration and produce a sufficient number of linear equations to fix all coefficients. The method using 2-particle cuts was introduced [7,8], where it was used to calculate the first infinite series of one-loop amplitudes in SYM, and later generalised to include 3-particle cuts in [9,11]. Finally, in [10] 4-particle cuts were shown to be an efficient tool to find coefficients of box functions. In particular they reduce the calculation of all one-loop amplitudes in $\mathcal{N} = 4$ SYM to a simple algebraic exercise.

In non-supersymmetric theories loop amplitudes contain additional rational terms, which do not have cuts in 4 dimensions. This problem can be elegantly solved by working in $D = 4 - 2\epsilon$ dimensions, since then also rational terms develop cuts and become amenable to unitarity techniques [28]. In [29] generalised unitarity techniques in D dimensions were shown to be a powerful tool to calculate complete one-loop amplitudes in non-supersymmetric theories like QCD. The price to pay is that we have to evaluate cuts in D dimensions; in practice this means that we have to make the internal particles massive with uniform mass μ and integrate over μ [28].

We now illustrate this technique for the simplest one-loop amplitude in pure YM which has four positive helicity gluons $\langle 1^+ 2^+ 3^+ 4^+ \rangle$. It turns out that in this case we only have to consider the 4-particle cut depicted in Figure 7; no other cuts are necessary. The 4-particle cut gives $\mu^4 \frac{[12][34]}{\langle 12\rangle\langle 34\rangle}$

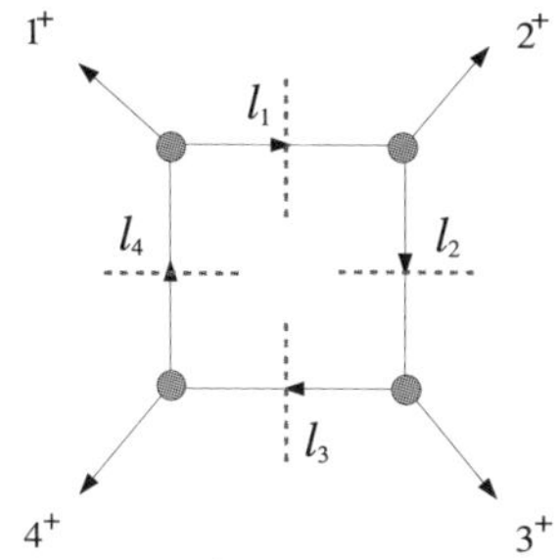

Fig. 7. The 4-particle cut of the one-loop $\langle 1^+2^+3^+4^+\rangle$ amplitude in Yang-Mills

where μ is the mass of the internal particle. This implies that the amplitude is proportional to a scalar box integral with μ^4 inserted in the integral $I_4[\mu^4] = -\epsilon(1-\epsilon)I_4^{D=8-2\epsilon} = -\frac{1}{6} + \mathcal{O}(\epsilon)$. Hence, the amplitude is purely rational, and we find

$$A_4(1^+,2^+,3^+,4^+) = -\frac{1}{6}\frac{[12][34]}{\langle 12\rangle\langle 34\rangle}\,. \tag{12}$$

In [29] this method was applied successfully to the remaining 4-point amplitudes $\langle 1^+2^+3^+4^-\rangle$, $\langle 1^-2^-3^+4^+\rangle$ and $\langle 1^-2^+3^-4^+\rangle$, and to the particular 5-point amplitude $\langle 1^+2^+3^+4^+5^+\rangle$. Interestingly, it turned out that for the amplitudes with one or more negative helicity gluons, only 3-particle and 4-particle cuts were needed in order to determine the amplitude. This method can be applied directly to more general amplitudes and the 5- and 6-point amplitudes are currently under investigation.

Acknowledgments

We would like to thank Bill Spence, James Bedford and Simon McNamara for very pleasant collaborations, and Adi Armoni, Zvi Bern, Emil Bjerrum-Bohr, Freddy Cachazo, Luigi Cantini, Lance Dixon, Dave Dunbar, Michael Green, Harald Ita, Valya Khoze, David Kosower, Paul Mansfield, Marco Matone, Christian Römelsberger and Sanjaye Ramgoolam for discussions. The work of AB is partially supported by a PPARC Special Project Grant. The research of GT is supported by an EPSRC Advanced Fellowship.

References

1. E. Witten, Commun. Math. Phys. **252**, 189 (2004), hep-th/0312171.
2. F. Cachazo, P. Svrček and E. Witten, JHEP **0409**, 006 (2004), hep-th/0403047.
3. A. Brandhuber, B. Spence and G. Travaglini, Nucl. Phys. B **706**, 150 (2005), hep-th/0407214.
4. J. Bedford, A. Brandhuber, B. Spence and G. Travaglini, Nucl. Phys. B **706**, 100 (2005), hep-th/0410280;
5. C. Quigley and M. Rozali, JHEP **0501** (2005) 053, hep-th/0410278.
6. J. Bedford, A. Brandhuber, B. Spence and G. Travaglini, Nucl. Phys. B **712**, 59 (2005), hep-th/0412108.
7. Z. Bern, L.J. Dixon, D.C. Dunbar and D.A. Kosower, Nucl. Phys. B **425** (1994) 217, hep-ph/9403226.
8. Z. Bern, L.J. Dixon, D.C. Dunbar and D.A. Kosower, Nucl. Phys. B **435** (1995) 59, hep-ph/9409265.
9. Z. Bern, L.J. Dixon and D.A. Kosower, Nucl. Phys. B **513** (1998) 3, hep-ph/9708239.
10. R. Britto, F. Cachazo and B. Feng, Nucl. Phys. B **725** (2005) 275, hep-th/0412103.
11. Z. Bern, L.J. Dixon and D.A. Kosower, Phys. Rev. D **72**, 045014 (2005).
12. S.J. Bidder, N.E.J. Bjerrum-Bohr, D.C. Dunbar and W.B. Perkins, Phys. Lett. B **612** (2005) 75, hep-th/0502028.
13. R. Britto, E. Buchbinder, F. Cachazo and B. Feng, Phys. Rev. D **72** (2005) 065012, hep-ph/0503132.
14. R. Britto, F. Cachazo, B. Feng and E. Witten, Phys. Rev. Lett. **94** (2005) 181602, hep-th/0501052.
15. K. Risager, JHEP **0512**, 003 (2005), hep-th/0508206.
16. P. Mansfield, JHEP **0603**, 037 (2006), hep-th/0511264.
17. A. Gorsky and A. Rosly, JHEP **0601**, 101 (2006), hep-th/0510111.
18. N. E. J. Bjerrum-Bohr, D. C. Dunbar, H. Ita, W. B. Perkins and K. Risager, JHEP **0601**, 009 (2006), hep-th/0509016.
19. A. Brandhuber, B. Spence and G. Travaglini, JHEP **0601** (2006) 142, hep-th/0510253.
20. R. P. Feynman, Acta Phys. Polon. **24** (1963) 697.
21. Z. Bern and G. Chalmers, Nucl. Phys. B **447**, 465 (1995), hep-ph/9503236.
22. D. A. Kosower, Nucl. Phys. B **552** (1999) 319, hep-ph/9901201.
23. L. J. Dixon, hep-ph/9601359.
24. D. A. Kosower and P. Uwer, Nucl. Phys. B **563** (1999) 477, hep-ph/9903515.
25. Z. Bern, V. Del Duca, W. B. Kilgore and C. R. Schmidt, Phys. Rev. D **60** (1999) 116001, hep-ph/9903516.
26. R.E. Cutkosky, J. Math. Phys. **1** (1960) 429.
27. R. J. Eden, P.V. Landshoff, D.I. Olive and J.C. Polkinghorne, *The Analytic S-Matrix*, Cambridge University Press, 1966.
28. Z. Bern and A.G. Morgan, Nucl. Phys. B **467** (1996) 479, hep-ph/9511336.
29. A. Brandhuber, S. McNamara, B. Spence and G. Travaglini, JHEP **0510** (2005) 011, hep-th/0506068

SIMILARITIES OF GAUGE AND GRAVITY AMPLITUDES

N. E. J. BJERRUM-BOHR, DAVID C. DUNBAR and HARALD ITA*

Department of Physics, Swansea University,
Swansea, SA2 8PP, Wales, UK
E-mail: {*n.e.j.bjerrum-bohr, d.c.dunbar, h.ita*}*@swansea.ac.uk*

We review recent progress in computations of amplitudes in gauge theory and gravity. We compare the perturbative expansion of amplitudes in $\mathcal{N}=4$ super Yang-Mills and $\mathcal{N}=8$ supergravity and find surprising similarities.

Keywords: Supergravity Models, Models of Quantum Gravity, General properties of perturbation theory in gauge theories.

1. Introduction

Perturbative gauge theory and gravity in four dimensions are quite dissimilar from a dynamical viewpoint. Gauge theory (*e.g.* pure Yang-Mills theory) is a renormalisable theory that is strongly coupled in the infrared and asymptotically free in the ultraviolet. Gravity on the other hand is a weakly coupled theory in the infrared but strongly coupled in the ultraviolet. By power counting, gravity in four dimensions is potentially a non renormalisable theory. Pure gravity scattering amplitudes are finite at one-loop with the first divergence occurring at two-loops [1].

Supersymmetry generally softens the UV behaviour in a quantum field theory. For example, maximally supersymmetric Yang-Mills is a finite theory [2] and supergravity theories have a finite S-matrix until at least three loops [3]. Although in four-dimensional power counting and counter-term arguments suggest that supergravity theories are non-renormalisable [4] this has, so far, not been tested by direct computations. Arguments based on power counting within unitary cuts suggest that the first counter term in maximal supergravity [5] is expected at five loops [6,7].

Recently, initiated by the duality between gauge theories and a twister string theory [8], there has been much progress in the computation of am-

*Presented by H. Ita at "Continuous Advances in QCD 2006". SWAT-06/481

plitudes in gauge theory. In this talk we discuss how these ideas may be applied to gravity calculations and the results thereof. We will first review the recent progress in computing physical on-shell tree amplitudes for gravity theories particularly focusing on the on-shell recursion relations [9,10] and the MHV-vertex construction [11–15]. Later we will discuss one-loop amplitudes. A surprising result is that the one-loop amplitudes of $\mathcal{N} = 4$ SYM and $\mathcal{N} = 8$ supergravity [16,17] occur to be expressible in terms of scalar box integral functions - despite the expectation from power counting. Supergravity multi-loop amplitudes are not directly addressed, however, the structure of amplitudes at tree-level and one-loop have, through factorisation and unitarity, important consequences on the structure of higher loop amplitudes.

2. Old and new techniques for gravity tree amplitudes

Graviton scattering amplitudes are extremely difficult to evaluate using conventional Feynman diagram techniques. In this section we review alternative methods: 1) the Kawai-Lewellen-Tye relations, 2) on shell recursion relations and 3) MHV vertex constructions.

1) Gravity amplitudes can be constructed through the Kawai, Lewellen and Tye (KLT)-relations [18] as squares of gauge theory amplitudes. The KLT relations are inspired by the naïve string theory relation

$$\text{closed string} \sim (\text{left-mover}) \times (\text{right-mover})\,, \tag{1}$$

and have the explicit form, up to five points,

$$\begin{aligned}
M_3^{\text{tree}}(1,2,3) &= -iA_3^{\text{tree}}(1,2,3)A_3^{\text{tree}}(1,2,3)\,,\\
M_4^{\text{tree}}(1,2,3,4) &= -is_{12}A_4^{\text{tree}}(1,2,3,4)A_4^{\text{tree}}(1,2,4,3)\,,\\
M_5^{\text{tree}}(1,2,3,4,5) &= is_{12}s_{34}\, A_5^{\text{tree}}(1,2,3,4,5)A_5^{\text{tree}}(2,1,4,3,5)\\
&\quad + is_{13}s_{24}\, A_5^{\text{tree}}(1,3,2,4,5)A_5^{\text{tree}}(3,1,4,2,5)\,,
\end{aligned}$$

where A_n^{tree} are the tree-level colour-ordered gauge theory partial amplitudes. We suppress factors of g^{n-2} in the A_n^{tree} and $(\kappa/2)^{n-2}$ in the M_3^{tree}.

The KLT relations are helpful in the calculation of gravity tree amplitudes, however they have some undesirable features. The factorisation structure is not manifest and the expressions do not tend to be compact, as the permutation sums grow rather quickly with n. In fact, the Berends,

Giele and Kuijf (BGK) form of the MHV gravity amplitude[19],

$$M_n^{\text{tree}}(1^-,2^-,3^+,\cdots,n^+) = -i\,\langle 1\,2\rangle^8 \times \left[\left(\frac{[1\,2]\,[n-2\,n-1]}{\langle 1\,n-1\rangle\,N(n)}\right)\times \right.$$
$$\left. \times\left(\prod_{i=1}^{n-3}\prod_{j=i+2}^{n-1}\langle i\,j\rangle\right)\prod_{l=3}^{n-3}(-[n|K_{l+1,n-1}|l\rangle) + \mathcal{P}(2,3,\cdots,n-2)\right],$$

is rather more compact than that of the KLT sum (as is the expression in Ref. 10.) In the above we use the definitions, $[k|K_{i,j}|l\rangle \equiv \langle k^+|\not{K}_{i,j}|l^+\rangle \equiv \langle l^-|\not{K}_{i,j}|k^-\rangle \equiv \langle l|K_{i,j}|k] \equiv \sum_{a=i}^{j}[k\,a]\,\langle a\,l\rangle$, and $N(n) = \prod_{1\le i<j\le n}\langle i\,j\rangle$. In terms of the above Weyl spinors we often use twistor variables $\lambda_i^a \equiv |k_i^+\rangle$ and $\bar\lambda_i^{\dot a} \equiv |k_i^-\rangle$. The MHV amplitudes for graviton scattering display a feature not shared by the Yang-Mills expressions, they depend not only on the holomorphic variables λ, but also on the anti-holomorphic $\bar\lambda$ variables (within the s_{ij} for the KLT expression).

2) In a recent computational approach for amplitudes, Britto, Cachazo, Feng and Witten[9] obtained on-shell recursion relations for trees. The recursion relations are based on factorisation properties of amplitudes and are thus applicable to a wide range of theories and in particular to gravity[9,10]. The technique is based on analytically shifting a pair of external legs, $\lambda_i \to \lambda_i + z\lambda_j, \bar\lambda_j \to \bar\lambda_j - z\bar\lambda_i$, and on determining the physical amplitude, $M_n(0)$, from the poles in the shifted amplitude, $M_n(z)$.This leads to a recursion relation of the form,

$$M_n(0) = \sum_\alpha \hat{M}_{n-k_\alpha+2}(z_\alpha)\times\frac{i}{P_\alpha^2}\times\hat{M}_{k_\alpha}(z_\alpha),$$

where the factorisation is only on these poles, z_α, where legs i and j are connected to different sub-amplitudes. An essential condition for the recursion relations is that the shifted amplitude $M_n(z)$ vanishes for large z. Whereas proven for gauge theory amplitudes a general proof (for arbitrary helicities[10]) in gravity is an open problem.

Recursion relations based on the analyticity in the complex plane can also be used at loop level both to calculate rational terms[20] and the coefficients of integral functions[21].

3) Finally, we would like to mention the CSW construction[11,22] of amplitudes and its generalisation to gravity[15]. In this approach MHV-amplitudes are treated as fundamental vertices and generic scattering amplitudes are expanded in terms of these MHV-vertices.

Considering the N^{s}MHV amplitude with n external legs. One would begin by drawing all diagrams which may be constructed using MHV vertices.

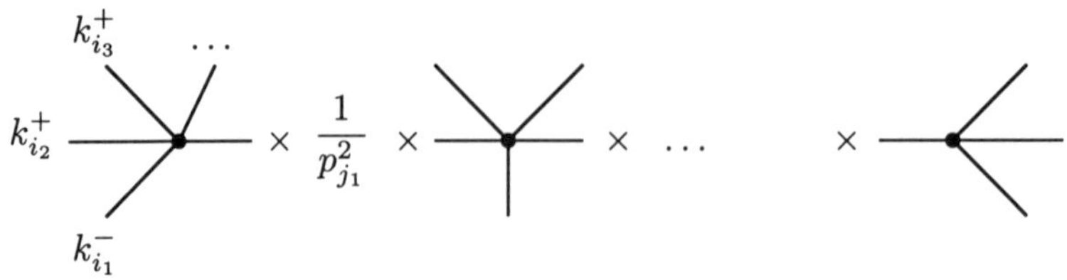

The contribution from each diagram is a product of $(s+1)$ MHV vertices and s propagators as indicated above.

The contribution of a given diagram to the total amplitude can be calculated by evaluating the product of MHV amplitudes and propagators,

$$M_N^s\big|_{\text{CSW-diagram}} = \left(\prod_{l=1,s+1} M_{N_l}^{\text{MHV}}(\hat{K}_l) \right) \prod_{j=1,s} \frac{i}{p_j^2},$$

where the propagators are computed on the set of momenta k_i and p_j, and the MHV vertices are evaluated at shifted momenta $\hat{k}_i$ and $\hat{p}_j$. The momenta k_i are external and the momenta p_j internal, and given by momentum conservation at each MHV-vertex. A key feature is the interpretation of the MHV amplitudes for internal legs. For Yang-Mills where the MHV vertices only depend on λ the correct interpretation is [11]

$$\lambda(p)_a = p_{a\dot{a}}\eta^{\dot{a}},$$

for an arbitrary reference spinor η. For gravity amplitudes we must also solve for $\bar{\lambda}(p)$ which is less obvious [23] and must be a function of the momentum of the negative helicity legs. It turns out that all spinors are uniquely defined in terms of the shifted momenta $\hat{k}_i$ and $\hat{p}_j$ if we demand that they are null vectors obeying momentum conservation at each vertex.

Explicitly they are given by shifting the negative helicity legs $\bar{\lambda}_i$ by

$$\bar{\lambda}_i \longrightarrow \bar{\lambda}_i + a_i\bar{\eta},$$

and leaving $\bar{\lambda}_j$ of the positive helicity legs k_{j^+} untouched. The $s+2$ parameters, a_{i^-} are uniquely fixed [15] by demanding a) overall momentum conservation, b) momentum conservation at each vertex and finally c) that the internal momenta, $\hat{p}_j$, are massless $\hat{p}_j^2 = 0$.

As an example of how the MHV-vertex constructions works for gravity, we can consider the n-point NMHV amplitude with three negative helicity legs m_i. The MHV-vertex approach gives the amplitude in the form,

$$\sum_{\text{perms},r} M^{\text{MHV}}(\hat{k}_1^-, k_4^+, \cdots k_r^+, \hat{p}^-) \times \frac{1}{p^2} \times M^{\text{MHV}}(\hat{k}_2^-, \hat{k}_3^- k_{r+1}^+, \cdots k_{n-3}^+, \hat{p}^+),$$

where the sum runs over diagrams involving all choices of $r > 0$ and all permutations of the negative and positive helicity legs. To illustrate the correct continuation, the three negative helicity $\bar{\lambda}$ must be shifted $\bar{\lambda}_i \to \bar{\lambda}_i + a_i\bar{\eta}$. Imposing the momentum constraints leaves us with a shift,

$$\bar{\lambda}_1 \longrightarrow \bar{\lambda}_1 + z\langle k_2, k_3\rangle\bar{\eta}\,,$$

together with the cyclic shifts of the other two legs. Momentum is conserved for any value of the parameter z. Requiring $\hat{p}^2 = 0$ then fixes z uniquely as $z = p^2/[\eta|p|k_1\rangle$ and the MHV vertex expansion is completely determined.

3. One-loop amplitudes in $N = 8$ supergravity

In a Yang-Mills theory, the loop momentum polynomial in a one-loop n-point diagram will generically be of degree $\leq n$. $\mathcal{N} = 4$ one-loop amplitudes exhibits considerable simplification and the loop momentum integral will be of degree $n-4$ [24,25]. Consequently, from a Passarino-Veltman reduction [26], the amplitudes can be expressed as a sum of scalar box integrals with rational coefficients,

$$A^{1-\text{loop}} = \sum_a c_a I_a^4\,. \tag{2}$$

Determining the amplitude then reduces to determining the rational coefficients c_a. Inspired by the duality in Ref. 8, considerable progress has recently been made in determining such coefficients, c_a, using a variety of methods based on unitarity [25,27–29].

For $\mathcal{N} = 8$ supergravity the equivalent power counting arguments [30] give a loop momentum polynomial of degree

$$2(n-4)\,,$$

which is consistent with Eq. (1). Reduction for $n > 4$ leads to a sum of tensor box integrals with integrands of degree $n-4$ which would then reduce to scalar boxes *and* triangle, bubble and rational functions,

$$M^{1-\text{loop}} = \sum_a c_a I_4^a + \sum_a d_a I_3^a + \sum_a e_a I_2^a + R\,,$$

where the I_3 are present for $n \geq 5$, I_2 for $n \geq 6$ and R for $n \geq 7$.

There is evidence that all one-loop amplitudes of $\mathcal{N} = 8$, like the $\mathcal{N} = 4$ amplitudes Eq. (2), can be expressed as a sum over scalar box integrals, the so called "no-triangle hypothesis" [16,17]. Firstly, in the few definite computations at one-loop level, triangle or bubble functions do not appear. The first computation was of the four-point amplitude [31] where only box functions appear (although this is consistent with power counting). Beyond this computations of the five and six point MHV-amplitudes yielded only scalar box-functions [32].

Secondly, the factorisation properties of physical amplitudes do not demand the presence of these functions. Since the four and five point amplitudes are triangle-free then in any factorisation limit of a higher point function the triangles must vanish. In this spirit an ansatz for the n-point MHV amplitude was constructed [32] entirely of box functions consistent in all soft limits.

Thirdly, one can check whether the amplitudes composed purely from box-functions precisely give the expected soft divergence in a n graviton amplitude [33],

$$M^{\text{one-loop}}_{[1,2,\ldots,n]} = ic_\Gamma \kappa^2 \left[\frac{\sum_{i<j} s_{ij} \ln[-s_{ij}]}{2\epsilon} \right] \times M^{\text{tree}}_{[1,2,\ldots,n]}.$$

In Ref. 17 the box coefficients were explicitly computed for the six-point NMHV amplitudes confirming the above claims.

The "no-triangle" hypothesis applies to one-loop amplitudes. However, by factorisation it should have implications beyond one-loop suggesting the UV behaviour of maximal supergravity may be significantly milder than expected from power counting.

References

1. G. 't Hooft and M. J. G. Veltman, Annales Poincare Phys. Theor. A **20** (1974) 69; M. H. Goroff and A. Sagnotti, Nucl. Phys. B **266** (1986) 709.
2. S.Mandelstam, Nucl.Phys. B **213**, 149 (1983).
3. M. T. Grisaru, P. van Nieuwenhuizen and J. A. M. Vermaseren, Phys. Rev. Lett. **37**, 1662 (1976); M. T. Grisaru, Phys. Lett. B **66** (1977) 75; E. Tomboulis, Phys. Lett. B **67** (1977) 417.
4. S. Deser, J. H. Kay and K. S. Stelle, Phys. Rev. Lett. **38**, 527 (1977); P. S. Howe and U. Lindstrom, Nucl. Phys. B **181**, 487 (1981).
5. E. Cremmer, B. Julia and J. Scherk, Phys. Lett. B **76**, 409 (1978); E. Cremmer and B. Julia, Phys. Lett. B **80**, 48 (1978).
6. Z. Bern, L.J. Dixon, D.C. Dunbar, M. Perelstein and J.S. Rozowsky, Nucl. Phys. B **530**, 401 (1998); Class. and Quant. Grav. **17**, 979 (2000).
7. P. S. Howe and K. S. Stelle, Phys. Lett. B **554**, 190 (2003).
8. E. Witten, Commun. Math. Phys. **252**, 189 (2004).
9. R. Britto, F. Cachazo and B. Feng, Nucl. Phys. B **715** (2005) 499; R. Britto, F. Cachazo, B. Feng and E. Witten, Phys. Rev. Lett. **94** (2005) 181602.
10. J. Bedford, A. Brandhuber, B. Spence and G. Travaglini; Nucl. Phys. B **721** (2005) 98; F. Cachazo and P. Svrcek. hep-th/0502160.
11. F. Cachazo, P. Svrcek and E. Witten, JHEP **0409**, 006 (2004).
12. A. Brandhuber, B. Spence and G. Travaglini, Nucl. Phys. B **706**, 150 (2005); C. Quigley and M. Rozali, JHEP **0501**, 053 (2005); J. Bedford, A. Brandhu-

ber, B. Spence and G. Travaglini, Nucl. Phys. B **706**, 100 (2005); A. Brandhuber, B. Spence and G. Travaglini, JHEP **0601**, 142 (2006).
13. G.Georgiou and V.V.Khoze, JHEP**0405**,070 (2004); J.B.Wu and C.J.Zhu JHEP**0409**,063(2004); X.Su and J.B.Wu. Mod. Phys. Lett. A **20** (2005) 1065
14. L. J. Dixon, E. W. N. Glover and V. V. Khoze, JHEP **0412**, 015 (2004); Z. Bern, D. Forde, D. A. Kosower and P. Mastrolia, hep-ph/0412167; S. D. Badger, E. W. N. Glover and V. V. Khoze. JHEP **0503** (2005) 023
15. N. E. J. Bjerrum-Bohr, D. C. Dunbar, H. Ita, W. B. Perkins and K. Risager, JHEP **0601**, 009 (2006).
16. Z. Bern, N. E. J. Bjerrum-Bohr and D. C. Dunbar, JHEP **0505** (2005) 056.
17. N.E.J. Bjerrum-Bohr, D.C. Dunbar and H. Ita, Phys. Lett B **621** (2005) 183; hep-th/0606268.
18. H. Kawai, D. C. Lewellen and S. H. H. Tye, Nucl. Phys. B **269**, 1 (1986).
19. F. A. Berends, W. T. Giele and H. Kuijf, Phys. Lett. B **211**, 91 (1988).
20. Z. Bern, L. J. Dixon and D. A. Kosower, Phys. Rev. D **73**, 065013 (2006); C. F. Berger et al hep-ph/0604195; hep-ph/0607014.
21. Z. Bern, N. E. J. Bjerrum-Bohr, D. C. Dunbar and H. Ita, JHEP **0511**, 027 (2005); hep-ph/0603187.
22. K. Risager, JHEP **0512**, 003 (2005).
23. S. Giombi, R. Ricci, D. Robles-Llana and D. Trancanelli, JHEP **0407**, 059 (2004); J. B. Wu and C. J. Zhu, JHEP **0407**, 032 (2004).
24. Z. Bern and D. A. Kosower, Phys. Rev. Lett. **66**, 1669 (1991); Nucl. Phys. B **379**, 451 (1992); Z. Bern and D. C. Dunbar, Nucl. Phys. B **379**, 562 (1992).
25. Z. Bern, L.J. Dixon, D.C. Dunbar and D.A. Kosower, Nucl. Phys. B **425**, 217 (1994).
26. G. Passarino and M. Veltman, Nucl. Phys. B **160**, 151, (1979).
27. Z. Bern, L.J. Dixon, D.C. Dunbar and D.A. Kosower, Nucl. Phys. B **435**, 59 (1995).
28. Z. Bern, V. Del Duca, L.J. Dixon and D. A. Kosower, Phys. Rev. D **71**, 045006 (2005); Z. Bern, L. J. Dixon and D. A. Kosower. Phys. Rev. D **72** (2005) 045014; R. Britto, F. Cachazo and B. Feng, Phys. Rev. D **71** (2005) 025012; R. Roiban, M. Spradlin and A. Volovich. Phys. Rev. Lett. **94** (2005) 102002; R. Britto, F. Cachazo and B. Feng. Nucl. Phys. B **725** (2005) 275; S. J. Bidder, D. C. Dunbar and W. B. Perkins, JHEP **0508** (2005) 055.
29. S. J. Bidder, N. E. J. Bjerrum-Bohr, L. J. Dixon and D. C. Dunbar, Phys. Lett. B **606**, 189 (2005); S. J. Bidder, N. E. J. Bjerrum-Bohr, D. C. Dunbar and W. B. Perkins, Phys. Lett. B **608** (2005) 151; Phys. Lett. B **612** (2005) 75; R. Britto, E. Buchbinder, F. Cachazo and B. Feng, Phys. Rev. D **72** (2005) 065012.
30. Z. Bern, D.C. Dunbar and T. Shimada, Phys. Lett. B **312**, 277, (1993); D.C. Dunbar and P.S. Norridge, Nucl. Phys. B **433**, 181 (1995).
31. M.B. Green, J.H. Schwarz and L. Brink, Nucl. Phys. B198:472 (1982).
32. Z. Bern, L. J. Dixon, M. Perelstein and J. S. Rozowsky, Nucl. Phys. B **546**, 423 (1999).
33. D. C. Dunbar and P. S. Norridge, Class. Quant. Grav. **14** (1997) 351.

BOOTSTRAPPING ONE-LOOP QCD AMPLITUDES*

CAROLA F. BERGER

Stanford Linear Accelerator Center
Stanford University, Stanford, California 94309 USA

We review the recently developed bootstrap method for the computation of high-multiplicity QCD amplitudes at one loop. The method combines (generalized) unitarity with on-shell recursion relations to determine the not cut-constructible, rational terms of these amplitudes. The bootstrap approach works for arbitrary configurations of gluon helicities and arbitrary numbers of external legs.

Keywords: On-shell recursion relations; QCD; One-loop amplitudes

1. Introduction

The Large Hadron Collider (LHC), which is scheduled to begin operation in 2007, will provide new insight into the main missing piece of the Standard Model, the origin of electroweak symmetry breaking, and potentially discover new physics beyond the Standard Model. Discoveries at the LHC require a thorough understanding of Standard Model processes, all of which have many particles in the final state. Many high-multiplicity processes need to be computed to as high accuracy as possible, which entails the computation of multi-leg amplitudes to at least one-loop order.

Here we review a recently developed method that combines on-shell recursion relations with (generalized) unitarity to compute one-loop multi-gluon amplitudes recursively, thereby "recycling" information from amplitudes with fewer legs[1–3]. The bootstrap method works for arbitrary helicity configurations. Moreover, in some cases, the recursion can be solved explicitly, yielding all-multiplicity expressions for one-loop gluon amplitudes[1–4].[a] Our bootstrap method relies only on factorization properties of the amplitudes as well as unitarity, and should therefore be amenable to an extension

*in collaboration with Zvi Bern, Lance Dixon, Darren Forde, and David Kosower
[a]See also D. Forde, these proceedings.

to amplitudes with external fermions and massive partons.

In the following, we use the spinor helicity formalism to express the amplitudes in terms of spinor inner products, $\langle j\, l\rangle = \bar{u}_-(k_j)u_+(k_l)$, $[j\, l] = \bar{u}_+(k_j)u_-(k_l)$, where $u_\pm(k)$ is a massless Weyl spinor with momentum k and positive or negative chirality, respectively, which we also write as, $\lambda_i \equiv u_+(k_i)$, $\tilde{\lambda}_i \equiv u_-(k_i)$. Furthermore, we strip off all color information, as well as coupling constants, and compute only the leading-color amplitudes. We employ a supersymmetric decomposition of the leading-color one-loop QCD amplitudes[6]. We write the QCD amplitude as a sum over contributions from $\mathcal{N} = 4$ and $\mathcal{N} = 1$ supersymmetric multiplets, which are completely cut-constructible, and thus computable via (generalized) unitarity, and a remaining non-supersymmetric $\mathcal{N} = 0$ contribution from a scalar running in the loop, which has both cut-constructible and rational parts. It is the latter part that we will compute via on-shell recursion relations. In the following we will suppress the $\mathcal{N} = 0$ superscript and write $A \equiv A^{\mathcal{N}=0}$ at one loop.

For further information and our notational conventions we refer to ref. 2 and the lecture notes 5.

2. On-shell recursion relations at tree level

Here we briefly review the on-shell recursion relations found in ref. 7 and proven in ref. 8 for tree level amplitudes. For further details we refer to these papers. The proof[8] of the tree-level relations employs a parameter-dependent shift of two of the external massless spinors, j and l,

$$[j,l\rangle: \qquad \tilde{\lambda}_j \to \tilde{\lambda}_j - z\tilde{\lambda}_l\,, \qquad \lambda_l \to \lambda_l + z\lambda_j\,. \qquad (1)$$

where z is a complex number. The corresponding momenta are then continued in the complex plane as well so that they remain massless, $k_j^2(z) = 0 = k_l^2(z)$, and overall momentum conservation is maintained.

An on-shell amplitude containing the momenta k_j and k_l then also becomes parameter-dependent, $A(z)$. At tree level, $A(z)$ is a rational function of z. The physical amplitude is given by $A(0)$.

We can then use Cauchy's theorem, $\frac{1}{2\pi i}\oint_C \frac{dz}{z}\, A(z) = 0$, where the contour C is taken around the circle at infinity, and the integral vanishes if the complex continued amplitude $A(z)$ vanishes as $z \to \infty$. Evaluating the integral as a sum of residues, we can then solve for the physical amplitude $A(0)$ to obtain,

$$A(0) = -\sum_{\text{poles }\alpha} \operatorname*{Res}_{z=z_\alpha} \frac{A(z)}{z}\,. \qquad (2)$$

As explained in ref. 8, at tree level $A(z)$ only has simple poles. These poles arise when shifting the propagators in the amplitude. Therefore, each residue in eq. (2) is given by factorizing the shifted amplitude on the poles in momentum invariants, so that at tree level,

$$A(0) = \sum_{\text{partitions}} \sum_{h} A_L^h(z = z_{rs}) \frac{i}{K_{r\cdots s}^2} A_R^{-h}(z = z_{rs}) , \tag{3}$$

where $h = \pm 1$ labels the helicity of the intermediate state, and the labels L and R denote amplitudes with fewer legs which the propagator $i/K_{r\ldots s}^2$ connects. The squared momentum associated with that pole, $K_{r\ldots s}^2$, is evaluated in the unshifted kinematics; whereas the on-shell amplitudes with fewer legs, A_L and A_R, are evaluated in kinematics that have been shifted by eq. (1) with $z = z_{rs}$, where the propagator has a pole. In the following, such shifted, on-shell momenta will be denoted by $k(z = z_{rs}) \equiv \hat{k}$. We have thus succeeded in expressing the n-point amplitude A in terms of sums over on-shell, but complex continued, amplitudes with fewer legs, which are connected by scalar propagators.

The proof reviewed above relies on two properties of the complex continued tree level amplitude:

- The amplitude only has simple poles, which correspond to *physical* multiparticle and collinear factorizations.
- The amplitude vanishes at infinity.

These properties do not depend on the specific details of the gauge theory under consideration. Therefore, at tree level, the application of the recursion relations eq. (3) has been extended beyond multi-gluon amplitudes, to the computation of amplitudes including fermions, scalars, and massive partons, of gravity amplitudes, and of Higgs amplitudes produced via a heavy quark loop, in the effective theory where the heavy quark has been integrated out. Further details and references can be found in ref. 9.

At loop level, however, these properties do not hold in general, and several new features arise:

- The (poly)logarithmic terms of the amplitude have spurious, unphysical, singularities, which however cancel in the full amplitude against corresponding terms in the rational part. These spurious singularities are already present in real kinematics. They the application of Cauchy's theorem because it requires to sum over all poles, whether physical or unphysical.

- The amplitude develops branch cuts in the complex plane that stem from the complex continuation of (poly)logarithms.
- The complex continued amplitude contains double poles $\sim \frac{\langle a\,b\rangle}{[a\,b]^2}$, and 'unreal' poles $\sim \frac{\langle a\,b\rangle}{[a\,b]}$. In real kinematics, expressions of this form have only a single pole, or are not singular, respectively, because the spinor products $\langle a\,b\rangle$ and $[a\,b]$ differ only by a phase. With the complex continuation eq. (1) these ratios of spinor products develop (double) poles. We term channels with such poles 'non-standard' channels.
- The amplitude does not vanish as $z \to \infty$.

For general helicity configurations, as explained in ref. 2, one can choose a pair of shifted momenta that avoids either non-standard channels or large-parameter contributions, but in general not both.

3. The bootstrap method

The bootstrap method developed in refs. 1, 2 systematically deals with the aforementioned complications. We use the term 'bootstrap' because we "assume that very general consistency criteria are sufficient to determine the whole theory completely"[10]. In the present case, the consistency criteria we employ are unitarity, to determine the branch cuts, and factorization, to obtain the rational remainder via on-shell recursion relations.

We begin by decomposing the amplitude into 'pure-cut' and 'rational' parts,

$$A_n(z) = c_\Gamma \Big[C_n(z) + R_n(z) \Big] , \tag{4}$$

where we have explicitly taken an ubiquitous one-loop factor c_Γ outside of $C_n(z)$ and $R_n(z)$, $c_\Gamma = \frac{1}{(4\pi)^{2-\epsilon}} \frac{\Gamma(1+\epsilon)\Gamma^2(1-\epsilon)}{\Gamma(1-2\epsilon)}$. The rational parts are defined by setting all logarithms, polylogarithms, and associated π^2 terms to zero. The pure-cut terms are the remaining terms, all of which must contain logarithms, polylogarithms, or π^2 terms. In the following we assume that the cut containing terms have been computed via (generalized) unitarity or other means (see, e. g. 11, 12, 13 and references therein), and derive an on-shell recursion for the rational remainder.

As mentioned above, the presence of spurious, unphysical singularities in the pure-cut parts complicates this task. These spurious poles cancel in the full amplitude. We will therefore 'complete the cut' by adding rational terms $\widehat{CR}$ to the pure-cut parts that eliminate these spurious poles from the beginning. Instead of the decomposition eq. (4) we now have,

$$A_n(z) = c_\Gamma \left[\widehat{C}_n(z) + \widehat{R}_n(z)\right], \tag{5}$$

where we have defined $\widehat{C} \equiv C + \widehat{CR}$ and $\widehat{R} = R - \widehat{CR}$. Of course, such a cut-completion is not unique. Also, when constructing a recursion relation for the rational remainder, we need to take into account that $\widehat{C}_n$ already contains rational terms from this cut-completion in order to avoid double counting. The resulting full amplitude is then unambiguous.

Since cut and rational parts factorize separately[1], we can now apply Cauchy's theorem to eq. (5). Note that $\widehat{C}_n(z)$ may have (spurious) contributions as $z \to \infty$, denoted by $\mathrm{Inf}\,\widehat{C}_n$, which we can subtract off, since we know $\widehat{C}_n(0)$ (and thus $\widehat{C}_n(z)$). We obtain,

$$A_n(0) = \mathrm{Inf}\, A_n + c_\Gamma \left[\widehat{C}_n(0) - \mathrm{Inf}\,\widehat{C}_n - \sum_{\text{poles }\alpha} \operatorname*{Res}_{z=z_\alpha} \frac{R_n(z)}{z} + O_n\right]. \tag{6}$$

Here, we have taken into account the fact that the full amplitude does not necessarily vanish at infinity under the chosen shift. This contribution is denoted by $\mathrm{Inf}\, A_n$. Its formal operator definition is the extraction of the constant z^0 term in a Laurent expansion of $A_n(z)$ around $z = \infty$.

Double counting between the rational terms in the cut completion $\widehat{C}_n$ and the rational remainder is avoided by the 'overlap terms' O_n,

$$O_n \equiv \sum_{\text{poles }\alpha} \operatorname*{Res}_{z=z_\alpha} \frac{\widehat{CR}_n(z)}{z}, \tag{7}$$

where in general every channel that gets shifted according to eq. (1) contributes to the sum, but for specific cut completions, individual channels may vanish.

Finally, the sum over residues of the rational part again results in a recursion relation, quite analogous to the tree-level case,

$$\begin{aligned} -\sum_{\text{poles }\alpha} \operatorname*{Res}_{z=z_\alpha} \frac{R_n(z)}{z} &\equiv R_n^D(k_1, \ldots, k_n) \\ &= \sum_{r,s} \sum_h \left\{ A_L^{\text{tree},h} \frac{i}{K_{r\ldots s}^2} R_R^{-h} + R_L^h \frac{i}{K_{r\ldots s}^2} A_R^{\text{tree},-h} \right. \\ &\qquad \left. + A_L^{\text{tree},h} \frac{i\mathcal{F}(K_{r\ldots s}^2)}{K_{r\ldots s}^2} A_R^{\text{tree},-h} \right\}. \end{aligned} \tag{8}$$

In the first two terms the scalar propagator connects the rational part of an on-shell loop amplitude with an on-shell tree amplitude, the last term corresponds to a one-loop correction to the propagator[14]. However, for a given shift, we may encounter non-standard channels with as yet unknown factorization behavior, which arise from two-particle channels with like-helicity gluons,

$$R_n^D \equiv R_n^{D,\,\mathrm{rec}} + R_n^{D,\,\mathrm{non-std}}\,, \tag{9}$$

where $R^{D,\,\mathrm{non-std}}$ involves $R_3(k_i^\pm, k_{i+1}^\pm, -\hat{K}_{i\,(i+1)})$ channels (i or $i+1 = \hat{\jmath}$ or $\hat{l}$ for a $[j,l\rangle$-shift)[b].

In summary, after attempting to construct a recursion relation for the rational terms, we have,

$$A_n(0) = \mathrm{Inf}\, A_n + c_\Gamma\big[\widehat{C}_n(0) - \mathrm{Inf}\,\widehat{C}_n + R_n^{D,\,\mathrm{rec}} + R_n^{D,\,\mathrm{non-std}} + O_n\big]\,, \tag{10}$$

which has two unknown contributions, the large-parameter contribution $\mathrm{Inf}\, A$, and the contribution from channels with not yet fully understood factorization behavior, $R^{D,\,\mathrm{non-std}}$. By using a pair of shifts in two *independent* complex parameters, we can construct a recursion relation which determines all unknown terms in eq. (10).

We use the primary shift eq. (1), and an auxiliary shift of two different legs, $[a,b\rangle$ in a different complex parameter w. We obtain the following recursion relations for the amplitude,

$$A_n(0) = \mathop{\mathrm{Inf}}_{[j,l\rangle} A_n + c_\Gamma\big[\widehat{C}_n(0) - \mathop{\mathrm{Inf}}_{[j,l\rangle} \widehat{C}_n + R_n^{D,\,\mathrm{rec}\,[j,l\rangle} + O_n^{[j,l\rangle}\big]\,, \tag{11}$$

$$A_n(0) = c_\Gamma\big[\widehat{C}_n(0) - \mathop{\mathrm{Inf}}_{[a,b\rangle} \widehat{C}_n + R_n^{D,\,\mathrm{rec}\,[a,b\rangle} + R_n^{D,\,\mathrm{non-std}\,[a,b\rangle} + O_n^{[a,b\rangle}\big]\,, \tag{12}$$

where we have indicated with additional superscripts which shift has been employed. Applying the primary shift eq. (1) to the auxiliary recursion (12), we can extract the large-parameter behavior of the *primary shift*,

$$\mathop{\mathrm{Inf}}_{[j,l\rangle} A_n = c_\Gamma\bigg[\mathop{\mathrm{Inf}}_{[j,l\rangle} \widehat{C}_n - \mathop{\mathrm{Inf}}_{[j,l\rangle}\Big(\mathop{\mathrm{Inf}}_{[a,b\rangle} \widehat{C}_n\Big) + \mathop{\mathrm{Inf}}_{[j,l\rangle} R_n^{D,\,\mathrm{rec}\,[a,b\rangle} + \mathop{\mathrm{Inf}}_{[j,l\rangle} O_n^{[a,b\rangle}\bigg]\,, \tag{13}$$

where now all terms on the right-hand side are either known or recursively constructible, *if* $\mathop{\mathrm{Inf}}_{[j,l\rangle} R_n^{D,\,\mathrm{non-std}\,[a,b\rangle} = 0$. A list of suitable pairs of shifts, as well as many examples, can be found in 2.

In summary, the bootstrap equation for the full amplitude is given by,

$$\boxed{A_n(0) = \mathop{\mathrm{Inf}}_{[j,l\rangle} A_n + c_\Gamma\bigg[\widehat{C}_n(0) - \mathop{\mathrm{Inf}}_{[j,l\rangle} \widehat{C}_n + R_n^{D,\,\mathrm{rec}\,[j,l\rangle} + O_n^{[j,l\rangle}\bigg]} \tag{14}$$

[b] $R_3(k_i^\pm, k_{i+1}^\mp, -\hat{K}_{i\,(i+1)}) = 0$, for either helicity of $\hat{K}_{i\,(i+1)}$, and $R_3(k_i^-, k_{i+1}^-, -\hat{K}) = 0$ if i or $i+1 = \hat{l}$, as well as $R_3(k_i^+, k_{i+1}^+, -\hat{K}) = 0$ if i or $i+1 = \hat{\jmath}$.

$\operatorname{Inf}_{[j,l\rangle} A_n$ is computed via eq. (13). All terms appearing in eq. (14) are either (assumed to be) known (the cut constructible part), or can be constructed recursively. Using this approach, we have computed many previously unknown amplitudes in ref. 2. We have tested all multiparticle and collinear factorization channels of these amplitudes. These tests are highly nontrivial, because any omitted or incorrect terms would spoil the factorization properties. Furthermore, at six points we have compared our results numerically to the results of refs. 15, 16 and found complete agreement.

4. Summary and outlook

Above, we have presented a method for the computation of one-loop multi-gluon amplitudes that combines information from unitarity, to determine the cut containing, (poly)logarithmic terms of the amplitude, and factorization, to obtain the rational remainder via on-shell recursion relations. The method is valid for arbitrary helicity configurations, and relies only on general properties of the gauge theory. We are therefore optimistic that it can be extended to include fermions, and massive partons.

In certain cases, the on-shell recursion relations can be solved explicitly. All-multiplicity results have been obtained for several multi-gluon amplitudes[1–4]. The study of the properties of these all-multiplicity amplitudes may reveal some more hidden structure of the underlying gauge theory.

Of course, the aforementioned work is just a starting point. It is desirable to extend the method to partons other than gluons, as well as fully automatize the computations. Furthermore, a better understanding of complex factorization properties in the non-standard channels would definitely be useful and might even lead to the applicability of recursion relations to higher loop order.

> "One of the most remarkable discoveries in elementary particle physics has been that of the existence of the complex plane."
>
> from J. Schwinger[17] (1970)

Acknowledgments

I thank Zvi Bern, Lance Dixon, Darren Forde, and David Kosower for a very fruitful collaboration. This work is supported by the Department of Energy, contract DE–AC02–76SF00515.

References

1. Z. Bern, L. J. Dixon, D. A. Kosower, Phys. Rev. D **71**, 105013 (2005); Phys. Rev. D **72**, 125003 (2005); Phys. Rev. D **73**, 065013 (2006).
2. C. F. Berger, Z. Bern, L. J. Dixon, D. Forde, D. A. Kosower, to appear in Phys. Rev. D (2006) [hep-ph/0604195].
3. C. F. Berger, Z. Bern, L. J. Dixon, D. Forde, D. A. Kosower, hep-ph/0607014.
4. D. Forde and D. A. Kosower, Phys. Rev. D **73**, 061701 (2006).
5. L. J. Dixon, in *QCD & Beyond: Proceedings of TASI '95*, ed. D. E. Soper (World Scientific, 1996) [hep-ph/9601359].
6. Z. Bern, L. J. Dixon, D. A. Kosower, Phys. Rev. Lett. **70**, 2677 (1993).
7. R. Britto, F. Cachazo, B. Feng, Nucl. Phys. B **715**, 499 (2005).
8. R. Britto, F. Cachazo, B. Feng, E. Witten, Phys. Rev. Lett. **94**, 181602 (2005).
9. F. Cachazo and P. Svrcek, PoS **RTN2005**, 004 (2005).
10. Bootstrapping. In *Wikipedia, The Free Encyclopedia*. Retrieved July 27, 2006, from `http://en.wikipedia.org/w/index.php?title=Bootstrapping`.
11. J. Bedford, A. Brandhuber, B. J. Spence, G. Travaglini, Nucl. Phys. B **712**, 59 (2005).
12. Z. Bern, N. E. J. Bjerrum-Bohr, D. C. Dunbar, H. Ita, JHEP **0511**, 027 (2005).
13. R. Britto, B. Feng, P. Mastrolia, Phys. Rev. D **73**, 105004 (2006).
14. Z. Bern and G. Chalmers, Nucl. Phys. B **447**, 465 (1995).
15. R. K. Ellis, W. T. Giele, G. Zanderighi, JHEP **0605**, 027 (2006).
16. Z. G. Xiao, G. Yang, C. J. Zhu, hep-ph/0607017.
17. J. Schwinger, *Particles, Sources, and Fields*, Volume 1, Reading, MA, 1970.

ON-SHELL RECURSION RELATIONS FOR N-POINT QCD*

DARREN FORDE

Service de Physique Théorique†*, CEA–Saclay*
F–91191 Gif-sur-Yvette cedex, France

We present on the use of on-shell recursion relations. These can be used not only for calculating tree amplitudes, including those with masses, but also to compute analytically the missing rational terms of one-loop QCD amplitudes. Combined with the cut-containing pieces calculated using a unitarity approach complete one-loop QCD amplitudes can be derived. This approach is discussed in the context of the adjacent 2-minus all-multiplicity QCD gluon amplitude.

Keywords: On-shell recursion relations, QCD, Proceedings.

1. Introduction

The forthcoming experimental program at CERN's Large Hadron Collider will place new demands on theoretical calculations. In order to reach the precision required by searches for and measurements of new physics, these processes need to be computed to next-to-leading order (NLO), which entails the computation of one-loop amplitudes. These are challenging calculations. State-of-the-art Feynman-diagrammatic computations have only recently reached six-point amplitudes[1] due to the large numbers of diagrams involved.

Feynman diagram techniques are not the only method for performing these needed one-loop contributions. Within the unitarity-based method[2–4] and related recent developments[5,6], one can decompose one-loop colour-ordered gluonic QCD amplitudes into pieces corresponding to $\mathcal{N} = 4$, $\mathcal{N} = 1$, and scalar contributions as $A_n = A_n^{\mathcal{N}=4} - 4A_n^{\mathcal{N}=1} + A_n^{\rm scalar}$. The supersymmetric contributions can be computed by performing the cut algebra strictly in four dimensions, with only the loop integrations computed

---

*Work in collaboration with C. F. Berger, Z. Bern, L. J. Dixon & D. A. Kosower.
†Laboratory of the *Direction des Sciences de la Matière* of the *Commissariat à l'Energie Atomique* of France.

in $D = 4 - 2\epsilon$ dimensions. Scalar-loop contributions require that the cut algebra, and the corresponding tree amplitudes fed into the unitarity machinery, also be computed in $D = 4 - 2\epsilon$ dimensions[4,7,8]. This makes the computation of these pieces somewhat more difficult than in the supersymmetric case and leads us to desire a more efficient approach.

At one loop, computing a scalar loop in $D = 4 - 2\epsilon$ dimensions is equivalent to computing a massive scalar loop in $D = 4$ dimensions, and then integrating over the mass with an appropriate weighting. The computation of tree-level amplitudes with massive scalars is thus of use in the unitarity method for computing massless loop amplitudes in non-supersymmetric gauge theories. On-shell recursion relations can be applied to calculate the necessary tree amplitudes[9,10]. These relations extend the tree-level on-shell recursion relations of Britto, Cachazo, Feng, and Witten[11]. The remarkable generality and simplicity of the proof of these recursion relations, requiring only Cauchy's theorem and a knowledge of the factorisation properties of the amplitudes, has enabled widespread application at tree level[12] and even at loop level[13–15].

Using massive scalars, although more straightforward than the unitarity method in $D = 4 - 2\epsilon$ dimensions, is still not the most efficient applicable technique. More efficient still is an updated version of the unitarity bootstrap technique[5]. This technique relies on first obtaining the cut-constructible parts of a desired amplitude — those terms containing polylogarithms, logarithms, and associated π^2 terms — via the unitarity method in $D = 4$. The missing rational terms that this process cannot capture are then derived using one-loop on-shell recursion relations[14]. This allows for a practical and systematic construction of the rational terms of loop amplitudes.

2. Recursive bootstrap approach

Before describing the extension of the on-shell recursion relations to loop processes we first give an overview of the tree level recursion relations including their application to massive theories. In the simplest case the recursion relations employ a parameter-dependent '$[j, l\rangle$' shift of two of the external massless spinors, j and l, in an n-point process,

$$[j,l\rangle : \qquad \tilde\lambda_j \to \tilde\lambda_j - z\tilde\lambda_l\,, \quad \lambda_j \to \lambda_j\,,$$
$$\lambda_l \to \lambda_l + z\lambda_j\,, \quad \tilde\lambda_l \to \tilde\lambda_l\,, \tag{1}$$

where z is a complex parameter. These shifted momentum then remain massless, $k_j^2(z) = k_l^2(z) = 0$, and overall momentum conservation

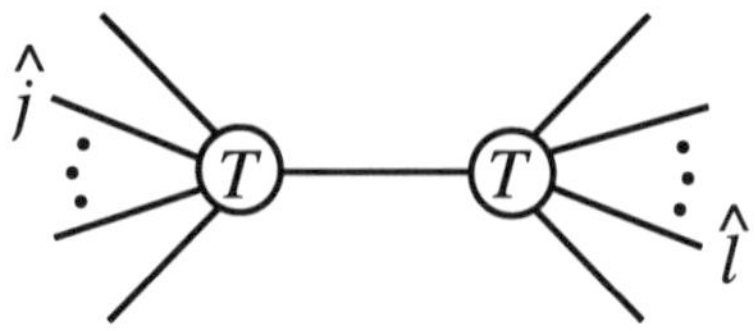

Fig. 1. Schematic representation of a tree-level recursive contribution to $A_n(0)$. The labels 'T' refer to tree vertices which are on-shell amplitudes. The momenta $\hat{\jmath}$ and $\hat{l}$ are shifted, on-shell momenta.

is maintained. Shifting massive particles is also possible although more involved[9,10]. An on-shell amplitude containing the momenta k_j and k_l then becomes parameter-dependent as well. Exploiting Cauchy's theorem to construct the analytic tree level function $A(z)$ from its residues and assuming that there is no contribution from the circular contour at infinity allows us to solve for the physical amplitude $A(0)$,

$$A(0) = - \sum_{\text{poles } \alpha} \operatorname*{Res}_{z=z_\alpha} \frac{A(z)}{z} \,. \tag{2}$$

The residues in eq. (2) may be obtained using the generic factorisation properties that any amplitude must satisfy[16]. The propagator in any factorised channel where the shifted legs j and l lie on opposite sides of the pole, as depicted in fig. 1, will be of the form $1/(K^2 - M^2 - z\langle j^-|\not{K}|l^-\rangle)$. Each pole therefore corresponds to a single factorised channel and hence evaluating the residues of all such poles results in an on-shell recurrence relation for $A(0)$ written schematically as

$$A(0) = \sum_{\text{channels}} \sum_{h=\pm} A(\ldots, \hat{\jmath}, \ldots, (-\hat{K})^h) \frac{i}{K^2 - M^2} A(\hat{K}^{-h}, \ldots, \hat{l}, \ldots) \,. \tag{3}$$

In both amplitudes the momenta are all on-shell including the intermediate momentum $\hat{K}$, which can be massive (i.e. $\hat{K}^2 = M^2$). Including massive external particles is therefore as straightforward as using, where necessary, massive propagators and amplitudes with the appropriate massive legs[9,10].

At loop level a number of new features arise. In particular, obtaining an on-shell recursion relation requires dealing with branch cuts, spurious singularities, and in some cases, the treatment of factorisation using complex momenta, which can differ from 'ordinary' factorisation using real momenta. An example of this, which applies also at tree level, is the vanishing of all three-point vertices in real momentum due to the constraints of momentum conservation. When using complex momentum this is no longer the case as for complex spinors $\overline{\lambda} \not\propto \tilde{\lambda}$, and so we must now include three-point amplitudes in the recursion relations. We must also contend with the possible appearance of double poles and unreal poles in two-particle channels with like-helicity gluons[13,14].

To set up a loop-level on-shell recursion we decompose the amplitude into 'pure-cut' and 'rational' pieces, $A_n(z) = c_\Gamma\,[C_n(z) + R_n(z)]$. The rational parts R_n are defined by setting all logarithms, polylogarithms, and associated π^2 terms to zero. It is then possible to show that the complete amplitude at one-loop is given by[14]

$$A_n(0) = \text{Inf}\, A_n + c_\Gamma \left[\widehat{C}_n(0) - \text{Inf}\, \widehat{C}_n + R_n^D + O_n \right], \tag{4}$$

where $\text{Inf}\, A_n$ is the potential contribution to the amplitude from large z, $\widehat{C}_n(0)$ is the completed-cut contribution, which can be calculated using unitarity based methods, $\text{Inf}\, \widehat{C}_n$ is the potential large-z spurious behaviour of the completed cut, which must be subtracted off, R_n^D are the recursive diagram contributions derived using an on-shell recursion relation, and the 'overlap' terms O_n remove double counting between the recursive diagrams and the rational terms that were added to complete the cuts.[a]

3. Solving recursion relations and all-multiplicity amplitudes

Our basic stratagem to derive the complete one-loop amplitude is therefore to calculate the cut-constructible pieces and then using these construct the overlap terms. The remaining R_n^D and $\text{Inf}\, A_n$ terms are then calculated using an on-shell recursion relation. Usually we will know the form of an amplitude only up to a certain number of negative helicity legs (for a mostly plus amplitude) and desire the form of the amplitude with one more negative helicity leg. On constructing a recursion relation though we will find that in some cases the recursion will contain an amplitude with the same number of negative helicity legs, though fewer positive. This is potentially problematic. For example consider the all-multiplicity one-loop amplitude $A_n^{\text{scalar}}(1^-, 2^-, 3^+, \ldots, n^+)$ the rational terms of this amplitude $R_n(1,2)$ will be given after a $[1,2\rangle$ shift by

$$\begin{aligned} R_n(1,2) &= A_s(1^-, 2^-, 3^+, ..., n^+) \\ &\quad + R_{n-1}(\hat{1}^-, \hat{K}_{23}^-, 4^+, ..., n^+) \frac{1}{K_{23}^2} A_3^{\text{tree}}((-\hat{K}_{23})^+, \hat{2}^-, 3^+) \,. \end{aligned} \tag{5}$$

[a]for a more detailed account of this see C. F. Berger's contribution in this proceedings.

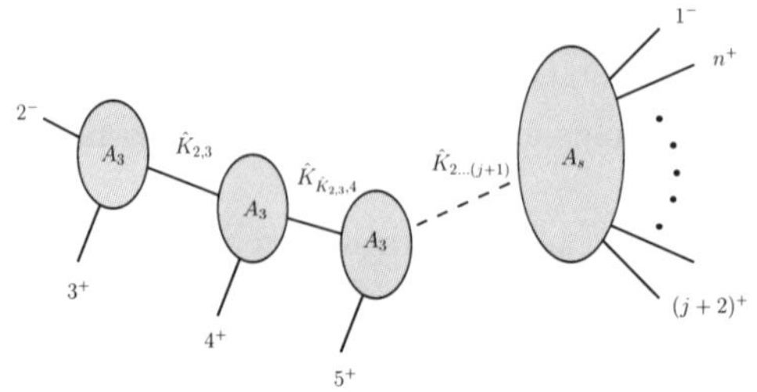

Fig. 2. A contribution to the "unwinding", the string of A_3 terms simplifies to $A_{j+1}(1^-, 2^-, 3^+, \ldots, (j+1)^+)/\langle 1\hat{K}_{2\ldots j+1}\rangle^2$.

Contained in A_s are terms which are already known; the one-loop amplitudes with one negative-helicity leg (which are completely rational) and the tree amplitudes that multiply them. The second term contains R_{n-1} which is the amplitude we are solving for but with one less positive helicity leg.

Our tactic to solve eq. (5) for $R_n(1,2)$ is to insert the left-hand side of eq. (5) into the right-hand side of eq. (5) repeatedly. At each insertion we find that our desired amplitude $R(1,2)$ appears on the right-hand side with one fewer positive-helicity leg, and multiplied by one more three-point gluon vertex and propagator, A_3^{tree}/K^2. This 'unwinding' of the amplitude continues until we have reduced the right-hand side of R_n (eq. (5)) down to, in this case, $R_4 = 0$ and a sum of terms that contain only known quantities (e.g. A_s and overlap terms O) multiplied by strings of A_3^{tree} vertices, a contributing example is shown in figure 2.

At each step of the unwinding we must choose new shifted momenta. We always choose to shift the two negative-helicity legs of R. For example, after the first step we choose $[\hat{1}, \hat{K}_{2,3}\rangle$ as the shifted legs. Similarly, when we perform a second insertion, of R_{n-2}, we choose the intermediate $\hat{K}$ momentum leg of the last shift and the previously shifted $\hat{1}$ leg.

After this "unwinding" each resulting term can be expressed schematically in the form

$$\left(\prod_{r=2}^{j+1} \frac{iA_{3r}^{\text{tree}}}{K_r^2}\right) A_s(1^-, \hat{K}^-, (j+2)^+, \ldots, n^+)\,. \tag{6}$$

The product of three point gluon vertices contained inside the brackets is equivalent to simply a tree amplitude divided by $\langle 1\,\hat{K}_{2\ldots(j+1)}\rangle^2$. Hence the recursion is solved as eq. (6) is written entirely in terms of objects we know

$$\frac{iA_{j+1}^{\text{tree}}(1^-, 2^-, \ldots, (j+1)^+)}{\langle 1\,\hat{K}_{2\ldots(j+1)}\rangle^2} A_s(1^-, \hat{K}^-_{2\ldots(j+1)}, (j+2)^+, \ldots, n^+)\,.$$

The complete unrenormalised scalar loop contribution is then given by

$$A_n^{\rm scalar} = c_\Gamma \left[\hat{C}_n + R_n^D \right] + \frac{1}{3} A_n^{\mathcal{N}=1 \text{ chiral}} + c_\Gamma \frac{2}{9} A_n^{\rm tree} , \tag{7}$$

as in this case Inf $A_n = 0$. The cut-completed contribution, $\hat{C}_n$, previously calculated from unitarity techniques is given in Ref. 15 and

$$\begin{aligned} R_n^D(1,2) = \sum_{m=2}^{n-3} \frac{i A_m^{\rm tree}(1^-, 2^-, \ldots, m^+)}{\langle 1\, \hat{K}_{2\ldots m} \rangle^2} \Big(& A_s(1^-, \hat{K}_{2\ldots m}^-, (m+1)^+, \ldots, n^+) \\ & + O_{n-m+2}(1^-, \hat{K}_{2\ldots m}^-, (m+1)^+, \ldots, n^+) \\ & + \widehat{CR}_{n-m+2}(1^-, \hat{K}_{2\ldots m}^-, (m+1)^+, \ldots, n^+) \Big) , \end{aligned} \tag{8}$$

were in the recursion for R_n^D both the overlap, O_n pieces, and the cut-completion $\widehat{CR}_n$ terms are included. Inserting the known forms of these terms into eq. (8) produces the result given in Ref. 15.

This "unwinding" technique also extends to other processes, so far this includes all-multiplicity massive scalar trees[10] and the all-multiplicity MHV one-loop gluonic QCD amplitude[14]. Hence the unitarity bootstrap approach provides a bright new outlook on the calculation of previously difficult to compute loop process needed to fully exploit the promise of the LHC.

References

1. A. Denner, S. Dittmaier, M. Roth and L. H. Wieders, Phys. Lett. B612:223 (2005) [hep-ph/0502063]; Nucl. Phys. B724:247 (2005) [hep-ph/0505042]; W. T. Giele and E. W. N. Glover, JHEP 0404:029 (2004) [hep-ph/0402152]; R. K. Ellis, W. T. Giele and G. Zanderighi, JHEP 0605:027 (2006) [hep-ph/0602185]; T. Binoth, G. Heinrich and N. Kauer, Nucl. Phys. B654:277 (2003) [hep-ph/0210023]; M. Kramer and D. E. Soper, Phys. Rev. D66:054017 (2002) [hep-ph/0204113]; T. Binoth, J. P. Guillet, G. Heinrich, E. Pilon and C. Schubert, JHEP 0510:015 (2005) [hep-ph/0504267].
2. Z. Bern, L. J. Dixon, D. C. Dunbar and D. A. Kosower, Nucl. Phys. B425:217 (1994) [hep-ph/9403226]; Z. Bern, L. J. Dixon and D. A. Kosower, Ann. Rev. Nucl. Part. Sci. 46:109 (1996) [hep-ph/9602280]; Z. Bern, L. J. Dixon and D. A. Kosower, Nucl. Phys. Proc. Suppl. 51C:243 (1996) [hep-ph/9606378]; Z. Bern, L. J. Dixon and D. A. Kosower, JHEP 0001:027 (2000) [hep-ph/0001001]; Z. Bern, L. J. Dixon and D. A. Kosower, Nucl. Phys. B513:3 (1998) [hep-ph/9708239]; Z. Bern, L. J. Dixon and D. A. Kosower, JHEP 0408:012 (2004) [hep-ph/0404293].
3. Z. Bern, L. J. Dixon, D. C. Dunbar and D. A. Kosower, Nucl. Phys. B435:59 (1995) [hep-ph/9409265].
4. Z. Bern and A. G. Morgan, Nucl. Phys. B467:479 (1996) [hep-ph/9511336].
5. Z. Bern, L. J. Dixon and D. A. Kosower, Nucl. Phys. B513:3 (1998) [hep-ph/9708239].

6. R. Britto, F. Cachazo and B. Feng, Nucl. Phys. B725:275 (2005) [hep-th/0412103]; A. Brandhuber, S. McNamara, B. Spence and G. Travaglini, JHEP 0510:011 (2005) [hep-th/0506068]; Z. Bern, L. J. Dixon and D. A. Kosower, JHEP 0408:012 (2004) [hep-ph/0404293]; Z. Bern, V. Del Duca, L. J. Dixon and D. A. Kosower, Phys. Rev. D71:045006 (2005) [hep-th/0410224]; R. Britto, E. Buchbinder, F. Cachazo and B. Feng, Phys. Rev. D72:065012 (2005) [hep-ph/0503132]; R. Britto, B. Feng and P. Mastrolia, Phys. Rev. D73:105004 (2006) [hep-ph/0602178].
7. Z. Bern, L. J. Dixon, D. C. Dunbar and D. A. Kosower, Phys. Lett. B394:105 (1997) [hep-th/9611127].
8. Z. Bern, L. J. Dixon and D. A. Kosower, JHEP 0001:027 (2000) [hep-ph/0001001]; Z. Bern, L. J. Dixon and D. A. Kosower, JHEP 0408:012 (2004) [hep-ph/0404293]; D. A. Kosower and P. Uwer, Nucl. Phys. B563: 477 (1999) [hep-ph/9903515]; Z. Bern, A. De Freitas and L. J. Dixon, JHEP 0203:018 (2002) [hep-ph/0201161]; A. Brandhuber, S. McNamara, B. Spence and G. Travaglini, JHEP 0510:011 (2005) [hep-th/0506068].
9. S. D. Badger, E. W. N. Glover, V. V. Khoze and P. Svrček, JHEP 0507:025 (2005) [hep-th/0504159]; S. D. Badger, E. W. N. Glover and V. V. Khoze, JHEP 0601:066 (2006) [hep-th/0507161]; C. Schwinn and S. Weinzierl, JHEP 0603:030 (2006) [hep-th/0602012]; P. Ferrario, G. Rodrigo and P. Talavera, Phys. Rev. Lett. 96:182001 (2006) [hep-th/0602043].
10. D. Forde and D. A. Kosower, Phys. Rev. D73:065007 (2006) [hep-th/0507292].
11. R. Britto, F. Cachazo and B. Feng, Nucl. Phys. B715:499 (2005) [hep-th/0412308]; R. Britto, F. Cachazo, B. Feng and E. Witten, Phys. Rev. Lett. 94:181602 (2005) [hep-th/0501052].
12. M.-X. Luo and C.-K. Wen, JHEP 0503:004 (2005) [hep-th/0501121]; Phys. Rev. D71:091501 (2005) [hep-th/0502009]; R. Britto, B. Feng, R. Roiban, M. Spradlin and A. Volovich, Phys. Rev. D71:105017 (2005) [hep-th/0503198]; K. J. Ozeren and W. J. Stirling, JHEP 0511:016 (2005) [hep-th/0509063]; M. Dinsdale, M. Ternick and S. Weinzierl, JHEP 0603:056 (2006) [hep-ph/0602204]; D. de Florian and J. Zurita, JHEP 0605:073 (2006) [hep-ph/0605291].
13. Z. Bern, L. J. Dixon and D. A. Kosower, Phys. Rev. D71:105013 (2005) [hep-th/0501240]; Z. Bern, L. J. Dixon and D. A. Kosower, Phys. Rev. D72:125003 (2005) [hep-ph/0505055].
14. Z. Bern, L. J. Dixon and D. A. Kosower, Phys. Rev. D73:065013 (2006) [hep-ph/0507005]; C. F. Berger, Z. Bern, L. J. Dixon, D. Forde and D. A. Kosower, hep-ph/0604195; C. F. Berger, Z. Bern, L. J. Dixon, D. Forde and D. A. Kosower, arXiv:hep-ph/0607014.
15. D. Forde and D. A. Kosower, Phys. Rev. D **73**, 061701 (2006) [arXiv:hep-ph/0509358].
16. M. L. Mangano and S. J. Parke, Phys. Rept. 200:301 (1991); L. J. Dixon, in *QCD & Beyond: Proceedings of TASI '95*, ed. D. E. Soper (World Scientific, 1996) [hep-ph/9601359]; Z. Bern and G. Chalmers, Nucl. Phys. B447:465 (1995) [hep-ph/9503236].

QCD ON-SHELL RECURRENCE RELATIONS FROM THE LARGEST TIME EQUATION

DIANA VAMAN and YORK-PENG YAO

Michigan Center for Theoretical Physics
Randall Laboratory of Physics, The University of Michigan
Ann Arbor, MI 48109-1120
email dvaman, yyao@umich.edu

We show how, by reassembling the tree level gluon Feynman diagrams in a convenient gauge, space-cone, we can explicitly derive the BCFW recursion relations. Moreover, the proof of the on-shell recursion relations hinges on an identity in momentum space which we show to be nothing but the Fourier transform of Veltman's largest time equation. Our approach lends itself to natural generalizations to include massive scalars and even fermions.

Keywords: QCD, Gauge symmetry

1. Introduction

QCD calculations are notoriously tedious if one is to follow the usual Feynman-Dyson expansion in some commonly used, such as Feynman, gauge. Over the past few years, great strides have been made to simplify such endeavors. The results for the complete amplitudes at the tree or one-loop level can be quite compact.

Following Witten's proposal for a description of perturbative Yang-Mills gauge theory as a string theory on twistor space [1], and subsequent proposal for an alternative to the usual Feynman diagrams in terms of the so-called maximally helicity violating (MHV) vertices [2], a new set of methods was available for the computation of QCD amplitudes. The latest advance in the form of recursion relations [3,4], in conjunction with the attendant rules for their construction, is particularly appealing. It is quite obvious from the flavor of such an approach that it bears on the cutting rules in field theory. In fact, some work at the one-loop level under the heading of cut-constructibility clearly points to the same origin [5].

As is well-known, unitarity of the S-matrix and the feasibility of an ordering of a sequence of space time points are intimately related. Indeed, the ordering need not be with respect to time, as is conventionally done. All that is essential in a perturbation series is that one must be able to separate the positive frequency and the negative frequency components in a propagator according to the signature of a certain linear combination of components Δx of the four vector between the two space-time points. For our purpose, a component $\eta \cdot \Delta x$ of the light-cone variables will be a convenient start, where η is a light-like vector. We shall rely on the existence of tubes of analyticity to continue such variables into the space cone, in order to incorporate a gauge condition for QCD. The resulting ordering is the equivalent of the largest time equations. The outcome, for QCD in particular, is that one factorizes a physical amplitude into products of physical amplitudes, with some momenta shifted but still on-shell. This is the content of the BCFW recursion relations [3,4]:

$$\mathcal{A}(P,\{P_i\},Q,\{Q_j\}) = \sum_{i,j} \mathcal{A}_L(\hat{P},\{P_i\}) \frac{1}{(P+\sum_i P_i)^2} \mathcal{A}_R(\hat{Q},\{Q_j\}),$$

where $\mathcal{A}_L, \mathcal{A}_R$ are lower n-point functions obtained by isolating two reference gluons with shifted momenta, $\hat{P} = P - z\eta$, $\hat{Q} = Q + z\eta$ with $\eta^2 = \eta \cdot P = \eta \cdot Q = 0$, on the two sides of the cut. The shifting is necessary in order to preserve energy-momentum conservation. We would like to take this opportunity to point out that in so far as factorization is concerned, the masses of the internal propagators have no bearing. However, the demand that the shifted momenta, which will be called reference momenta, should be on-shell will force these external momenta to be light-like.

We now turn to the important step of gauge fixing. In order to facilitate natural cancellation of terms at every level of a QCD calculation, the gauge that is most convenient for us is the space-cone gauge [6]. A crucial advantage of this gauge is that when we shift the momenta to obtain recursion relations, the dependence on momenta of the vertices will not be affected. Thereupon the factorization of the amplitudes is the same as that in a scalar theory. It is this special attribute which makes the program manageable.

2. The space-cone gauge fixed Yang-Mills action

Consider the four-dimensional Yang-Mills gauge theory, with the Lagrangian

$$L = -\frac{1}{8} Tr(\partial^a A^b - \partial^b A^a + i[A^a, A^b])^2 . \qquad (1)$$

According to the spinor helicity formalism, null vectors can be decomposed into a product of two commuting spinors (twistors):

$$P^{\alpha\dot{\beta}} = p^{\alpha} p^{\dot{\beta}} \equiv |p\rangle\,[p|\,. \tag{2}$$

Moreover, we can use the twistors to define a basis on the space of four-vectors

$$P = p^{+}\,|+\rangle[+| + p^{-}\,|-\rangle[-| + p\,|-\rangle[+| + \bar{p}\,|+\rangle[-|\,. \tag{3}$$

As shown by Chalmers and Siegel, [6], in the spinor helicity formalism, a powerful simplification is achieved in the Feynman diagramatics by choosing the space-cone gauge:

$$a = 0\,, \tag{4}$$

followed by the elimination of the "auxiliary" component $\bar{a}$ from its equation of motion. The gauge fixed Lagrangian has now only two scalar degrees of freedom

$$\mathcal{L} = \mathrm{Tr}\Bigg[\frac{1}{2}a^{+}\Box a^{-} - i\left(\frac{\partial^{-}}{\partial}a^{+}\right)[a^{+},\partial a^{-}]$$

$$-i\left(\frac{\partial^{+}}{\partial}a^{-}\right)[a^{-},\partial a^{+}] + [a^{+},\partial a^{-}]\frac{1}{\partial^{2}}[a^{-},\partial a^{+}]\Bigg]\,. \tag{5}$$

Choosing the space-cone gauge amounts to selecting two of the external momenta to be the reference null vectors for defining a twistor basis: $|+\rangle[+|, |-\rangle[-|$, such that the space-cone gauge fixing is equivalent to $N \cdot A = 0$, where the null vector N is equal to $|+\rangle[-|$. The other ingredient which is needed in converting the essentially scalar Feynman diagrams arising from the gauge fixed Lagrangian (5) into definite helicity gluon Feynman diagrams is inserting external line factors

$$\epsilon_{+}{}^{+} = \frac{[-p]}{\langle +p\rangle}, \qquad \epsilon_{-}{}^{-} = \frac{\langle +p\rangle}{[-p]}\,. \tag{6}$$

for the positive, respectively negative helicity external gluons. The helicities of the internal lines/virtual gluons are accounted for by the the scalar Lagrangian: a "$+-$" helicity internal line corresponds to a $a^{+}a^{-}$ propagator, and vice versa.

3. The causality ("largest time") equations

As stated in the Introduction, we shall show that the recursion relations are rooted in the largest time equation. To this end, we briefly revisit here the causality equations as derived by Veltman[7], but appropriately rewriting them in a light-cone frame.

First, we introduce the following set of rules:

-duplicate the Feynman diagram 2^N times, for N vertices, by adding circles around vertices in all possible ways;

-each vertex can be circled or not; a circled vertex will bring a factor of i, and an uncircled vertex will bring a factor of $(-i)$;

-the propagator between two uncircled vertices is $\Delta(x-y)$, while the propagator between two circled vertices is the complex conjugate $\Delta^*(x-y)$;

-the propagator between a circled x_k and an uncircled x_l is $\Delta^+(x_k - x_l)$, while the propagator between an uncircled x_k and a circled x_l is $\Delta^-(x_k - x_l)$.

Clearly, the uncircled Feynman diagram is the usual one, while the fully circled diagram corresponds to its complex conjugate.

The largest time equation states that the sum of all 2^N circled Feynman diagram vanishes:

$$F(x_i) + F^*(x_i) + \mathbf{F}(x_i) = 0\,, \tag{7}$$

where $F(x_i)$ stands for the usual Feynman diagram, $F^*(x_i)$ is its complex conjugate, and $\mathbf{F}(x_i)$ is the sum of $2^N - 2$ diagrams in which at least one vertex is circled and at least one is uncircled.

Other causality equations can be obtained by singling out 2 vertices, x_k and x_l. Let us assume $x_k^+ < x_l^+$. Then, one has

$$\theta((x_l - x_k)^+)(F(x_i) + \mathbf{F}(k, x_i)) = 0\,, \tag{8}$$

where $\mathbf{F}(k, x_i)$ is the sum of all diagrams with k uncircled, but at least one other vertex circled. Similarly, one has

$$\theta((x_k - x_l)^+)(F(x_i) + \mathbf{F}(l, x_i)) = 0\,, \tag{9}$$

By adding these two equations one finds

$$F(x_i) = -\mathbf{F}(k, l, x_i) - \theta((x_l - x_k)^+)\mathbf{F}(k, \mathbf{l}, x_i) - \theta((x_k - x_l)^+)\mathbf{F}(\mathbf{k}, l, x_i))\,, \tag{10}$$

where $\mathbf{F}(k, l, x_i)$ is the sum of all diagrams with neither k, l circled, but at least one other vertex circled, $\mathbf{F}(k, \mathbf{l}, x_i)$ is the sum of all amplitudes with k uncircled , but l circled and finally, $\mathbf{F}(\mathbf{k}, l, x_i))$ has k circled and l uncircled.

The momentum space propagators between two vertices with both vertices uncircled is $-i/(p^2 - i\epsilon)$, the complex conjugate expression if the two vertices are circled, and $2\pi\delta^+(p^2)$ if the momentum flows between an uncircled and a circled vertex. We shall see that *the η-shift of two of the external momenta, required in the on-shell recurrence relations, is a consequence of* *the momentum inflow associated with the step-functions which arise in the largest time equation*

$$\theta((x_l - x_k)^+) = \int \frac{dz}{2\pi i(z - i\epsilon)} e^{iz\eta\cdot(x_l - x_k)} . \tag{11}$$

4. Reassembling Feynman diagrams into BCFW recursion relations

The Feynman diagrams are those following from the space-cone gauge-fixed Lagrangian (5). To set the stage for the recursion relations involving arbitrary tree level gluon n-point functions, we begin our investigation with the lowest ones.

Let us consider the 4-point function $(1234) = (+--+)$, with 1 and 2 selected the reference gluons. One of the crucial observation of this paper is that shifting the reference gluon momenta by $\eta = |2\rangle\,[1| = |+\rangle\,[-|$ changes only their $\bar{p}$ component. More clearly, the shifted momenta are

$$\widehat{P_1} = P_1 + z\eta, \qquad \widehat{P_2} = P_2 - z\eta, , \tag{12}$$

or, in components,

$$\hat{p}_1^- = 1, \qquad \hat{\bar{p}}_1 = -z[21]\langle 21\rangle, \qquad \hat{p}_2^+ = 1, \qquad \hat{\bar{p}}_2 = z[21]\langle 21\rangle . \tag{13}$$

This is particularly important, since the vertices in the space-cone gauge turn out to be independent of $\bar{p}$, as it can be seen by inspecting the Lagrangian (5).

We conclude that any shift of the external momenta as in (12) is of no consequence for the vertex-dependence of any Feynman diagram, and leaves an imprint only over the internal line propagators.

The 4-point function in the space-cone gauge is given by a single Feynman diagram

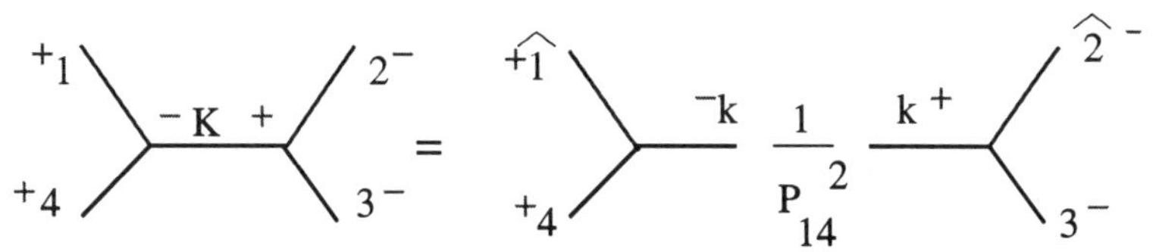

Fig. 1. Factorization of the 4-point function

which is represented by the left hand side of Fig.1. The factorization of the

4-point function is now manifest at the level of the Feynman diagram: there is no difference between the 3-point function and 3-point vertex (which is insensitive to the shift of the momenta of the reference gluons 1 and 2), other than the multiplication by external line polarizations. Moreover, because we have essentially a scalar field theory, one can insert freely factors $\epsilon_k^+ \epsilon_k^-$ (=1).

The 5-point function is the first non-trivial example in which we invoke an identity rooted in the largest time equation. The factorization arises at

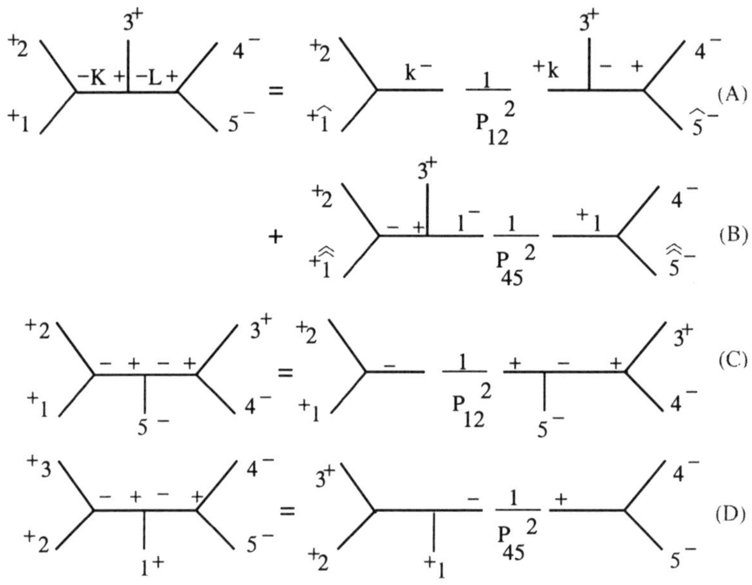

Fig. 2. Factorization of the 5-point function

the level of the Feynman diagrams, as indicated in Fig.2. The equality of the two sides of the last two diagrams is obvious from the fact that the vertices on the left and right hand side of Fig.2 are the same, irrespective of having shifted the external reference momenta in order to put the cut line on shell. Moreover, the propagators on the right and left hand side of the bottom two diagrams coincide as well.

The attentive reader could observe that B+D add up to the $(+++-)$ 4-point function, and so they vanish. Thus indeed, the BCFW recursion relation amounts to including only the terms A and C, corresponding to a factorization of the $(+++--)$ amplitude into $(++-)(++--)$ amplitudes.

To show that the top Feynman diagram equals A+B, once we factored out the vertices (which are the same on the left and right side of Fig.2), amounts to an identity between propagators:

$$\frac{1}{P_{12}^2}\frac{1}{P_{45}^2} = \frac{1}{P_{12}^2}\frac{1}{P_{4\hat{5}}^2} + \frac{1}{P_{\hat{\hat{1}}2}^2}\frac{1}{P_{45}^2}, \tag{14}$$

where we defined the shifted reference gluon external momenta by $P_{\hat{1}} = P_1 + \hat{z}\eta$, $P_{\hat{5}} = P_5 - \hat{z}\eta$, $P_{\hat{\hat{1}}} = P_1 + \hat{\hat{z}}\eta$, $P_{\hat{\hat{5}}} = P_5 - \hat{\hat{z}}\eta$ and where the null space-cone vector is $\eta = |+\rangle\,[-| \equiv |5\rangle\,[1|$. The variables $\hat{z}, \hat{\hat{z}}$ are such that we put the internal lines K, L, respectively, on-shell. Next, we observe that in terms of the z-shifts (14) becomes a trivial algebraic identity

$$\frac{1}{\hat{z}\hat{\hat{z}}} = \frac{1}{(\hat{\hat{z}}-\hat{z})\hat{z}} - \frac{1}{(\hat{\hat{z}}-\hat{z})\hat{\hat{z}}}. \tag{15}$$

5. The recursion relations and the largest time equation

The deeper meaning behind the partial fractioning identity (15) is uncovered by rewriting the Fourier-transformed of the rhs of (14) in position space, and comparing the result with the largest time equation (10) for the same Feynamn diagram. From (10), the largest time equation reads

$$\Delta(x_1-x_2)\Delta(x_2-x_3) = \Delta^-(x_1-x_2)\Delta^+(x_2-x_3)$$
$$+\theta((x_1-x_3)^+)\Big(\Delta^+(x_1-x_2)\Delta(x_2-x_3) - \Delta^*(x_1-x_2)\Delta^+(x_2-x_3)\Big)$$
$$+\theta((x_3-x_1)^+)\Big(\Delta(x_1-x_2)\Delta^-(x_2-x_3) - \Delta^-(x_1-x_2)\Delta^*(x_2-x_3)\Big), \tag{16}$$

where the two vertices which are being singled out (whose coordinates are x_1 and x_3) correspond to the external reference gluons.

This equals the Fourier-transform of the rhs of (14) up to the following two extra terms: $\theta((x_1-x_3)^+)\Delta^+(x_1-x_2)\Delta^+(x_2-x_3)$ and $\theta((x_3-x_1)^+)\Delta^-(x_1-x_2)\Delta^-(x_2-x_3)$. However, these terms are zero as the product of the three distributions has zero support.

Before closing this section, we need to add a few remarks. First, we have to notice that in order to write largest time equation a real η is required, since it is only for such values that we know how to order a sequence of space-time points in $\eta \cdot x_i$. However, now that we have the largest time equation, let us Fourier-transform it into momentum space, appropriate for a physical process under consideration and keeping η as a variable. From the Fourier-transformed step functions we obtain a set of shifted momenta. We then analytically complexify η. For tree level, this is certainly possible and justifiable, because the dependence on η is only through some algebraic functions of propagators. At the loop level, we need to invoke the analysis of axiomatic field theorists [8], which states that there are tubes of analyticity to allow this extension and to lead to complexified unitarity relations. We now identify these propagators with the ones which we need in the space-cone gauge to carry on with the analysis.

6. The general case

To exploit the full generality of the problem, we derive an identity satisfied by the momentum space scalar propagators working under the assumption that we deal with massive propagators, with arbitrary masses.

We single out two external momenta which do not land on the same vertex as reference vectors and call them p_a and p_b. For a tree graph, there is a unique path through some of the internal lines which connects p_a to p_b. The factorization procedure is to cut these q_i successively by shifting them by $z\eta$. The on-shell conditions $\bar{q}_i^2 + m_i^2 = 0, \;\; \bar{q}_i \equiv q_i - z\eta$ will give us a set of solutions, points in the complex plane, namely $z_i = \frac{q_i^2+m_i^2}{2\eta \cdot q_i}$, More precisely stated, the factorization amounts to splicing the graph into a sum of products of two on-shell graphs with shifted momenta $\{p_a - z_i\eta, \ldots, \bar{q}_i\}$ and $\{-\bar{q}_i, \cdots, p_b + z_i\eta\}$, where $\cdots$ stand for the other momenta in the left graph segment and similarly for those in the right graph segment, with the propagator $\frac{1}{q_i^2+m_i^2}$ as the partition. Imposing the condition that the shifted momenta remain on-shell requires that $p_a \cdot \eta = p_b \cdot \eta = 0$, in other words, p_a, p_b must be null.

The identity which we want to establish is

$$\frac{1}{q_1^2+m_1^2}\frac{1}{q_2^2+m_2^2}\cdots\frac{1}{q_{n-1}^2+m_{n-1}^2}$$

$$=\frac{1}{q_1^2+m_1^2}\frac{1}{(q_2-z_1\eta)^2+m_2^2}\cdots\frac{1}{(q_{n-1}-z_1\eta)^2+m_{n-1}^2}$$

$$+\frac{1}{(q_1 - z_2\eta)^2 + m_1^2}\frac{1}{q_2^2 + m_2^2}\cdots\frac{1}{(q_{n-1} - z_2\eta)^2 + m_{n-1}^2} + \ldots$$

$$+\frac{1}{(q_1 - z_{n-1}\eta)^2 + m_1^2}\cdots\frac{1}{(q_{n-2} - z_{n-1}\eta)^2 + m_{n-2}^2}\frac{1}{q_{n-1}^2 + m_{n-1}^2}, \quad (17)$$

and its proof hinges on yet another partial fractioning formula $\oint \frac{dz}{z(z-z_1)(z-z_2)\ldots(z-z_n)} = 0$. Next we notice that eqn. (17) (with massless propagators) is precisely the identity needed to reassemble a generic tree level gluon Feynman diagram into lower on-shell amplitudes, as shown in Section 4. The combinatorics work out properly to reproduce the BCFW recursions.

Furthermore, the arguments presented in Section 5, relating the momentum space identity (17) to the Fourier transform of the corresponding largest time equation (10), with the singled out vertices corresponding to those of the external reference gluons, can be easily carried through.

This completes our purely field theoretical proof of the BCFW recursion relations. In the process, we have identified the underlying principle behind them in the form of the largest time equation. Using the same method, is is not difficult to establish on-shell recurrence relations for minimally coupled (massless or massive) scalars and even fermions [12].

References

1. E. Witten, Commun. Math. Phys. **252**, 189 (2004) [arXiv:hep-th/0312171].
2. F. Cachazo, P. Svrcek and E. Witten, JHEP **0409**, 006 (2004) [arXiv:hep-th/0403047].
3. R. Britto, F. Cachazo and B. Feng, Nucl. Phys. B **715**, 499 (2005) [arXiv:hep-th/0412308].
4. R. Britto, F. Cachazo, B. Feng and E. Witten, Phys. Rev. Lett. **94**, 181602 (2005) [arXiv:hep-th/0501052].
5. Z. Bern, L. J. Dixon, D. C. Dunbar and D. A. Kosower, Nucl. Phys. B **435**, 59 (1995) [arXiv:hep-ph/9409265].
6. G. Chalmers and W. Siegel, Phys. Rev. D **59**, 045013 (1999) [arXiv:hep-ph/9801220].
7. M. J. G. Veltman, Physica **29**, 186 (1963).
8. R. F. Streater and A. S. Wightman, *PCT, Spin & Statics and all that*, The Mathematical Physics Monograph Series, ed. W. A. Benjamin, Inc (1964).
9. S. D. Badger, E. W. N. Glover, V. V. Khoze and P. Svrcek, JHEP **0507**, 025 (2005) [arXiv:hep-th/0504159].

10. S. D. Badger, E. W. N. Glover and V. V. Khoze, arXiv:hep-th/0507161.
11. Z. Bern, L. J. Dixon and D. A. Kosower, Phys. Rev. D **71**, 105013 (2005) [arXiv:hep-th/0501240].
12. D. Vaman and Y. P. Yao, JHEP **0604**, 030 (2006) [arXiv:hep-th/0512031].

SECTION 11

LIGHT CONE

Convener J. Hiller

THE NUCLEON ELECTRIC DIPOLE MOMENT IN LIGHT-FRONT QCD

S. GARDNER

Department of Physics and Astronomy, University of Kentucky,
Lexington, KY, USA
E-mail: gardner@pa.uky.edu

I present an exact relationship between the electric dipole moment and anomalous magnetic moment of the nucleon in the light-front formalism of QCD and consider its consequences.

1. Introduction

Interpreting the electric dipole moments of leptons and baryons as constraints on fundamental, CP-violating Lagrangian parameters of various extensions of the Standard Model (SM) gives key insight into TeV-scale physics. In this contribution I report on work in collaboration with Stan Brodsky and Dae Sung Hwang [1], in which we sharpen the connection between the computed neutron electric dipole moment and fundamental CP violation by comparing its hadronic matrix element to that of the anomalous magnetic moment. In the context of the SM, and, specifically, of the Cabbibo-Kobasyashi-Maskawa (CKM) mechanism of CP violation, the assessed values of the neutron electric dipole moment (EDM) have been disparate, ranging from $d^n_{\rm CKM} \simeq 10^{-32}$e-cm [2,3] arising from a $\pi - N$ loop calculation in a chiral Lagrangian treatment, to $d^d_{\rm KM} \simeq 10^{-34}$e-cm [4,5] for the EDM of the d-quark itself, computed to three-loop precision in leading-logarithmic approximation. These EDMs are much too small to be experimentally observable, so that the marked disparity is actually of little consequence. However, if we restrict ourselves to effective CP-violating operators of dimension five or less, the method of QCD sum rules can be employed to compute the EDM of the neutron [6,7], yielding a value for d^n, induced by a QCD $\bar{\theta}$-term, e.g., commensurate in size with that of the chiral estimate [8], with a surety of $\sim 50\%$ [6]. The evaluation of d^n and d^p, and the errors therein, is also important to interpreting the ^2H EDM [9]. Here we analyze the nucleon electric dipole moment in the light-front formalism of QCD [1], to the end of realizing an independent test of the methods used to compute d^N — and of their assessed errors.

2. Electromagnetic form factors in light-front QCD

Our study of the electric dipole form factor $F_3(q^2)$ in the light-front formalism of QCD complements earlier studies of the Dirac and Pauli form factors [10]. The Pauli and electric dipole form factors emerge from the spin-flip matrix elements of the electromagnetic current $J^\mu(0)$:

$$\langle P', -S_z | J^\mu(0) | P, S_z \rangle = \bar{U}(P', -\lambda) \left[\frac{i}{2M} \sigma^{\mu\alpha} \right.$$
$$\left. \times \left(F_2(q^2) + iF_3(q^2)\gamma_5 \right) q_\alpha \right] U(P, \lambda) , \quad (1)$$

where $U(P, \lambda)$ is a Dirac spinor for a nucleon of momentum P and helicity λ, with $S_z = \lambda/2$. Recall that the anomalous magnetic moment κ and the electric dipole moment d are given by $\kappa = (e/2M)[F_2(0)]$ and $d = (e/M)[F_3(0)]$. We find a close connection between κ and d [1], as long anticipated [11]. Working in the $q^+ = 0$ frame, with $q = (q^+, q^-, \mathbf{q}_\perp) = (0, -q^2/P^+, \mathbf{q}_\perp)$ and $P = (P^+, P^-, \mathbf{P}_\perp) = (P^+, M^2/P^+, \mathbf{0}_\perp)$, in the interaction picture for $J^+(0)$, and in the assumed simple vacuum of the light-front formalism, we find, noting $q^{R/L} \equiv q^1 \pm iq^2$,

$$\frac{1}{2M} \begin{pmatrix} F_2(q^2) \\ -iF_3(q^2) \end{pmatrix} = \sum_a \int [\mathrm{d}x][\mathrm{d}^2\mathbf{k}_\perp] \sum_j e_j \, \frac{1}{2} \left[-\frac{1}{q^L} \psi_a^{\uparrow *}(x_i, \mathbf{k}'_{\perp i}, \lambda_i) \right.$$
$$\left. \times \psi_a^{\downarrow}(x_i, \mathbf{k}_{\perp i}, \lambda_i) \pm \frac{1}{q^R} \psi_a^{\downarrow *}(x_i, \mathbf{k}'_{\perp i}, \lambda_i) \, \psi_a^{\uparrow}(x_i, \mathbf{k}_{\perp i}, \lambda_i) \right] , \quad (2)$$

where $\mathbf{k}'_{\perp j} = \mathbf{k}_{\perp j} + (1 - x_j)\mathbf{q}_\perp$ for the struck constituent j and $\mathbf{k}'_{\perp i} = \mathbf{k}_{\perp i} - x_i \mathbf{q}_\perp$ for each spectator ($i \neq j$). The electric dipole form factor $F_3(q^2)$ vanishes if the usual light-front wave functions are employed, so that we must learn how parity- and time-reversal-violating effects can be included in the light-cone framework.

3. Discrete symmetries on the light front and a relation for the electric dipole moment

We construct parity $\mathcal{P}_\perp$ and time-reversal $\mathcal{T}_\perp$ in the light-front formalism [1] by noting that these operations should act on the $k_\perp$ of a free particle alone, so that $|\mathbf{k}_\perp|$, k^+, and k^- remain unchanged. We choose $\mathcal{P}_\perp$ so that the components of a vector d^μ transform as $d^1 \to -d^1$, $d^2 \to d^2$, or $d^{R,L} \to -d^{L,R}$, and $d^\pm \to d^\pm$. With this $\mathcal{P}_\perp$ is an unitary operator, though it flips the spin as well. We find that $F_2(q^2)$ is even and $F_3(q^2)$ is odd under $\mathcal{P}_\perp$. The choice of $\mathcal{T}_\perp$ is predicated by that for $\mathcal{P}_\perp$; a momentum vector q^μ transforms as $q^{R,L} \to -q^{L,R}$ and $q^\pm \to q^\pm$ under $\mathcal{T}_\perp$, so that the position vector $x^\mu \equiv (x^+, x^-, x^L, x^R) \to (-x^+, -x^-, x^R, -x^L)$. With this we find that $\mathcal{T}_\perp$ is antiunitary, but it does not flip the spin. With the charge-conjugation operator $\mathcal{C}$ defined in the usual way, we note that all scalar fermion bilinears are invariant under $\mathcal{C}\mathcal{P}_\perp\mathcal{T}_\perp$ as needed. We find that $\mathrm{Re}(F_2)$ and $\mathrm{Im}(F_3)$ are even and $\mathrm{Re}(F_3)$ and $\mathrm{Im}(F_2)$ are odd under $\mathcal{T}_\perp$. Thus to realize a non-zero electric dipole form factor, we must include a $\mathcal{T}_\perp$- and $\mathcal{P}_\perp$-odd parameter β_a in the light-front wave function $\psi_a^{S_z}(x_i, \mathbf{k}_{\perp i}, \lambda_i)$; namely, $\psi_a^{S_z}(x_i, \mathbf{k}_{\perp i}, \lambda_i) = \phi_a^{S_z}(x_i, \mathbf{k}_{\perp i}, \lambda_i)\exp(i\lambda\beta_a)$, where $\phi_a^{S_z}(x_i, \mathbf{k}_{\perp i}, \lambda_i)$ is both $\mathcal{P}_\perp$- and $\mathcal{T}_\perp$- invariant. We assume $\mathcal{C}\mathcal{P}_\perp$ is broken at scales much larger than those of interest, so that $M^2_{\mathrm{CP}} \gg q^2$ and that any q^2-dependence in β_a can be neglected. With this we find, for a Fock component a [1],

$$[F_3(q^2)]_a = (\tan\beta_a)[F_2(q^2)]_a \quad \text{and} \quad d_a = 2\kappa_a\beta_a \quad \text{as} \quad q^2 \to 0\,, \tag{3}$$

since β_a is small. Thus the EDM and the anomalous magnetic moment of the nucleon should both be computed with a given method, to test for consistency. If the method employed is unable to confront the empirical anomalous magnetic moments successfully, it cannot be trusted to predict the electric dipole moments reliably. We note in the case of the QCD sum rule method, that the computed anomalous magnetic moments, as long known, are in good agreement with experiment, namely, $\kappa^n_{\mathrm{th}} = -2$ and $\kappa^p_{\mathrm{th}} = +2$ [12].

4. Implications

We now consider some specific consequences of Eq. (3). In what follows we consider the EDM induced through a QCD $\bar{\theta}$-term only. In a quark–scalar-diquark, $q(qq)_0$, model of the nucleon a single $\mathcal{P}_\perp$- and $\mathcal{T}_\perp$-violating parameter β^N suffices to characterize d^N. Since δL_{CP} is isoscalar, $\beta^n = \beta^p$, and we can employ the empirical anomalous magnetic moments $\kappa^n = -1.91$ and $\kappa^p = 1.79$, in units of μ_N, to estimate $(d^n + d^p)/(d^p - d^n) = (\kappa^n + \kappa^p)/(\kappa^p - \kappa^n) \approx -0.12/3.70 \approx -0.03$. The isoscalar electric dipole moment of the nucleon is extremely small. This is in accord with the chiral Lagrangian estimate, for which it is zero — the relevant diagrams are mediated by a $\pi - N$ loop, which is logarithmically enhanced as $M_\pi \to 0$ [8].

Our estimate can be compared to the QCD sum rule calculation, for which $(d^n + d^p)/(d^p - d^n) \approx -0.3$ [9], which is much larger. We note that the QCD sum rule method also predicts a zero isoscalar magnetic moment [12]; the method is less successful in reproducing a quantity which suffers partial cancellation. The ^2H magnetic moment is determined not only by the sum of $d^n + d^p$ but also by a CP-violating meson-exchange current — the former is estimated to be numerically larger [9]. The efficacy of a EDM measurement in a particular system in bounding $\bar{\theta}$ is determined by the size of the coefficient multiplying $\bar{\theta}$. The larger the coefficient, the better the bound on $\bar{\theta}$, for a given experimental limit. Were $d^n + d^p$ smaller, the bound on $\bar{\theta}$ from a putative ^2H EDM measurement would weaken.

5. Summary

In summary, we have analyzed the electromagnetic form factors in the light-front formalism of QCD, extending the earlier Drell-Yan-West-Brodsky framework to the analysis of $\mathcal{P}_\perp$ and $\mathcal{T}_\perp$-odd observables. We have used the light-front formalism to find a general equality between the anomalous magnetic and electric dipole moments. The relation holds for spin-1/2 systems, in general: it is not specific to the neutron and is independent of the mechanism of CP violation. An earlier study noting the importance of the simultaneous study of the muon's electric dipole and anomalous magnetic moments is given in Ref. [13]. The relation we derive implies that both the EDM and anomalous magnetic moment of the spin-1/2 system of interest should be calculated in a given model, to test for consistency. Ultimately, this can lead to sharpened constraints on models containing non-CKM sources of CP violation.

Acknowledgments

S.G. thanks S.J. Brodsky and D.-S. Hwang for a most enjoyable collaboration and acknowledges the support of the U.S. Department of Energy under contract no. DE-FG02-96ER40989.

References

1. S. J. Brodsky, S. Gardner, and D. S. Hwang, *Phys. Rev.* **D73**, 036007 (2006).
2. M. B. Gavela, A. Le Yaouanc, L. Oliver, O. Pene, J. C. Raynal, and T. N. Pham, *Phys. Lett.* **B109**, 215 (1982).
3. I. B. Khriplovich and A. R. Zhitnitsky, *Phys. Lett.* **B109**, 490 (1982).
4. I. B. Khriplovich, *Phys. Lett.* **B173**, 193 (1986).

5. A. Czarnecki and B. Krause, *Phys. Rev. Lett.* **78**, 4339 (1997).
6. M. Pospelov and A. Ritz, *Phys. Rev. Lett.* **83**, 2526 (1999). Note arXiv:hep-ph/9904483v3.
7. M. Pospelov and A. Ritz, *Annals Phys.* **318**, 119 (2005) and references therein.
8. R. J. Crewther, P. Di Vecchia, G. Veneziano and E. Witten, *Phys. Lett.* **B88**, 123 (1979) [*Erratum-ibid.* **B91**, 487 (1980)].
9. O. Lebedev, K. A. Olive, M. Pospelov, and A. Ritz, *Phys. Rev.* **D70**, 016003 (2004).
10. S. D. Drell and T. M. Yan, *Phys. Rev. Lett.* **24**, 181 (1970); G. B. West, *Phys. Rev. Lett.* **24**, 1206 (1970); S. J. Brodsky and S. D. Drell, *Phys. Rev.* **D22**, 2236 (1980).
11. I. I. Y. Bigi and N. G. Uraltsev, *Nucl. Phys.* **B353**, 321 (1991).
12. B. L. Ioffe and A. V. Smilga, *Nucl. Phys.* **B232**, 109 (1984).
13. J. L. Feng, K. T. Matchev, and Y. Shadmi, *Nucl. Phys.* **B613**, 366 (2001).

MASSES AND BOOST-INVARIANT WAVE FUNCTIONS OF HEAVY QUARKONIA FROM THE LIGHT-FRONT HAMILTONIAN OF QCD

STANISŁAW D. GŁAZEK

Institute of Theoretical Physics, Warsaw University,
Hoża 69, 00-681 Warsaw, Poland
E-mail: stglazek@fuw.edu.pl
www.fuw.edu.pl/~stglazek

A new scheme for calculating masses and boost-invariant wave functions of heavy quarkonia is developed in a light-front Hamiltonian formulation of QCD. Only the simplest approximate version with one flavor of quarks and an ansatz for the mass gap for gluons is discussed. The resulting spectra look reasonably good in view of the crude approximations made in the simplest version.

Keywords: bottomonium, charmonium, constituent, quark, gluon

1. Motivation for the LF Hamiltonian approach to QCD

The method for calculating masses and wave functions of heavy quarkonia that is reported here stems from the program of a weak-coupling expansion for Hamiltonians in light-front (LF) QCD[1]. The LF form of dynamics was discovered by Dirac[2,3] and continues to excite imagination of physicists[4]. Many authors have rediscovered LF dynamics. A famous example concerns application to hard exclusive processes[5]. Review articles provide other references concerning LF formalism[6,7]. One reason of the great interest is that the field quantization on the front hyperplane leads to 7 kinematical generators of the Poincaré group, instead of only 6 kinematical generators in the standard form (3 momentum and 3 angular momentum operators). Another reason is that the vacuum problem in the LF formulation of quantum field theory appears intriguingly different from the standard version. The same two reasons propelled also the development of the renormalization group procedure for effective particles (RGPEP) that is the basis of the method discussed here[8]. But there are two more reasons.

The first is that the LF Fock space of free bare particles can be introduced before one constructs the concept of a quantum field operator and builds Hamiltonian interaction terms for the bare quanta using such field operators[9]. This is useful when one attempts to mathematically define a theory of quarks and gluons that never appear as incoming or outgoing particles in scattering experiments but exist inside hadrons. Proceeding in this order, one can regulate the interaction terms in the LF Hamiltonian in a boost-invariant way. The regularization is accomplished using the relative transverse momenta and fractions of total "plus" momentum that the bare particles in interaction are carrying[8]. The transverse and "plus" momenta are defined with respect to the direction of the front hyperplane, the latter conventionally defined by the condition $x^+ = x^0 + x^3 = 0$ in a frame of reference in which the front is moving along z-axis (x^3), extending in the transverse directions of coordinates $x^\perp$. The transverse momenta of the particles are denoted by $k^\perp$ and their "plus" momenta by $k^+ = k^0 + k^3$. The regulated interaction Hamiltonian for bare particles is invariant with respect to boosts along the z-axis and two additional boost-like transformations that can change transverse momenta to arbitrary values. It is also invariant with respect to two translations in the $\perp$ directions, translation in the x^- direction, and rotations around the z-axis (typically directed along the beam, a dominant momentum transfer, or a suitable combination thereof depending on a scattering experiment, but in a complete theory the choice should not matter). Thus, the Hamiltonian has the same structure in a large class of frames of reference (7 dimensional). Consequently, one does not need to construct Hamiltonian counterterms that restore boost symmetry when one tries to quantitatively explain the mechanism by which masses, spins, and other quantum numbers of hadrons are formed. Most attractively, the basic Hamiltonian has the same structure in the rest frame of a hadron, where the constituent picture works[10], and in the infinite momentum frame, where the parton model works[11]. The LF Hamiltonian approach raises hopes for conceptual and quantitative explanation of the constituent and parton models in a single and complete formulation of QCD.

The second reason is that one can take advantage of the concept of potentials acting at a distance between relativistic quarks and gluons, a feast not conceivable in the standard approach that is defined using objects distributed on a space-like hyperplane in space-time. Every interaction between two objects located at different points of such a hyperplane corresponds to a dynamical effect spreading faster than light and has to eventually cancel out in observables (this happens in perturbative QED but it is not clear how it may happen in non-perturbative QCD). Therefore, it is common in the standard approach to consider only action and use local Lagrangian densities for field variables in a path-integral formula for transition amplitudes. Geometrical ideas such as strings and other nonlocal objects in multidimensional spaces are then used to regulate and explain the interaction terms in the Lagrangians. An additional argument for the Lagrangian approach is that it can incorporate variation of the metric in space-time and, hopefully, illuminate the problem of connection between particle dynamics and gravity[13]. But if one leaves gravity aside as too weak to be of an immediate concern at the scale of hadronic binding mechanism, it is useful to observe that the LF Hamiltonian at $x^+ = 0$ can contain potential terms that act between particles separated by arbitrarily large distances and such interactions can obey the rule that dynamical effects do not spread faster than light. Namely, when the bare point-like particles have the same transverse positions, the four-dimensional space-time interval between them is zero no matter how large is their separation in the direction of x^-. In fact, the LF counterpart of the Coulomb interaction between two particles 1 and 2 on the LF is proportional to $|x_1^- - x_2^-|$ when $x_1^\perp = x_2^\perp$, and otherwise vanishes. Precisely this type of interaction leads to a model of confinement in a 1+1 dimensional theory[14]. It is clear that the LF Hamiltonians are very singular when transverse distances between charged point-like particles tend to zero, and the singular terms can involve entire functions of the x^- distances between the particles.

Both reasons described above indicate that one needs a powerful ultraviolet renormalization technique for Hamiltonians in order to develop LF QCD (note that the Wilsonian concept of universality could help in identifying effective Hamiltonians irrespectively of many details in setting up the initial bare theory). A new technique has been invented[15,16] and adopted in a general scheme of weak coupling expansion in LF QCD[1]. More recently, the Hamiltonian approach has been redesigned in the form of RGPEP[8]. The results reported below are obtained using RGPEP and an ansatz for a mass-gap for gluons.

2. Binding above threshold in heavy quarkonia

Since there is not enough room here to thoroughly explain the RGPEP in application to heavy quarkonia[17] in comparison to other approaches, only the main steps are indicated. The central puzzle is how a systematic treatment of QCD can produce binding of quarks above threshold. QED describes quantum binding only below the mass threshold. So, how can binding above the threshold emerge in a relativistic quantum theory?

One begins from the Lagrangian for QCD with one flavor of quarks

$$\mathcal{L} = \bar{\psi}(i\not{D} - m)\psi - \frac{1}{4}F^{\mu\nu a}F^a_{\mu\nu} \,. \tag{1}$$

A canonical LF procedure in gauge $A^+ = 0$ produces a Hamiltonian with many terms (constraint equations are solved explicitly)[5]

$$\begin{aligned} H_{can} &= H_{\psi^2} + H_{A^2} + H_{\psi A\psi} + H_{(\psi\psi)^2} \\ &\quad + H_{A^3} + H_{A^4} + H_{\psi AA\psi} + H_{[\partial AA]^2} + H_{[\partial AA](\psi\psi)} \,. \end{aligned} \tag{2}$$

Each of these terms is an integral of the corresponding Hamiltonian density over the LF hyperplane with $x^+ = 0$, $H = \int dx^- d^2x^\perp \mathcal{H}$. For example,

$$\mathcal{H}_{\psi^2} = \frac{1}{2}\bar{\psi}\gamma^+\frac{-\partial^{\perp\,2} + m^2}{i\partial^+}\psi\,, \qquad \mathcal{H}_{A^2} = -\frac{1}{2}A^\perp(\partial^\perp)^2 A^\perp\,, \tag{3}$$

$$\mathcal{H}_{\psi A\psi} = g\,\bar{\psi}\not{A}\psi\,, \qquad \mathcal{H}_{(\psi\psi)^2} = \frac{1}{2}g^2\,\bar{\psi}\gamma^+t^a\psi\frac{1}{(i\partial^+)^2}\bar{\psi}\gamma^+t^a\psi\,,\ etc. \tag{4}$$

The fields at $x^+ = 0$ are expanded into creation and annihilation operators for bare quarks and gluons, the measure is $[k] = dk^+d^2k^\perp/(16\pi^3k^+)$:

$$\psi = \sum_{\sigma c}\int [k]\left[\chi_c u_{k\sigma} b_{k\sigma c} e^{-ikx} + \chi_c v_{k\sigma} d^\dagger_{k\sigma c} e^{ikx}\right]\,, \tag{5}$$

$$A^\mu = \sum_{\sigma c}\int [k]\left[t^c \varepsilon^\mu_{k\sigma} a_{k\sigma c} e^{-ikx} + t^c \varepsilon^{\mu *}_{k\sigma} a^\dagger_{k\sigma c} e^{ikx}\right]\,, \tag{6}$$

c stands for color, σ for spin. The bilinear terms in $\mathcal{H}$ provide kinetic energies for the bare particles and the terms with more than two fields provide interactions that are regulated[17] as indicated in Section 1. The regularization implies appearance of counterterms, H_{CT}, that restore the dynamics that was cut off by the regularization. The full regulated Hamiltonian, $H = [H_{can} + H_{CT}]_{reg}$ provides the initial condition for RGPEP (RGPEP is also used to determine H_{CT})[8]. The main step is to replace the canonical operators b, d, and a, or their hermitean conjugates in Eqs. (5) and (6), commonly denoted by q_{can}, by unitarily equivalent operators that create or annihilate effective particles corresponding to the renormalization group parameter λ, $q_\lambda = U_\lambda\, q_{can}\, U_\lambda^\dagger$, so that $q_\infty = q_{can}$ and $dH_\lambda/d\lambda = [\mathcal{T}_\lambda, H_\lambda]$, where $\mathcal{T}_\lambda = dU_\lambda/d\lambda\, U_\lambda^\dagger$. Given the initial condition $H_\infty = [H_{can} + H_{CT}]_{reg}$, one can systematically evaluate the Hamiltonian $H_\lambda = H_\infty + \int_\infty^\lambda ds[\mathcal{T}_s, H_s]$ in perturbation theory. H_λ is equal to H but it is expressed in terms of operators creating and annihilating effective particles of size $1/\lambda$ with respect to strong interactions. Since H_λ is expressed in terms of the creation and annihilation operators for effective quarks and gluons instead of the bare canonical ones, it contains different interaction terms, including new effective potentials. For λ on the order of hadronic masses, the effective particles are expected to correspond to the constituent quarks and gluons that are used to describe hadrons in particle tables[10]. The perturbative procedure for evaluating H_λ is safe from genuine infrared singularities because the RGPEP generator $\mathcal{T}_\lambda$ is designed to exclude the possibility that energy denominators in perturbation theory are significantly smaller than λ.

A quarkonium eigenvalue problem for the QCD Hamiltonian H_λ, $(P^+ H_\lambda - P^{\perp\,2})|P\rangle = M^2|P\rangle$, is solved by first eliminating the eigenvalues P^+ and $P^\perp$ of three kinematical momentum operators P^+_λ and $P^\perp_\lambda$ (these operators are also provided by RGPEP[18]) and obtaining an eigenvalue equation for the quarkonium mass M (the center-of-mass motion is eliminated from the eigenvalue problem exactly). Still, the eigenstate $|P\rangle$ is built from the virtual effective particles in the LF Fock space and carries four-momentum P with $P^- = (M^2 + P^{\perp\,2})/P^+$. In terms of the effective quark-antiquark, quark-antiquark-gluon, and other components:

$$|P\rangle = |Q_\lambda \bar{Q}_\lambda\rangle + |Q_\lambda \bar{Q}_\lambda g_\lambda\rangle + \dots\,. \tag{7}$$

This expansion may converge, in distinction from the expansion of the same state into canonical bare-particle sectors, because interactions in H_λ are limited to momentum transfers smaller than λ by the form factors f_λ that appear in all interaction vertices in H_λ. The form factors are introduced through the generator $\mathcal{T}_\lambda$ of RGPEP. In the effective-particle basis, the Hamiltonian H_λ takes a matrix form

$$[H_\lambda] = \begin{bmatrix} \cdot & \cdot & \cdot \\ \cdot & T_3 + V_3 & Y \\ \cdot & Y^\dagger & T_2 + V_2 \end{bmatrix} \to H_{2+3} = \begin{bmatrix} T_3 + \mu^2_{ansatz} & Y \\ Y^\dagger & T_2 + V_2 \end{bmatrix}, \tag{8}$$

in which dots denote couplings with sectors with more than 3 effective constituents, T refers to kinetic energy terms, V to potentials, Y to emission of effective gluons by quarks, and 2 and 3 to the Fock components with 2 and 3 effective particles. The arrow indicates a truncation of the system to sectors $|Q_\lambda \bar{Q}_\lambda\rangle$ and $|Q_\lambda \bar{Q}_\lambda g_\lambda\rangle$ only, which is done at the price of introducing an ansatz for the gluon mass gap,

$$\mu^2_{ansatz} = \left(1 - \frac{\alpha^2}{\alpha_s^2}\right)\mu^2\,. \tag{9}$$

The ansatz is so designed that when the coupling constant α (this is the effective coupling at some small scale λ) is extrapolated to a realistically strong value α_s, the ansatz will be removed and the true QCD interactions can be recovered order-by-order in the weak coupling expansion in α. The gap function μ^2 is inserted in order to model the effect of all the non-abelian Coulomb potentials, V_3, that act in sector $|Q_\lambda \bar{Q}_\lambda g_\lambda\rangle$ and the interactions that produce couplings to additional sectors with more constituents (the dots). It is very unlikely that the first approximation in QCD should be $\mu^2 = 0$. But if $\mu^2 \neq 0$, the resulting dynamics in the $|Q_\lambda \bar{Q}_\lambda\rangle$ sector is described in the leading order in α by the eigenvalue equation $H_{Q\bar{Q}\lambda}|P\rangle = M^2|P\rangle$, where the effective quark-antiquark Hamiltonian has the form (qualitatively)

$$H_{Q\bar{Q}\lambda} = T_2 + V_2 + Y^\dagger \frac{1}{T_3 + \mu^2} Y \,. \tag{10}$$

The main point is that the gluon emission and absorption produces diverging (for small q_z) terms of the form $f_\lambda \frac{4m^2}{q_z^2} \frac{\mu^2}{q^2+\mu^2} f_\lambda$, in which the momentum transfer $\vec{q}$ approaches zero. This happens also in the quark self-interaction terms. The net effect is positive, lifting the quark energy above threshold. In addition, the factor dependent on μ^2 becomes 1 for small $\vec{q}$ irrespectively of the details of the ansatz for μ^2. The final result[17] is a harmonic oscillator potential that appears as a leading correction to the color Coulomb interaction at typical distances between the quarks (the Coulomb term appears with the Breit-Fermi spin factors). Technically, it is the harmonic oscillator term that leads to the binding above threshold, $M > 2m$, where m is the mass ascribed to the quarks. Such effect is absent in positronium in QED because there are no Coulomb-like interactions between photons and electrons and no mass-gap for photons .

3. Masses and wave functions in the crudest approximation

The resulting eigenvalue equation for quarkonium wave function can be solved numerically and the mass spectrum depends on the choice of the coupling constant α and quark mass m at some value of λ. The Breit-Fermi terms include three-dimensional δ-functions that are smeared and made finite by the presence of the form factors f_λ. If one assumes $\alpha_{M_Z} \sim 0.12$, the RGPEP evolution with one flavor of quarks in the same Hamiltonian scheme[19] gives $\alpha \sim 0.326$ at $\lambda \sim 3.7$ GeV. Table 1 shows masses of $b\bar{b}$ quarkonia obtained for $\alpha = 0.326$ and $m = 4857$ MeV at $\lambda = 3699$ MeV, adjusted to fit masses of $\chi_1(1P)$ and $\chi_1(2P)$. The pattern of differences in the 4th column agrees with expectations in the new scheme. All details concerning this calculation and results for some other quarkonia can be found elsewhere[20]. The oscillator frequency corresponding to Table 1 is $\omega = 182$ MeV. The bottom line is that the realistic value of α is near 1/3 and

Table 1. Qualitative illustration of bottomonium masses.

meson name	calculation (MeV)	experiment (MeV)	difference (MeV)
$\Upsilon 10865$	10725	10865	-140
$\Upsilon 10580$	10464	10580	-116
$\Upsilon 3S$	10382	10355	27
$\chi_2 2P$	10276	10269	7
$\chi_1 2P$	10256	10256	0
$\chi_0 2P$	10226	10232	-6
$\Upsilon 2S$	10012	10023	-11
$\chi_2 1P$	9912	9912	-1
$\chi_1 1P$	9893	9893	0
$\chi_0 1P$	9865	9859	5
$\Upsilon 1S$	9551	9460	91
$\eta_b 1S$	9510	9300	210

the new Hamiltonian approach to QCD can be further studied in a weak coupling expansion in the case of heavy quarkonia, including many effects in the complex relativistic color dynamics of virtual quarks and gluons.

References

1. K. G. Wilson et al., *Phys. Rev. D* **49**, 6720 (1994).
2. P. A. M. Dirac, *Rev. Mod. Phys.* **21**, 392 (1949).
3. P. A. M. Dirac, in *Mathematical Foundations of Quantum Theory*, Ed. A. R. Marlow (Academic Press, New York, 1978).
4. K. G. Wilson, *Nucl. Phys. Proc. Suppl.* **140**, 3 (2005).
5. G. P. Lepage, S. J. Brodsky, Phys. Rev. D**22**, 2157 (1980).
6. J. Kogut, L. Susskind, Phys. Rep. C**8**, 75 (1973).
7. S. J. Brodsky, H. C. Pauli, S. S. Pinsky, Phys. Rept. **301**, 299 (1998).
8. S. D. Głazek, *Acta Phys. Polon. B* **29**, 1979 (1998).
9. S. Weinberg, *The Quantum Theory of Fields*, (Cambridge University Press, Cambridge, 1995).
10. S. Eidelman et al., *Phys. Lett. B* **592**, 1 (2004).
11. R. P. Feynman, *Photon-Hadron Interactions* (Benjamin, New York, 1972).
12. K. G. Wilson, Phys. Rev. D**10**, 2445 (1974).
13. M. B. Green, J. H. Schwarz, E. Witten, *Superstring Theory*, (Cambridge University Press, Cambridge, 1987).
14. G. 't Hooft, Nucl. Phys. B**75**, 461 (1974).
15. S. D. Głazek and K. G. Wilson, *Phys. Rev. D* **48**, 5863 (1993).
16. S. D. Głazek and K. G. Wilson, *Phys. Rev. D* **49**, 4214 (1994).
17. S. D. Głazek, Phys. Rev. D**69**, 065002 (2004).
18. S. D. Głazek, T. Masłowski, *Phys. Rev. D* **65**, 065011 (2002).
19. S. D. Głazek, *Phys. Rev. D* **63**, 116006 (2001).
20. S. D. Głazek and J. Młynik, IFT/08/06, hep-th/0606235.

LATTICE FORMULATION OF QCD "NEAR THE LIGHT CONE"

D. GRÜNEWALD*

Institut für Theoretische Physik, Universität Heidelberg,
D-69120 Heidelberg, Germany
E-mail: d.gruenewald@tphys.uni-heidelberg.de
www.thphys.uni-heidelberg.de

Non-perturbative physics on the light cone is investigated in a Hamiltonian lattice framework. We use near light cone coordinates and perform a limiting procedure onto the light cone. Such a formulation is natural in order to describe high energy scattering. It contains an additional parameter η which represents the distance to the light cone and is varying the energy. The QCD vacuum is planned to be generated by a quantum diffusion Monte Carlo algorithm. In order to minimize the algorithmic variance, a guidance wave functional close to the exact ground state is required. We present a solution for the ground state corresponding to the dominant part of the Hamiltonian in the light cone limit.

Keywords: Light cone; quantum chromodynamics; lattice.

1. Introduction

Already in 1949, Dirac[1] has shown that there are only three basically different parameterizations of space time which are not related by a Lorentz transformation, namely the instant form, the point form and the front form which is parameterized by light cone (LC) coordinates. These parameterizations differ by the hypersphere on which classical and quantum fields are initialized. Since physical observables must not depend on the parameterization of space time one chooses, it is a pure matter of convenience and often a matter of calculability which framework is used. For example, the LC coordinates are natural to describe high energy scattering. The meson-meson scattering amplitude is in principle governed by the correlation function of two Wegner-Wilson loops lying on the LC[2]. Furthermore, a model independent, non-perturbative, quantum field theoretical definition

*Supported by the EU project EU RII3-CT-2004-506078

of the jet-quenching parameter $\widehat{q}$ was given recently[3], which relates the logarithm of the thermal expectation value of a light-like Wegner-Wilson loop to $\widehat{q}$. The jet-quenching parameter is the essential parameter used in parton energy loss models which successfully describe the large suppression of hadron spectra, its dependence on the centrality and orientation to the reaction plane in heavy ion collisions. In LC coordinates, the Wilson Loops along the light cone are lying along the coordinates axis which is easy to implement. This makes the LC coordinates physically very appealing.

Of course, the vacuum expectation values of the Wilson loops involve non-perturbative physics. The only tool in order to obtain fully non-perturbative results from a first principle calculation is lattice gauge theory. Therefore, the method of discretized light cone quantization combined with a transversal lattice has been pursued[4,5], because the standard lattice approach[6] fails for the front form. Both, the Euclidean path integral formulation as well as the Hamiltonian formulation of lattice gauge theory in the front form yield severe numerical problems. The standard tool of lattice gauge theory, the Monte Carlo sampling of the Euclidean path integral, does not apply for the LC framework, because the Euclidean LC action has complex parts. Therefore the integrand of the path integral is not positive definite and it can not be interpreted as a probability density. Similar problems arise for QCD at finite baryonic density which is often referred to as the sign problem. So far, there has not been found a convenient solution. Concerning LC coordinates, the problem originates from the fact that the chromoelectric field does not only appear quadratically in the action but also linearily due to off-diagonal terms in the metric. In the Hamiltonian formulation one is faced with the problem, that the LC Hamiltonian is highly non-local due to the resolution of a constraint equation. This non-locality yields a bad numerical convergence.

We propose to circumvent the problem of the highly non-local Hamiltonian by using near light cone (NLC) coordinates[7,8]. Using the NLC Hamiltonian, the basic idea is to produce the ground state wave functional of NLC lattice gauge theory with a quantum diffusion Monte Carlo algorithm for fixed distance to the LC. This wave function can be used to measure expectation values of various operators. The corresponding expectation values on the LC are obtained via a limiting procedure.

2. Near light cone coordinates

The definition of NLC coordinates is closely related to the definition of LC coordinates. The temporal coordinate x^+ and the longitudinal spatial coordinate x^- are given by

$$x^+ = \frac{1}{\sqrt{2}}\left[\left(1+\frac{\eta^2}{2}\right)x^0 + \left(1-\frac{\eta^2}{2}\right)x^3\right]$$
$$x^- = \frac{1}{\sqrt{2}}\left[x^0 - x^3\right] \quad (1)$$

and the transversal coordinates x^k are unchanged. The parameter η controls the distance to the LC. In the limit $\eta \to 0$ the definition of LC coordinates is recovered and for $\eta = 1$ the temporal coordinate is proportional to x^0, the ordinary temporal coordinate. Therefore, by varying η in the interval $[0, 1]$ the NLC coordinates allow for a smooth interpolation between equal time theories and the front form. The transition from Minkowski to NLC coordinates corresponds to a boost in longitudinal direction with boost factor β, where β is given by

$$\beta = \frac{1-\eta^2/2}{1+\eta^2/2} \quad (2)$$

plus an additional linear transformation which is not element of the Lorentz group. For NLC coordinates, the pure gluonic action for $SU(2)$ is given by

$$S = \sum_a \int d^4x \left[\frac{1}{2}F^a_{+-}(x)^2 + \sum_k \left(F^a_{+k}(x)F^a_{-k}(x) + \frac{\eta^2}{2}F^a_{+k}(x)^2\right)\right.$$
$$\left. - \frac{1}{2}F^a_{12}(x)^2\right], \quad (3)$$

where $F^a_{\mu\nu}(x)$ are the components of the standard non-abelian field strength tensor

$$F^a_{\mu\nu}(x) = \partial_\mu A^a_\nu(x) - \partial_\nu A^a_\mu(x) + g f^{abc} A^b_\mu(x) A^c_\nu(x). \quad (4)$$

The analytical continuation to imaginary temporal coordinate yields a complex Euclidean action due to the $F^a_{+k}F^a_{-k}$-term which couples the transversal chromo-electric field to the transversal chromo-magnetic field. Therefore, a Monte-Carlo sampling of the Euclidean path integral fails. Nevertheless, for the NLC coordinates it is more advantageous to switch to a Hamiltonian formulation. The transition to a Hamiltonian formulation is standard. One defines the momenta conjugate to the gauge fields as the following functional derivatives of the Lagrange density

$$\Pi^a_\mu \equiv \frac{\delta \mathcal{L}}{\delta\left(\partial_+ A^a_\mu\right)} \quad (5)$$

and performs a Legendre transformation of the Lagrangean. The Hamiltonian is given by

$$\widehat{H} = \frac{1}{2}\sum_a \int d^3x \left[\widehat{\Pi}^a_-(\vec{x})^2 + \widehat{F}^a_{12}(\vec{x})^2 + \sum_{k=1}^{2} \frac{1}{\eta^2} \left(\widehat{\Pi}^a_k(\vec{x}) - \widehat{F}^a_{-k}(\vec{x}) \right)^2 \right]. \quad (6)$$

Here, we have chosen the axial gauge $A^a_+ = 0$ which is quite natural because the temporal gauge field A^a_+ is not dynamical, i.e. there is no temporal derivative appearing in the Lagrange function. It acts like a Lagrange multiplier in the Lagrangean. But, in order to cover the full dynamics, one has to impose the constraint equation represented by that Lagrange multiplier, the Gauss' law, on physical states

$$\left(\widehat{D}^{ab}_- \widehat{\Pi}^b_-(\vec{x}) + \sum_{k=1}^{2} \widehat{D}^{ab}_k \widehat{\Pi}^b_k(\vec{x}) \right) |\Psi\rangle = 0 \quad \forall\, \vec{x}, a. \quad (7)$$

Here $\widehat{D}^{ab}_i$ denotes the covariant derivative operator in i-th direction. The Gauss' law operator commutes with the Hamiltonian, i.e. the Gauss' law operator is a constant of motion. In order to find the ground state numerically, one introduces a spatial lattice, $SU(2)$-link matrices representing the gauge connections between the lattice sites and momentum operators acting on these link matrices[9,10], so that in leading order of the lattice spacing the continuum Hamiltonian is obtained in the naive continuum limit. The ground state is found by using the fact that the time evolution operator, analytically continued to imaginary times, is a projector onto the ground state, i.e.

$$|\Psi_0\rangle = \lim_{\tau \to \infty} e^{-(\widehat{H}-E_0)\tau} |\Phi_t\rangle, \quad (8)$$

where $|\Phi_t\rangle$ is an arbitrary trial state, with non-vanishing overlap to the exact ground state $|\Psi_0\rangle$ and E_0 is the ground state energy. By switching to an imaginary temporal coordinate the Schrödinger equation turns into a diffusion equation. Due to the fact that the Gauss' law operator is a constant of motion, it is sufficient to choose the trial state $|\Phi_t\rangle$ gauge invariant. Then, the ground state is in the physical sector as well because time evolution leaves the subspace of physical states invariant. On the lattice, a guided diffusion Monte-Carlo algorithm for $SU(2)$ implementing the projection on the ground state has been developed for equal time theories[11] and may be extended to the NLC framework. One uses the Trotter formula to write the time evolution operator as product of infinitesimal time evolution operators, each evolving the system by a timestep $\Delta\tau$, and inserts in between a complete set of link operator eigenstates

$$\mathbf{1} = \int \mathcal{D}x \; \Phi_{\rm g}(\widehat{x})^{-1} \, |x\rangle \, \langle x| \, \Phi_{\rm g}(\widehat{x}). \tag{9}$$

Here, x denotes an entire configuration of link matrices and in addition, the unit operator is extended with the guidance wave functional Φ_g. After evaluating the resulting matrix elements in link space, the expression is computed by a Monte-Carlo integration. The inclusion of the guidance wave functional improves the convergence of the algorithm since the Monte-Carlo is lead to regions of the configuration space where the guidance wave functional is large. The quality of the guidance wave functional controls the convergence of the algorithm. Therefore, it is very important to have guidance wave functionals which are very close to the exact ground state wave functional. In the next section we compute a possible candidate.

3. The guidance wave functional

In order to find an approximate solution for the ground state wave functional, we concentrate on the dominant part of the Hamiltonian Eq. 6 in the $\eta \to 0$ limit, namely

$$\widehat{H}_0 = \frac{1}{2}\frac{1}{\eta^2} \sum_a \sum_{k=1}^{2} \int d^3x \left(\widehat{\Pi}^a_k(\vec{x}) - \widehat{F}^a_{-k}(\vec{x}) \right)^2 . \tag{10}$$

We introduce a spatial box with boxlength L on each side and periodic boundary conditions. The gauge fields and the conjugate momenta obey the following commutation relations

$$\left[\widehat{\Pi}^a_k(\vec{x}), \widehat{A}^b_{k'}(\vec{x}') \right] = -\mathrm{i}\delta^{ab}\delta_{kk'}\delta^{(3)}(\vec{x} - \vec{x}'). \tag{11}$$

All other commutators vanish. In order to simplify the calculation, we fix the gauge to $A^a_-(\vec{x}) = A^3_-(\vec{x}_\perp)$. Therefore, $A^a_-(\vec{x})$ is gauged away up to the zero mode in longitudinal minus direction which is chosen to lie in $SU(2)$ 3-direction. In contrast to $A^a_-(\vec{x}) = 0$-gauge, this gauge is compatible with periodic boundary conditions in the box. We want to express the solution of the Schrödinger equation in terms of the field strength tensors. Therefore, we compute the commutator of the momentum operator with the field strength tensor:

$$\begin{aligned} \left[\widehat{\Pi}^a_k(\vec{x}), \widehat{F}^b_{-k'}(\vec{x}') \right] &= -\mathrm{i}\delta^{ab}\delta_{kk'}\partial^{x'}_- \delta^{(3)}(\vec{x} - \vec{x}') \\ &\quad -\mathrm{i}gf^{abc}A^c_-(\vec{x}_\perp)\delta_{kk'}\delta^{(3)}(\vec{x} - \vec{x}') \, . \end{aligned} \tag{12}$$

Due to the appearance of a derivative in minus direction inside of the field strength tensor, the commutator couples in minus direction. In order to decouple the longitudinal coordinate it is convenient to perform a discrete Fourier transformation and write the Hamiltonian in terms of the Fourier coefficients. An appropriate transformation of Bogolubov type brings the Hamiltonian into a form which may be interpreted as a sum over 1-d harmonic oscillators. Therefore, the ground state can be read off easily. Having the ground state in terms of the Fourier coefficients, it is easy to express it in terms of the field strength tensors by performing the inverse transformation. Finally, the ground state in the box expressed in terms of the field strength tensors is given by

$$\Psi_0 = \exp\left\{ \sum_k \sum_{a,b} \int d^3x d^3y \widehat{F}^a_{-k}(\vec{x}) g^{ab}(\vec{x},\vec{y}) \widehat{F}^b_{-k}(\vec{y}) \right\} \tag{13}$$

with

$$g^{ab}(\vec{x},\vec{y}) \equiv -\frac{1}{2}\left[\frac{1}{L}\sum_{n\neq 0} \widetilde{g}^{ab}(n,\vec{x}_\perp) e^{ik_n(x^- - y^-)}\right] \delta^{(2)}\left(\vec{x}_\perp - \vec{y}_\perp\right), \tag{14}$$

$$k_n \equiv \frac{2\pi}{L} n \tag{15}$$

and

$$\widetilde{g}^{ab}(n,\vec{x}_\perp) \equiv \begin{cases} \frac{\delta^{ab}|k_n| - i\,\mathrm{sign}(k_n) g f^{ab3} A^3_-(\vec{x}_\perp)}{|k_n|^2 - g^2 A^3_-(\vec{x}_\perp)^2} & \text{if } a,b \in \{1,2\} \\ \frac{1}{|k_n|} & \text{if } a,b = 3 \end{cases} . \tag{16}$$

For ease of numerical implementability, we turn to $A_-(\vec{x}) = 0$-gauge. Of course, this choice of gauge violates the periodic boundary conditions. Nevertheless, it is a convenient guidance wave functional. Then, $g^{ab}(\vec{x},\vec{x}')$ simplifies to

$$g^{ab}(\vec{x},\vec{y}) = \frac{\delta^{ab}}{4\pi} \log\left\{2\left[1-\cos\left(2\pi\frac{x^- - y^-}{L}\right)\right]\right\} \delta^{(2)}\left(\vec{x}_\perp - \vec{y}_\perp\right). \tag{17}$$

In order to put it on the computer, one has to discretize the integrals and express the field strength tensors in terms of $SU(2)$-plaquette variables. Gauge invariance is achieved by multiplying with appropiate strings of gauge links connecting in minus direction.

4. Conclusions

We propose to investigate the non-perturbative features of QCD in the front form by a lattice simulation of the NLC Hamiltonian with fixed distance η to the LC. The interesting observables like Wilson loop correlation functions or Wilson loop expectation values on the LC are obtained by a limiting procedure afterwards. In contrast to LC coordinates, quantization with NLC coordinates does not yield further constraint equations beside the gauss law. The gauss law is used as an additional initial condition due to the fact that it is a constant of motion. Therefore, in contrast to the LC Hamiltonian, the NLC Hamiltonian remains local and a quantum diffusion Monte-Carlo algorithm is a promising tool in order to generate the QCD ground state.

In order to reduce the algorithmic variances it is advantageous to introduce a guidance wave functional which has large overlap with the exact ground state, i.e. which is an approximation to the vacuum wave functional. We have analytically computed the ground state of $\widehat{H}_0$, the leading term in the light cone ($\eta \to 0$) limit, for a box with periodic boundary conditions and for the gauge in which the longitudinal field A_- is given by its zero mode in longitudinal direction.

This is a promising starting point for the investigation of non-perturbative physics in the front form in the framework of lattice gauge theory.

References

1. P. A. M. Dirac, "Forms Of Relativistic Dynamics," Rev. Mod. Phys. **21** (1949) 392.
2. E. R. Berger and O. Nachtmann, Eur. Phys. J. C **7** (1999) 459 [arXiv:hep-ph/9808320].
3. H. Liu, K. Rajagopal and U. A. Wiedemann, arXiv:hep-ph/0605178.
4. M. Burkardt, Phys. Rev. D **49** (1994) 5446 [arXiv:hep-th/9312006].
5. S. Dalley and B. van de Sande, Phys. Rev. D **59** (1999) 065008 [arXiv:hep-th/9806231].
6. K. G. Wilson, Phys. Rev. D **10** (1974) 2445.
7. E. V. Prokhvatilov and V. A. Franke, Sov. J. Nucl. Phys. **49** (1989) 688 [Yad. Fiz. **49** (1989) 1109].
8. F. Lenz, M. Thies, K. Yazaki and S. Levit, Annals Phys. **208** (1991) 1.
9. J. B. Kogut and L. Susskind, Phys. Rev. D **11** (1975) 395.
10. M. Creutz, Phys. Rev. D **15** (1977) 1128.
11. S. A. Chin, O. S. Van Roosmalen, E. A. Umland and S. E. Koonin, Phys. Rev. D **31** (1985) 3201.

NEW PERSPECTIVES FOR QCD FROM AdS/CFT

S. J. BRODSKY

Stanford Linear Accelerator Center
Stanford University, Stanford, California 94309
E-mail: sjbth@slac.stanford.edu

The AdS/CFT correspondence between conformal field theory and string states in an extended space-time has provided new insights into not only hadron spectra, but also their light-front wavefunctions. We show that there is an exact correspondence between the fifth-dimensional coordinate of anti-de Sitter space z and a specific impact variable ζ which measures the separation of the constituents within the hadron in ordinary space-time. This connection allows one to predict the form of the light-front wavefunctions of mesons and baryons, the fundamental entities which encode hadron properties and scattering amplitudes. A new relativistic Schrödinger light-cone equation is found which reproduces the results obtained using the fifth-dimensional theory.

1. The conformal approximation to QCD

One of the most interesting recent developments in hadron physics has been the use of Anti-de Sitter space holographic methods in order to obtain a first approximation to nonperturbative QCD. The essential principle underlying the AdS/CFT approach to conformal gauge theories is the isomorphism of the group of Poincare' and conformal transformations $SO(4,2)$ to the group of isometries of Anti-de Sitter space $SO(1,5)$. The AdS metric is

$$ds^2 = \frac{R^2}{z^2}(\eta^{\mu\nu} dx_\mu dx^\mu - dz^2)$$

which is invariant under scale changes of the coordinate in the fifth dimension $z \to \lambda z$ and $dx_\mu \to \lambda dx_\mu$. Thus one can match scale transformations of the theory in $3+1$ physical space-time to scale transformations in the fifth dimension z. The amplitude $\phi(z)$ represents the extension of the hadron into the fifth dimension. The behavior of $\phi(z) \to z^\Delta$ at $z \to 0$ must match the twist dimension of the hadron at short distances $x^2 \to 0$. As shown by Polchinski and Strassler[1], one can simulate confinement by imposing the condition $\phi(z = z_0 = \frac{1}{\Lambda_{QCD}})$. This approach, has been successful in reproducing general properties of scattering processes of QCD bound states[1,2], the low-lying hadron spectra[3,4], hadron couplings and chiral symmetry breaking[4,5], quark potentials in confining backgrounds[6] and pomeron physics[7].

It was originally believed that the AdS/CFT mathematical tool could only be applicable to strictly conformal theories such as $\mathcal{N} = 4$ supersymmetry. However, if one considers a semi-classical approximation to QCD with massless quarks and without particle creation or absorption, then the resulting β function is zero, the coupling is constant, and the approximate theory is scale and conformal invariant. Conformal symmetry is of course broken in physical QCD; nevertheless, one can use conformal symmetry as a *template*, systematically correcting for its nonzero β function as well as higher-twist effects. For example, "commensurate scale relations"[8] which relate QCD observables to each other, such as the generalized Crewther relation[9], have no renormalization scale or scheme ambiguity and retain a convergent perturbative structure which reflects the underlying conformal symmetry of the classical theory. In general, the scale is set such that one has the correct analytic behavior at the heavy particle thresholds[10].

In a confining theory where gluons have an effective mass, all vacuum polarization corrections to the gluon self-energy decouple at long wavelength. Theoretical[11] and phenomenological[12] evidence is in fact accumulating that QCD couplings based on physical observables such as τ decay[13] become constant at small virtuality; *i.e.*, effective charges develop an infrared fixed point in contradiction to the usual assumption of singular growth in the infrared. The near-constant behavior of effective couplings also suggests that QCD can be approximated as a conformal theory even at relatively small momentum transfer. The importance of using an analytic effective charge[14] such as the pinch scheme[15,16] for unifying the electroweak and strong couplings and forces is also important[17]. Thus conformal symmetry is a useful first approximant even for physical QCD.

2. Hadronic spectra in AdS/QCD

Guy de Teramond and I[18,3] have recently shown how a holographic model based on truncated AdS space can be used to obtain the hadronic spectrum of light quark $q\bar{q}$, qqq and gg bound states. Specific hadrons are identified by the correspondence of the amplitude in the fifth dimension with the twist dimension of the interpolating operator for the hadron's valence Fock state, including its orbital angular momentum excitations. An interesting aspect of our approach is to show that the mass parameter μR which appears in the string theory in the fifth dimension is quantized, and that it appears as a Casimir constant governing the orbital angular momentum of the hadronic constituents analogous to $L(L+1)$ in the radial Schrödinger equation.

As an example, the set of three-quark baryons with spin 1/2 and higher is described in AdS/CFT by the Dirac equation in the fifth dimension[18]

$$\left[z^2\,\partial_z^2 - 3z\,\partial_z + z^2\mathcal{M}^2 - \mathcal{L}_\pm^2 + 4\right]\psi_\pm(z) = 0. \tag{1}$$

The constants $\mathcal{L}_+ = L+1$, $\mathcal{L}_- = L+2$ in this equation are Casimir constants which are determined to match the twist dimension of the solutions with arbitrary relative orbital angular momentum. The solution is

$$\Psi(x,z) = Ce^{-iP\cdot x}\left[\psi(z)_+\; u_+(P) + \psi(z)_-\; u_-(P)\right], \tag{2}$$

with $\psi_+(z) = z^2 J_{1+L}(z\mathcal{M})$ and $\psi_-(z) = z^2 J_{2+L}(z\mathcal{M})$. The physical string solutions have plane waves and chiral spinors $u(P)_\pm$ along the Poincaré coordinates and hadronic invariant mass states given by $P_\mu P^\mu = \mathcal{M}^2$. A discrete four-dimensional spectrum follows when we impose the boundary condition $\psi_\pm(z = 1/\Lambda_{\rm QCD}) = 0$: $\mathcal{M}^+_{\alpha,k} = \beta_{\alpha,k}\Lambda_{\rm QCD}$, $\mathcal{M}^-_{\alpha,k} = \beta_{\alpha+1,k}\Lambda_{\rm QCD}$, with a scale-independent mass ratio[3]. Figure 1(a) shows the predicted orbital spectrum of the nucleon states and Fig. 1(b) the Δ orbital resonances. The spin 3/2 trajectories are determined from the corresponding Rarita-Schwinger equation. The data for the baryon spectra are from S. Eidelman *et al.*[19] The internal parity of states is determined from the SU(6) spin-flavor symmetry. Since only one parameter, the QCD mass scale Λ_{QCD}, is

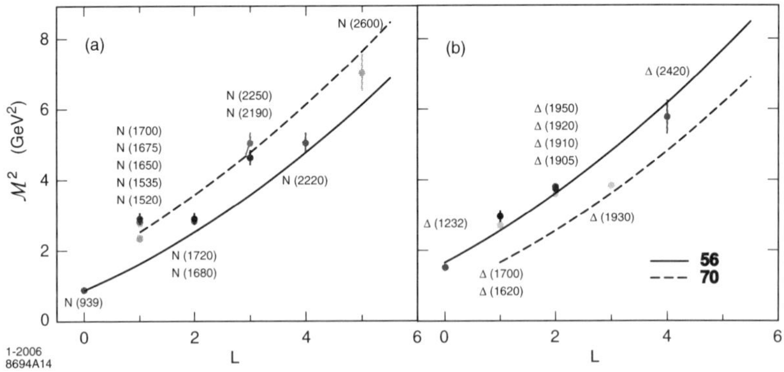

Fig. 1. Predictions for the light baryon orbital spectrum for $\Lambda_{QCD} = 0.25$ GeV. The **56** trajectory corresponds to L even $P = +$ states, and the **70** to L odd $P = -$ states.

introduced, the agreement with the pattern of physical states is remarkable. In particular, the ratio of Δ to nucleon trajectories is determined by the ratio of zeros of Bessel functions. The predicted mass spectrum in the truncated space model is linear $M \propto L$ at high orbital angular momentum, in contrast to the quadratic dependence $M^2 \propto L$ in the usual Regge parametrization.

3. Hadron wavefunctions in AdS/QCD

One of the important tools in atomic physics is the Schrödinger wavefunction; it provides a quantum mechanical description of the position and spin coordinates of nonrelativistic bound states at a given time t. Similarly, it is an important goal in hadron and nuclear physics to determine the wavefunctions of hadrons in terms of their fundamental quark and gluon constituents. The dynamics of higher Fock states such as the $|uudq\overline{Q}\rangle$ fluctuation of the proton is nontrivial, leading to asymmetric $s(x)$ and $\overline{s}(x)$ distributions, $\overline{u}(x) \neq \overline{d}(x)$, and intrinsic heavy quarks $c\overline{c}$ and $b\overline{b}$ which have their support at high momentum[20] . Color adds an extra element of complexity: for example there are five-different color singlet combinations of six 3_C quark representations which appear in the deuteron's valence wavefunction, leading to the hidden color phenomena[21].

An important example of the utility of light-front wavefunctions in hadron physics is the computation of polarization effects such as the single-spin azimuthal asymmetries in semi-inclusive deep inelastic scattering, representing the correlation of the spin of the proton target and the virtual photon to hadron production plane: $\vec{S}_p \cdot \vec{q} \times \vec{p}_H$. Such asymmetries are time-reversal odd, but they can arise in QCD through phase differences in different spin amplitudes. In fact, final-state interactions from gluon exchange between the outgoing quarks and the target spectator system lead to single-spin asymmetries in semi-inclusive deep inelastic lepton-proton scattering which are not power-law suppressed at large photon virtuality Q^2 at fixed x_{bj}.[22] In contrast to the SSAs arising from transversity and the Collins fragmentation function, the fragmentation of the quark into hadrons is not necessary; one predicts a correlation with the production plane of the quark jet itself. Physically, the final-state interaction phase arises as the infrared-finite difference of QCD Coulomb phases for hadron wave functions with differing orbital angular momentum. The same proton matrix element which determines the spin-orbit correlation $\vec{S} \cdot \vec{L}$ also produces the anomalous magnetic moment of the proton, the Pauli form factor, and the generalized parton distribution E which is measured in deeply virtual Compton scattering. Thus the contribution of each quark current to the SSA is proportional to the contribution $\kappa_{q/p}$ of that quark to the proton target's anomalous magnetic moment $\kappa_p = \sum_q e_q \kappa_{q/p}$.[22,23] The HERMES collaboration has recently measured the SSA in pion electroproduction using transverse target polarization.[24] The Sivers and Collins effects can be separated using planar correlations; both contributions are observed to contribute, with values not in disagreement with theory expectations.[24,25]

We have recently shown that the amplitude $\Phi(z)$ describing the hadronic state in AdS_5 can be precisely mapped to the light-front wavefunctions $\psi_{n/h}$ of hadrons in physical space-time[18], thus providing a relativistic description of hadrons in QCD at the amplitude level. The light-front wavefunctions are relativistic and frame-independent generalizations of the familiar Schrödinger wavefunctions of atomic physics, but they are determined at fixed light-cone time $\tau = t + z/c$—the "front form" advocated by Dirac—rather than at fixed ordinary time t. Formally, the light-front expansion is constructed by quantizing QCD at fixed light-cone time [26] $\tau = t + z/c$ and forming the invariant light-front Hamiltonian: $H_{LF}^{QCD} = P^+P^- - \vec{P}_\perp^2$ where $P^\pm = P^0 \pm P^z$.[27] The momentum generators P^+ and $\vec{P}_\perp$ are kinematical; *i.e.*, they are independent of the interactions. The generator $P^- = i\frac{d}{d\tau}$ generates light-cone time translations, and the eigen-spectrum of the Lorentz scalar H_{LF}^{QCD} gives the mass spectrum of the color-singlet hadron states in QCD together with their respective light-front wavefunctions. For example, the proton state satisfies: $H_{LF}^{QCD} \mid \psi_p\rangle = M_p^2 \mid \psi_p\rangle$.

Our approach shows that there is an exact correspondence between the fifth-dimensional coordinate of anti-de Sitter space z and a specific impact variable ζ in the light-front formalism which measures the separation of the constituents within the hadron in ordinary space-time. We derived this correspondence by noticing that the mapping of $z \to \zeta$ analytically transforms the expression for the form factors in AdS/CFT[1] to the exact Drell-Yan-West expression in terms of light-front wavefunctions. In the case of a two-parton constituent bound state the correspondence between the string amplitude $\Phi(z)$ and the light-front wave function $\widetilde{\psi}(x, \mathbf{b})$ is expressed in closed form[18]

$$\left|\widetilde{\psi}(x,\zeta)\right|^2 = \frac{R^3}{2\pi}\, x(1-x)\, e^{3A(\zeta)}\, \frac{\left|\Phi(\zeta)\right|^2}{\zeta^4}, \tag{3}$$

where $\zeta^2 = x(1-x)\mathbf{b}_\perp^2$. Here $b_\perp$ is the impact separation and Fourier conjugate to $k_\perp$. The variable ζ, $0 \leq \zeta \leq \Lambda_{\rm QCD}^{-1}$, represents the invariant separation between point-like constituents, and it is also the holographic variable z in AdS; *i.e.*, we can identify $\zeta = z$. The prediction for the meson light-front wavefunction is shown in Fig. 2. We can also transform the equation of motion in the fifth dimension using the z to ζ mapping to obtain an effective two-particle light-front radial equation

$$\left[-\frac{d^2}{d\zeta^2} + V(\zeta)\right] \phi(\zeta) = \mathcal{M}^2 \phi(\zeta), \tag{4}$$

with an effective potential $V(\zeta) = -(1 - 4L^2)/4\zeta^2$ in the conformal limit. The solution to (4) is $\phi(z) = z^{-\frac{3}{2}}\Phi(z) = Cz^{\frac{1}{2}} J_L(z\mathcal{M})$. This equation reproduces the AdS/CFT solutions. The lowest stable state is determined by the Breitenlohner-Freedman bound[28]. The mass eigenvalues are determined by the boundary conditions at $\phi(z = 1/\Lambda_{\rm QCD}) = 0$ are given in terms of the roots of the Bessel functions: $\mathcal{M}_{L,k} = \beta_{L,k}\Lambda_{\rm QCD}$. Normalized LFWFs follow from (3)

$$\widetilde{\psi}_{L,k}(x,\zeta) = B_{L,k}\sqrt{x(1-x)} J_L\left(\zeta\beta_{L,k}\Lambda_{\rm QCD}\right) \theta\big(z \le \Lambda^{-1}_{\rm QCD}\big), \tag{5}$$

where $B_{L,k} = \pi^{-\frac{1}{2}}\Lambda_{\rm QCD}\ J_{1+L}(\beta_{L,k})$. The resulting wavefunctions (see: Fig. 2) display confinement at large inter-quark separation and conformal symmetry at short distances, reproducing dimensional counting rules for hard exclusive processes and the scaling and conformal properties of the LFWFs at high relative momenta in agreement with perturbative QCD. The hadron form factors can be predicted from overlap integrals in AdS

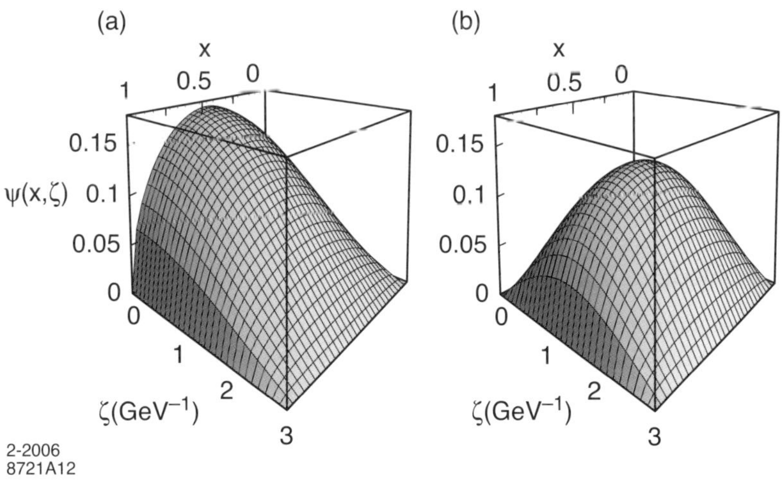

Fig. 2. AdS/QCD Predictions for the light-front wavefunctions of a meson.

space[1] or equivalently by using the Drell-Yan West formula in physical space-time. The prediction for the pion form factor is shown in Fig. 3.

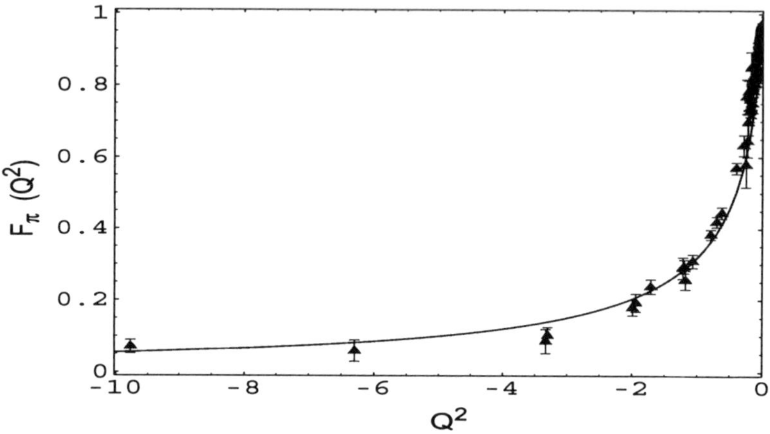

Fig. 3. AdS/QCD Predictions for the pion form factor. Here $\Lambda_{QCD} = 0.2$ GeV

Since they are complete and orthonormal, these AdS/CFT model wavefunctions can be used as an initial ansatz for a variational treatment or as a basis for the diagonalization of the light-front QCD Hamiltonian. We are now in fact investigating this possibility with J. Vary and A. Harindranath. The wavefunctions predicted by AdS/QCD have many phenomenological applications ranging from exclusive B and D decays, deeply virtual Compton scattering and exclusive reactions such as form factors, two-photon processes, and two body scattering[29]. A connection between the theories and tools used in string theory and the fundamental constituents of matter, quarks and gluons, has thus been found.

Acknowledgments

Work supported by the Department of Energy under contract number DE–AC02–76SF00515. This talk is based on work done in collaboration with Guy de Teramond.

References

1. J. Polchinski and M. J. Strassler, Phys. Rev. Lett. **88**, 031601 (2002) [arXiv:hep-th/0109174].
2. S. J. Brodsky and G. F. de Teramond, Phys. Lett. B **582**, 211 (2004) [arXiv:hep-th/0310227].

3. G. F. de Teramond and S. J. Brodsky, Phys. Rev. Lett. **94**, 201601 (2005) [arXiv:hep-th/0501022].
4. J. Erlich, E. Katz, D. T. Son and M. A. Stephanov, Phys. Rev. Lett. **95**, 261602 (2005) [arXiv:hep-ph/0501128].
5. S. Hong, S. Yoon and M. J. Strassler, arXiv:hep-ph/0501197.
6. H. Boschi-Filho, N. R. F. Braga and C. N. Ferreira, Phys. Rev. D **73**, 106006 (2006) [arXiv:hep-th/0512295].
7. H. Boschi-Filho, N. R. F. Braga and H. L. Carrion, Phys. Rev. D **73**, 047901 (2006) [arXiv:hep-th/0507063].
8. S. J. Brodsky and H. J. Lu, Phys. Rev. D **51**, 3652 (1995) [arXiv:hep-ph/9405218].
9. S. J. Brodsky, G. T. Gabadadze, A. L. Kataev and H. J. Lu, Phys. Lett. B **372**, 133 (1996) [arXiv:hep-ph/9512367].
10. S. J. Brodsky, G. P. Lepage and P. B. Mackenzie, Phys. Rev. D **28**, 228 (1983).
11. R. Alkofer, C. S. Fischer and F. J. Llanes-Estrada, Phys. Lett. B **611**, 279 (2005) [arXiv:hep-th/0412330].
12. S. J. Brodsky, S. Menke, C. Merino and J. Rathsman, Phys. Rev. D **67**, 055008 (2003) [arXiv:hep-ph/0212078].
13. S. J. Brodsky, J. R. Pelaez and N. Toumbas, Phys. Rev. D **60**, 037501 (1999) [arXiv:hep-ph/9810424].
14. S. J. Brodsky, M. S. Gill, M. Melles and J. Rathsman, Phys. Rev. D **58**, 116006 (1998) [arXiv:hep-ph/9801330].
15. M. Binger and S. J. Brodsky, arXiv:hep-ph/0602199.
16. J. M. Cornwall and J. Papavassiliou, Phys. Rev. D **40**, 3474 (1989).
17. M. Binger and S. J. Brodsky, Phys. Rev. D **69**, 095007 (2004) [arXiv:hep-ph/0310322].
18. S. J. Brodsky and G. F. de Teramond, Phys. Rev. Lett. **96**, 201601 (2006) [arXiv:hep-ph/0602252].
19. S. Eidelman *et al.* [Particle Data Group], Phys. Lett. B **592**, 1 (2004).
20. S. J. Brodsky, arXiv:hep-ph/0004211.
21. S. J. Brodsky, C. R. Ji and G. P. Lepage, Phys. Rev. Lett. **51**, 83 (1983).
22. S. J. Brodsky, D. S. Hwang and I. Schmidt, Phys. Lett. B **530**, 99 (2002) [arXiv:hep-ph/0201296].
23. M. Burkardt, Nucl. Phys. Proc. Suppl. **141**, 86 (2005) [arXiv:hep-ph/0408009].
24. A. Airapetian *et al.* [HERMES Collaboration], Phys. Rev. Lett. **94**, 012002 (2005) [arXiv:hep-ex/0408013].
25. H. Avakian and L. Elouadrhiri [CLAS Collaboration], AIP Conf. Proc. **698**, 612 (2004).
26. P. A. M. Dirac, Rev. Mod. Phys. **21**, 392 (1949).
27. S. J. Brodsky, H. C. Pauli and S. S. Pinsky, Phys. Rept. **301**, 299 (1998) [arXiv:hep-ph/9705477].
28. P. Breitenlohner and D. Z. Freedman, Annals Phys. **144**, 249 (1982).
29. For a review and additional references, see S. J. Brodsky, arXiv:hep-ph/0412101.

SUPERSYMMETRIC TWO-DIMENSIONAL QCD AT FINITE TEMPERATURE

J.R. HILLER

Department of Physics
University of Minnesota-Duluth
Duluth, MN 55812 USA
E-mail: jhiller@d.umn.edu

Light-cone coordinates and supersymmetric discrete light-cone quantization are used to analyze the thermodynamics of two-dimensional supersymmetric quantum chromodynamics with a Chern–Simons term in the large-N_c approximation. This requires estimation of the entire spectrum of the theory, which is done with a new algorithm based on Lanczos iterations. Although this work is still in progress, some preliminary results are presented.

Keywords: supersymmetry, quantum chromodynamics, finite temperature, density of states

1. Introduction

Recent work[1] has shown that thermodynamic properties can be computed for large-N_c supersymmetric theories. The approach is based on light-cone coordinates[2] and the numerical technique of supersymmetric discrete light-cone quantization (SDLCQ).[3,4] Here we consider two-dimensional supersymmetric quantum chromodynamics with a Chern–Simons term (SQCD-CS),[5] dimensionally reduced from three dimensions.

Light-cone coordinates[2] are defined by the time variable, $x^+ = (t + z)/\sqrt{2}$, and spatial components, $\underline{x} = (x^-, \vec{x}_\perp)$, where $x^- \equiv (t - z)/\sqrt{2}$ and $\vec{x}_\perp = (x, y)$. The light-cone energy and momentum are given by $p^- = (E - p_z)/\sqrt{2}$ and $\underline{p} = (p^+ \equiv (E + p_z)/\sqrt{2}, \vec{p}_\perp = (p_x, p_y))$, respectively.

For field theories quantized in terms of these coordinates, the standard numerical technique is discrete light-cone quantization (DLCQ).[6,7] Space is restricted to a light-cone box $-L < x^- < L$, $-L_\perp < x, y < L_\perp$ with periodic boundary conditions. Momentum is then discretized as $p_i^+ \rightarrow \frac{\pi}{L} n_i$, $\mathbf{p}_{i\perp} \rightarrow (\frac{\pi}{L_\perp} n_{ix}, \frac{\pi}{L_\perp} n_{iy})$ with n_i, n_{ix} and n_{iy} all integers. The limit $L \rightarrow \infty$ is exchanged for a limit in terms of the integer (harmonic) resolution[6]

$K \equiv \frac{L}{\pi} P^+$ for fixed total momentum P^+. Because the n_i are positive, the number of particles is limited to no more than K. Integrals are replaced by discrete sums.

SDLCQ[3,4] is a special form of DLCQ that preserves at least part of the supersymmetry algebra

$$\{Q^+, Q^+\} = 2\sqrt{2}P^+, \quad \{Q^-, Q^-\} = 2\sqrt{2}P^-, \quad \{Q^+, Q^-\} = -4P_\perp. \tag{1}$$

Instead of discretizing the Hamiltonian P^- directly, the supercharge Q^- is discretized, and P^- is computed from the algebra as

$$P^-_{\rm SDLCQ} = \frac{1}{2\sqrt{2}} \{Q^-, Q^-\} \neq P^-_{\rm DLCQ}. \tag{2}$$

For ordinary DLCQ, one recovers supersymmetry only in the infinite resolution limit.

After a brief summary of the SQCD-CS theory, we show how thermodynamic quantities can be constructed from the partition function. This requires knowledge of the spectrum of the theory, which we obtain numerically with an iterative Lanczos algorithm. Some preliminary results are presented and future work discussed.

2. Supersymmetric QCD

We consider a dimensional reduction from 2+1 to 1+1 dimensions of $\mathcal{N} = 1$ supersymmetric quantum chromodynamics with a Chern Simons term (SQCD-CS).[5] The action is

$$\begin{aligned} S = \int d^3x \mathrm{Tr} \Big\{ & -\frac{1}{4} F_{\mu\nu}F^{\mu\nu} + D_\mu \xi^\dagger D^\mu \xi + i\bar{\Psi} D_\mu \Gamma^\mu \Psi \\ & -g \left[\bar{\Psi}\Lambda\xi + \xi^\dagger \bar{\Lambda}\Psi \right] + \frac{i}{2} \bar{\Lambda}\Gamma^\mu D_\mu \Lambda \\ & + \frac{\kappa}{2} \epsilon^{\mu\nu\lambda} \left[A_\mu \partial_\nu A_\lambda + \frac{2i}{3} g A_\mu A_\nu A_\lambda \right] + \kappa \bar{\Lambda}\Lambda \Big\} , \end{aligned} \tag{3}$$

where the adjoint fields are the gauge boson A_μ (gluons) and a Majorana fermion Λ (gluinos) and the fundamental fields are a Dirac fermion Ψ (quarks) and a complex scalar ξ (squarks). The Chern–Simons coupling, κ, has the effect of providing a mass for the adjoint fields. The covariant derivatives are

$$\begin{aligned} D_\mu \Lambda &= \partial_\mu \Lambda + ig[A_\mu, \Lambda] , \quad D_\mu \xi = \partial_\mu \xi + ig A_\mu \xi , \\ D_\mu \Psi &= \partial_\mu \Psi + ig A_\mu \Psi . \end{aligned} \tag{4}$$

The fields transform according to

$$\delta A_\mu = \frac{i}{2}\bar{\varepsilon}\Gamma_\mu \Lambda\,, \qquad \delta\Lambda = \frac{1}{4}F_{\mu\nu}\Gamma^{\mu\nu}\varepsilon\,,$$
$$\delta\xi = \frac{i}{2}\bar{\varepsilon}\Psi\,, \qquad \delta\Psi = -\frac{1}{2}\Gamma^\mu \varepsilon D_\mu \xi. \tag{5}$$

We reduce to 1+1 dimensions by assuming the fields to be independent of the transverse coordinate x. We define fermion components by

$$\Lambda = \left(\lambda, \tilde{\lambda}\right)^T, \qquad \Psi = \left(\psi, \tilde{\psi}\right)^T, \qquad Q = \left(Q^+, Q^-\right)^T. \tag{6}$$

There are constraints, which in light-cone gauge ($A^+ = 0$) are written

$$\partial_- \tilde{\lambda} = -\frac{ig}{\sqrt{2}}\left([A^2, \lambda] + i\xi\psi^\dagger - i\psi\xi^\dagger\right), \tag{7}$$

$$\partial_- \tilde{\psi} = -\frac{ig}{\sqrt{2}}A^2\psi + \frac{g}{\sqrt{2}}\lambda\xi - \kappa\lambda/\sqrt{2}, \;\; \partial_-^2 A^- = gJ\,, \tag{8}$$

with

$$J \equiv i[A^2, \partial_- A^2] + \frac{1}{\sqrt{2}}\{\lambda, \lambda\} + \kappa\partial_- A^2 - ih\partial_-\xi\xi^\dagger + i\xi\partial_-\xi^\dagger + \sqrt{2}\psi\psi^\dagger\,. \tag{9}$$

The reduced supercharge is

$$\begin{aligned} Q^- = g\int dx^- \Big\{ & 2^{3/4}\left(\mathrm{i}[A^2, \partial_- A^2] - \kappa\partial_- A^2 + \frac{1}{\sqrt{2}}\{\lambda, \lambda\}\right)\frac{1}{\partial_-}\lambda \\ & -\frac{1}{\sqrt{2}}\left(i\sqrt{2}\xi\partial_-\xi^\dagger - i\sqrt{2}\partial_-\xi\xi^\dagger + 2\psi\psi^\dagger\right)\frac{1}{\partial_-}\lambda \\ & -2\left(\xi^\dagger A^2\psi + \psi^\dagger A^2 \xi\right)\Big\}\,. \end{aligned} \tag{10}$$

In the large-N_c approximation, there are only single-trace Fock states, the mesons

$$\bar{f}^\dagger_{i_1}(k_1) a^\dagger_{i_1 i_2}(k_2)\ldots b^\dagger_{i_n i_{n+1}}(k_{n-1})\ldots f^\dagger_{i_p}(k_n)|0\rangle \tag{11}$$

and glueballs

$$\mathrm{Tr}[a^\dagger_{i_1 i_2}(k_1)\ldots b^\dagger_{i_n i_{n+1}}(k_n)]|0\rangle, \tag{12}$$

where $\bar{f}^\dagger_i$ and $f^\dagger_i$ create fundamental partons and $a^\dagger_{ij}$ and $b^\dagger_{ij}$ create adjoint partons. Either type of state could be a boson or a fermion.

The theory possesses a useful Z_2 symmetry[8]

$$a_{ij}(k, n^\perp) \to -a_{ji}(k, n^\perp), \;\; b_{ij}(k, n^\perp) \to -b_{ji}(k, n^\perp), \tag{13}$$

which further divides the Fock space between states with even and odd numbers of gluons. We then diagonalize in each sector separately.

3. Finite temperature

From the partition function, $Z = e^{-p_0/T}$, we compute the bosonic free energy

$$\mathcal{F}_B = \frac{VT}{\pi} \sum_{n=1}^{\infty} \int_{M_n}^{\infty} dp_0 \frac{p_0}{\sqrt{p_0^2 - M_n^2}} \ln\left(1 - e^{-p_0/T}\right). \tag{14}$$

and the fermionic free energy

$$\mathcal{F}_F = -\frac{VT}{\pi} \sum_{n=1}^{\infty} \int_{M_n}^{\infty} dp_0 \frac{p_0}{\sqrt{p_0^2 - M_n^2}} \ln\left(1 + e^{-p_0/T}\right). \tag{15}$$

The total free energy, once expanded in logarithms and p_0 integrals are performed, is given by

$$\mathcal{F}(T, V) = -\frac{(K-1)\pi}{4} VT^2 - \frac{2VT}{\pi} \sum_{n=1}^{\infty} \sum_{l=0}^{\infty} M_n \frac{K_1\left((2l+1)\frac{M_n}{T}\right)}{(2l+1)}. \tag{16}$$

The sum over l is well approximated by the first few terms. We represent the sum over n as an integral over a density of states ρ: $\sum_n \rightarrow \int \rho(M) dM$. The density is approximated by a continuous function, and the integral $\int dM$ is computed by standard numerical techniques.

4. Lanczos algorithm for density of states

The discrete density of states is $\rho(M^2) = \sum_n d_n \delta(M^2 - M_n^2)$, where d_n is the degeneracy of the mass eigenvalue M_n. The density can be written in the form of a trace over the evolution operator $e^{-iP^- x^+}$:

$$\rho(M^2) = \frac{1}{4\pi P^+} \int_{-\infty}^{\infty} e^{iM^2 x^+/2P^+} \mathrm{Tr} e^{-iP^- x^+} dx^+. \tag{17}$$

We approximate the trace as an average over a random sample of vectors[9]

$$\rho(M^2) \simeq \frac{1}{S} \sum_{s=1}^{S} \rho_s(M^2), \tag{18}$$

with ρ_s a local density for a single vector $|s\rangle$, defined by

$$\rho_s(M^2) = \frac{1}{4\pi P^+} \int_{-\infty}^{\infty} e^{iM^2 x^+/2P^+} \langle s| e^{-iP^- x^+} |s\rangle dx^+. \tag{19}$$

The sample vectors $|s\rangle$ can be chosen as random phase vectors;[10] the coefficient of each Fock state in the basis is a random number of modulus one.

We approximate the matrix element $\langle s|e^{-iP^- x^+}|s\rangle$ by Lanczos iterations.[11] Let D be the length of $|s\rangle$, and define $|u_1\rangle = \frac{1}{\sqrt{D}}|s\rangle$ as the initial Lanczos vector. The matrix element $\langle u_1|e^{-iP^- x^+}|u_1\rangle$ can be approximated by the $(1,1)$ element of the exponentiation of the Lanczos tridiagonalization of P^-. Let P_s^- be the tridiagonal Lanczos matrix. It can be exponentiated by first diagonalizing it:

$$P_s^- \vec{c}_j^{\,s} = \frac{M_{sj}^2}{2P^+} \vec{c}_j^{\,s}, \tag{20}$$

such that $P_s^- = U\Lambda U^{-1}$, with $U_{ij} = (c_j^s)_i$ and $\Lambda_{ij} \equiv \delta_{ij} \frac{M_{sj}^2}{2P^+}$. The $(1,1)$ element is given by

$$\left(e^{-iP_s^- x^+}\right)_{11} = \sum_n |(c_n^s)_1|^2 e^{-iM_{sn}^2 x^+/2P^+}. \tag{21}$$

The local density can now be estimated by

$$\rho_s(M^2) \simeq \sum_n w_{sn} \delta(M^2 - M_{sn}^2), \tag{22}$$

where $w_{sn} \equiv D|(c_n^s)_1|^2$ is the weight of each Lanczos eigenvalue. Only the extreme Lanczos eigenvalues are good approximations to eigenvalues of the original P^-. The other Lanczos eigenvalues provide a smeared representation of the full spectrum.

From this density of states, we compute the cumulative distribution function (CDF), $N(M^2) = \int^{M^2} dM^2 \rho(M^2)$ as an average

$$N(M^2) \simeq \frac{1}{S} \sum_s N_s(M^2), \tag{23}$$

of local CDFs

$$N_s(M^2) \equiv \int^{M^2} dM^2 \rho_s(M^2) \simeq \sum_n w_{sn} \theta(M^2 - M_{sn}^2). \tag{24}$$

The convergence of the approximation is dependent on the number of Lanczos iterations per sample, as well as the number S of samples. Test runs indicate that taking 20 samples is sufficient. The number of Lanczos iterations needs to be on the order of 1000 per sample; using only 100 leaves errors on the order of 1-2%.

5. Preliminary results

Some preliminary results are presented in the accompanying figures. Figure 1 compares the numerical results for the CDF to analytic results, which can be obtained when the Yang-Mills coupling is zero. The numerical results are quite good, with only one noticeable deviation, at large M^2, a region where the discrete spectrum is sparse. Figure 2 shows the free energy at particular values of the Yang–Mills coupling.

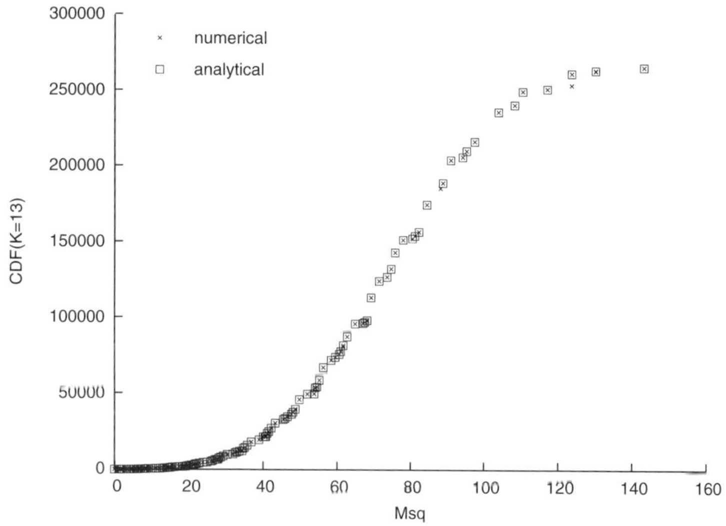

Fig. 1. Cumulative distribution function for resolution $K = 13$ in the analytically solvable case of zero Yang–Mills coupling. The numerical and analytic solutions are compared.

6. Future work

Additional work is in progress to complete this study of finite temperature properties of two-dimensional SQCD. Beyond this particular effort, one can consider finite-N_c effects, with baryons and mixing of mesons and glueballs, and the full three-dimensional theory. As these techniques mature, analysis of four-dimensional theories can be considered.

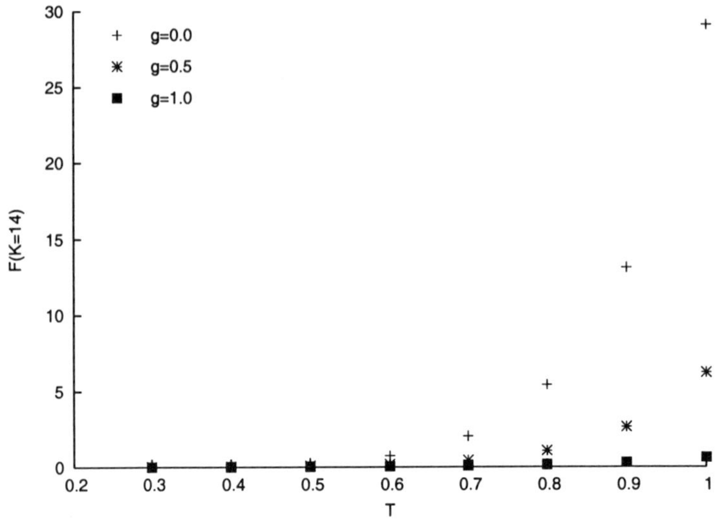

Fig. 2. Free energy at fixed Yang–Mills coupling g as a function of temperature T for resolution $K = 14$.

Acknowledgments

The work reported here was done in collaboration with S. Pinsky, Y. Proestos, N. Salwen, and U. Trittmann and supported in part by the US Department of Energy and the Minnesota Supercomputing Institute.

References

1. J. R. Hiller, Y. Proestos, S. Pinsky, and N. Salwen, *Phys. Rev. D* **70**, 065012 (2004).
2. P. A. M. Dirac, *Rev. Mod. Phys.* **21**, 392 (1949).
3. Y. Matsumura, N. Sakai, and T. Sakai, *Phys. Rev. D* **52**, 2446 (1995).
4. O. Lunin and S. S. Pinsky, AIP Conf. Proc. **494**, 140 (1999).
5. J. R. Hiller, S. S. Pinsky, and U. Trittmann, *Nucl. Phys. B* **661**, 99 (2003); *Phys. Rev. D* **67**, 115005 (2003).
6. H.-C. Pauli and S. J. Brodsky, *Phys. Rev. D* **32**, 1993 (1985); 2001 (1985).
7. S. J. Brodsky, H.-C. Pauli, and S. S. Pinsky, *Phys. Rep.* **301**, 299 (1997).
8. D. Kutasov, *Nucl. Phys. B* **414**, 33 (1994).
9. R. Alben, M. Blume, H. Krakauer, and L. Schwartz, *Phys. Rev. B* **12**, 4090 (1975); A. Hams and H. De Raedt, *Phys. Rev. E* **62**, 4365 (2000).
10. T. Iitaka and T. Ebisuzaki, *Phys. Rev. E* **69**, 057701 (2004).
11. J. Jaklič and P. Prelovšek, *Phys. Rev. B* **49**, 5065(R) (1994); M. Aichhorn, M. Daghofer, H. G. Evertz, and W. von der Linden, *Phys. Rev. B* **67**, 161103(R) (2003).

POLE APPROXIMATION FOR PION ELECTROMAGNETIC FORM FACTOR WITHIN LIGHT-FRONT DYNAMICS

J. P. B. C. DE MELO[a,b] and J. S. VEIGA[a]

[a]*Centro de Ciências Exatas e Tecnológicas, Universidade Cruzeiro do Sul, 08060-070, São Paulo, SP, Brazil*

[b]*Instituto de Física Teórica, Universidade Estadual Paulista, 01405-900, São Paulo, SP, Brazil*

T. FREDERICO

Departamento de Física, Instituto Tecnológico de Aeronáutica, 12228-900, São José dos Campos, SP, Brazil

E. PACE

Dipartimento di Fisica, Università di Roma "Tor Vergata" and Istituto Nazionale di Fisica Nucleare, Sezione Tor Vergata, Via della Ricerca Scientifica 1, I-00133, Roma, Italy

G. SALMÉ

Istituto Nazionale di Fisica Nucleare, Sezione Roma I, P.le A. Moro 2, I-00185, Roma, Italy

We carefully investigate the reliability of the propagator Pole Approximation, i.e. the approximation of retaining only the propagator poles in the evaluation of the Mandelstam covariant expression for the electromagnetic current of the pion. Different frames are analyzed, in order to find the most suitable one for calculating the pion form factor within the proposed approximation. It turns out that the approximation is more accurate in the frame where q^+ is maximal. The relevance of the Pole Approximation is briefly discussed in view of calculations of hadron form factors based on wave functions generated by dynamical models.

Keywords: light-front, quark model, electromagnetic form factor

1. Introduction

Light-front field theory starts with the paper by Dirac in 1949, when he proposed different forms to describe relativistic systems [1]. After Dirac, Light-front field theory was applied to calculated many process in relativistic nuclear physics and particle physics (see the reviews by Brodsky, Pinsky, Pauli [2] and Harindranath [3] for details). In parallel, the description of the electromagnetic properties of hadronic systems within the light-front dynamics framework, based on the introduction of a valence bound state wave function, has found some success [4–9]. In the light-front dynamics the bound state wave functions are defined on the hyper-surface $x^+ = x^0 + x^3 = 0$. These wave functions are covariant under kinematical front-form boosts, due to the stability of Fock-state decomposition under these boosts [10,11].

In recent works, (see, e.g., 12, 13), a third path for evaluating hadron electromagnetic (em) form factors has been proposed. The Mandelstam covariant expression for the matrix elements of the current has been calculated replacing the quark-hadron vertex function in the valence range by a hadron wave function obtained in a dynamical model, able to reproduce the energy spectrum. In those works the analytical structure of quark-hadron vertex function has been neglected, retaining only the poles from the constituent propagators. In this contribution we will study the reliability of such an approximation within an analytic, covariant model for the pion em form factor [14].

2. Electromagnetic pion form factor

In general, the pion em form factor is given by

$$P^\mu F_\pi(q^2) \ = \langle \pi(p')|J^\mu|\pi(p)\rangle, \tag{1}$$

where $P = p+p'$, $q = p'-p$ and J^μ is the electromagnetic current operator. It is expressed in terms of the quark fields q_f and charge e_f (f is the flavor of the quark field): $J^\mu = \sum_f e_f \bar{q}_f \gamma_\mu q_f$.

In the impulse approximation the em form factor is given by:

$$P^\mu F_\pi(q^2) = \int \frac{d^4k}{(2\pi)^4} Tr\Big[\gamma^5 S(k-p')\gamma^\mu S(k-p)\gamma^5 S(k)\Big]\Gamma(k,p')\Gamma(k,p) \tag{2}$$

where $\Gamma(k,p)$ is the pion-quark vertex function. Here, we will show calculations for a vertex function with a symmetric form, previously used in Ref. 14:

$$\Gamma(k,p) = \left[\frac{N}{(k^2 - m_R^2 + \imath\epsilon)} + \frac{N}{((p-k)^2 - m_R^2 + \imath\epsilon)}\right]. \tag{3}$$

The pion em form factor is extracted from the J^+_π current using Eq. (2):

$$F_\pi(q^2) = \frac{N_c\, m^2}{f_\pi^2(p^+ + p'^+)} \int \frac{dk^- dk^+ d^2k_\perp}{2(2\pi)^4} \frac{\Gamma(k,p')\Gamma(k,p)}{k^+(k^- - \frac{f_1 - \imath\epsilon}{k^+})} \tag{4}$$

$$Tr\left[\frac{(\not k + m)\gamma^5(\not k - \not p' + m)\gamma^+(\not k - \not p + m)\gamma^5]}{(p^+ - k^+)(p^- - k^- - \frac{f_2 - \imath\epsilon}{p^+ - k^+})(p'^+ - k^+)(p'^- - k^- - \frac{f_3 - \imath\epsilon}{(p'^+ - k^+)})}\right],$$

where, $f_1 = k_\perp^2 + m^2$, $f_2 = (p-k)_\perp^2 + m^2$ and $f_3 = (p'-k)_\perp^2 + m^2$. The k^+ momentum integration has two contributions: (i) $0 < k^+ < p^+$ and (ii) $p^+ < k^+ < p'^+$, where $p'^+ = p^+ + q^+$. The first interval (i) in the k^+ integration includes the contribution of the valence component of the wave function and the second one (ii) corresponds to the contribution of the pair term [17].

The pion valence wave function for the symmetric vertex is given by [14]:

$$\Psi(k^+, \vec{k}_\perp) = \left[\frac{N}{(1-x)(m_\pi^2 - M^2(m^2, m_R^2))} + \frac{N}{x(m_\pi^2 - M^2(m_R^2, m^2))}\right] \times \frac{p^+}{m_\pi^2 - M_0^2}.$$

where $x = k^+/p^+$ is the fraction of the momentum carried by the quark. The function M is defined as

$$M^2(m^2, m_R^2) = \frac{k_\perp^2 + m^2}{x} + \frac{(p-k)_\perp^2 + m_R^2}{(1-x)} - p_\perp^2\,. \tag{5}$$

The free square mass of the quark-antiquark system is $M_0^2 = M^2(m^2, m^2)$ and the normalization constant N is found from the condition $F_\pi(0) = 1$.

The parameters of the model are the quark mass $m_q = 0.220$ GeV, the regulator $m_R = 0.60$ GeV mass and the experimental pion mass $m_\pi = 0.140$ GeV [14]. Our choice of the regulator mass value fits the pion decay constant $f_\pi^{exp} = 92.4$ MeV.

3. Propagator pole approximation

Taking into account only the poles coming from the quark propagators in Eq. (5), the pion em form factor is calculated from: i) the residue of the pole of the spectator quark on-minus-shell (*valence* contribution), and ii) a contribution coming from the pair production mechanism (*non-valence* contribution). The pole approximation (PA) is compared with the full result in different reference frames, labeled by different values of the plus component of the momentum transfer. In order to accomplish the freedom of changing the frames, let us parametrize the momentum transfer as follows: $q^+ = -q^- = \sqrt{-q^2}\sin\alpha$, $q_x = \sqrt{-q^2}\cos\alpha$ and $q_y = 0$.

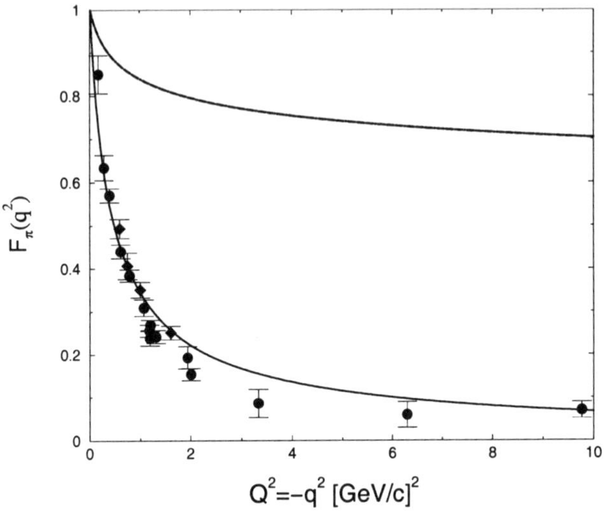

Fig. 1. Pion electromagnetic form factor vs square momentum transfer $Q^2 = -q^2$ for $\alpha = 0$. Lower solid curve: full calculation. Upper solid curve: pion Pole Approximation. Experimental data from the compilation of Baldini et al. [18]

The poles in the k^- integration contributing to the pion em current in the triangle diagram from the quark propagators, in the PA, correspond to two intervals in the k^+ integration: the first is (*i*) $0 < k^+ < p^+$ and in this case, the pole is $k_1^- = (k_\perp^2 + m^2/k^+$ (valence region) and, the second one is (ii) $p^+ < k^+ < p'^+$, and we consider the pole $k_3^- = p^- - (p'-k)_\perp^2 + m^2)/(p'^+ - k^+)$ (pair term region).

The em pion form factor within the PA is given by

$$F_\pi^{(PA)}(q^2) = F_\pi^{(I)}(q^2)|_{k_1^-} + F_\pi^{(II)}(q^2)|_{k_3^-}. \tag{6}$$

In the PA, the contributions coming from the poles of the quark-pion vertex are neglected. We have used three values of $\alpha = 0^o$ (Drell-Yan condition, $q^+ = 0$), 45° and 90°.

In particular, the em form factor in the Drell-Yan frame reduces to the valence contribution, namely

$$F_\pi^{(PA)}(q^2) = F_\pi^{(I)}(q^2)|_{k_1^-} = \frac{m^2 N_c}{p^+ f_\pi^2} \int \frac{d^2 k_\perp}{2(2\pi)^3} \int_0^1 dx \left[k_{on}^- p^{+2} + \frac{1}{4} x p^+ q^2 \right] \times \frac{\Psi_f^*(x, k_\perp) \Psi_i(x, k_\perp)}{x(1-x)^2}, \qquad (7)$$

where $k_{on}^- = k_\perp^2 + m^2/k^+$.

In Figs. 1, 2 and 3, the pion form factor calculated within the PA for $\alpha = 0, 45^o$ and 90°, respectively, is presented. Results obtained considering both the valence contribution only ($F_\pi^{(I)}(q^2)|_{k_1^-}$) and valence plus nonvalence contributions, Eq. (6), are shown. The comparison with the exact result, Eq. (5), indicates that in the Drell-Yan frame PA works very badly, while in the frame where the momentum transfer has no transverse components at all PA appears remarkably effective, in particular at high momentum transfer. In this kinematical region, the pair production mechanism dominates, that can be related to the absorption of a $q\bar{q}$ pair by a single constituent, that has a given longitudinal-momentum distribution.

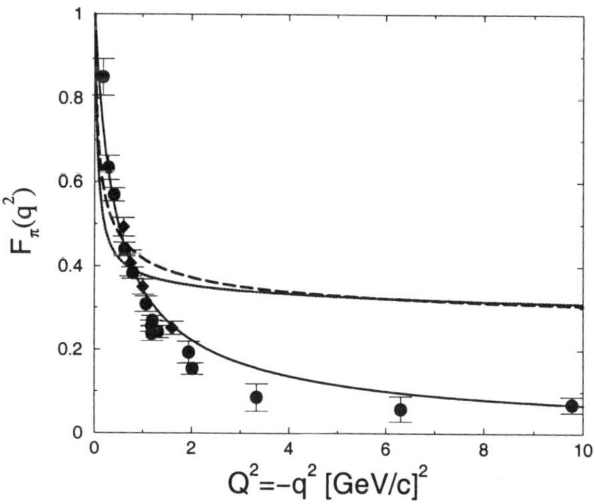

Fig. 2. Pion electromagnetic form factor vs $Q^2 = -q^2$ for $\alpha = 45^o$. Lower solid curve: full calculation. Upper solid curve: valence contribution in Pole Approximation. Dashed curve: valence + nonvalence contribution in Pole Approximation. Experimental data from the compilation of Baldini et al. [18]

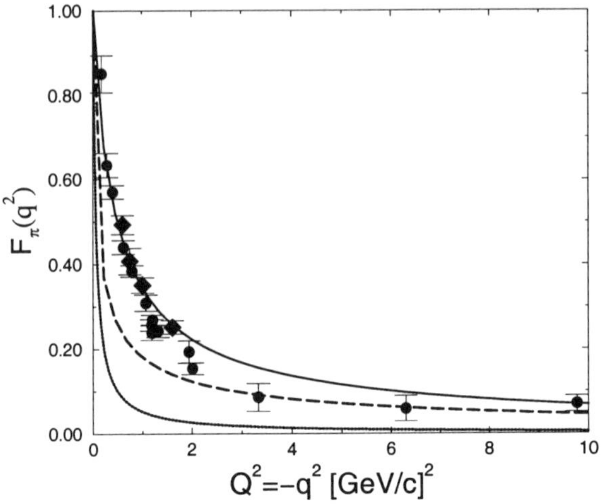

Fig. 3. Pion electromagnetic form factor vs $Q^2 = -q^2$ for $\alpha = 90^o$. Upper solid curve: full calculation. Lower solid curve: valence contribution in Pole Approximation. Dashed curve: valence + nonvalence contribution in Pole Approximation. Experimental data from the compilation of Baldini et al. [18]

4. Conclusions

The pion em form factor is calculated for q^2 up to 10 $[GeV/c]^2$ with the covariant symmetric model for the quark-pion vertex [14]. The exact calculation is compared with an approximate evaluation of the form factor in frames with different values of q^+. In the propagator Pole Approximation, investigated in this contribution, only the poles originated by the quark propagators are taken into account, while the poles of the quark-hadron vertex are disregarded in the analytical integration over the light-front energy. We found that in the frame where q^+ is maximal and the pair term dominates, the approximation is able do describe qualitatively well the exact results.

Acknowledgments

This work was partially supported by Fundação de Amparo à Pesquisa do Estado de São Paulo, Conselho Nacional de Pesquisas and by Ministero dell'Istruzione, dell'Università e della Ricerca.

References

1. P. A. M. Dirac, Rev. Mod. Phys. **21** (1949) 392.
2. S. J. Brodsky, H.-C. Pauli, and S. S. Pinsky, Phys. Rep. **301** (1998) 299.
3. An Introduction to Light-Front Dynamics for Pedestrian, A. Harindranath, hep -ph/9612244; A. Harindranath, Pramana **55** (2000) 241.
4. M. V. Terentév, Sov. J. Nucl. Phys. **24** (1976) 106; L. A. Kondratyuk and M.V. Terentév, ibid. **31** (1980) 561.
5. Z. Dziembowski and L. Mankiewicz, Phys. Rev. Lett. **58** (1987) 2175.
6. F.Cardarelli, I. L. Grach, I. M. Narodetskii, G. Salmé and E. Pace, Phys. Lett. B **357** (1995) 267.
7. J. P. B. C. de Melo and T. Frederico, Phys. Rev. C **55** (1997) 2043.
8. J. P. B. C. de Melo, H. W. L. Naus and T. Frederico, Phys. Rev. **C59** (1999) 2278.
9. J. P. B. C. de Melo, T. Frederico, H. W. L. Naus and P. U. Sauer, Nucl. Phys. A **660** (1999) 219.
10. R. J. Perry, A. Harindranath, K. G. Wilson, Phys. Rev. Lett. **65** (1990) 2959.
11. W. R. B. de Araújo, J. P. B. C. de Melo and T. Frederico, Phys. Rev. C **52** (1995) 2733.
12. J. P. B. C. de Melo, T. Frederico, G. Salmé and E. Pace, Phy. Lett. B **581** (2004) 75.
13. J. P. B. C. de Melo, T. Frederico, E. Pace and G. Salmé, Phys. Rev. D **73** (2006) 074013.
14. J. P. B. C. de Melo, T. Frederico, E. Pace and G. Salmé, Nucl. Phys. A **707** (2002) 399; ibid., Braz. J. Phys. **33** (2003) 301.
15. F. M. Lev, E. Pace and G. Salmè, Nucl. Phys. **A 641**, 229 (1998); Few-Body Syst. Suppl. 10, 135 (1998); Phys. Rev. Lett. **83**, 5250 (1999); Phys. Rev. **C 62**, 064004 (2000); Nucl. Phys. **A 663**, 365 (2000); E. Pace and G. Salmè, Nucl. Phys. **A 684**, 487 (2001); **A 689**, 411 (2001).
16. B. L. G. Bakker, H.-M. Choi and C. R. Ji, Phys. Rev. D **63** (2001) 074014.
17. J. P. B. C. de Melo, J.H.O. Sales, T. Frederico and P. U. Sauer, Nucl. Phys. A **631** (1998) 574c.
18. R. Baldini, et al., Eur. Phys. J. C **11** (1999) 709; ibid., Nucl. Phys. A **666-667** (2000) 3.

A SUM RULES CALCULATION IN THE LIGHT-CONE REPRESENTATION

GARY MCCARTOR

Department of Physics, SMU
Dallas, Texas 75275, U.S.A.
E-mail: mccartor@mail.physics.smu.edu

In the light-cone representation, vacuum effects are correctly introduced into the dynamics by the inclusion of induced operators. Recently, some of the induced operators in QCD have been derived. The knowledge of these operators makes it possible to perform calculations of the propertied of QCD resonances using essentially the QCD sum rules approach; but with the inclusion of the induced operators providing the nonperturbative physics rather than the operator product expansion that is used in the standard sum rules approach.

Keywords: light-cone; sum rules; QCD; resonances

1. A Sum Rule Calculation for the ρ Meson

In this section I shall briefly review the QCD sum rules calculation on the ρ meson as described in Ref. 1. The main point of that article is that in the resonance region there are peaks in the cross section — the resonances; but perturbative QCD gives no peaks; so nonperturbative effects must correct that deficiency.

The author of Ref. 1 examines the problem of the ρ resonance by making a calculation of the polarization operator, defined in Ref. 1 as:

$$\Pi_{\mu\nu}(q^2) = (q_\mu q_\nu - q^2 g_{\mu\nu})\Pi(q^2) \tag{1}$$

where $\Pi_{\mu\nu}(q^2)$ is the usual polarization tensor. The author of Ref. 1 makes use of the standard sum rule for the polarization operator:

$$\Pi(Q^2) = \frac{1}{\pi}\int ds \frac{\Im\Pi(s)}{s+Q^2} + Const.\ ;\ Q^2 = -q^2 \tag{2}$$

The author of Ref. 1 then calculates both sides of the sum rule. Using the right hand side he calculates an "experimental" value at positive Q^2 by first fitting the data with two Breit-Wigner terms (that gives quite a good fit to the data in the neighborhood of the ρ). He then performs the integral on the right hand side of the sum rule. The author of Ref. 1 then calculates a theoretical value for the polarization operator in two steps: he includes a perturbative calculation; and tries to include nonperturbative physics through the operator product expansion. In that way, the author of ref. 1 produces a formula that includes several terms in the perturbative expansion and several terms in the nonperturbative expansion. As a technical point, he Borel transforms everything to try to improve convergence and so, presents the answer as a function of the Borel parameter, M, rather than Q^2. Although several terms are given, only the first term in the perturbative expansion and the nonperturbative term that is proportional to the chiral condensate are of much importance. These terms are:

$$\frac{4}{9}\left(\ln\left[\frac{M^2}{e^{\gamma}\Lambda^2}\right]\right)^{-1}\left\{1+\dots\right\} - x_{4q}\frac{448\pi^3}{81M^6}\alpha\left(\langle\bar{\psi}\psi\rangle\right)^2 + \dots \tag{3}$$

The above formula agrees quite well with the "experimental" formula, at least over the range of values of M where the author of Ref. 1 considers the theoretical calculation to be reasonably accurate; in particular, that range includes a peak.

2. Induced operators

Light-cone quantization usually requires the solution to differential constraint relations. In particular — the case that will concern us — if the theory includes fermions, the field, ψ_- always satisfies a differential constraint relation. In QCD, for the case of massless quarks and the case of light-cone gauge, the equation is

$$\mathrm{i}\partial_-\psi_- = \mathrm{i}\gamma_\perp \cdot D_\perp\gamma^0\psi_+ \ . \tag{4}$$

The general solution to this equation is given by:

$$\psi_- = \psi_-^0(x^+, x^\perp) + \int dx^-\gamma_\perp \cdot D_\perp\gamma^0\psi_+ \ . \tag{5}$$

Here, $\int$ is the antiderivative which just replaces e^{ikx} with $\frac{1}{ik}\mathrm{e}^{ikx}$ in the Fourier expansion of the field. In many cases the field, $\psi_-^0(x^+, x^\perp)$ is taken to be zero, but that is not the correct solution: $\psi_-^0(x^+, x^\perp)$ is not zero and, in particular, vectors created by the action of $\psi_-^0(x^+, x^\perp)$ on the bare vacuum dress the bare vacuum to form the physical vacuum [2,3]. That the bare vacuum must be dressed in that way is required by gauge invariance.

Although modes from $\psi_-^0(x^+, x^\perp)$ do dress the vacuum, and that dressing has physical consequences, $\psi_-^0(x^+, x^\perp)$ does not create physical particles from the vacuum. The operators that create physical particles from the vacuum are the usual light-cone operators that do not involve $\psi_-^0(x^+, x^\perp)$ or any other integration constant field (in light-cone gauge, the field A^- satisfies a second order differential constraint relation and therefore involves two integration-constant fields; those fields are not zero and they have physical consequences [4], but they do not create physical particles from the vacuum). Thus, while physical states do contain quanta from $\psi_-^0(x^+, x^\perp)$, it is only as they inherit those quanta from the vacuum. That structure is quite special and it allows the development of an effective theory that uses only the usual light-cone subspace, the space that results from applying physical operators to the bare vacuum. It is an exact effective theory in that the spectrum of the effective theory is the same as that of the full theory and the eigenstates of the effective theory are the projections of the eigenstates of the full theory onto the space of the effective theory. However, the dynamics of the effective theory includes induced operators: operators that would not be present if the physical vacuum were truly simple[2]. From the point of view of the usual light-cone quantization, which ignores the integration-constant fields, the induced operators appear as noncanonical operators; but they are not so in the sense that if the theory is quantized at equal time using standard methods, the dynamics will automatically include the induced operators.

Not all of the induced operators in QCD are known, but some of them were recently derived [2]. Here we shall focus on one of those that is now known; we conjecture that it is the most important of the induced operators that is associated with the fact that the physical vacuum breaks chiral symmetry. The derivation of the operator is rather complicated but the result is easy to state: the interaction has the same structure as an ordinary QCD three point vertex (two quarks and a gluon); the spin of the quark is flipped and the gluon absorbs all the longitudinal momentum of the quark; the transverse momentum as shared in all possible ways as in the usual vertex. The fact that there exists a spin-flip interaction even at zero quark mass is a signal of chiral symmetry breaking. The vertex is shown in figure 1.

3. A sum rules calculation

If we knew all of the induced operators in QCD it would be possible to perform a calculation, in principle equivalent to the sum rules calculation described in section 1., but with the nonperturbative physics included through

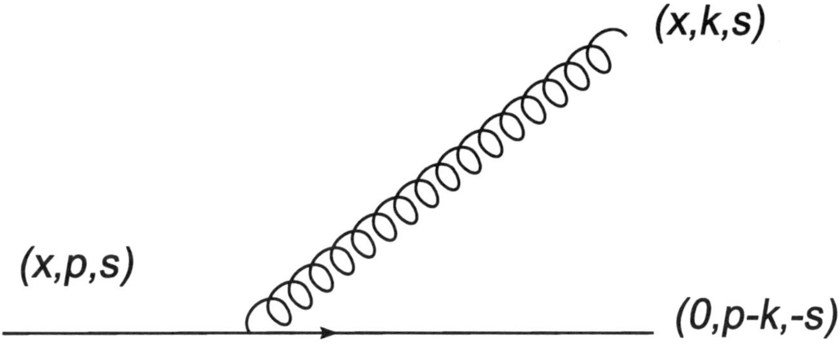

Fig. 1. The vertex that results from the induced interaction considered in this paper.

the inclusion of the induced operators rather than through the use of the operator product expansion. If both calculations were complete and exact, presumably the results would agree. Since both will actually be approximate, there may not be exact agreement but they should be approximations to the same answer.

To make a sum rules calculation for the ρ meson we shall simply calculate as in perturbation theory but shall include the induced operator described in the previous section. The calculation has the appearance of a perturbative calculation but it is not since it includes the induced operator which carries information about nonperturbative physics. The situation is in some ways similar to the standard sum rules calculation in that both expansions look like perturbative expansions, but in the case of the usual sum rules calculation the condensates carry nonperturbative information, whereas in the light-cone calculation the coefficients of the induced operators carry nonperturbative information. In each case there is at least one constant that is fit to data rather than calculated from first principles. In the case of the induced operator, the coefficient is, in principle, calculable; but performing such a calculation would require techniques beyond those that we currently have available.

We are currently performing the calculation just outlined. The perturbative part of the calculation will be the same as in the equal time case and we can just take the result from ref. 1. (the only important term is the first term in eq. (3)). We must add to that result the terms contributed by the induced operator. The graphs that give the lowest order contributions that include the induced operator are shown in figures 2-4. These are light-cone graphs, not Feynman graphs, so different possible orderings of the vertices have to be considered.

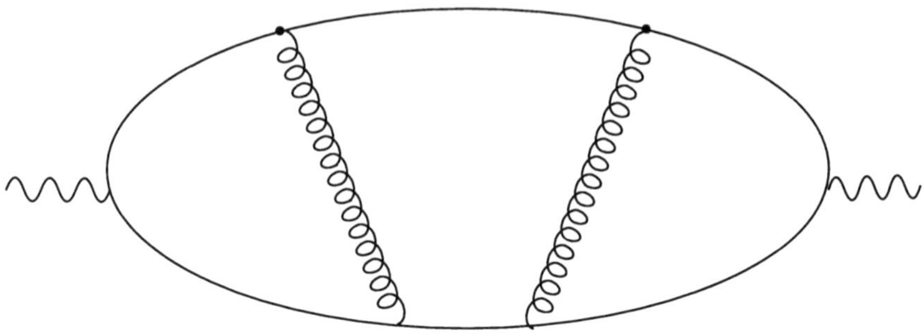

Fig. 2. A graph that includes contributions from the induced operator; the induced vertices are marked with dots.

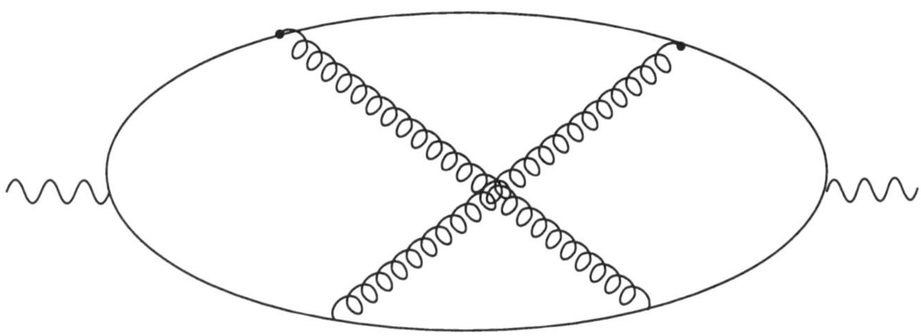

Fig. 3. A graph that includes contributions from the induced operator; the induced vertices are marked with dots.

Even though the calculation of the graphs containing induced vertices is not yet complete, we can say something about the results we will eventually get. Our answer will be of the form:

$$\frac{4}{9}\left(\ln\left[\frac{M^2}{e^{\gamma}\Lambda^2}\right]\right)^{-1}\left\{1+\ldots\right\} \pm \frac{C}{M^2}\alpha + \ldots \tag{6}$$

Comparing this equation to equation (3) we see that our nonperturbative term goes like M^{-2}, whereas the corresponding term in Eq. (3) goes like M^{-6}. The constant C is a parameter that, at the moment, we have to fit to data. With that freedom it turns out that, in spite of the quite different

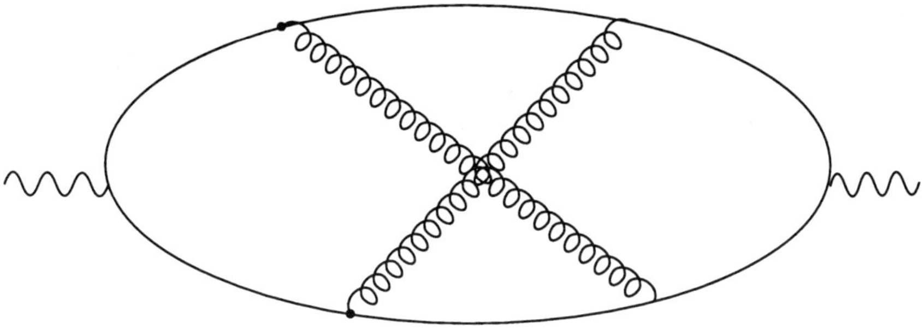

Fig. 4. A graph that includes contributions from the induced operator; the induced vertices are marked with dots.

powers of M, our curve will fit the "experimental" curve about as well as Eq. (3) if the sign in our equation turns out to be negative. If the sign turns out to be positive we will have no peak and no reasonable fit to the data. Even if we do get a negative sign, and therefore a good fit to the data, the fact that we have the freedom to fit C will prevent us from drawing any very strong conclusions. In order to form a strong impression that the induced operators are working properly it will be necessary to calculate a number of processes with all the parameters fixed, as has been done in the usual sum rules approach.

4. Summary

In the light-cone representation, the effects of a structured vacuum that breaks symmetries present in the dynamical equations are properly included in the dynamics by the inclusion of induced operators. For the case of QCD, some of the induced operators are known. We conjecture that, of those operators that are associated with the fact that the QCD vacuum breaks chiral symmetry, one of the known operators is the most important operator in determining the properties of low-mass bound states and resonances . The fact that the nonperturbative vacuum structure is encoded in the induced operators means that it should be possible to examine the properties of QCD resonances by performing a sum rules calculation, similar to the usual QCD sum rules calculations, but with the inclusion of the induced operators replacing the use of the operator product expansion. Such calculations are now underway.

References

1. M. Shifman, Snapshots of Hadrons,in *ITP Lectures in Particle Physics and Field Theory* (World Scientific, Singapore, 1999).
2. S. Dalley and G. McCartor, *Annals Phys.* **321**, 402 (2006).
3. G. McCartor, *Few-Body Systems* **36**, 181 (2005).
4. Y. Nakawaki and G. McCartor, *Prog.Theor.Phys.* **115**, 425 (2006).

SECTION 12

PAGES OF THE PAST

A BRIEF HISTORY OF FTPI

STEPHEN GASIOROWICZ

William I. Fine Theoretical Physics Institute,
University of Minnesota
116 Church Str. S.E. Minneapolis, MN 55455
USA

The creation of the Theoretical Physics Institute is a direct consequence of Bill Fine's lifelong interest in the physical sciences. In the early 1980's Bill Fine and Steve Gasiorowicz had a number of conversations on the current advances in the understanding of the fundamental laws of physics. In this context, Bill asked the question of how he could contribute to the advancement of physics, and responded very warmly to the suggestion of the formation of an institute devoted to theoretical physics. The response of the University of Minnesota to the first proposals for such an institute were so cautious that it did not seem likely that these ideas would bear fruit. It was therefore critical that Gloria Lubkin entered the picture at this time. She pointed out that the proposals themselves were on too small a scale, and that it was necessary to bring in the top levels of the University administration into the planning. She suggested bringing in Leo Kadanoff as spokesman and potential director to give reality to the proposal. In the summer of 1986, during a festive and intense get together in Minneapolis, Bill and Leo, with strong support by Chuck Campbell, outgoing head of the School of Physics and Astronomy, and Marvin Marshak, his successor, persuaded then-President Ken Keller of the merits of building a Theoretical Physics Institute at the University. Building on a very generous pledge by Bill Fine, the University committed itself to matching Bill's gift to create two chairs (subsequently split into three), and to provide permanent funding of a magnitude to support an active, vibrant institute. The Theoretical Physics Institute (TPI) officially came into existence at the end of January 1987. Steve Gasiorowicz was appointed acting director, and during the period 1987-89 conducted a vigorous search for a director. There was no predetermined choice of fields that were to be covered, but the goal of the TPI was to become a center of theoretical physics research of the highest quality, so as to make a permanent impact on that field. Right from the start it was envisaged that the first goal of the director would be the recruitment of a first-rate faculty to the TPI. The faculty, guided by the director and the newly created oversight committee, were to set the direction of activity at the TPI. The TPI would serve as a center of attraction for postdocs and long- and short-term visitors, and to help in the rapid dissemination of new directions in research through timely workshops in a variety of fields of physics.

William I. Fine, 1999

In 1989 Larry McLerran, whose research activity centered on high energy theoretical physics, became the first director of the TPI. It was decided at that time that the activities at the TPI would be evenly split between high energy physics and condensed matter physics. Very early in his tenure, Larry, in close collaboration with Gloria Lubkin, who was very active on the oversight committee, decided to take advantage of the unique opportunities presented by the break-up of the Soviet Union. In short order, five extremely creative and productive physicists were recruited from the former Soviet Union. These included Arkady Vainshtein, Boris Shklovskii, Leonid Glazman, Mikhail Shifman and Mikhail Voloshin. This group, together with later additions of Anatoly Larkin and Keith Olive form the permanent membership of TPI. Through their research, their intensive interaction with the international theoretical physics community, and their role as mentors to postdocs and visitors, they made the impact on the field that was hoped for when TPI was first conceived. Their research has received world-wide recognition and has been honored by the award of a number of prizes. Anatoly Larkin, one of the world leaders in the field of condensed matter theory was awarded the Hewlett Packard Europhysics Prize, the London Prize in Low Temperature Physics, the Onsager Prize in Theoretical Statistical Physics and the Bardeen Prize for Superconductivity. Leonid Glazman and Keith Olive are holders of McKnight Presidential Chairs, Boris Shklovskii is the recipient of the Landau Prize. Arkady Vainshtein was awarded the Sakurai Prize and the Pomeranchuk Prize, Mikhail Voloshin was awarded the Sakurai Prize, and Mikhail Shifman was awarded the Sakurai Prize and the Lilienfeld Prize.

In 2002, at the time of the fifteenth anniversary of the founding of TPI, the Institute officially changed its name to the William I. Fine Theoretical Physics Institute (FTPI), to honor Bill Fine for his critical role in the creation of the Institute. Sadly, Bill Fine died shortly after he received this recognition, on May 18, 2002. Anatoly Larkin, deeply involved in research and the teaching of theoretical physicists died suddenly on August 4, 2005. His death is a great loss to FTPI and the world-wide community of physicists.

SLOUCHING TOWARDS THE STANDARD MODEL

STEPHEN GASIOROWICZ

William I. Fine Theoretical Physics Institute,
University of Minnesota, 116 Church Str. S.E.
Minneapolis, MN 55455, USA

Next year, on January 31, 2007, FTPI will have its twentieth anniversary. It was born in an era in which elementary particle physics had reached a level of completeness and maturity that could not have been foreseen when I received my Ph.D. in 1952. The path that led to the Standard Model has been documented in various places: Nobel Prize acceptance speeches, various semi-popular books and in a very complete book, *The Second Creation.*[a] All of these reports suggest an unwavering line that led from confusion to the final triumph. They resemble history book accounts of battles, which seldom deal with all the messy details and diversions that are seen on the ground, but do not appear in the grand summaries. This essay contains some reminiscences of the messy sidelines that preoccupied many theorists while the field ultimately advanced to the establishment of the Standard Model. Since this is a personal view, it is not a history; it covers my impressions from the time I came to the Radiation Laboratory in Berkeley in 1952 to the very beginnings of the Standard Model, that is, 1967 or so.

I came to the Radiation Laboratory in Berkeley in 1952, having graduated from UCLA, then at the very beginning of its rise in excellence in this field. UCLA had no high-energy experimental program when I was there, so that when I came to Berkeley, I was particularly struck by the intensive interaction between theory and experiment. The junior staff members in the theory group, in addition to doing their own research, served as a conduit between the experimentalists and theorists, inside or outside of Berkeley. Theorists had to deal with small, incremental steps in teasing out some facts to determine what particle physics was about. The emphasis

---

[a]Robert P. Crease and Charlers C. Mann, *The Second Creation,* (MacMillan, 1986); Reprint edition: Rutgers University Press, 1996.

was clearly on phenomenology. The year 1952 was also the beginning of the preprint era, so that communication among a subset of particle physicists became very rapid. The Radiation Laboratory in Berkeley was the recipient of all preprints, and communicating outside results at seminars and Journal-clubs became a definite part of our job as postdocs. This gave me something of a perspective on what the thinking was in other research centers in the U.S. Contact with the outside world was somewhat limited. Europe was just recovering and there was little work done in phenomenological high-energy physics. CERN was still in the early planning stages. Communication with Japan was also limited. The first visitors, shortly after I came to Berkeley were Nambu and Kinoshita on the way to the Institute of Advanced Study, but it took a few more years till a steady exchange program developed with the Japanese community. There was essentially no contact with physics in the Soviet Union, and we were not even allowed to send preprints there when I first arrived. These restrictions had to be taken seriously, since the laboratory was very security conscious. [b] In part, this was because there was some classified work going on in Berkeley and because there was not yet a total barrier between Berkeley and Livermore. It was possible to get hold of Russian journals and to arrange for translations of individual articles. When I became interested in the work on field theory by Landau and collaborators, I tried to take advantage of this service, but the translations did not use physics language and were almost unreadable. I made an abortive attempt to learn Russian. Fortunately, the translations soon got better.

The influence of the "Founding Fathers" was fading. The new stars were Schwinger, Feynman and Dyson, soon to be followed by Gell-Mann, Goldberger, Chew, Low and Yang. Among the rising stars, Karplus and Ruderman were at Berkeley, Drell was at Stanford and we saw a lot of them. The senior theorists, Serber and Wick left in part because of the U.C. Regents' imposed loyalty oath, as did Panofsky.

In contrast to today, there was no lack of unexplained experimental results to confront theorists. The problem was that there was not yet any

---

[b] In order to get hired, I needed a Q-clearance, a top security clearance. I still recall that I was required to provide the addresses of all the places that I had lived in the fifteen years between 1937 and the time of application 1952. Since this was the period during which my family escaped from Danzig to Warsaw then, in September 1939 from Warsaw, to Lwow, then Bucharest, Istanbul, Baghdad, Karachi, Delhi Mussoorie, Dehra Dun and Bombay, and then the U.S, I created what was probably the largest file among Q-clearance applicants.

coherent basis on which to build. Only in quantum electrodynamics was theory well established, and in excellent agreement with experiment. Renormalization theory had been tested in calculations of the Lamb shift, positronium decay and $(g-2)$ for electrons.[c] In Berkeley the emphasis was on what we would call the strong interactions. In the late 40's the "holy grail" in theory had been the form of the nuclear potential, with the deuteron selected to play the role of the "hydrogen atom." The discovery that the deuteron had a quadrupole moment, and low energy nucleon-nucleon scattering, showed that this potential was very complicated. The Yukawa idea of meson exchange was generally accepted. When this was combined with the requirement of renormalizability, the only acceptable meson was a pseudoscalar one. (The fact that pions were pseudoscalar particles was established experimentally in 1951.) The idea of I-spin conservation, proposed by Heisenberg and by Kemmer in the 30's implied that there was a triplet of the mesons (pions). This meant that the interaction Lagrangian had to be of the form

$$L_{\text{int}} = g \int d^4x\, \bar{\psi}(x)\, \gamma_5\, \tau_a \psi(x)\, \phi^a(x)\,.$$

Perturbation theory was useless, since rough estimates set $g^2/4\pi$ in the range of 10-20. The best that one could do was a selective summation of certain classes of Feynman graphs. A variant of the Bethe-Salpeter equation (which was also proposed by Gell-Mann and Low, Nambu and Schwinger) was used by Maurice Levy to arrive at a nuclear potential that showed promise. Both Oppenheimer and Bethe were strong proponents of this approach, but a closer look at the approximations made showed that they were quite unjustified, and the problem of nuclear forces as a guide to the strong interactions died quietly. Dick Arnowitt, who was also a post-doc at the Radiation Laboratory, and I, put one of the many nails in the coffin.

A number of cyclotrons started operating in the 1950's with energies such that pions could be produced. Beams of pions were then directed at targets. Initially the targets consisted of complex nuclei (e.g. in photographic emulsions), so that little could be learned other than some cross sections. Wigner described these efforts as "the scattering of mesons by morons." Soon, however, Fermi started working with hydrogen targets. At the energies available then, only elastic scattering was possible. This was a process that could well be described by nonrelativistic quantum mechanics.

[c]The impressive fourth order calculation by Karplus and Kroll contained an error which was corrected by A. Petermann, and by C. Summerfield, five years later.

Fermi obtained data and carried out a conventional phase-shift analysis. The I-spin quantum numbers were $1/2$ and $3/2$, and because of the low energies, the total angular momentum also was limited to $1/2$ and $3/2$. The $(3/2, 3/2)$ phase shift rose rapidly, while the others were small. This result was in some way anticipated. In the limiting case of the nucleon mass much larger than the pion mass, the pseudoscalar coupling reduces to a P-wave coupling of the pions to the spin of the nucleon. In the strong coupling limit, people had already seen the existence of an isobar with the quantum numbers $(3/2, 3/2)$, and indeed, the Fermi analysis soon showed that there was a resonance in the pion-nucleon system with these quantum numbers. Unfortunately, the theory gave totally wrong numbers for the S-wave scattering amplitudes, and theorists started looking at the structure of the scattering matrix in a less model-dependent way. Progress was made possible by the formulation of the scattering amplitude in terms of matrix elements of commutators of pion field operators. This is generally associated with the elegant work of Lehmann, Symanzik and Zimmermann, but it was developed, somewhat less generally, by Low, Gell-Mann, Goldberger, Thirring and was also known to Nambu and to students of Schwinger. Chew and Low used this relation to study the scattering of pions by a fixed source (no recoil, no pairs). They limited the intermediate states in the evaluation of the commutator matrix elements to states containing no more than one pion, and obtained a nonlinear equation for the scattering amplitude. The equation gave strong support to the dominance of the $I = J = 3/2$ resonance, (called the Δ) and it also led to satisfactory predictions of photoproduction of pions. More interesting, the general formulation of scattering amplitudes in terms of the commutators allowed for a clear statement of causality, necessary for the formulation of dispersion relations. The Kramers-Kronig relation expressing the forward scattering amplitude of light in terms of an integral involving its absorptive part relied on a simple mathematical formulation of causality. For massless particles (photons) a sharp wavefront made this trivial: nothing happens before the incident particle arrives at the target. It was causality that led to the analyticity properties of the scattering amplitude in the frequency of the photon. With massive projectiles there was no sharp wavefront, and there were problems with defining causality. The difficulty was solved by Gell-Mann, Goldberger and Thirring who pointed out that the vanishing of the commutator of two field operators when their separation was spacelike was a proper formulation of causality. This allowed for the construction of dispersion relations for the forward scattering amplitude of pions by nucleons. The absorptive part of the elastic scattering amplitude could be formally expressed using the completeness of the intermediate states in the evaluation of the commutator of the field operators. In the forward direction, the absorptive part was directly proportional to the total cross section. An important tool was the notion of *crossing symmetry.* Gell-Mann and Goldberger noticed that for each graph in pion-nucleon scattering, there was a corresponding graph with the incoming and outgoing pion lines interchanged. This obvious remark had not been made before, and it played an increasingly important role in the study of the structure of scattering amplitudes.[d] The one-particle intermediate states led to poles in the scattering amplitudes, and the residues provided a unique way of defining renormalized coupling constants. The analyticity properties of the scattering amplitudes, easy to prove for the scattering of light, were finally established, even in the non-forward directions. With this result, it was possible to obtain relations involving individual partial waves, and this work, carried out by Chew, Goldberger, Low and Nambu completed the pion-nucleon scattering program.[e]

Around 1957 analyticity properties (usually assumed[f]) and unitarity started being used for other scattering matrix elements. Experimental developments prompted these changes. For example, the Hofstadter experiments on electron proton scattering carried out at Stanford showed that the proton had an extended structure and gave data on the proton electric and magnetic form factors, both for the $I = 0$ and the $I = 1$ part of the electromagnetic current. Dispersion analyses of the form factors (using assumptions about the analyticity in the square of the momentum transfer) led to a new direction in hadronic physics. Chew, Karplus, Zachariasen and I looked at the form factor and assumed analyticity in the momentum

---

[d] My attachment to crossing symmetry had one unfortunate effect. A graduate student at Berkeley showed me a soluble model that he had devised. I pointed out to him that it was flawed, because it lacked crossing symmetry. Because of this, the Lee model is not known as the Stapp-Lee model! Ruderman and I later found out that the ghosts in the Lee model are a direct consequence of the absence of crossing symmetry.

[e] The results were summarized by Low at the International Congress on Theoretical Physics held in Seattle in the fall of 1956. The talk is printed in the Review of Modern Physics, April 1957. I wrote a review of all the material on relativistic 2-particle reactions for Fortschritte der Physik while I was at the Max Planck Institute in Munich in 1959-60. During my absence from Berkeley I missed the excitement connected with Regge theory in the scattering matrix theory.

[f] Much effort had gone into establishing mathematical proofs. In many cases, the best one could do was by the use of perturbation theory. More often, practitioners followed Goldberger's proof using *reductio ad absurdum.* "Suppose the proposed analyticity properties were wrong. But that would be absurd..."

transfer variable. Parallel work by Federbush, Goldberger and Treiman led to similar conclusions for the form factor, but they did not stop at this vertex. Their analogous discussion of the pion decay form factor led them after some unconvincing acrobatics to the Godberger-Treiman relation. This relation, as I will note later, led to a whole new development in particle theory.

In order to get a fit to the Hofstadter data for the $I = 1$ (isovector) form factor, the two-pion intermediate state (with $I = 1$ and $J = 1$) in the dispersion analysis had to be boosted, and this led to the prediction of an $I = J = 1$ particle, the ρ meson,[g] which treated as a one-particle intermediate state whose mass and coupling to the two mesons and the photon could be adjusted to fit the data. This only had meaning if the ρ meson also took part in other processes. The large hydrogen bubble chamber was just coming into operation in Berkeley at the time, and people started looking for evidence of ρ-production. This was soon found in the study of

$$p + \pi^- \rightarrow n + \pi^+ \pi^- \, .$$

The two outgoing pions, if they were the decay products of a virtual ρ, would have a strong enhancement in their invariant mass at the mass of the ρ. This was indeed found to be the case, and the mass of the ρ, as well as its coupling to the two pions was determined. The study of enhancements of this type was actually just an extension of the finding of the Δ in the final state in elastic scattering of pions on nucleons. The success led to the proposals that other classes of intermediate states (e.g. three pions for $I = 0, J = 1$) could be dominated by resonances.

A whole industry of finding new "resonances" or, as they were soon called "particles" developed in Berkeley and other laboratories. In Berkeley the tool was a large bubble-chamber and an almost completely automated method of scanning the photographic plates from the bubble chamber. Theoretical methods were developed for finding their spins and parities, and large tables of new particles, with their decay modes soon appeared. Such particles could be exchanged among other particles (just like the pion is exchanged between nucleons in nucleon-nucleon scattering), and by concentrating only on exchanges of particles, quite a good understanding of inelastic reactions, produced at ever increasing energies, was obtained. By treating resonances as "particles" and assuming that these dominated channels associated with their quantum numbers, the S matrix took on the form

[g]The prediction was made by Nambu, and subsequently by Frazer and Fulco.

described by generalized tree graphs. There were some brilliant solutions to technical problems. For example, the exchange of particles with spin $S \geq 1$ led to violations of unitarity, and this problem was solved by an application of Regge's work on the analytic continuation of angular momentum.[h] Effectively, the high energy contribution of an amplitude in which a spin J was exchanged, E^J, actually found to have the form $E^{J(t)}$ where t was the momentum transfer. For t values at the mass of the particle, $J(t)$ took on the spin value for that particle, while at $t = 0$ the value of $J(0)$ never exceeded 1. Interestingly $J(t)$ was experimentally determined to be a linear function of t, at least for the experimentally accessible resonances, and this led to a great deal of work in fitting the rapidly growing amount of data. Although quantum field theory provided a background, strong interaction physics became more and more dominated by S-matrix theory[i] which had acquired a more definite structure. The amplitudes had to satisfy all the necessary analyticity properties, they had to obey crossing symmetry, satisfy unitarity conditions and they had to obey Regge behavior at high energies.

Although it was not possible to satisfy all of these requirements, the replacement of all intermediate states by appropriate resonances (treated as single-particle states) allowed for an algebraic treatment of scattering amplitudes. Critical to this approach was the observation of a form of *duality*. A scattering amplitude, for example for the process

$$\pi + \pi \rightarrow \pi + \omega$$

could be written as a sum of terms in which the incident two pions form resonances (an infinite number of them according to Regge theory), which then decayed into a pion and an ω, or equivalently pion-pion and pion-omega lines exchanging an infinite number of Regge trajectories. This duality was used to connect decay widths of resonances with Regge trajectory parameters, but the most dramatic result was the discovery by Veneziano that he could write down a form of the scattering amplitude for the above process that satisfied all of the S-matrix conditions (with unitarity expressed in terms of one-particle intermediate states only). The Veneziano formula was soon generalized to describe any n-particle process. This was probably the

[h]The extension to particle physics was pioneered by Mandelstam and Chew, two of the leading figures in S-matrix theory at the time.

[i]Field Theory came into disrepute partly as a result of the work of the Ladau group, who found what would now be called "Asymptotic slavery" in Quantum Electrodynamics, and partly because in the US, at least, it was viewed as useless for the strong interactions. Nevertheless, most theorists still used Feynman graphs for guidance.

high point of S-matrix theory.[j]

Although the study of weak interactions was not a high priority in Berkeley at the time, a bridge from the strong interactions to the weak ones came from the study of the "strange" particles.

In the early 50's attention was beginning to be paid to new particles, originally found in cosmic rays, that had the peculiar property of being produced with typical "strong interaction" cross sections, but that decayed at rates many orders of magnitude smaller than expected. In 1953, Gell-Mann (and independently Nishijima and Nakano) introduced a new quantum number, "strangeness," which allowed for the classification of the new particles, the prediction of yet undiscovered partners and which brought order to what was for a while a totally confused picture. Strange particles were not a center of attention in experimental physics in Berkeley at the time, and most of the early accelerator results came from experiments done at the Brookhaven Cosmotron. An interesting recollection: Mal Ruderman spent the summer of 1954 at Brookhaven and on his return reported on some data on the production and the decay of Λ^0s in a cloud chamber. The data indicated that there was an asymmetry between the number of pions emitted up vs. down relative to the production plane (i.e. relative to the direction of the Λ spin, and Yang had noted that this must be an experimental error, since if this were taken seriously, *the pion would have no parity.* Somehow, none of us absorbed this comment till two years later.

Still, the new particles became a preoccupation of the theorists. Two papers stirred particular interest: one, by Gell-Mann and Pais on the K_1, K_2 prediction struck all of us by its beauty and simplicity, and subsequently by the experimental consequences associated with regeneration. The other one was Dalitz's analysis of the spin and parity of one of the new particles, the τ. This led to the assignment of spin-parity 0^- to the τ. The "other" particle, the θ decayed into two pions, and its spin-parity were determined to be 0^+. These two particles had the same mass, within experimental errors. There was some speculation about a new symmetry, "parity-doubling" which would account for this particular phenomenon, but, like supersymmetry today, a lot of particles were missing. The question of parity nonconservation in the decay that led to the decay of a single particle (the K) into three and two pions, respectively, was very much in the air. However, it was only Lee and Yang who came up with the experimental require-

[j]This work led directly to the old "strong interaction" string theory, and the rest is history.

ments for testing this possibility: you must measure a correlation that is pseudoscalar.[k] By the end of 1956 the first results were in: parity was not conserved in weak interactions. When the first reports of the Wu experiment reached the community in December 1956, an unprecedented growth of experimental activity led, within a year, to a complete clarification of the phenomenological structure of the weak interaction Hamiltonian. Perhaps the most detailed formulation of that structure was of the current-current form given by Feynman and Gell-Mann

$$H_W = \frac{G_F}{\sqrt{2}} J^\alpha J^\dagger_\alpha \,.$$

These authors identified the nonleptonic part of the *vector* current is the charge-changing component of the conserved strong interaction I-spin current.[l] This conjecture was checked by the magnetic contribution to beta decay of N^{12} and B^{12} to C^{12}. Effectively this meant that further progress in the weak interactions, especially those of the strange particles became a strong interaction problem. Attempts to treat this Hamiltonian as the effect of an exchange of massive vector bosons faltered on the grounds that no such particles had been observed, and that the data indicated that if they existed, they would have to have masses larger than 2 GeV. The weak interaction community was too cautious in their speculations!

It was again Gell-Mann who properly identified the axial current as a partner of the I-spin vector current, resulting from a new symmetry, in which the "charges" corresponding to the currents obeyed commutation rules of the form[m]

$$[V_\alpha\,, V_\beta] = [A_\alpha\,, A_\beta] = i\varepsilon_{\alpha\beta\gamma}V^\gamma\,, \quad [A_\alpha\,, A_\beta] = i\varepsilon_{\alpha\beta\gamma}A^\gamma\,.$$

---

[k]A very low key presentation of this was given by Yang at the Seattle Congress. That may account for the fact that the experimental frenzy of discoveries did not materialize before the most difficult of possible parity-nonconservation experiments was done by Mme Wu and collaborators and reported at the very end of 1956. I was one of the scientific secretaries writing up the talks and, like everybody else, missed the importance of Yang's final comments. (Also printed in Reviews of Modern Physics, April 1957).

[l]The fact that the current was of the form (vector − axial) was first conjectured by Marshak and Sudarshan.

[m]The first model to show this chiral SU(2)×SU(2) structure appeared in a paper by Gell-Mann and Levy, whose purpose was to construct models that satisfied the condition that the divergence of the axial current be proportional to the canonical pion field. This so called PCAC (partially conserved axial current) condition yielded a transparent derivation of the Goldberger–Trieman relation connecting the pion decay constant with the strong pion-nucleon coupling. The original GT derivation was totally unacceptable, consisting of unjustified approximations, but the relation was good to better than 20% accuracy.

The *current algebra* resulting from these relations became a major research program in the early 60's. Probably of the order of 1000 papers were written on a variety of sum rules following from these relations, but the most dramatic one was due to Adler and Weisberger, who independently found the magnitude of the enhancement of the axial current due to the strong interactions.

Although progress in determining the spins and parities of the strange particles was slow, it was clear that since the baryonic particles, the Λ^0, $\Sigma^{\pm,0}$, $\Xi^{0,-}$ particles differed from the nucleon mass by 1–2 pion masses, there should be a relation between all of them. Assuming that the strange hadrons also had the same spins and parities as the nucleons and pions, respectively, Gell-Mann proposed a "global symmetry" model of the strong couplings in 1957. The model was meant to provide a structure to be compared with future experiments, and I don't believe that Gell-Mann took the model very seriously. There is one reason for mentioning it here: In the text there is the sentence "The symmetry properties of the model may be correct, even though the use of field theory is unjustified. For this reason, an analysis purely in terms of the symmetry group of the theory is in order." The first sentence signaled the new philosophy that crept into strong interaction physics, pointing in the direction of a decade of S-matrix theory. The second part showed that the author had not actually done this group-theoretical analysis. Lee and Yang pointed out that the symmetry group was quite complicated, but even before that Gell-Mann had evidently had looked into group theory and in 1961 came up with the "Eightfold Way," equivalently, the SU(3) of flavor. This was discovered simultaneously by a then yet unknown Israeli physicist, Yuval Ne'eman. Gell-Mann also proposed that the symmetry-breaking term transformed as the octet representation of SU(3). On the basis of this, the $I = J = 3/2$ resonance which had to belong to the **10** representation of SU(3) had to have a partner of strangeness $S = -3$ (isosinglet), with charge -1, which, following the symmetry-breaking proposal, had to have a mass in the vicinity of 1680 MeV. The predicted particle, the Ω^- was discovered almost exactly where expected, and the SU(3) symmetry of the then known hadronic world became completely established. I should remark that this was the first occasion that group theory became a necessary tool for phenomenologists and even for experimentalists. Until then, the only group one had to deal with was SU(2), and this was, for all practical purposes equivalent to angular momentum. I recall giving a series of lectures on a graphical approach to SU(3) at the Argonne National Laboratory, and was surprised at the large number of reprint requests that came from experimentalists. Within a year Gell-Mann, and independently George Zweig proposed that the known octets and decuplets of the hadrons were composites of fundamental particles, named quarks by Gell-Mann. Not everybody took quarks seriously at the beginning,[n] but there was a body of work[o] which treated quarks effectively as 1/3 of a nucleon, and studied the "nuclear physics" of three-quark states.

The field theory of quarks became of interest as a by-product of current algebra. This work was stimulated by a development that occurred at the end of the period that I am describing. Bjorken studied the cross section for deep inelastic electron scattering, the process

$$e^- + p \rightarrow e^- + X$$

where all possible states X are summed over. The cross section depends on the commutator of electromagnetic currents, and data, first presented in 1968 agreed with Bjorken's use of free-field commutators for the evaluation of the cross section. The search for theories for which this was a good approximation ultimately led Quantum Chromodynamics, whose credentials were confirmed by the discovery of *Asymptotic Freedom*, a property restricted to non-Abelian gauge theories, first described by Yang and Mills in 1954. This work originally had little impact,[p] except for a very stimulating paper by Sakurai that appeared in 1960. He proposed that the strong interactions involve the coupling of hadrons to Yang–Mills vector bosons, and derived a number of interesting consequences. Although he had to mutilate the theory by adding a mass term for the vector bosons, which destroyed gauge invariance and made the theory un-renormalizable, it led to a number of models, described by effective Lagrangians, which included broken SU(3), PCAC, chiral symmetry and vector mesons.[q] It was in puzzling

[n] In my book *Elementary Particle Physics* (J. Wiley, 1966), which more or less summarized the state of the field by 1965, I stated "A search for new particles of this type (charge $e/3$) has so far proved unsuccessful. In view of this failure and the difficulty of inventing a mechanism which would bind a quark and an antiquark, or three quarks, but not two quarks, for example, we will not discuss the quark model further." This was reasonable but wrong. The only redeeming feature of this judgment is that in a footnote, I mention that the use of quark fields in the construction of vector and axial currents is of some importance.

[o] In my memory I associate it with the names of Morpurgo and Dalitz, but I am sure that there were many other contributors to this effort.

[p] Bludman conjectured that the "intermediate vector bosons" proposed by a number of people as mediating the coupling of various hadrons and leptons to the weak currents be members of the Yang–Mills vector particles.

[q] Don Geffen and I reviewed this program in Reviews of Modern Physics in 1969. Before the year was out, the particle physics community was deeply involved in the Standard

about the construction of such a model that Weinberg realized that the proper application of his work was to the weak interactions. This work led to the electroweak unification. The solution to the mass problem turned out to be the Higgs mechanism.[r]

The journey to the inclusion of quarks, to the degradation of SU(3) from a basic symmetry to an almost accidental consequence of the magnitude of the quark masses (flavor), and the rise of the *color* SU(3), is beyond the scope of this essay.

Model, and chiral symmetry became more of an off-shoot of QCD.

[r]Another personal note: I spent 1959-60 in Munich at the Max Planck Institute. My preoccupation was dispersion theory, but I did have a long talk with Heisenberg about the Goldstone problem. Although Heisenberg was focused on his "World Formula," he very clearly indicated that there was no problem, since in the presence of long-range interactions the massless boson would disapear, playing the role of making the forces short-range.

PARADISE LOST

SASHA MIGDAL

CEO & President, Magic Works, LLC,
 # 3322, Empire State Building, New York City, NY 10019
E-mail: sasha.migdal@gmail.com

Initiation into Landau School of Thought started with the famous Theoretical Minimum Exam. This was a sequence of increasing tests of that rare mix of knowledge, durability and passion, which was necessary to be accepted and survive the school.

I was lucky to be initiated by Landau himself, just one year before the tragic automobile accident, after which Landau never recovered.

My friend Sasha Polyakov and I came to Landau in the spring of 1961. We came with zero combined life experience and infinite combined self-confidence, after some success in Mathematical *Olympiads*[a] where we shared the First Prize.

Lev Landau

[a] *Olympiads* (or math competitions) were organized on a regular basis at every level of the Soviet educational system, beginning in school districts, through city competitions, and finally at the national level. National prize winners were praised by the media just as winners of the national spelling-bee competitions are praised in the US. -Editors' note

Landau had that childish expression on his face, a mixture of surprise, impatience and arrogance, as if he was saying "Let me see if you are as dumb as I think you are."

He placed us in two different offices in "Kapichnik"(Kapitsa Institute for Physical Problems), he gave us some math problems and started pacing between these offices, looking over our shoulders with curiosity and making nasty remarks about every line we wrote.

I felt as if my lord had come to earth and crucified me with his own hands. One of my problems was to integrate a rational function. The textbooks recommended the method of variation of constants. One was supposed to represent the ratio $P(x)/Q(x)$ as

$$\frac{P(x)}{Q(x)} = R(x) + \sum_i \frac{c_i}{x - a_i} \tag{1}$$

with a_i being roots of $Q(x)$, and obtain coefficients c_i as residues at $x = a_i$, and similarly for coefficients of polynomial $R(x)$, as residues at $x = \infty$.

Kapitsa Institute in Moscow

But I have never read those textbooks. In fact I was very bad at reading textbooks all my life. I preferred to do it myself and then ask my educated friends whether my solution was correct.

I had just a few minutes to invent my own method before Dau would have run out of his short patience. I do not remember now how I solved that problem, but I do remember very well that I did. I still hear his voice in my head: "Well, you finally got a solution, invented a wheel, you should have known such triviality in a first place."

Sashka had a similar experience in his cell, after which we were pulled together in Dau's office and he looked into our eyes without smiling. "Well kids, you passed, but you have to really work on your mathematical skills. A theoretician cannot live with skills like that. You have to know complex analysis, PDE, group theory, all nice and easy stuff. As for everything else they will teach you in university, take it easy, just flow around (обтекайте)."

I came back home totally humiliated. My father Arkady, former student, friend and rival of Landau, was waiting for me. He was initiated in the 1930s without even being tested by Theor-Minimum.

Arkady Migdal and Sasha Migdal in the early 1960s

"Congratulations—you passed the first Landau Exam."He hugged me, but I pushed back. "What are you talking about, Landau was cursing us, he said that a theoretician cannot live with a technique like that, he was mocking the way we solved problems."

"He honored you, idiots, by treating 15-year old boys as grown-up theoreticians! Do you really think the grown-up theoretician could live with your mathematical technique? Go and learn group theory like he suggested, and always take good advice regardless how insulting it may sound. By the way, he called me up right after you two left his office, he was very excited. He said you were so smart, you reminded him of himself when he was young."

Landau did more than that to help us. He wrote a letter recommending us to be accepted into the University. This letter actually helped that miracle to happen. There were two miracles, in fact. First, young kids were allowed to attend the Physical Technical Institute [b] entrance exams without graduating from high school. Second, Jews were actually accepted into PhysTech, which was next to impossible.

Sasha Polyakov, Vitya Zharinov, Seriozha Gurvitz, and Sasha Migdal in mid-1960s during the Military Training

The Landau letter was so good that it was never shown to us until a few years ago. We lived in a tough country during tough times, when kids were not supposed to be spoiled by being praised. I suppose now this letter cannot spoil us anymore.

After Landau's death Isaak Khalatnikov and a few other Apostles created the Landau Institute of Theoretical Physics. It was in the late 1960s when the Stalin's Fear was starting to fade, and the Iron Curtain was starting to rust a little. KGB still ruled the country,[c] but it was already

[b] The Moscow Institute for Physics and Technology, in Dolgoprudny, about 15 miles to the northwest of Moscow. This is an elite Russian institution analogous to MIT. PhysTech (Физтех) is the Russian abbreviation. – Editors' note.

[c] KGB is the Russian-language abbreviation for State Security Committee, the principal secret police agency. The Soviet secret police changed acronyms many times. It started out as the Cheka, and then became the GPU, the OGPU, the NKVD, the MGB, and, finally – since 1954 till the demise of the Soviet Union in 1990 – the KGB. The term KGB is also used in a more general sense to refer to the Soviet State Security organization

playing more sophisticated games, not necessarily the deadly ones.

Isaak Khalatnikov is truly a remarkable man, whose immense contribution to Theoretical Physics is underestimated in my view. He was involved in seminal work with Landau and Abrikosov, where they first discovered the famous "zero charge"problem, and laid the ground work for the modern quest for a consistent filed theory. But his life achievement is the creation and leadership of the Landau Institute, which played such an important role in the history of physics during the 20th century.

Khalat was a genius of political intrigue. Marrying into the Inner Circle of the Soviet System (his wife Valya is the daughter of a legendary Revolution hero), he used all his connections and all the means to achieve his secret goal. To assemble the best brains and let them think freely.

On the surface, his pitch to the party went as follows.

Landau with Apostles. First row, right to left: E. Lifshitz, L. Landau, I. Khalatnikov, A. Abrikosov, unknown

"The West is attacking us for anti-Semitism. The best way to counter this slander is to create and Institute, where Jews are accepted, allowed to travel abroad and generally look happy. This can be a very small Institute, by standards of the Atomic Project, it will have no secret military research,

since its foundation as the Cheka in 1917. –Editors' note

it will cost you very little, but it will help “Разрядка" (Détente). These Jews will be so happy, they will tell all their Jewish friends in the West how well they live. If they will not, it is after all, us who decide which one goes abroad and which one stays home. They are smart kids, they will figure out which side of the toast is buttered.”

As I put it, Khalat sold half of his soul to the devil and used the money to save the other half. I truly respect him for that, now once I learned what it takes to create a startup and try to protect it against the hostile world.

As many crazy plans that came before it, this plan really worked. The best brains were assembled in Landau Institute, they were given a chance to happily solve problems without being forced to eat political shit like the whole country, and yes they did sometimes travel abroad, they did make friends in the West.

In a way the plan worked too well, we became so worldly and free that we could no longer be controlled. Needless to say, our friends in the West became closer to us than our curators in KGB.

To some extent, KGB played similar games in other areas of the Soviet Culture as well. The Soviet Union improved its image in the eyes of the West, however at the same time the whole country, starting with KGB, got spoiled and seduced by Western influence. The Cold War was lost by the USSR not just because of an economical burden of arms race, but also because of a loss of fighting spirit as a result of careless flirting with the West.

Political games and Cold War was the least of our worries in the 1960s and ’70s. I joined Landau Institute in 1969, a year after its inception, after defending the PhD [d] on the scale invariant Reggeon Field Theory. When I look back at that time, it turns out to be a time of great discoveries in solid state physics (the main specialty of Landau Institute) as well as in the elementary particle physics.

Both streams of discoveries were fed by the realization of a remarkable analogy between these two fields, allowing for cross-fertilization. This analogy was foreseen by great Julian Schwinger, who noted that inverse temperature $\beta = 1/T$ in statistical mechanics is equivalent to imaginary

---

[d] The academic hierarchy in Russia follows the German rather than the Anglo-American pattern. An approximate Russian equivalent of PhD in the US is the so called *candidate* degree. This is the degree the author received in 1969. The highest academic degree, doctoral, is analogous to the German *Habilitation*. The doctoral dissertation is usually presented at a mature stage of the academic career; only a fraction of the *candidate* degree holders make it to the doctoral level. –Editors’ note

time in quantum theory, as it follows from the comparison of the partition functions

$$Z = \mathrm{Trace}\,(\exp(-\beta\, H))$$

in statistics versus

$$Z = \mathrm{Trace}\,(\exp(-i\, t\, H))$$

in quantum theory.

At the time I entered the field in the 1960s, the analogy was barely noted, and its implications were not known at all. I learned about this analogy from the remarkable physicist Tolya Larkin, who pioneered it in the USSR with Valya Vaks. The whole discipline of the Euclidean field theory, which does not make a distinction between quantum and statistical applications and uses notions from both, was yet to be born.

I guess the missing link was a realization that this "statistical" imaginary time is the same as Minkowski's imaginary time of special relativity. This is the same observation that drove Steven Hawking to interpret the imaginary time of the black hole as inverse temperature. Good ideas are so scarce they come back again and again in different disguise.

Werner Heisenberg

Moreover, the Field Theory was pronounced dead by Heisenberg and Landau. As Landau put it: "The Lagrangean Field Theory is dead and should be buried, with all the proper honors of course."

Landau's motivation was the "zero charge," which indicated inconsistency of all known field theories except one. The one exception was still in its infancy and could not yet stand and defend itself. That one exception was of course the Yang-Mills theory, which was born in the fifties and never taken seriously until the seventies. In the 1970s it was finally quantized and shown to be free of the "zero charge" problem. Now it is the theory of all elementary forces except gravity.

Heisenberg's motivation was even more ambitious. This is an example of how one great leader can block the way to the whole army by falling down in a narrow pass. He dared to go one step further from his celebrated uncertainty principle and declare the physics must only study observable quantities. His own approach was to study so called S matrix. S matrix is a collection of transition amplitudes between various observable incoming states, such as an anti-proton flying towards the hydrogen atom and the observable out-coming states, such as beams of electrons, positrons, photons and mesons.

Pretty much like the medieval Scholastic Magisters who were extremely inventive in defending the church dogmas and blocking the way to experimental science, some great minds in the 1960s developed the S matrix dogma with great perfection and skill before it was buried down in the 1970s after discovery of quarks and asymptotic freedom.

Left to right: Arkady Migdal, Sidney Drell, Vladimir Gribov, Sasha Migdal, Mark Strickman, Leonid Frankfurt, Lev Okun, unknown. Late 1970s. We are playing push-game (forcing opponent to lose balance).

It turned out — quite unfortunate for physics — one could deduce a lot about S matrix on purely phenomenological grounds without ever asking heretical questions like "what is inside?" One could not of course even attempt to compute proton mass or its magnetic moment or explain anything about properties of so-called resonances — short living subatomic particles. Scattering amplitudes were the only numbers to be considered and computed.

In a way this reactionary idea was a truly revolutionary one. For the first time since Galileo the quest for the structure of matter was stopped on philosophical grounds. There is nothing inside, total nuclear democracy! Everything consists of everything else. Do not ask whether there was a rabbit inside the hat, you are only allowed to compute how far it will jump and in what direction.

It is a bitter irony of history that such a restriction on a free thought was imposed by a German scientist and so widely accepted in Russia in the second half of the 20th century. My physics teachers Gribov and Okun were respected as liberals and free thinkers. They were followers of the great Landau, but still they would not even talk to me about Yang-Mills theory because it was "unobservable."

For the whole two years 1964 and 1965 JETP refused to publish our work with Sasha Polyakov "Spontaneous Symmetry Breaking of Strong Interaction and Absence of Massless Particles." This is where we correctly argued that vector mesons of the Yang-Mills Theory must acquire mass by absorbing zero-mass Goldstone particles. We were stomped to the ground at every seminar we tried to present our work at. The most disturbing thing was that nobody would even argue with us on the subject. The mere mention of "spontaneous symmetry breaking" caused healthy laughter, which ended the conversation. Independently this effect was discovered and published by Higgs and rightfully is called the Higgs phenomenon.

This was the first, and very useful lesson of the danger of independent thinking. There were many more of them proving the same point: "не высовывайся" – do not stick out.

I eventually realized that I was born dissident and maverick, never to be part of any crowd or achieve any social success. I cannot be satisfied when I reach some harmony with the world. I have to abandon everything and go further.

As for the Lagrangean field theory, so respectfully buried by Heisenberg and Landau, my good friend Sasha Zamolodchikov (another Sasha from the Landau Institute) summarized it like this: "They buried the Lagrangean

matician N.N. Bogoliubov took the microphone and said literally the following: "There is no mathematical difference, but when some young people want to use a fancy word they call it conformal symmetry."

David Gross and Sasha Migdal in Aspen

Obviously, his ignorant lieutenants misinformed him, and he did not bother to look up for himself what was the conformal symmetry.

I could not stand it any longer, I raised the hand to give everybody a brief introduction to Conformal Symmetry (naturally, no one invited any of us, suckers, to speak at such an important International Conference, but I was allowed into the audience). Vigilant Organizers of the conference igno-

Soviet delegation in Aspen, CO., 1976

red my raised hand, the break was quickly announced, so that my indignant cry: "15 parameters!" went apparently unnoticed. (By the way, somebody told me recently that he heard that cry and wondered for years what could that mean, until he learned conformal symmetry).

Here is the interesting part. I came home and said to my father, "Look, what a fool N.N. really is," and then I told him the story of 15 parameters. My father laughed with me, surely he knew what conformal symmetry was about. Then he said something remarkable. "You know, Sasha, there are two kinds of intellect. The first kind helps you to say smart things. But the second kind helps you to do smart things. N.N. used to have great intellect of the first kind, but later he switched to the second kind. Do you think he cares about parameters of conformal group? He is involved in Big Science, where political truth is more important than scientific truth. You would do yourself some good by borrowing the second kind of intellect form N.N."

N.N. Bogolyubov

I wish I would know how to follow this wise advice! Fortunately, in my new life there are lawyers around to zip my mouth when it is necessary to keep it shut.

I feel obliged to say something about the atmosphere in the Landau Institute, how we worked, how we discussed things, how we had fun.

It was more like a Gentlemen's Club than like any other research institute I ever knew. We worked at home the whole week, mostly alone, sometimes meeting at someone's apartment or talking by phone. My students Volodya Kazakov and Ivan Kostov practically lived at my small apartment when we were developing our matrix models of quantum gravity.

These models were another example how original thinking gets you in trouble. They were just too simple as a solution for quantum gravity problem, which at that time was attacked without much success by sophisticated methods of string theory. Everybody was hypnotized by celebrated words of Ed Witten: "String Theory cannot be solved by methods of the 20th century mathematics."

Matrix models were plain vulgar, they started by postulating that fluctuating curved space of quantum gravity can be made from small identical equilateral simplexes, dynamically reconnecting with each other, and computing the partition function (sum over all these reconnections) by good old combinatorial methods. Naturally, nobody believed our dumb solutions for almost a decade, until Knizhnik, Polyakov and Zamolochikov heroically reproduced some of these formulas, using all the heavy weaponry of string theory and conformed field theory. We got out share of ostracism before that happened.

Volodya Kazakov and Sasha Migdal, Paris, early 1990s

As I bitterly commented on it: "Ed was right as always, 2D string theory was solved by methods of the 19th century mathematics."

Every friday we were obliged to go to Chernogolovka,[e] and sit at the

[e]Chernogolovka is a small town located 50 kilometers (30 miles) northwest of Moscow. That's where the Landau Institute headquarters were located, along with other Soviet research institutions founded after 1956.

General Seminar. This was the time to exchange all the news and gossips, to get your paycheck, and to fill out some paperwork for trips abroad (if you were entitled for those). Also there were informal seminars in one of a few small rooms owned by the Landau Institute. This was where the new Physics was born and discussed, at those shabby blackboards with rough chalk, disintegrating in your hands and leaving almost no marks on the board.

I must also say something about the style of the Landau seminar, which was quite unique. The best analogy would be the dog hunt, with the speaker as a wolf and the rest of us as a dog pack. The role of the Hunter was initially reserved for Landau, but after his death it was vacant which added to the chaos of the hunt.

The seminar lasted forever, we had nothing better to do than to hunt each other in quest for truth. No limits, nor any mercy, not to mention good manners or political correctness was allowed. The resulting stress and high emotions were relieved by a good drink, where it all ended.

My naive attempt to revive this Landau style of discussions (without the drinking part of course) in my undergrad class at Princeton ended with a disaster. When I said to them at the opening lecture, "I will try to make you think, which is quite painful. But, as we say in Russia, "suffering purifies," half of the class dropped with indignant child abuse complaints to the Dean of the Faculty.

Local parade is passing by the Landau Institute in Chernogolovka

The remaining half stayed and suffered all the way. One of them, a nice Chinese student told me later at the exam, "I forgot the shortcut you showed us at your lecture, so I solved this problem the hard way, but this suffering purified me Professor!"

"Oh yes it did, "thought I with an inner smile," "Long live Landau!"

The large facade in this picture is misleading, in fact this is the place which was given to us on fridays for seminars. The rest of the friday as well as the rest of the week the Institute occupied only eight to ten small rooms on the first floor of the annex, not visible in this picture. We had a lot of fun in the Landau Institute, at least in the 1970s when it was one family, where everyone knew all the secrets of everyone else.

We did not really suffer that much from our isolation, the preprints came regularly by mail, everyone who traveled abroad, dumped all the latest news on the rest of us after coming back.

In the eighties it was all starting to get sour, nothing lasts forever in this world. Some of us turned out to be more equal than others. Our Western friends, regardless of how hard they tried to help us in isolation, could do nothing with the laws of a free market. If you are not present to explain and defend your ideas they will get stolen or simply ignored and reinvented.

Giorgio Parisi

As Giorgio Parisi, another hero of the statistical field theory put it, "All the credit goes to the *last* one to make an important discovery." He made quite a lot of those discoveries himself, the most important in my view being, the stochastic quantization, which explained how the Bohm's idea of hidden variables can be implemented by introducing extra 'stochastic' time τ and adding two forces, a quantum friction $\partial X/\partial\tau$ and Gaussian random force $\eta(X,\tau)$ to quantize Newton's equations:

$$\frac{\partial X}{\partial \tau} + \frac{\delta S}{\delta X} = \eta(X,\tau),$$

$$\langle \eta(X,\tau)\, \eta(X',\tau') \approx \delta(X - X')\, \delta(\tau - \tau'). \tag{2}$$

My own feeling is that this may be the way to go beyond quantum theory with its postulates of linear superposition of amplitudes of histories and resulting infinities coming from the quantum sea of zero point fluctuations. What if, in fact, there actually is an extra dimension, where some natural forces are present at Planck scale, imitating stochastic noise at observable scales? Maybe God does not play dice after all? Besides, nothing is precisely linear in our world, why should the superposition principle be an exception?

Here it is, an approximation coming about after averaging over hidden variables $\eta(X,\tau)$.

We had some wild parties in Chernogolovka, with rivers of vodka, lots of dances, flirtation, sometimes ending in fistfights. We were young, talented, brazen, careless and free. I have never been so free in my life since then. When I came into the big real world where I happily live now, I realized that one cannot live without responsibilities, but we had almost no responsibilities back then, in the golden seventies. Money meant very little, nobody had any money by modern standards, but all the good things of life were free back then, or so it seemed to us.

Yasha Sinai, Sasha Polyakov and Volodya Zakharov at the camp before Peak Energia in Fan Mountains, 1980s

We traveled to Siberia, we hiked in the mountains, we carelessly risked our lives and thought nothing of it. I remember Volodya Zakharov, Sasha Polyakov, Tolya Larkin, Yasha and Lena Sinai and myself hiking for two or three weeks in the Fan Mountains in what is now Tajikistan. It was less dangerous back then, but still pretty wild. We talked and laughed a lot in our tents after walking a whole day under the hot sun in those high mountains.

I remember one episode, when Volodya Zakharov, Sasha Polyakov and me left the rest of the team and went over the pass to get some food supplies from aul (the mountain village). We made it over the pass by late night, and camped at a safe distance from the village (those aul dogs would easily tear your apart if you enter uninvited). In the morning we had to make a cautious contact with locals and ask them to give us some food in exchange for alcohol and medicine (like I said, money did not count at those times).

Volodya Zakharov, Ira Payusova, Sasha Migdal and Sasha Polyakov at the summit of the Peak Energia (18,373 feet above sea level) in Fan Mountains, 1980s

I was awaken by screams of Sasha Polyakov who was the first to discover that we were all covered by huge hairy earwigs ("ukhovertka"). Each measuring an inch or more, they looked quite scary, and they were biting us. After defeating this army, we started laughing and immediately wrote a poem describing the feat. It was joint creation of Zakharov and myself, it sounds wildly funny in Russian. Unfortunately, I do not know how to translate it in English. Not to mention the rhymes, the whole mindset was totally Russian, to be more precise it was the mindset of the Declining Soviet Empire. Here it is, for the Russian speaking readers:

Был он в Риме и в Нью-Йорке
И нигде не унывал,
Но напали уховёртки
На него меж Фанских скал.

Эти гады скользки, вёртки,
И, отбросив ложный стыд,
Поляков на малой горке
Как ошпаренный, кричит:

Эта тварь мерзка, двухвоста,
Я от ужаса горю,
Заползает в ухо просто,
Выползет через ноздрю!

И висит, вцепившись за нос,
Обкусает шею, грудь,
А не то – залезет в анус,
Иль ещё куда-нибудь.

Весь рюкзак зауховёрчен:
Он метался и рыдал,
Будет провиант испорчен –
Просыпайся, друг Мигдал!

Оный друг, пренебрегая
Низших тварей суетой,
Под арчой лежал, зевая,
В равновесьи со средой.

Но, восстав, как Ангел Смерти,
Пробудившийся Мигдал
Мигом всё разуховертил.
В назиданье же сказал:

"Не пугайся, друг, двухвосток,
Все двухвостки – ерунда!
От двуногих вертихвосток
Больше сраму и вреда..."

I made this photo — thumbing the nose to KGB, leaving USSR for good (1988). I left my heart there though

Am I missing these crazy, happy, funny, exciting times? Sure I am, but they are gone to never come back. I am quite happy with the world I live in now. It also had its share of craze and excitement in the nineties during the Internet revolution. I feel that there is more craze and excitement ahead of us in the near future. Alas, it hardly will be Physics again!

SECTION 13

GLIMPSES OF THE CONFERENCE

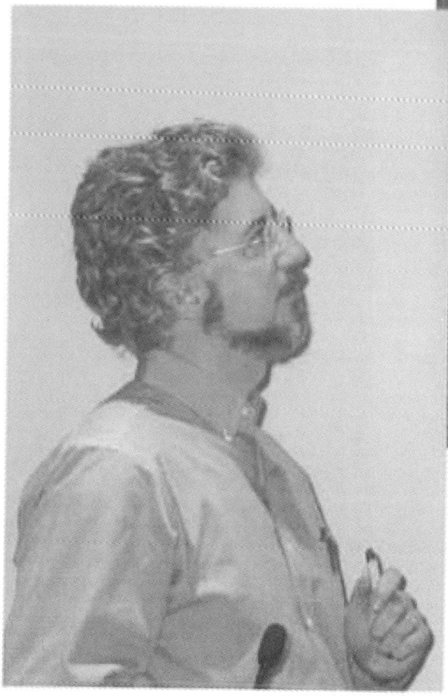

LIST OF THE PARTICIPANTS OF CAQCD-06

Thomas Appelquist, Yale University
Adi Armoni, University of Wales, Swansea
Roberto Auzzi, University of Minnesota
Ian Balitsky, Old Dominion University
Andrei Belitsky, Arizona State University
Carola Berger, Stanford Linear Accelerator Center
Zvi Bern, Univ. California Los Angeles
Stefano Bolognesi, The Niels Bohr Institute
Nora Brambilla, University of Milan
Andreas Brandhuber, Queen Mary University, London
Vladimir Braun, University of Regensburg
Stan Brodsky, Stanford Linear Accelerator Center
Thomas Cohen, University of Maryland
Pietro Colangelo, Instituto Nazionale di Fisica Nucleare, Bari
Andrzej Czarnecki, University of Alberta
Fulvia De Fazio, Instituto Nazionale di Fisica Nucleare, Bari
Guy de Teramond, Universidad de Costa Rica
Sergiy Dubynskiy, University of Minnesota
Gerald Dunne, University of Connecticut
Georgi Dvali, New York University
Johanna Erdmenger, Max Planck Institute for Physics, Munich
Joshua Erlich, William-Mary College
Angelo Raffaele Fazio, Universidad Nacional de Colombia
Darren Forde, Spht CEA-Saclay
Hilmar Forkel, University of Heidelberg
Tobias Frederico, Instituto Tecnologico de Aeronautica, Brazil
Gregory Gabadadze, New York University
Susan Gardner, University of Kentucky
Einan Gardi, Cambridge University, UK
Steve Gasiorowicz, University of Minnesota
Joel Giedt, University of Minnesota

Stanislaw Glazek, Warsaw University,
Leonid Glozman, University of Graz
Jose Goity, Hampton University
Daniel Gruenewald, Universität Heidelberg
Elena Gubankova, Massachusetts Institute of Technology
Amihay Hanany, Massachusetts Institute of Technology
John Hiller, University of MinnesotaDuluth
Johannes Hirn, University of Valencia
Ralf Hofmann, Universität Heidelberg
Yuichi Hoshino, Kushiro National College of Technology
Harald Ita, University of Wales, Swansea
Robert Jaffe, Massachusetts Institute of Technology
David Kaplan, Institute for Nuclear Theory, Seattle
Joe Kapusta, University of Minnesota
Alexander Khodjamirian, University of Siegen
Igor Klebanov, Princeton University
Evgeni Kolomeitsev, University of Minnesota
Kenichi Konishi, University of Pisa
Gregory Korchemsky, University of Paris XI
Cris Korthals Altes, CNRS, Marseille
Alex Kovner, University of Connecticut
Richard Lebed, Arizona State University
Laurent Lellouch, CNRS, Marseille
Frieder Lenz, Universität Erlangen-Nürnberg
Heinrich Leutwyler, University of Bern
Michael Lublinsky, University of Connecticut
Thomas Mannel, University of Siegen
Gary McCartor, Southern Methodist University
Max Metlitski, University of British Columbia
Yannick Meurice, University of Iowa
Agnes Mocsy, Research Center RIKEN-BNL
Rajamani Narayanan, Florida International University
Giuseppe Nardulli, University of Bari
Matthias Neubert, Cornell University
Maribel Nunez-Valdez, University of Minnesota
Keith Olive, University of Minnesota
Joao Pacheco B. C. de Melo, Universidad Cruzeiro do Sul, Brazil
Leopoldo Pando Zayas, University of Michigan
Andrei Parnachev, Rutgers University

Jan Martin Pawlowski, University of Heidelberg
Marco Peloso, University of Minnesota
Alexey Petrov, Wayne State University
Erich Poppitz, University of Toronto
Maxim Pospelov, Perimeter Institute
Parmar Rajendrasinh, Science College Modasa, Gujarat, India
Anatoly Radyushkin, Jefferson Lab
Ralf Rapp, Texas A&M University
Adam Ritz, University of Victoria
Radu Roiban, Pennsylvania State University
Serge Rudaz, University of Minnesota
Pedro Ruiz-Femenia, Max Planck Institut für Physik. Munich
Norisuke Sakai, Tokyo Institute of Technology
Francesco Sannino, Niels Borh Institute
Veronica Sanz, University of Granada & CAFPE
Yudi Santoso, University of Victoria
Mikhail Shifman, University of Minnesota
Edward Shuryak, SUNY, Stony Brook
Dam Thanh Son, University of Washington
Jacob Sonnenschein, Tel Aviv University
Vassilis Spanos, University of Minnesota
Marcus Spradlin, University of Michigan
Misha Stephanov, University of Illinois, Chicago
Matt Strassler, University of Washington
Shigeki Sugimoto, Kyoto University
Peter Svrcek, Stanford University
Bayram Tekin, Middle East Technical University
Tonnis ter Veldhuis, Macalester College, St. Paul
Charles Thorn, University of Florida
David Tong, Cambridge University, UK
Gabriele Travaglini, Queen Mary University, London
Arkady Tseytlin, Ohio State University
Mithat Unsal, Boston University
Nikolai Uraltsev, University of Milan
Arkady Vainshtein, University of Minnesota
Antonio Vairo, University of Milan
Diana Vaman, University of Michigan
Liliana Velasco-Sevilla, University of Minnesota
Jacobus Verbaarschot, SUNY, Stony Brook

Mikhail Voloshin, University of Minnesota
Larry Yaffe, Washington University
Aleksander Yelnykov, Virginia Tech
Eric Zhitnitsky, University of British Columbia, Canada
Roman Zwicky, IPPP-University of Durham